Lecture Notes in Computer Science   9394

Commenced Publication in 1973
Founding and Former Series Editors:
Gerhard Goos, Juris Hartmanis, and Jan van Leeuwen

More information about this series at http://www.springer.com/series/7407

Wiebe van der Hoek · Wesley H. Holliday
Wen-fang Wang (Eds.)

Logic, Rationality, and Interaction

5th International Workshop, LORI 2015
Taipei, Taiwan, October 28–31, 2015
Proceedings

 Springer

Editors
Wiebe van der Hoek
Department of Computer Science
University of Liverpool
Liverpool
UK

Wen-fang Wang
Institute of Philosophy of Mind and Cognition
National Yang-Ming University
Taipei
Taiwan

Wesley H. Holliday
Department of Philosophy
Group in Logic and the Methodology
  of Science
University of California, Berkeley
Berkeley, CA
USA

ISSN 0302-9743
Lecture Notes in Computer Science
ISBN 978-3-662-48560-6
DOI 10.1007/978-3-662-48561-3

ISSN 1611-3349   (electronic)

ISBN 978-3-662-48561-3   (eBook)

Library of Congress Control Number: 2015950894

LNCS Sublibrary: SL1 – Theoretical Computer Science and General Issues

Printed on acid-free paper

Springer-Verlag GmbH Berlin Heidelberg is part of Springer Science+Business Media
(www.springer.com)

Preface

This volume contains the papers presented at LORI-5, the 5th International Workshop on Logic, Rationality and Interaction, held during October 28–31, 2015, in Taipei, Taiwan, and hosted by the Department of Philosophy of National Taiwan University and the Institute of Philosophy of Mind and Cognition of National Yang-Ming University.

There were 62 submissions to LORI-5. Each submission was reviewed by at least two, and on average three, Program Committee (PC) members. The committee decided to accept 32 full papers and seven abstracts for poster presentations.

The topics covered in this program represent well the span and depth that have become a trademark of the LORI workshop series, where logic interfaces with disciplines as diverse as game theory and decision theory, philosophy and epistemology, linguistics, computer science, and artificial intelligence. The technical program of the workshop was further enriched with invited talks by Maria Aloni, Branden Fitelsen, Joseph Halpern, Churn-Jung Liau, Fenrong Liu, and Eric Pacuit.

The LORI series took off with a first event, LORI-1, hosted in August 2007 by Beijing Normal University in Beijing. That event was a great success, providing an effective platform for Chinese and non-Chinese logicians to meet and exchange research ideas. The wish to perpetuate such a platform led to four later editions: LORI-2, hosted by Southwest University in Chongqing; LORI-3, hosted by Sun Yat-sen University in Guangzhou; LORI-4, held at Zhejiang University in Hangzhou; and of course LORI-5 in Taipei, of which this book collects the proceedings. A history of the series can be accessed at www.golori.org.

As Organizing Committee and PC chairs we would like to thank the authors of all submitted papers for their submissions and the PC members and external reviewers for a truly outstanding job under extremely tight time constraints. The program owes the greatest debt to their contribution. Our activity was further supported by the indefatigable work of Fenrong Liu and Johan van Benthem on the general organization of LORI. As to the local organization, LORI-5 would not have taken place without the tremendous amount of work put in by Eric Peng. A great thanks for organizing LORI-5 should also go to the Department of Philosophy at the National Taiwan University and the Institute of Philosophy of Mind and Cognition at National Yang-Ming University. We would also like to acknowledge the use of EasyChair, which has been a fantastic tool for both organizing the reviewing process and creating these proceedings.

Special thanks for sponsorship goes to the College of Liberal Arts at National Taiwan University and National Yang-Ming University, to the Ministry of Science and Technology of Republic of China (Taiwan), and to the National Committee of the Republic of China for the Division of Logic, Methodology and Philosophy of Science. Last but not least, thanks to the School of EEE&CS at the University of Liverpool, UK, for financially supporting the proceedings of LORI-5.

August 2015   Wiebe van der Hoek
                                                              Wesley H. Holliday
                                                              Wen-fang Wang

Organization

Program Committee

Natasha Alechina	University of Nottingham, UK
Sergei Artemov	City University of New York, USA
Guillaume Aucher	University of Rennes, France
Alexandru Baltag	University of Amsterdam, The Netherlands
Patrick Blackburn	University of Roskilde, Denmark
Thomas Bolander	Technical University of Denmark, Denmark
Jan Broersen	Utrecht University, The Netherlands
Fabrizio Cariani	Northwestern University, USA
Nina Gierasmczuk	University of Amsterdam, The Netherlands
Patrick Girard	The University of Auckland, New Zealand
Valentin Goranko	Stockholm University, Sweden
Davide Grossi	University of Liverpool, UK
Andreas Herzig	Université Paul Sabatier, France
Wesley H. Holliday	University of California, Berkeley, USA
John Horty	University of Maryland, USA
Tomohiro Hoshi	Stanford University, USA
Linton I-Chi Wang	National Chung Cheng University, Taiwan
Thomas Icard	Stanford University, USA
Gerhard Jäger	University of Bern, Switzerland
Makoto Kanazawa	National Institute of Informatics, Japan
Barteld Kooi	University of Groningen, The Netherlands
Jérôme Lang	Université Paris Dauphine, France
Hannes Leitgeb	Ludwig-Maximilians-Universität München, Germany
Beishui Liao	Zhejiang University, China
Churn-Jung Liau	Academia Sinica, Taiwan
Hanti Lin	University of California, Davis, USA
Fenrong Liu	Tsinghua University, China
Emiliano Lorini	Université Paul Sabatier, France
John-Jules Meyer	Utrecht University, The Netherlands
Larry Moss	Indiana University, USA
Sara Negri	University of Helsinki, Finland
Eric Pacuit	University of Maryland, USA

Daniele Porello	Institute of Cognitive Science and Technology, Italy
Brian Renne	University of Amsterdam, The Netherlands
Hans Rott	University of Regensburg, Germany
Olivier Roy	University of Bayreuth, Germany
Joshua Sack	University of Amsterdam, The Netherlands
Jeremy Seligman	University of Auckland, New Zealand
Sonja Smets	University of Amsterdam, The Netherlands
Martin Stokhof	University of Amsterdam, The Netherlands
Kaile Su	Griffith University, Australia
Jakub Szymanik	University of Amsterdam, The Netherlands
Allard Tamminga	University of Groningen, The Netherlands
Paolo Turrini	Imperial College, UK
Wiebe van der Hoek	University of Liverpool, UK
Leon van der Torre	University of Luxembourg, Luxembourg
Hans van Ditmarsch	LORIA, France
Robert van Rooij	University of Amsterdam, The Netherlands
Bow-Yaw Wang	Institute of Information Science, Taiwan
Wen-fang Wang	National Yang-Ming University, Taiwan
Yanjing Wang	Peking University, China
Gregory Wheeler	Ludwig-Maximilians-Universität München, Germany
Michael Wooldridge	University of Oxford, UK
Thomas Ågotnes	University of Bergen, Norway

Additional Reviewers

Ines Crespo	Hu Liu
Peter Fritz	Bastien Maubert
Petar Iliev	Katsuhiko Sano
Dominik Klein	Cristián Santibáñez Yáñez
Chanjuan Liu	Francois Schwarzentruber

Contents

Full Papers

X   Contents

Poster Papers

Sabotage Modal Logic:
Some Model and Proof Theoretic Aspects

Guillaume Aucher[1], Johan van Benthem[2], and Davide Grossi[3]

[1] University of Rennes 1 – INRIA
guillaume.aucher@irisa.fr
[2] University of Amsterdam, Stanford University, Tsinghua University
J.vanBenthem@uva.nl
[3] University of Liverpool
D.Grossi@liverpool.ac.uk

Abstract. We investigate some model and proof theoretic aspects of sabotage modal logic. The first contribution is to prove a characterization theorem for sabotage modal logic as the fragment of first-order logic which is invariant with respect to a suitably defined notion of bisimulation (called sabotage bisimulation). The second contribution is to provide a sound and complete tableau method for sabotage modal logic. We also chart a number of open research questions concerning sabotage modal logic, aiming at integrating it within the current landscape of logics of model update.

1 Introduction

Sabotage modal logic (SML) [6] expands the standard modal language with an edge-deletion modality $\blacklozenge\varphi$ whose intended reading is "after the deletion of at least one edge in the frame it holds that φ". As such it can be viewed as the modal logic of arbitrary edge deletion. Although it inspired several later formalisms in the dynamic epistemic logic tradition [7] (e.g., *graph modifiers logic* [3], *memory logic* [1], *swap logic* [2], *arrow update logic* [13]), and is directly related to recent work in theoretical computer science (e.g., [16,12]) and learning theory (e.g., [11]) it remains a rather under-investigated logic.

The only work focusing specifically on SML is, to the best of our knowledge, [15,14] where the undecidability of the satisfiability problem and the complexity of the model-checking problem of SML are established. Among the open questions concerning SML, that work points to the lack of a notion of bisimulation characteristic for SML. The present article addresses such question and can be regarded as an application of standard techniques and methods of modal correspondence theory [5] to sabotage modal logic. The article provides as well a sound and complete tableau method for SML. This contributes to the proof theory of SML, which has rather been neglected so far. In pursuing our investigations, the article establishes a few related model-theoretic results and aims at putting SML 'on the map' of current research in dynamic epistemic logic.

© Springer-Verlag Berlin Heidelberg 2015
W. van der Hoek et al. (Eds.): LORI 2015, LNCS 9394, pp. 1–13, 2015.
DOI: 10.1007/978-3-662-48561-3_1

Outline of the article. Section 2 introduces SML and what is thus far known of its properties. That is the starting point of the article. Section 3 introduces a notion of bisimulation for SML—called sabotage bisimulation—and Section 4 characterizes SML as the fragment of first-order logic (FOL) which is invariant for sabotage bisimulation. Section 5 provides a sound and complete tableau method for sabotage modal logic. Section 6 concludes with some open research questions.

2 Preliminaries

In this section, we introduce the syntax and semantics of SML, recapitulate some key results from [14], and present a standard translation for SML (to FOL).

2.1 Syntax

Let **P** be a countable set of propositional atoms. The set of formulae of the sabotage modal language $\mathcal{L}^s$ is defined by the following grammar in BNF:

$$\mathcal{L}^s : \varphi ::= p \mid \neg\varphi \mid (\varphi \wedge \varphi) \mid \Diamond\varphi \mid \blacklozenge\varphi$$

where $p \in \mathbf{P}$. The remaining set of Boolean connectives $\{\vee, \rightarrow\}$ and the modal operators $\Box$ and $\blacksquare$ can be defined in the standard way. The formula $\bot$ is an abbreviation for the formula $p \wedge \neg p$ (for a chosen $p \in \mathbf{P}$) and $\top$ is an abbreviation for $\neg\bot$. The iteration of n sabotage operators or modalities will sometimes be denoted by $\blacklozenge_n$ and $\Diamond_n$, respectively, $\blacksquare_n$ and $\Box_n$. To save parenthesis, we use the following ranking of binding strength: $\blacksquare, \blacklozenge, \Box, \Diamond, \neg, \wedge, \vee, \rightarrow$.

A natural measure of syntactic complexity for sabotage formulae is given by their *sabotage depth* [14]. Let $\varphi \in \mathcal{L}^s$. The *sabotage depth* of φ, written $sd(\varphi)$, is inductively defined as follows: $sd(\top) = sd(p) := 0$, $sd(\neg\varphi) := sd(\varphi)$, $sd(\varphi_1 \wedge \varphi_2) := \max\{sd(\varphi_1), sd(\varphi_2)\}$, $sd(\Diamond\varphi) := sd(\varphi)$ and $sd(\blacklozenge\varphi) := sd(\varphi) + 1$.

2.2 Semantics

We will be working with standard Kripke models $\mathcal{M} = (W, R, V)$ where: W is a non-empty set; $R \subseteq W \times W$; and $V : \mathbf{P} \longrightarrow 2^W$. The pair (W, R) is called a frame, and is denoted by $\mathcal{F}$.

Such structures will be also interpreted as models for the binary fragment of FOL with equality[1] denoted $\mathcal{L}^1$. Sometimes we will use the following FOL terminology/notation. We say that a model $\mathcal{M}$ *satisfies* a formula $\varphi(x) \in \mathcal{L}^1$ (or a set $\Gamma(x) \subseteq \mathcal{L}^1$) with one free variable x under the assignment of w to x if and only if φ (respectively Γ) is true of w, in symbols, $\mathcal{M} \models \varphi(x)[w]$ (respectively, $\mathcal{M} \models \Gamma(x)[w]$). We say that a model $\mathcal{M}$ *realizes* a set $\Gamma(x) \subseteq \mathcal{L}^1$ with one free variable x (i.e., a type) if and only if there exists an element $w \in W$ such that $\mathcal{M} \models \Gamma(x)[w]$.

---

[1] We refer the reader to [4, Ch. 2.4].

The satisfaction relation for $\mathcal{L}^s$ is defined as usual for the atomic and Boolean cases, and for the standard modalities. For the sabotage modality it is as follows:

$$(W, R, V), w \models \blacklozenge\varphi \Longleftrightarrow \exists(w', w'') \in R \text{ s.t. } (W, R \setminus \{(w', w'')\}, V), w \models \varphi \quad (1)$$

In other words, $\blacklozenge\varphi$ is satisfied by a pointed model if and only if there exist two R-related (possibly identical) states such that, once the edge between these two states is removed from R, φ holds at the same evaluation state. The notions of validity and logical consequence are defined as usual.

We say that two pointed models $(\mathcal{M}, w)$ and $(\mathcal{M}', w')$ are *sabotage-related* (notation, $(\mathcal{M}, w) \xrightarrow{\blacklozenge} (\mathcal{M}', w')$) if and only if: $w' = w$; $W' = W$; $R' = R \setminus \{(w'', w''')\}$ for some $w'', w''' \in W$; $V' = V$. The set $\mathbf{r}(\mathcal{M}, w) = \{(\mathcal{M}', w') \mid (\mathcal{M}, w) \xrightarrow{\blacklozenge} (\mathcal{M}', w')\}$ denotes the set of all models which are sabotage-related to a given pointed model $(\mathcal{M}, w)$. Similarly, $\mathbf{r}^n(\mathcal{M}, w) = \{(\mathcal{M}', w') \mid (\mathcal{M}_1, w_1) \xrightarrow{\blacklozenge} (\mathcal{M}_2, w_2) \xrightarrow{\blacklozenge} \ldots \xrightarrow{\blacklozenge} (\mathcal{M}_{n+1}, w_{n+1}) \ \& \ (\mathcal{M}_1, w_1) = (\mathcal{M}, w) \ \& \ (\mathcal{M}_{n+1}, w_{n+1}) = (\mathcal{M}', w')\}$ denotes the set of all models which are related to $(\mathcal{M}, w)$ by a $\xrightarrow{\blacklozenge}$-path of length n. Finally, $\mathbf{r}^*(\mathcal{M}, w) = \{(\mathcal{M}', w') \mid (\mathcal{M}, w) \xrightarrow{\blacklozenge}^* (\mathcal{M}', w')\}$ denotes the set of all pointed models which are reachable from $(\mathcal{M}, w)$ by the reflexive and transitive closure of $\xrightarrow{\blacklozenge}$. We will often drop the reference to a given point in the model, which will be clear by the context, and simply write $\mathcal{M} \xrightarrow{\blacklozenge} \mathcal{M}'$ instead of $(\mathcal{M}, w) \xrightarrow{\blacklozenge} (\mathcal{M}', w')$.

The set of sabotage modal formulae which are satisfied by a pointed model $(\mathcal{M}, w)$, i.e., the *sabotage modal logic theory* of w in $\mathcal{M}$, is denoted $\mathbb{T}^s(\mathcal{M}, w)$. We say that two pointed models $(\mathcal{M}, w)$ and $(\mathcal{M}', w')$ are *sabotage modally equivalent* —notation: $(\mathcal{M}, w) \leadsto_s (\mathcal{M}', w')$—if and only if they satisfy the same sabotage modal formulae, that is, they have the same sabotage modal logic theory.

SML can express properties that are beyond the reach of standard modal logic. An example is the property "there are at most n successors" with $1 \leq n$ (see also Example 1 below):

$$\exists^{\leq n} y \, (xRy). \quad (2)$$

This property can be expressed in SML with:

$$\Box\bot \lor \bigvee_{1 \leq i \leq n} \blacklozenge_i\Box\bot. \quad (3)$$

A dual formula expresses the property "there are at least n successors". In fact, SML can even define frames up to isomorphisms. Indeed, one can easily show that the formula $\Diamond\top \land \Box\Diamond\top \land \blacksquare\Box\bot$ is true in a model if and only if its underlying frame consists of one reflexive point.

Finally, let us recapitulate the findings of [15,14]. They are proved with a multi-modal version of SML, but all our results and methods are easily generalizable to this multi-modal setting.

Theorem 1 ([15,14]). *The model-checking problem of* SML *is PSPACE-complete.* SML *lacks the finite model property, the tree-model property and its satisfiability problem is undecidable.*

2.3   A Standard Translation for SML

A standard translation for SML was first sketched in the technical report [15]. In this section we describe such translation and its correctness in detail. This is essential to prepare the later sections of the article.

Setting up the Translation. In order to define a standard translation from the language of SML to the free variable fragment of FOL with equality one needs to keep track of the changes that the sabotage operators introduce in the model.

This can be achieved by indexing the standard translation with a set E consisting of pairs of variables. The idea is that when the standard translation is applied to the outermost operator of a given formula, this set is empty. As the analysis proceeds towards inner operators, each sabotage operator $\blacklozenge$ in the formula will introduce a new pair of variables in E, which will be bound by an existential quantifier. Here is the formal definition:

Definition 1 (Standard Translation for SML). *Let E be a set of pairs (y,z) of variables—edges—and x be a designated variable. The translation ST_x^E : $\mathcal{L}^s \longrightarrow \mathcal{L}^1$ is recursively defined as follows:*

$$ST_x^E(p) = P(x)$$
$$ST_x^E(\bot) = x \neq x$$
$$ST_x^E(\neg\varphi) = \neg ST_x^E(\varphi)$$
$$ST_x^E(\varphi_1 \wedge \varphi_2) = ST_x^E(\varphi_1) \wedge ST_x^E(\varphi_2)$$

$$ST_x^E(\Diamond\varphi) = \exists y \left(xRy \wedge \bigwedge_{(v,w)\in E} \neg(x = v \wedge y = w) \wedge ST_y^E(\varphi) \right)$$

$$ST_x^E(\blacklozenge\varphi) = \exists y,z \left(yRz \wedge \bigwedge_{(v,w)\in E} \neg(y = v \wedge z = w) \wedge ST_x^{E\cup\{(y,z)\}}(\varphi) \right)$$

The key clauses concern $\Diamond$-formulae and $\blacklozenge$-formulae. Let us start with the latter. Formula $\blacklozenge\varphi$ is translated as the first order formula stating the following: that there exists some R-edge denoted by (y,z); that such edge is different from any edge possibly denoted by the pairs in E; that the translation of φ should now be carried out with respect to the set $E \cup \{(y,z)\}$; and that this translation is realized at x.

As to the former clause, it says that formula $\Diamond\varphi$ is translated as the first order formula with x free, which states the existence of a state y accessible from x via an edge which is different from all the edges in the set E, and that the translation of φ is realized at y.

Setting up the translation like this allows one to book-keep the removal of edges via E. The removal of edges is handled by imposing the existence of states which are different from the ones reachable via the 'removed' edges. In other words edge removal is simulated by imposing the existence of edges which are then not used to interpret inner modal operators.

It is important to notice the following feature of the translation. Depending on the chosen E, ST^E can possibly yield formulae with several free variables, e.g.: $ST_x^{(v,w)} \Diamond p = \exists y \, (xRy \wedge \neg (x = v \wedge y = w) \wedge p)$. However, if ST^E is applied to a formula φ by setting $E = \emptyset$, that is to say, if the translation is initiated with an empty E, then, at each successive application of ST^E to subformulae of φ, the variables occurring in W will be bound by some quantifiers introduced at previous steps. For any φ, $ST_x^\emptyset(\varphi)$ yields a FOL formula with only x free.

Correctness of the Translation. We prove now the correctness of the translation proposed in Definition 1.

Theorem 2. *Let $\mathcal{M}, w$ be a pointed model and $\varphi \in \mathcal{L}^s$:*

$$\mathcal{M}, w \models \varphi \Longleftrightarrow \mathcal{M} \models ST_x^\emptyset(\varphi)[w]$$

Proof (Sketch). By induction on the structure of φ. We omit the Boolean and modal cases. The case for the sabotage operator $\blacklozenge$ is proven by the following series of equivalences:

$$\mathcal{M}, w \models \blacklozenge\varphi \Longleftrightarrow \mathcal{M}, w \xrightarrow{\blacklozenge} \mathcal{M}', w \models \varphi \qquad \text{semantics of } \blacklozenge \text{ (1)}$$
$$\Longleftrightarrow \mathcal{M}, w \xrightarrow{\blacklozenge} \mathcal{M}' \models ST_x^\emptyset(\varphi)[w] \qquad \text{IH}$$
$$\Longleftrightarrow \mathcal{M} \models \exists y, z \left(yRz \wedge ST_x^{\{(y,z)\}}(\varphi)[w] \right) \quad \text{sem. of } \blacklozenge \text{ (1) and Def. 1}$$
$$\Longleftrightarrow \mathcal{M} \models ST_x^\emptyset(\blacklozenge\varphi)[w] \qquad \text{Def. 1} \quad \square$$

We conclude the section with the following observation:

Proposition 1. SML *is not contained in any fixed variable fragment of* FOL.

Proof. We show SML contains formulae that are not definable in any fixed variable fragment of FOL. Consider the above FOL formulae with counting quantifier of Expression (2) with $1 \leq n$. Clearly, for each integer n, Expression (2) is definable in FOL (without counting quantifiers) using a fixed number of variables. But no fixed variable fragment can define Expression (2) for all integers n. Since Expression (2) is equivalent to (3) it follows that although SML is FOL-definable (Corollary 2) it is not definable in any fixed variable fragment of FOL. $\quad \square$

3   Bisimulation for SML

In this section, we introduce a notion of bisimulation for SML.

3.1   Sabotage Bisimulation

Definition 2 (s-bisimulation). *Let $\mathcal{M}_1 = (W_1, R_1, V_1)$ and $\mathcal{M}_2 = (W_2, R_2, V_2)$ be two Kripke models. A non-empty relation $Z \subseteq \mathbf{r}^*(\mathcal{M}_1, w) \times \mathbf{r}^*(\mathcal{M}_2, v)$ is an s-bisimulation between the two pointed models $(\mathcal{M}_1, w)$ and $(\mathcal{M}_2, v)$—notation, $Z : (\mathcal{M}_1, w) \leftrightarrow_s (\mathcal{M}_2, v)$—if the following conditions are satisfied:*

Atom: *If* $(\mathcal{M}_1, w)Z(\mathcal{M}_2, v)$ *then* $\mathcal{M}_1, w \models p$ *iff* $\mathcal{M}_2, v \models p$, *for any atom* p.

Zig$_\Diamond$: *If* $(\mathcal{M}_1, w)Z(\mathcal{M}_2, v)$ *and there exists* $w' \in W_1$ *s.t.* wR_1w' *then there exists* $v' \in W_2$ *s.t.* vR_2v' *and* $(\mathcal{M}_1, w')Z(\mathcal{M}_2, v')$;

Zag$_\Diamond$: *If* $(\mathcal{M}_1, w)Z(\mathcal{M}_2, v)$ *and there exists* $v' \in S_s$ *s.t.* vR_1v' *then there exists* $w' \in W_1$ *s.t.* wR_1w' *and* $(\mathcal{M}_1, w')Z(\mathcal{M}_2, v')$;

Zig$_\blacklozenge$: *If* $(\mathcal{M}_1, w)Z(\mathcal{M}_2, v)$ *and there exists* $\mathcal{M}_1'$ *such that* $(\mathcal{M}_1, w) \overset{\blacklozenge}{\rightarrow} (\mathcal{M}_1', w)$, *then there exists* $\mathcal{M}_2'$ *such that* $(\mathcal{M}_2, v) \overset{\blacklozenge}{\rightarrow} (\mathcal{M}_2', v)$ *and* $(\mathcal{M}_1', w)Z(\mathcal{M}_2', v)$;

Zag$_\blacklozenge$: *If* $(\mathcal{M}_1, w)Z(\mathcal{M}_2, v)$ *and there exists* $\mathcal{M}_2'$ *such that* $(\mathcal{M}_2, v) \overset{\blacklozenge}{\rightarrow} (\mathcal{M}_2', v)$, *then there exists* $\mathcal{M}_1'$ *such that* $(\mathcal{M}_1, w) \overset{\blacklozenge}{\rightarrow} (\mathcal{M}_1', w)$ *and* $(\mathcal{M}_1', w)Z(\mathcal{M}_2', v)$.

We write $(\mathcal{M}_1, w) \leftrightarroweq_s (\mathcal{M}_2, v)$ *if there exists an s-bisimulation* Z *s.t.* $(\mathcal{M}_1, w) Z(\mathcal{M}_2, v)$.

It is worth spending a few words about Definition 2. The notion of s-bisimulation strengthens the standard modal bisimulation with the 'zig' and 'zag' conditions for the sabotage modality. Just like the sabotage modality is an 'external' modality so is s-bisimulation an 'external' notion of bisimulation. Standard bisimulation keeps the model fixed and changes the evaluation point along the accessibility relation of the Kripke model, s-bisimulation keeps the evaluation point fixed and changes the model by picking one among the sabotage-accessible ones.

3.2   Bisimulation and Modal Equivalence in SML

We first show that s-bisimulation implies SML equivalence.

Proposition 2 ($\leftrightarroweq_s \subseteq \leftrightsquigarrow_s$). *For any two pointed models* $(\mathcal{M}_1, w)$ *and* $(\mathcal{M}_2, v)$ *it holds that:* $(\mathcal{M}_1, w) \leftrightarroweq_s (\mathcal{M}_2, v) \implies (\mathcal{M}_1, w) \leftrightsquigarrow_s (\mathcal{M}_2, v)$.

Proof. The proof is by induction on the syntax of φ. Assume $(\mathcal{M}_1, w_1)Z(\mathcal{M}_2, w_2)$. **Base:** The Atom clause of Definition 2 covers the case of atoms and nullary operators. **Step:** The Boolean cases are as usual. The Zig$_\Diamond$ and Zag$_\Diamond$ clauses of Definition 2 take care of $\Diamond$-formulae in the standard way. As to $\blacklozenge$-formulae, assume $\mathcal{M}_1, w_1 \models \blacklozenge\varphi$. By the semantics of $\blacklozenge$ we have that $\mathcal{M}_1 \overset{\blacklozenge}{\rightarrow} \mathcal{M}_1', w \models \varphi$ and, by clause Zig$_\blacklozenge$ of Definition 2, it follows that $\mathcal{M}_2 \overset{\blacklozenge}{\rightarrow} \mathcal{M}_2'$ and $(\mathcal{M}_1', w)Z(\mathcal{M}_2', v)$. By IH we conclude that $\mathcal{M}_2', v \models \varphi$ and, consequently, $\mathcal{M}_2, v \models \blacklozenge\varphi$. Similarly, from $\mathcal{M}_2, v \models \blacklozenge\varphi$ we conclude $\mathcal{M}_1, w \models \blacklozenge\varphi$ by clause Zag$_\blacklozenge$ of Definition 2. $\square$

Just like for the standard modal language, the converse of Proposition 2 can be proven under the assumption that the models at issue are ω-saturated. Before introducing such notion let us fix some notation. Given a finite set Y, the expansion of $\mathcal{L}^1$ with a finite set of constants Y is denoted $\mathcal{L}_Y^1$, and the expansion of a Kripke model $\mathcal{M}$ to $\mathcal{L}_Y^1$ is denoted $\mathcal{M}_Y$.[2]

Definition 3 (ω-saturation). *A model* $\mathcal{M} = (W, R, V)$ *is* ω-saturated *if, and only if, for every* $Y \subseteq W$ *such that* $|Y| < \omega$, *the expansion* $\mathcal{M}_Y$ *realizes every set* $\Gamma(x)$ *of* $\mathcal{L}_Y^1$-*formulae whose finite subsets* $\Gamma'(x) \subseteq \Gamma(x)$ *are all realized in* $\mathcal{M}_Y$.

---

[2] For more on ω-saturation we refer the reader to [4, Ch. 2] and [9, Ch. 2].

Intuitively, a model $\mathcal{M}$ is ω-saturated if for any set of formulae $\Gamma(x, y_1, \ldots, y_n)$ over a finite set of variables, once some interpretation of $y_1, \ldots, y_n$ is fixed to, e.g., $w_1, \ldots, w_n$, and all finite subsets of $\Gamma(x)[w_1, \ldots, w_n]$ are realizable in $\mathcal{M}$, then the whole of $\Gamma(x)[w_1, \ldots, w_n]$ is realizable in $\mathcal{M}$. From a modal point of view, Definition 3 requires that if for any subset of Γ there are accessible states satisfying it at the evaluation point, then there are accessible states satisfying the whole of Γ at the evaluation point. This is precisely the property used in the proof of the following proposition.

Proposition 3 ($\rightsquigarrow_s \, \subseteq \, \leftrightarrows_s$)**.** *For any two ω-saturated pointed models $(\mathcal{M}_1, w_1)$ and $(\mathcal{M}_2, w_2)$ it holds that: $(\mathcal{M}_1, w_1) \rightsquigarrow_s (\mathcal{M}_2, w_2) \implies (\mathcal{M}_1, w_1) \leftrightarrows_s (\mathcal{M}_2, w_2)$.*

Proof. It suffices to show that $\rightsquigarrow_s$ is an s-bisimulation (Definition 2). **Base:** The condition Atom is straightforwardly satisfied. **Step:** The proof for conditions Zig$_\diamond$ and Zag$_\diamond$ proceeds as usual for basic modal languages. We prove that the condition Zig$_\blacklozenge$ is satisfied. Assume $(\mathcal{M}_1, w_1) \rightsquigarrow_s (\mathcal{M}_2, w_2)$ and $(\mathcal{M}_1, w_1) \xrightarrow{\blacklozenge} (\mathcal{M}_1', w_1)$. We show that there exists $(\mathcal{M}_2', w_2)$ such that $(\mathcal{M}_2, w_2) \xrightarrow{\blacklozenge} (\mathcal{M}_2', w_2)$ and $(\mathcal{M}_1', w_1) \rightsquigarrow_s (\mathcal{M}_2', w_2)$. We have that for any finite $\Gamma \subseteq \mathbb{T}^s(\mathcal{M}_1', w_1)$ the following sequence of equivalences holds:

$$\mathcal{M}_1, w_1 \models \blacklozenge \bigwedge \Gamma \Longleftrightarrow \mathcal{M}_2, w_2 \models \blacklozenge \bigwedge \Gamma$$

$$\Longleftrightarrow \mathcal{M}_2 \models ST_x^\emptyset \left(\blacklozenge \bigwedge \Gamma \right) [w_2]$$

$$\Longleftrightarrow \mathcal{M}_2 \models \exists y, z \left(yRz \wedge ST_x^{\{(y,z)\}} \left(\bigwedge \Gamma \right) \right) [w_2]$$

The first equivalence holds by the assumption of sabotage equivalence between $(\mathcal{M}_1, w_1)$ and $(\mathcal{M}_2, w_2)$. The second one follows by Theorem 2 and the third one by Definition 1. From this, by ω-saturation of $\mathcal{M}_2$ we can conclude that:

there are $y, z \in \mathcal{M}_2$ such that yRz and $\mathcal{M}_2 \models ST_x^{\{(y,z)\}} (\mathbb{T}^s(\mathcal{M}_1', w_1)) [w_2]$.

By Theorem 2 there exists then a model $\mathcal{M}_2'$ such that $\mathcal{M}_2 \xrightarrow{\blacklozenge} \mathcal{M}_2'$ and $\mathcal{M}_2' \models ST_x^\emptyset (\mathbb{T}^s(\mathcal{M}_1', w_1)) [w_2]$. By Theorem 2 we conclude that $(\mathcal{M}_1', w_1) \rightsquigarrow_s (\mathcal{M}_2', w_2)$, which completes the proof of the Zig$_\blacklozenge$ clause. In the same way it can be proven that also the condition Zag$_\blacklozenge$ is satisfied. $\square$

We have thus established a match between sabotage modal equivalence and sabotage bisimulation for the class of ω-saturated models.

4   Characterization of SML by Invariance

In this section, we characterize SML as the one free variable fragment of FOL which is invariant under s-bisimulation.[3]

---

[3] Recall that the standard translation $ST^\emptyset$ of a sabotage modal logic formula always produces a FOL formula with only one free variable.

Theorem 3 (Characterization of SML by s-bisimulation Invariance). *An $\mathcal{L}_1$-formula is equivalent to the translation of an $\mathcal{L}^s$ formula if, and only if, it is invariant for sabotage bisimulation.*

Proof. [Left to right] This direction follows from Proposition 2. [Right to left] We proceed as customary. Let $\varphi \in \mathcal{L}^1$ with one free variable x. Assume that φ is invariant under s-bisimulation and consider the following set:

$$\mathbb{C}^s(\varphi) = \{ST_x^\emptyset(\psi) \mid \psi \in \mathcal{L}^s \text{ and } \varphi \models ST_x^\emptyset(\psi)\}.$$

The result follows from these two claims:

(i) If $\mathbb{C}^s(\varphi) \models \varphi$ then φ is equivalent to the translation of an $\mathcal{L}^s$-formula.
(ii) It holds that $\mathbb{C}^s(\varphi) \models \varphi$, i.e., for any pointed model $\mathcal{M}, w$: if $\mathcal{M} \models \mathbb{C}^s(\varphi)[w]$ then $\mathcal{M} \models \varphi[w]$.

As to (i). Assume that $\mathbb{C}^s(\varphi) \models \varphi$. From the deduction and compactness theorems of FOL we have that $\models \bigwedge \Gamma \to \varphi$ for some finite $\Gamma \subset \mathbb{C}^s(\varphi)$. The converse holds by the definition of $\mathbb{C}^s(\varphi)$: $\models \varphi \to \bigwedge \Gamma$. We thus have that $\models \varphi \leftrightarrow \bigwedge \Gamma$ proving the claim.

As to (ii). Take a pointed model $\mathcal{M}, w$ such that $\mathcal{M} \models \mathbb{C}^s(\varphi)[w]$ and consider its sabotage modal theory $\mathbb{T}^s(\mathcal{M}, w)$. Now consider the set $\Sigma = ST_x^\emptyset(\mathbb{T}^s(\mathcal{M}, w)) \cup \{\varphi\}$. We proceed by showing that:

(a) Σ is consistent;
(b) $\mathcal{M} \models \varphi[w]$, thus proving claim (ii).

To prove (a) assume, towards a contradiction, that Σ is inconsistent. By the compactness of FOL we then obtain that $\models \varphi \to \neg \bigwedge \Gamma$ for some finite $\Gamma \in \Sigma$. But then, by the definition of $\mathbb{C}^s(\varphi)$, we have that $\neg \bigwedge \Gamma \in \mathbb{C}^s(\varphi)$, and hence $\neg \bigwedge \Gamma \in ST_x^\emptyset(\mathbb{T}^s(\mathcal{M}, w))$ which is impossible as $\Gamma \subset ST_x^\emptyset(\mathbb{T}^s(\mathcal{M}, w))$.

Now we will prove (b). As Σ is consistent, it can be realized by a pointed model, which we call $\mathcal{M}', w'$. Observe, first of all, that $\mathcal{M}, w \rightsquigarrow_s \mathcal{M}', w'$ as they both have the same sabotage modal theory. Now take two ω-saturated elementary extensions $(\mathcal{M}_\omega, w)$ and $(\mathcal{M}'_\omega, w')$ of $(\mathcal{M}, w)$ and $(\mathcal{M}', w')$. That such extensions exist can be proven by a chain construction argument (see [9, Proposition 3.2.6]). By the invariance of FOL under elementary extensions, since $\mathcal{M}' \models \varphi[w]$ (by the construction of Σ) we can conclude that $\mathcal{M}'_\omega \models \varphi[w]$. From this, by the assumption that φ is invariant for s-bisimulation and Proposition 3, we conclude that $\mathcal{M}_\omega \models \varphi(x)[w]$ and again, by elementary extension, that $\mathcal{M} \models \varphi(x)[w]$, which establishes claim (ii) and completes the proof. □

Definable and undefinable properties in SML. So which FOL properties belong to the fragment identified by Theorem 3 and which ones do not? We provide examples of SML-definable and undefinable (at model level) properties.

Example 1 (Counting successors). Consider the FOL property "there exist at most n successors" (2). This property is not bisimulation invariant, but it is invariant with respect to sabotage bisimulation. It is therefore definable in SML (by formula (3)).

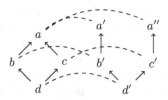

Fig. 1. Two s-bisimilar models (s-bisimulation rendered by the dashed lines). At state d the property "all successors have one same successor" is true. It fails at state d'.

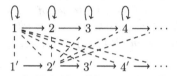

Fig. 2. Sabotage bisimulation between two frames (dashed lines). Only the part of the s-bisimulation relation originating in points 1 and $2'$ is depicted. The top frame is $\mathcal{F} = \langle \mathbb{N}, \geq \rangle$ (transitive edges are omitted) and the bottom model is $\mathcal{F}' = \langle \mathbb{N}, > \rangle$ (transitive edges are omitted).

Example 2 (Confluence). Consider the FOL property "all successors have one same successor". The property is not invariant for sabotage bisimulation. It is therefore not definable in SML. See Figure 1 for an illustration.

Example 3 (Reflexive states). Consider the FOL property xRx. This property is not invariant with respect to sabotage bisimulation. To witness this fact take two pointed models built on the set of natural numbers (with 0) where the point of evaluation is set at 0 and the accessibility relations are: on the first model the *greater or equal* relation (hence reflexive), and on the second one the strictly greater relation (hence irreflexive). That is: $\mathcal{M} = \langle \mathbb{N}, \geq \rangle$ and $\mathcal{M}' = \langle \mathbb{N}, > \rangle$. We have that $\langle \mathcal{M}, 0 \rangle \leftrightarrow_s \langle \mathcal{M}', 0 \rangle$. Figure 2 depicts (part of) a relation which is a (standard) bisimulation Z between the two models and which in addition has the property that any edge deletion on one model can be 'mirrored' on the other model obtaining pointed models that are still connected by Z (recall Definition 2). In particular observe that the deletion of a reflexive edge in $\mathcal{M}$ at point i can be 'mirrored' by the deletion of edge $(i, i+1)$ in $\mathcal{M}'$ (note that the accessibility relations are transitive in both models). However, $\mathcal{M} \models xRx[0]$ and $\mathcal{M}' \not\models xRx[0]$. Property xRx is therefore not definable in SML.

5 Tableau Method for SML

Since SML is not invariant under (standard) bisimulation [14], it is clear that the sort of reduction argument normally used to obtain sound and complete axiom systems for logics of model update (cf. [10]) can not be applied as that would imply an embedding of SML into logic K. It is therefore natural to attempt a semantics-driven approach to the proof theory of SML, like a tableau method.

Moreover, SML does not have the tree-model property: there are specific SML formulae satisfied in Kripke models whose underlying frames can *not* be trees. For example, the formula $\Diamond\top \wedge \Box\Diamond\top \wedge \blacksquare\Box\bot$ is true in a model if and only if its underlying frame consists of one reflexive point. Hence, the labeled tableau system for logic K has to be adapted for SML.

Definition 4 (Label, Labeled Formula and Relation Term). *Let S be an infinite set whose elements are called* labels. *An extended label is an expression of the form ℓ^E where $\ell \in S$ and E is a finite set of pairs of S. A labeled formula is an expression of the form $(\ell^E \; \varphi)$ where ℓ^E is a label and $\varphi \in \mathcal{L}^s$. A relation term is an expression of the form $(R\,\ell_1\,\ell_2)$ where $\ell_1, \ell_2 \in S$.*

Input: A formula $\varphi \in \mathcal{L}^s$.

Output: A tableau $\mathcal{T}$ for φ: each branch may be infinite, finite and labeled open, or finite and labeled closed.

1. Initially, $\mathcal{T}$ is a tree consisting of a single root node labeled with $(\ell^0 \; \varphi)$.
2. Repeat the following steps as long as possible:
   (a) *Choose* a branch which is neither closed nor open and choose a labeled formula $(\ell^E \; \psi)$ (or a pair of labeled formula $(\ell^E \; \psi)$ and relation term $(R\,\ell_1\,\ell_2)$) not selected before on this branch.
   (b) *Apply* the appropriate tableau rule of Figure 4 to $(\ell^E \; \psi)$ (or the pair $(\ell^E \; \psi)$, $(R\,\ell_1\,\ell_2)$):
      – if the tableau rule is rule $\neg\wedge$ (or rules $\Diamond$, $\blacklozenge$), add two successor nodes (resp. $n+1$, n successor nodes) to the branch labeled with the instantiations of the denominators of that rule,
      – otherwise, add a unique successor node labeled with the instantiation of the denominator(s) of that rule.
   (c) i. *Label* by $\times$ (*closed*) the (new) branches which contain two labeled formulae $(\ell^E \; p)$ and $(\ell^F \; \neg p)$ (where E and F may possibly be different sets) or two labeled formulae $(\ell^E \; \varphi)$ and $(\ell^E \; \neg\varphi)$.
      ii. *Label* by $\odot$ (*open*) the (new) branches where there are no more formulae to decompose.

Fig. 3. Construction of a tableau.

Definition 5 (Tableau). *A (labeled) tableau is a tree whose nodes are labeled with labeled formulae or relation terms. The tableau tree for a formula is constructed as shown in the algorithm of Figure 3. In the tableau rules of Figure 4, the formulae above the horizontal lines are called* numerators *and those below are called* denominators. *A tableau closes when all its branches are closed. A branch is open when it is infinite or it terminates in a leaf labeled open.*

$$\frac{(\ell^E \ \varphi \wedge \psi)}{(\ell^E \ \varphi) \ (\ell^E \ \psi)} \ \wedge \qquad \frac{(\ell^E \ \neg(\varphi \wedge \psi))}{(\ell^E \ \neg\varphi) \mid (\ell^E \ \neg\psi)} \ \neg\wedge \qquad \frac{(\ell^E \ \neg\neg\varphi)}{(\ell^E \ \varphi)} \ \neg\neg$$

$$\frac{(\ell_1^E \ \neg\Diamond\varphi) \quad (R \ \ell_1 \ \ell_2)}{(\ell_2^E \ \neg\varphi)} \ \neg\Diamond \qquad \frac{(\ell^E \ \neg\blacklozenge\varphi) \quad (R \ \ell_1 \ \ell_2)}{(\ell^{E\cup\{(\ell_1,\ell_2)\}} \ \neg\varphi)} \ \neg\blacklozenge$$

where $(\ell_1, \ell_2) \notin E$ in both rules above.

$$\frac{(\ell^E \ \Diamond\varphi)}{(R \ \ell \ \ell_1)(\ell_1^E \ \varphi) \mid \ldots \mid (R \ \ell \ \ell_n)(\ell_n^E \ \varphi) \mid (R \ \ell \ \ell_{n+1})(\ell_{n+1}^E \ \varphi)} \ \Diamond$$

where $\{\ell_1, \ldots, \ell_n\}$ are all the labels occurring in the current branch such that $(\ell, \ell_i) \notin E$ for all $i \in \{1, \ldots, n\}$ and ℓ_{n+1} is a 'fresh' label not occurring in the current branch.

$$\frac{(\ell^E \ \blacklozenge\varphi)}{(R \ \ell_1 \ \ell_1')(\ell^{E\cup\{(\ell_1,\ell_1')\}} \ \varphi) \mid \ldots \mid (R \ \ell_n \ \ell_n')(\ell^{E\cup\{(\ell_n,\ell_n')\}} \ \varphi)} \ \blacklozenge$$

where $\{(\ell_1, \ell_1'), \ldots, (\ell_n, \ell_n')\} := (M \times M) \cup \{(\ell_+, \ell_{++})\} \setminus E$, with M the set of labels occurring in the current branch to which we add a 'fresh' label ℓ_*, and (ℓ_+, ℓ_{++}) is a pair of 'fresh' and distinct labels.

Fig. 4. Tableau rules.

The construction of a tableau may not necessarily terminate (see Example 4). This is in line with the fact that the satisfiability problem of SML is undecidable. Nevertheless, a tableau closes only if the construction terminates. Note that if we remove the rules for sabotage we obtain a sound and complete tableau method for logic K which is somewhat non-standard (and computationally demanding).

Theorem 4 (Soundness and Completeness). *Let* $\varphi \in \mathcal{L}^s$*. If* φ *is unsatisfiable, then the tableau for* φ *closes (completeness). If the tableau for* φ *closes then* φ *is unsatisfiable (soundness).*

Example 4. In Figure 5, on the right, we display the execution of the tableau method of Figure 3 on the formula $\Diamond\top \wedge \Box\Diamond\top \wedge \blacksquare\Box\bot$. We obtain a single open branch (labeled with $\odot$) from which we can extract a model whose frame is a single reflexive point. This formula is thus satisfiable, and in fact only in this frame. In Figure 5, on the left, we show that the tableau construction may not necessarily terminate by exhibiting an infinite branch in the tableau for the formula $\Diamond\top \wedge \blacksquare\Diamond\top$. Even if the formula holds in pointed models having at least two successors, our tableau method does not terminate with this formula as input and produces a pointed model with infinitely many successors.

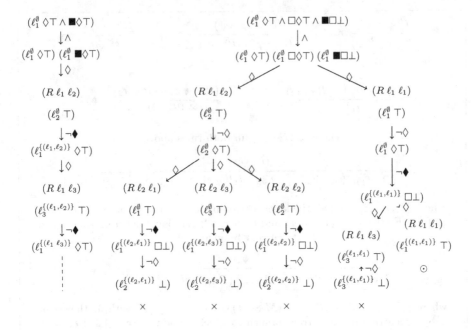

Fig. 5. An infinite branch in the tableau for $\Diamond\top \wedge \blacksquare\Diamond\top$ (*left*), and tableau for $\Diamond\top \wedge \Box\Diamond\top \wedge \blacksquare\Box\bot$ (*right*).

6 Conclusions and Future Work

We have touched upon some model theoretic aspects of SML and fleshed out the theory of a standard translation for SML, which was only sketched in [15]. We have studied such translation together with a notion of bisimulation tailored to SML thereby establishing a novel characterization theorem for the logic. We have also provided the first proof system for SML in the form of a sound and complete tableau method.

SML remains a rather under-investigated formalism and many natural questions are still open. We conclude by mentioning a few. First, it is unclear to what extent standard techniques of modal correspondence theory (see [4, Ch. 3]) are applicable to SML. In particular, can the Sahlqvist theorem be extended to SML? Second, the set of valid formulae of SML is not closed under uniform substitution (e.g., $p \leftrightarrow \blacksquare p$). Is the set of schematic validities of SML decidable? Is it axiomatizable? Third, SML is not a well-behaved logic (recall Theorem 1). The fact that edge deletion is arbitrary seems to be the key feature that sets SML apart from better behaved logics in the dynamic epistemic logic landscape where deletions, even of a very general kind, are definable (e.g., [8]). Are there natural restrictions on the semantics of SML (e.g., 'localized' edge deletion) which yield better behaved variants? Finally, the notion of sabotage bisimulation suggests a natural operationalization of equivalence in terms of model comparison games. How such games relate to the original sabotage game of [6] and what further insights they can give into SML are also worthwhile lines of research.

Acknowledgments. For this research Davide Grossi was supported in part by NWO (VENI grant 639.021.816), and in part by EPSRC (grant EP/M015815/1). Guillaume Aucher was supported in part by AFR (grant TR-PDR BFR08-056).

References

1. Areces, C., Fervari, R., Hoffmann, G.: Moving arrows and four model checking results. In: Ong, L., de Queiroz, R. (eds.) WoLLIC 2012. LNCS, vol. 7456, pp. 142–153. Springer, Heidelberg (2012)
2. Areces, C., Fervari, R., Hoffmann, G.: Swap logic. Logic Journal of the IGPL 22(2), 309–332 (2014)
3. Aucher, G., Balbiani, P., Fariñas del Cerro, L., Herzig, A.: Global and local graph modifiers. Electronic Notes in Theoretical Computer Science 231, 293–307 (2009)
4. Blackburn, P., de Rijke, M., Venema, Y.: Modal Logic. Cambridge University Press, Cambridge (2001)
5. van Benthem, J.: Modal Logic and Classical Logic. Monographs in Philosophical Logic and Formal Linguistics. Bibliopolis (1983)
6. van Benthem, J.: An essay on sabotage and obstruction. In: Hutter, D., Stephan, W. (eds.) Mechanizing Mathematical Reasoning. LNCS (LNAI), vol. 2605, pp. 268–276. Springer, Heidelberg (2005)
7. van Benthem, J.: Logical Dynamics of Information and Interaction. Cambridge University Press (2011)
8. van Benthem, J., Liu, F.: Dynamic logic of preference upgrade. Journal of Applied Non-Classical Logic 17(2) (2007)
9. Chang, C.C., Keisler, H.J.: Model Theory. Studies in Logic and the Foundations of Mathematics. North-Holland (1973)
10. van Ditmarsch, H., Kooi, B., van der Hoek, W.: Dynamic Epistemic Logic. Synthese Library Series, vol. 337. Springer (2007)
11. Gierasimczuk, N., Kurzen, L., Velázquez-Quesada, F.R.: Learning and teaching as a game: a sabotage approach. In: He, X., Horty, J., Pacuit, E. (eds.) LORI 2009. LNCS (LNAI), vol. 5834, pp. 119–132. Springer, Heidelberg (2009)
12. Gruener, S., Radmacher, F., Thomas, W.: Connectivity games over dynamic networks. Theoretical Computer Science 498, 46–65 (2013)
13. Kooi, B., Renne, B.: Arrow update logic. Review of Symbolic Logic 4(4) (2011)
14. Löding, C., Rohde, P.: Model checking and satisfiability for sabotage modal logic. In: Pandya, P.K., Radhakrishnan, J. (eds.) FSTTCS 2003. LNCS, vol. 2914, pp. 302–313. Springer, Heidelberg (2003)
15. Löding, C., Rohde, P.: Solving the sabotage game is PSPACE-hard. Technical report, Department of Computer Science RWTH Aachen (2003)
16. Radmacher, F., Thomas, W.: A game theoretic approach to the analysis of dynamic networks. Electronic Notes in Theoretical Computer Science 200(2), 21–37 (2008)

Game Theoretical Semantics
for Paraconsistent Logics

Can Başkent

Department of Computer Science, University of Bath, England
can@canbaskent.net, www.canbaskent.net/logic

1 Introduction

Game theoretical semantics suggests a very intuitive approach to formal semantics. *The semantic verification game* for classical logic is played by two players, *verifier* and *falsifier* who we call Heloise and Abelard respectively. The goal of Heloise in the game is to verify the truth of a given formula in a given model whereas for Abelard it is to falsify it. The rules are specified syntactically based on the form of the formula. During the game, the given formula is broken into subformulas step by step by the players. The game terminates when it reaches the propositional literals and when there is no move to make. If the game ends up with a propositional literal which is true in the model in question, then Heloise wins the game. Otherwise, Abelard wins. Conjunction is ssociated with Abelard, disjunction with Heloise. That is, when the main connective is a conjunction, it is Abelard's turn to choose and make a move, and similarly, disjunction yields a choice for Heloise. The negation operator switches the roles of the players: Heloise becomes the falsifier, Abelard becomes the verifier. The major result of this approach states that Heloise has a winning strategy *if and only if* the given formula is true in the given model. The semantic verification game and its rules are shaped by classical logic and consequently by its restrictions. In this work, we first observe how the verification games change in non-classical, especially propositional paraconsistent logics, and give Hintikka-style game theoretical semantics for them. We will obtain games in which winning strategies for players are not necessary and sufficient conditions for truth values of the formulas.

Game theoretical semantics (GTS, for short) was largely popularized by Hintikka and Helsinki School researchers even though earlier pointers to similar ideas can be found in Parikh [12]. An overview of the field and its relation to various epistemic and scientific topics can be found in [15]. Moreover, [9,14,15] provide extensive surveys of GTS. A game theoretical concept of truth and its relation to winning strategies were investigated by [3]. Pietarinen considered various non-classical issues including partiality and non-competetive games within the framework of GTS with some connections to the Kleene logic without focusing on particular (paraconsistent) logics [13,16,23]. Hintikka and Sandu discussed non-classicality in GTS also without specifically offering any insight on paraconsistency [9,14]. Tulenheimo studied languages with two negation signs, which can bear some resembles to paraconsistent ideas on weak and strong negations [26]. Additionally, there

© Springer-Verlag Berlin Heidelberg 2015
W. van der Hoek et al. (Eds.): LORI 2015, LNCS 9394, pp. 14–26, 2015.
DOI: 10.1007/978-3-662-48561-3_2

were some technical work discussing the intersection of GTS and intuitionism including some work on type-theoretical foundations [21]. An epistemic, first-order extension of GTS, called "Independence-Friendly" logic, was suggested by Hintikka and Sandu relating GTS to Henkin quantifiers [8,11]. Some discussions on intuitionism from the viewpoint of GTS are worth noting. Tennant argued that some aspects of GTS do not work intuitionistically [25]. Similarly, Hintikka noted that the law of excluded middle may not hold in some instances since the lack of a winning strategy for a player does not entail the existence of a winning strategy for the other player [7]. However, Hintikka himself, perhaps with the exception of independence-friendly logic, is not very clear on GTS and intuitionism, especially when it comes to negation [25]. GTS relates directly to various issues in programming languages, yet, this will not be our focus here.

In this work, we consider propositional paraconsistent logics. We define paraconsistent logic as any formal system that does *not* satisfy the explosion principle: $\varphi, \neg\varphi \vdash \psi$ for any φ, ψ. There exists a wide variety of paraconsistent logics, and there are numerous ways to construct them [5,18,19]. Apart from its proof-theoretical definition, paraconsistency can also be described semantically suggesting that in paraconsistent logic some formulas and their negations can both be true.

Apart from studying the underlying logic, GTS can also be approached from a game theoretical perspective. It is then worthwhile to consider verification games where i) Abelard and Heloise both may win, ii) Abelard and Heloise both may lose, iii) Heloise may win, Abelard may not lose, iv) Abelard may win, Heloise may not lose, v) There is a tie, vi) There is an additional player, vii) Players do not take turns. Such different possibilities can occur, for instance, when both p and $\neg p$ are true, so that both players can have winning strategies. We can also imagine verification games with additional truth values and additional players beyond verifiers and falsifiers, and also construct games where players may play simultaneously.

This paper investigates the logical conditions which entail such game theoretical conditions, and aims at filling the gap in the literature between GTS and paraconsistency. In what follows, we consider a variety of well-known paraconsistent logics, offer a game semantics for them and observe how different logics generate different verification games. This is also important philosophically especially when winning strategies are seen as constructive proofs for truth in an intuitionistic sense or when they are seen as verifications [3]. Therefore, by focusing on inconsistent formulas and associated winning strategies, we offer (constructive) proofs for inconsistencies (cf. appendix) and expand the computational discussions on the connection between proofs, strategies and truth.

2   Game Semantics for Logic of Paradox

Logic of paradox (LP, for short) introduces an additional truth value P, called *paradoxical*, which intuitively stands for both true and false [17].

LP is a conservative extension of the classical logic, thus preserves the classical truth. The logics LP and Kleene's three valued system K3 have the same truth tables. However, they differ on the truth values that they preserve in valid inferences, and how they read P. It is read as *over-valuation* in LP and as *under-valuation* in K3. The truth values that are preserved in validities are called *designated truth values* [20]. In LP, it is the set $\{T, P\}$; in K3, it is the set $\{T\}$. Designated truth values can be thought of as extensions of the classical notion of truth. Even if the truth tables of two logics are the same, different sets of designated truth values produce different sets of validities, thus different logics. For instance, $p \vee \neg p$ is a theorem in LP, but not in K3.

	$\neg$
T	F
F	T
P	P

$\wedge$	T	P	F
T	T	P	F
P	P	P	F
F	F	F	F

$\vee$	T	P	F
T	T	T	T
P	T	P	P
F	T	P	F

Fig. 1. *The truth table for LP and K3.*

We stipulate that the introduction of the third truth value requires an additional player that we call *Astrolabe* after Abelard and Heloise's son. Astrolabe is the *paradoxifier* in the game forcing the game to an end with P.

In GTS for LP, the first problem is to determine the turns of the players at each connective. For instance, if the formula $T \wedge P$ is considered, the problem becomes evident. In this quick game, if we assume that it is Abelard's turn then he will not have a move that can bring him a win. From the truth table, it can be seen that the formula evaluates to P, so Astrolabe can be expected to have a winning strategy. In order to make it possible, then, Astrolabe must be allowed to make a move at a conjunction, too. Similarly, if $F \vee P$ is considered, which evaluates to P, Eloise cannot make a move that can bring him a win, and Astrolabe needs to be given a turn to make a move to win the game. Therefore, we associate disjunction with Heloise and Astrolabe, and conjunction with Abelard and Astrolabe. This modification introduces parallel play where the players may make moves in a parallel, concurrent fashion. In the case of a negation, Heloise and Abelard will switch their role, and Astrolabe will keep his role as P is a fixed-point for negation in LP. Astrolabe's role always remains as the paradoxifier.

Let us now formally define GTS for LP following the terminology in [14]. First, we take the language $\mathcal{L}$ of propositional logic with its standard signature. A model M is a tuple (S, v) where S is a non-empty domain on which the game is played, and the valuation function v assigns the terms in $\mathcal{L}$ to truth values in the logic. For simplicity, we assume $\mathcal{L}$ does not have $\rightarrow$ nor $\leftrightarrow$. We define the verification game as a tuple $\Gamma = (\pi, \rho, \sigma, \delta)$ where π is the set of players, ρ is the set of well-defined game rules, σ is the set of positions, and δ is the set of designated truth values. The set of positions is determined by the subformulas of the given formula and remains unchanged in the logics we discuss as they use the same propositional syntax. We embed the turn function at the positions into the rules of the game for simplicity. A *semantic verification game* is defined as $\Gamma(M, \varphi)$ for a game Γ, model M and a formula $\varphi \in \mathcal{L}$. A *strategy* for a player is a set of rules that tells him which move to make at each position where it is his turn. A *winning strategy* is the one that guarantees a win for the player regardless of the moves of the opponent(s). A winning strategy for a player does not necessarily entail the lack of a winning strategy for the opponent(s). Let us now reconsider the following example before determining the π and ρ for LP.

Example 1. Consider the formula $(P \wedge T) \vee (P \wedge F)$ which evaluates to P. In this game, Astrolabe has a winning strategy: at each end-node ($P \wedge T$ and $P \wedge F$), he selects P. Here, we also observe that Abelard being stuck at some states (such as $P \wedge T$) does not necessarily entail a win for neither of the other players.

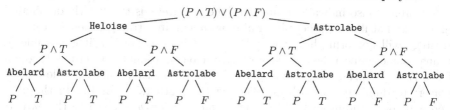

We call the verification game for LP as GTS^{LP}. GTS^{LP} is a non-zero sum verification game where more than one player may have a winning strategy, and making the opponent lose does not necessarily entail that it is a win for the player himself. Also, as we shall see, in GTS^{LP} admitting winning strategies does not necessarily entail the truth value of the formula in question.

Definition 1. *The tuple $\Gamma_{\mathrm{LP}} = (\pi, \rho, \sigma, \delta)$ is an LP verification game for LP where $\pi = \{Astrolabe, Heloise, Abelard\}$, σ is as in classical logic, δ is $\{T, P\}$ and ρ is given as follows inductively for a game $\Gamma_{\mathrm{LP}}(M, \varphi)$.*
— If φ is atomic, the game terminates, and Heloise wins if φ is true, Abelard wins if φ is false and Astrolabe wins if φ is paradoxical,
— if $\varphi = \neg\psi$, Abelard and Heloise switch roles, Astrolabe keeps his role, and the game continues as $\Gamma_{\mathrm{LP}}(M, \psi)$,
— if $\varphi = \chi \wedge \psi$, Abelard and Astrolabe choose between χ and ψ simultaneously,
— if $\varphi = \chi \vee \psi$, Heloise and Astrolabe choose between χ and ψ simultaneously.

Correctness theorem for GTS^{LP} follows.

Theorem 1. *In a GTS^{LP} verification game $\Gamma_{\mathrm{LP}}(M, \varphi)$*
— Heloise has a winning strategy if φ is true in M,
— Abelard has a winning strategy if φ is false in M,
— Astrolabe has a winning strategy if φ is paradoxical in M.

LP distinguishes different trues and falses: trues that are only true (T), falses that are only false (F), and trues that are also false (P) and falses that are also ture (P). In GTS, this carries over to games allowing Astrolabe making moves alongside Heloise and Abelard. In, GTS^{LP} there are winning strategies that causes a loss for the opponent, and there are winning strategies do not. Additionally, there are winning strategies that cannot guarantee the logical truth of formulas. A game for $P \wedge F$ illustrate this point, where both Abelard and

Astrolabe can have a winning strategy. But, this does not directly say anything about the truth value of $P \wedge F$. Therefore, in GTS^{LP}, the immediate connection between the existence of winning strategies and truth values becomes slightly more complicated as the following theorem identifies.

Theorem 2. *In a GTS^{LP} verification game $\Gamma_{\mathrm{LP}}(M, \varphi)$,*
— *If Heloise has a winning strategy, then φ is true in M;*
— *If Abelard has a winning strategy, then φ is false in M;*
— *If Astrolabe has a winning strategy, but not the other players, then φ is paradoxical in in M.*

Theorem 2 also indicates that Astrolabe's strategy is the strictly dominated in a sense that if some other player also has a winning strategy, then Astrolabe's strategy will not bring him a win. Based on this observation, it is possible to change some game rules in order to give a biconditional correctness theorem for GTS^{LP} by prioritizing some players over the others. This will allow some players to dominate the others reflecting the truth table for LP. In this new and extended reading of GTS^{LP}, such a *move priority* is given to the *parents* (Abelard and Heloise), they are let to play first, then Astrolabe makes his move. This extension prevents parallel moves and incorporates winning strategies into the game rules. These additional rules are given as follows.

1. For propositional letters and negation, the rules are as before.
2. Disjunction belongs to Heloise and Astrolabe; conjunction belongs to Abelard and Astrolabe.
3. If Heloise (resp. Abelard) has a winning strategy in the sub-game they choose, the game proceeds with her (resp. his) move.
4. Otherwise, Astrolabe makes a move.

Example 2. Let us consider the formula in Example 1. Given $(P \wedge T) \vee (P \wedge F)$, Heloise first attempts to choose either of them only to realize that she does not have a winning strategy in either of the sub-games with $P \wedge T$ or $P \wedge F$. So, she cannot make a move, and it becomes Astrolabe's turn. Astrolabe chooses $P \wedge T$. Now, Abelard attempts to choose either P or T only to realize that neither brings him a win. So, he cannot make a move. Astrolabe makes a move, chooses P, and wins - this is Astrolabe's winning strategy. If Astrolabe chose $P \wedge F$, then first Abelard would make a move and choose F for a win. Yet, Abelard still does not have a winning strategy in this game.

As we mentioned earlier, such a twist on GTS^{LP} is *ad-hoc*. It incorporates possessing winning-strategies, which is a meta-logical condition, into game rules, which are supposed to be syntactic. This modification naturally provides a biconditional Theorem 1 at the expense of violating the pure syntacticality of the game rules, resulting in completely *ad-hoc* game rules.

3   Game Semantics for First-Degree Entailment

Semantic evaluations are generally thought of as *functions* from logical formulas to truth values. This ensures that each and every formula is assigned a *unique* truth value. However, it is possible to replace the valuation function with a valuation *relation* which can produce multiple truth values for logical formulas. The system obtained in this manner is called *First-degree entailment* (FDE, for short), and is due to Dunn [1,6].

For the given propositional language $\mathcal{L}$, the valuation relation **r** is defined on $\mathcal{L} \times \{0, 1\}$. By $\varphi \mathbf{r} \emptyset$, we will denote the situation where φ is not related to any truth value. By $\varphi \mathbf{r} \{0, 1\}$, we denote the situation when φ is related to both truth values. FDE is a paraconsistent (inconsistency-tolerant) and paracomplete (incompleteness-tolerant) logic. For formulas $\varphi, \psi \in \mathcal{L}$, the valuation **r** is defined inductively as follows.

- $\neg \varphi \mathbf{r} 1$ iff $\varphi \mathbf{r} 0$
- $\neg \varphi \mathbf{r} 0$ iff $\varphi \mathbf{r} 1$
- $(\varphi \wedge \psi) \mathbf{r} 1$ iff $\varphi \mathbf{r} 1$ and $\psi \mathbf{r} 1$
- $(\varphi \wedge \psi) \mathbf{r} 0$ iff $\varphi \mathbf{r} 0$ or $\psi \mathbf{r} 0$
- $(\varphi \vee \psi) \mathbf{r} 1$ iff $\varphi \mathbf{r} 1$ or $\psi \mathbf{r} 1$
- $(\varphi \vee \psi) \mathbf{r} 0$ iff $\varphi \mathbf{r} 0$ and $\psi \mathbf{r} 0$

Notice that LP can be obtained from FDE by imposing a restriction on FDE that no formula gets the truth value $\emptyset$. We denote the GTS for FDE as $\mathrm{GTS}^{\mathrm{FDE}}$.

What does the relational semantics correspond to in verification games? If the truth value P in LP can intuitively be thought of as both true and false, and if this allows concurrent moves in $\mathrm{GTS}^{\mathrm{LP}}$, then the same approach works in $\mathrm{GTS}^{\mathrm{FDE}}$ as well. In FDE, unlike LP, formulas can have no truth value which suggests that neither Heloise nor Abelard may have a winning strategy. Also, in FDE, both players can have winning strategies. We define the verification games for FDE in the standard fashion as follows.

Definition 2. *The tuple $\Gamma_{\mathrm{FDE}} = (\pi, \rho, \sigma, \delta)$ is a FDE verification game where $\pi = \{Heloise, Abelard\}$, σ is as in classical logic, δ is $\{T\}$ and ρ is given as follows inductively for a game $\Gamma_{\mathrm{FDE}}(M, \varphi)$.*

— *If φ is atomic, the game terminates, and Heloise wins if $\varphi \mathbf{r} 1$, Abelard wins if $\varphi \mathbf{r} 0$, neither wins if $\varphi \mathbf{r} \emptyset$,*
— *if $\varphi = \neg \psi$, players switch roles, and the game continues as $\Gamma_{\mathrm{FDE}}(M, \psi)$,*
— *if $\varphi = \chi \wedge \psi$, Abelard and Heloise choose between χ and ψ simultaneously,*
— *if $\varphi = \chi \vee \psi$, Abelard and Heloise choose between χ and ψ simultaneously.*

The above rules determines the turn function for the $\mathrm{GTS}^{\mathrm{FDE}}$ which suggests that both players make moves at all binary connectives. A simple example can be helpful.

Example 3. Consider the formula $p \wedge (q \vee r)$ where $p \mathbf{r} \{0, 1\}$, $q \mathbf{r} \emptyset$ and $r \mathbf{r} 0$. Then, this formula evaluates to 0. In the verification game, Abelard first chooses $q \vee r$, and then chooses r. Alternatively, he can also choose p as his winning strategy, yet this also gives Heloise a win. This is also another case where existence of winning strategies do not guarantee the truth value of the formula in question.

The correctness theorem for $\mathrm{GTS}^{\mathrm{FDE}}$ is given as follows.

Theorem 3. *In a game $\Gamma_{\mathrm{FDE}}(M, \varphi)$, we have the following:*
— *Heloise has a winning strategy if $\varphi \mathbf{r} 1$,*
— *Abelard has a winning strategy if $\varphi \mathbf{r} 0$,*
— *Either of the players or none of the players has a winning strategy if $\varphi \mathbf{r} \emptyset$.*

The connection between FDE and LP can further be explicated as follows.

Corollary 1. *For an LP model M and a formula φ, let M' be the model obtained from M by maintaining the same carrier set and replacing the valuation function of LP with the valuation relation of FDE as follows: $T \mapsto 1$, $F \mapsto 0$ and $P \mapsto \{0, 1\}$. If Heloise or Abelard has a winning strategy in $\Gamma_{\text{LP}}(M, \varphi)$, then Heloise or Abelard has a winning strategy in $\Gamma_{\text{FDE}}(M', \varphi)$ respectively. If only Astrolabe has a winning strategy in $\Gamma_{\text{LP}}(M, \varphi)$, then both Heloise and Abelard have winning strategies in $\Gamma_{\text{FDE}}(M', \varphi)$.*

The converse of Corollary 1 is not true. In GTS^{FDE}, for a game $T \wedge F$, both Abelard and Heloise have winning strategies. Yet, in LP for a game $T \wedge F$, Astrolabe does not have a winning strategy.

The lack of biconditional correctness theorem for GTS^{FDE} can be seen more clearly once LP is considered as a restricted case of FDE.

4   Game Semantics for A Relevant Logic

Relevant logics define negation differently by resorting to possible worlds modalizing the negation operator. The idea is due to Routley and Routley, and we will focus on their logic [22]. A *Routley model* is a structure $(W, \#, v)$ where W is a set of possible worlds, $\#$ is a map from W to itself, and v is a valuation function defined in the standard way. In this system, the semantics for disjunction and conjunction is local, whereas for negation, possible worlds are needed.

$$v(w, \neg\varphi) = 1 \quad \text{iff} \quad v(\#w, \varphi) = 0$$
$$v(w, \varphi \wedge \psi) = 1 \quad \text{iff} \quad v(w, \varphi) = 1 \text{ and } v(w, \psi) = 1$$
$$v(w, \varphi \vee \psi) = 1 \quad \text{iff} \quad v(w, \varphi) = 1 \text{ or } v(w, \psi) = 1$$

We call Routleys' system RR, and denote its GTS as GTS^{RR}. Notice that if $\#w = w$, then we have the classical truth conditions. Further connections between RR and FDE or LP can be found in [19]. We define semantical games in RR as $\Gamma_{\text{RR}}(M, \varphi, w)$ where M, φ are as before, and $w \in W$ is a possible world.

Definition 3. *The tuple $\Gamma_{\text{RR}} = (\pi, \rho, \sigma, \delta)$ is a RR verification game where $\pi = \{\text{Heloise, Abelard}\}$, σ is in the form of (φ, w) for $\varphi \in \mathcal{L}$ and $w \in W$, δ is $\{T\}$ and ρ is given as follows inductively for a game $\Gamma_{\text{RR}}(M, \varphi, w)$ where w is a possible world.*
– If φ is atomic, the game terminates, and Heloise wins if φ is true, Abelard wins if φ is false,
– if $\varphi = \neg\psi$, the players switch roles, and the game continues as $\Gamma_{\text{RR}}(M, \psi, \#w)$,
– if $\varphi = \chi \wedge \psi$, Abelard chooses between χ and ψ,
– if $\varphi = \chi \vee \psi$, Heloise chooses between χ and ψ.

The correctness theorem is given as follows.

Theorem 4. *In a game $\Gamma_{\text{RR}}(M, \varphi, w)$, Heloise has a winning strategy if φ is true, and Abelard has a winning strategy if φ is false.*

The converse of Theorem 4 is not correct as the # operator can create inconsistencies. In order to see this, let $w \models \neg\varphi$ and $w' \models \varphi$. If $\#(w) = w'$, then by definition φ is both true and false at w' satisfying an inconsistency.

5 Translating Games

It is possible to give a translation between three-valued logics and modal logic S5 [10]. Modal logic S5 is defined as a system (W, R, V) where W is a non-empty set, R is an equivalence relation on $W \times W$ and V is the valuation.

Now, we give a translation of LP (and K3) into S5 via GTS. The translation is built on the following observation: "In an S5-model there are three mutually exclusive and jointly exhaustive possibilities for each atomic formula p: either p is true in all possible worlds, or p is true in some possible worlds and false in others, or p is false in all possible worlds" [10].

Given the propositional language $\mathcal{L}$, we extend it with the modal symbols $\Box$ and $\Diamond$ and close it under the standard rules to obtain the modal language $\mathcal{L}_M$. GTS for modal logic is well-known. "Diamond" formulas are assigned to Heloise whereas the "Box" formulas are assigned to Abelard. Also, similar to the RR, formulas in $\mathcal{L}_M$ are associated with a possible world, and when a move is made from a modal formula, the next possible world is determined by R.

The translations $\mathsf{Tr}_{LP} : \mathcal{L} \mapsto \mathcal{L}_M$ and $\mathsf{Tr}_{K3} : \mathcal{L} \mapsto \mathcal{L}_M$ for LP and K3 respectively are given as follows where p is a propositional variable [10].

$$
\begin{array}{l|l}
\mathsf{Tr}_{LP}(p) = \Diamond p & \mathsf{Tr}_{LP}(\varphi \wedge \psi) = \mathsf{Tr}_{LP}(\varphi) \wedge \mathsf{Tr}_{LP}(\psi) \\
\mathsf{Tr}_{K3}(p) = \Box p & \mathsf{Tr}_{K3}(\varphi \wedge \psi) = \mathsf{Tr}_{K3}(\varphi) \wedge \mathsf{Tr}_{K3}(\psi) \\
\mathsf{Tr}_{LP}(\neg\varphi) = \neg\mathsf{Tr}_{K3}(\varphi) & \mathsf{Tr}_{LP}(\varphi \vee \psi) = \mathsf{Tr}_{LP}(\varphi) \vee \mathsf{Tr}_{LP}(\psi) \\
\mathsf{Tr}_{K3}(\neg\varphi) = \neg\mathsf{Tr}_{LP}(\varphi) & \mathsf{Tr}_{K3}(\varphi \vee \psi) = \mathsf{Tr}_{K3}(\varphi) \vee \mathsf{Tr}_{K3}(\psi)
\end{array}
$$

The translation is a co-induction, and it generates fully modalized formulas. As the authors underlined, for fully modalized formulas in S5, a formula is true somewhere in an S5 model if and only if it is true everywhere in the model. This fact is due to the frame properties of S5 [10].

Given $\Gamma_{\mathrm{LP}} = (\pi, \rho, \sigma, \delta)$, we define $\Gamma_{\mathrm{S5}} = (\pi', \rho', \sigma', \delta')$ as follows: $\pi' = \{\text{Heloise, Abelard}\}$, ρ and σ' are the rules and positions of verifications games of S5, and $\delta' = \{1\}$. The correctness of the translation for LP is as follows.

Theorem 5. *Let $\Gamma_{\mathrm{LP}}(M, \varphi)$ be given. Then,*
– if Heloise has a winning strategy in $\Gamma_{\mathrm{LP}}(M, \varphi)$, then she has a winning strategy in $\Gamma_{\mathrm{S5}}(M, \mathsf{Tr}_{LP}(\varphi))$,
– if Abelard has a winning strategy in $\Gamma_{\mathrm{LP}}(M, \varphi)$, then he has a winning strategy in $\Gamma_{\mathrm{S5}}(M, \mathsf{Tr}_{LP}(\varphi))$,
– if only Astrolabe has a winning strategy in $\Gamma_{\mathrm{LP}}(M, \varphi)$, then both Abelard and Heloise have winning strategies in $\Gamma_{\mathrm{S5}}(M, \mathsf{Tr}_{LP}(\varphi))$.

For an LP valuation v, and a model M of S5, v and M are said to be Tr_{LP}-*equivalent* if for all $\varphi \in \mathcal{L}$ we have (i) $1 \in v^*(\varphi) \Leftrightarrow M \models_{S5} \mathsf{Tr}_{LP}(\varphi)$, and (ii) $0 \in v^*(\varphi) \Leftrightarrow M \not\models_{S5} \mathsf{Tr}_{K3}(\varphi)$, where v^* is the (truth table) function based on v

that maps *formulas* to truth values of LP. Based on various results in [10], we now prove the following, the converse of Theorem 5.

Theorem 6. *Let M be an S5 model, $\varphi \in \mathcal{L}$ with an associated verification game $\Gamma_{S5}(M, \varphi)$. Then, there exists an LP model M' and a game $\Gamma_{LP}(M', \varphi)$ where,*
— *if Heloise has a winning strategy for $\Gamma_{S5}(M, \varphi)$ at each point in M, then Heloise has a winning strategy in $\Gamma_{LP}(M', \varphi)$,*
— *if Abelard has a winning strategy for $\Gamma_{S5}(M, \varphi)$ at each point in M, then Abelard has a winning strategy in $\Gamma_{LP}(M', \varphi)$,*
— *if Heloise or Abelard has a winning strategy for $\Gamma_{S5}(M, \varphi)$ at some points but not all in M, then Astrolabe has a winning strategy in $\Gamma_{LP}(M', \varphi)$.*

For an application of Theorem 5, consider the formula $p \vee q$ where p and q have the truth values P, F respectively in LP. Then, $\mathsf{Tr}(p \vee q) = \Diamond p \vee \Diamond q$ where p, q have the truth values $\{T, F\}, \{F\}$ respectively in S5. Based on Theorem 5, we expect both players to have winning strategies. First, Heloise has a winning strategy in this game if she chooses p. Also, notice that all possible moves of Heloise brings Abelard a win without him even not making any moves, due to the truth values of p, q. Thus, both players have winning strategies in this game.

6   Conclusion

Giving a full picture of GTS for all paraconsistent logics goes beyond the limits of this article. Some well-studied logics such as da Costa's C-systems and LFIs (*Brazilian School*), 4-valued Belnap logic, the modal extensions of the logics we presented, and the *preservationist* approach (*Canadian School*) are the natural next steps of this project [4,5,2,24].

The current work can be seen as a case for logical pluralism. The classical GTS is essentially a very narrow and limited case with many additional and auxiliary game theoretical and logical presuppositions. Once those assumptions are set aside (or at least questioned) for various reasons, GTS turns out to be expressive enough for a variety of non-classical logics as we have exemplified.

References

1. Anderson, A.R., Belnap, N.D.: First degree entailments. Mathematische Annalen 149, 302–319 (1963)
2. Belnap, N.D., Stell, T.B.: The Logic of Questions and Answers. Yale University Press (1976)
3. Boyer, J., Sandu, G.: Between proof and truth. Synthese 187, 821–832 (2012)
4. da Costa, N.C.A.: On the theory of inconsistent formal systems. Notre Dame Journal of Formal Logic 15(4), 497–510 (1974)
5. da Costa, N.C.A., Krause, D., Bueno, O.: Paraconsistent logics and paraconsistency. In: Jacquette, D. (ed.) Philosophy of Logic, vol. 5, pp. 655–781. Elsevier (2007)
6. Michael Dunn, J.: Intuitive semantics for first-degree entailments and 'coupled trees'. Philosophical Studies 29(3), 149–168 (1976)

7. Hintikka, J.: The Principles of Mathematics Revisited. Cambridge University Press (1996)
8. Hintikka, J., Sandu, G.: Information independence as a semantical phenomenon. In: Fenstad, J.E., Frolov, I.T., Hilpinen, R. (eds.) Logic, Methodology and Philosophy of Science VIII, pp. 571–589. Elsevier (1989)
9. Hintikka, J., Sandu, G.: Game-theoretical semantics. In: van Benthem, J., ter Meulen, A. (eds.) Handbook of Logic and Language, pp. 361–410. Elsevier (1997)
10. Kooi, B., Tamminga, A.: Three-valued logics in modal logic. Studia Logica 101(5), 1061–1072 (2013)
11. Mann, A.L., Sandu, G., Sevenster, M.: Independence-Friendly Logic. Cambridge University Press (2011)
12. Parikh, R.: D structures and their semantics. In: Gerbrandy, J., Marx, M., de Rijke, M., Venema, Y. (eds.) JFAK. UvA (1999), http://www.illc.uva.nl/j50/. ILLC
13. Pietarinen, A.: Logic and coherence in the light of competitive games. Logique et Analyse 43, 371–391 (2000)
14. Pietarinen, A., Sandu, G.: Games in philosophical logic. Nordic Journal of Philosophical Logic 4(2), 143–173 (2000)
15. Pietarinen, A.-V.: Games as formal tools versus games as explanations in logic and science. Foundations of Science 8(4), 317–364 (2003)
16. Pietarinen, A.-V.: Semantic games in logic and epistemology. In: Rahman, S., Gabbay, D., van Bendegem, J.P. (eds.) Logic, Epistemology, and the Unity of Science, pp. 57–103. Kluwer (2004)
17. Priest, G.: The logic of paradox. Journal of Philosophical Logic 8, 219–241 (1979)
18. Priest, G.: Paraconsistent logic. In: Gabbay, D., Guenthner, F. (eds.) Handbook of Philosophical Logic, vol. 6, pp. 287–393. Kluwer (2002)
19. Priest, G.: Paraconsistency and dialetheism. In: Gabbay, D.M., Woods, J. (eds.) Handbook of History of Logic, 1st edn., vol. 8, pp. 129–204. Elsevier (2007)
20. Priest, G.: An Introdiction to Non-Classical Logic. Cambridge University Press (2008)
21. Ranta, A.: Propositions as games as types. Synthese 76(3), 377–395 (1988)
22. Routley, R., Routley, V.: The semantics of first degree entailment. Noûs 6(4), 335–359 (1972)
23. Sandu, G., Pietarinen, A.: Partiality and games: Propositional logic. Logic Journal of the IGPL 9(1), 107–127 (2001)
24. Schotch, P., Brown, B., Jennings, R.: On Preserving: Essays on Preservationism and Paraconsistent Logic. University of Toronto Press (2009)
25. Tennant, N.: Games some people would have all of us play. Philosophia Mathematica 6(3), 90–115 (1998)
26. Tulenheimo, T.: Classical negation and game-theoretical semantics. Notre Dame Journal of Formal Logic 55(4), 469–498 (2014)

Appendix: Proofs

Proof (Proof of Theorem 1). We start with the case for Heloise. We proceed by induction on φ. Let φ be true in M.

If φ is a propositional letter p which is true in M, then Heloise wins the game by definition, hence has a winning strategy.

Let $\varphi = \neg\psi$. Then, ψ is false. By the game rules, now the game continues where Heloise is the falsifier. By the induction hypothesis (for falsifier), Heloise the falsifier has a winning strategy for ψ. Then, she has a winning strategy as the verifier for φ.

Now, let φ be a conjunction of the form $\chi \wedge \psi$. Since, φ is assumed to be true, the only way to make it true is to have χ and ψ both true. Then, by the induction hypothesis, Heloise has a winning strategy for both χ and ψ. Then, for φ, Abelard and Astrolabe make moves. Yet, whichever move they make (whichever of χ or ψ they choose), Heloise will have a winning strategy. Thus, for φ, she has a winning strategy: whatever move Abelard and Astrolabe make, she has a win.

Let φ be a disjunction of the form $\chi \vee \psi$. Then, by the induction hypothesis, Heloise has a winning strategy for either χ or ψ whichever is true. Then, choosing the true disjunct is her winning strategy at φ, independent from whatever Astrolabe chooses.

The case for Abelard is almost identical to that of Heloise's, hence skipped.

For Astrolabe, we first assume that the given formula φ is paradoxical in M. If φ is a propositional letter p which is paradoxical in M, then Astrolabe has a winning strategy by definition. Similarly, if $\varphi = \neg\psi$, then, ψ is paradoxical, too. By the game rules, Astrolabe's rule remains the same. By the induction hypothesis, he has a winning strategy for ψ, and thus for for φ by simply maintaining the same role and the strategy, and proceeding with ψ.

For $\varphi = \chi \wedge \psi$. Since φ is assumed to be paradoxical, we only have two options for χ and ψ: (1) either one of them has the truth value P and the other has the truth value T, (2) both have the truth value P. Therefore, Astrolabe has winning strategy for at least one of χ and ψ, by the induction hypothesis. Then, for φ, Astrolabe chooses the conjunct that has the truth value P for which he has a winning strategy already. This forms his winning strategy for φ, independent from whatever move Abelard makes.

If $\varphi = \chi \vee \psi$, then we have two options as well: (1) one of the disjuncts has the truth value P and the other one has the truth value F, (2) both have the truth value P. By a similar argument Astrolabe has a winning strategy for either case.

Proof (Proof of Theorem 2). The proof is by induction on φ for each player, and the cases for Heloise and Abelard are very similar to the classical case. Now, assume that for φ, only Astrolabe has a winning strategy. The cases for propositional variables and negation are as above, hence skipped.

Now, let $\varphi = \chi \wedge \psi$. If only Astrolabe has a winning strategy, this means, Astrolabe has a winning strategy for either of the conjuncts (as he can choose whichever he likes), say χ without loss of generality. Then, by the induction hypothesis, χ is paradoxical. Since Abelard does not have a winning strategy, by Theorem 1, then neither of the conjuncts is false. Thus, by the truth table φ is forced to be paradoxical as χ is paradoxical. Otherwise, if Abelard had a winning strategy, and if one of the conjuncts was F, then $P \wedge F$ would return F, not P disproving the claim. This is the reason why only Astrolabe is supposed to have a winning strategy.

The case for disjunction for Astrolabe is very similar.

Proof (Proof of Theorem 3). We start with the case for Heloise. Suppose $\varphi r1$. The cases for propositional variables and negation are immediate. Let $\varphi = \chi \wedge \psi$.

If φ**r1**, then we have both χ**r1** and ψ**r1**. By the induction hypothesis, Heloise has winning strategies for both χ and ψ. Thus, she has a winning strategy for φ. For the failure of the reverse direction, assume that Heloise has a winning strategy, that is, to choose χ (without loss of generality). Assume further that, Abelard has a winning strategy as well, that is, to choose ψ. Then, by the indiction hypothesis χ**r1** and ψ**r0** which forces φ**r0**. Heloise's case for disjunction is very similar.

The interesting case is for $\emptyset$. Now, assume φ**r$\emptyset$**. If φ is a propositional variable, by definition, no player wins. If $\varphi = \neg\psi$, then ψ**r$\emptyset$**, and by the induction hypothesis, no player has a winning strategy.

Let $\varphi = \chi \wedge \psi$. Then, we have two options: (1) both χ**r$\emptyset$** and ψ**r$\emptyset$**, or (2) χ**r1** and ψ**r$\emptyset$** (without loss of generality). If the prior one is the case, by the induction hypothesis, no player has a winning strategy for χ or ψ. Thus, no player has a winning strategy for φ. If the latter is the case, then Heloise can have a winning strategy for φ as she can make a move at a conjunction which forms her winning strategy for φ. Dually, if $\varphi = \chi \vee \psi$, then, we have two options: (1) both χ**r$\emptyset$** and ψ**r$\emptyset$**, or (2) χ**r0** and ψ**r$\emptyset$** (without loss of generality). If the prior one is the case, by the same argument as above, no player has a winning strategy for φ. If the latter is the case, as Abelard can make a move at a disjunction and choose χ, then he can have a winning strategy for φ.

Proof (Proof of Corrolary 1). The first part about Heloise and Abelard follows from Theorem 2 and Theorem 3. In other words, if Heloise has a winning strategy in an LP game, then the formula is true in LP by Theorem 2. The translation then translates T of LP to 1 of FDE. Then, by Theorem 3, Heloise has a winning strategy in the FDE game. The argument is similar for Abelard.

If only Astrolabe has a winning strategy for the LP game for φ, then by Theorem 2, φ is paradoxical. By the translation, then φ is related to both 0 and 1 in FDE. By Theorem 3, then both Heloise and Abelard has winning strategies in the FDE game.

Proof (Proof of Theorem 4). The proof is by induction on φ. Let us see the case for Heloise at w. The case for Abelard is very similar hence will be skipped.

If φ is a propositional letter p. Then, if p is true then, by definition, Heloise has a winning strategy.

Let $\varphi = \neg\psi$. Then the game continues at $\#$ for ψ with switched roles, where $v(\#w, \psi) = 0$. Thus Heloise becomes falsifier. Then, by the induction hypothesis (for Abelard), the falsifier has a winning strategy for the game at $\#w$ for ψ. Thus, Heloise has a winning strategy at w for $\neg\psi$ which forms her winning strategy for φ. The cases for conjunction and disjunction are as expected thus omitted.

Proof (Proof of Theorem 5). The theorem is given for LP and S5. Yet, a similar theorem for K3 and S5 can also be given. We will assume the correctness of such a theorem for this proof as the translation co-depends on both LP and K3.

Assume that Heloise has a winning strategy for φ in LP. Let us proceed by induction on φ. If φ is a propositional letter p, then p is true in LP. Then, it translates to S5 as $\Diamond p$, which is a turn for Heloise. Then, the game in S5 starts by Heloise with $\Diamond p$, and she makes a move to p for which she has a winning strategy.

For $\varphi = \neg\psi$, suppose Heloise has a winning strategy for $\neg\psi$ in LP. By the translation, she has a winning strategy for $\neg\mathsf{Tr}_{K3}(\psi)$ in S5. So, by the assumed similar theorem for K3 and S5, Abelard has a winning strategy in S5 for $\mathsf{Tr}_{K3}(\psi)$. Then, in S5 Heloise has a winning strategy for $\neg\mathsf{Tr}_{K3}(\psi)$ which is $\mathsf{Tr}_{LP}(\neg\psi)$. Thus, Heloise has a winning strategy for $\mathsf{Tr}_{LP}(\varphi)$ in S5.

The cases for conjunction and disjunction are immediate. Also the case for Abelard is very similar, hence skipped. The case for Astrolabe is interesting.

Assume that only Astrolabe has a winning strategy for φ in LP. As the first step of the induction, assume $\varphi = p$ for a propositional variable p. So, p is paradoxical. The translation of p into S5 is $\Diamond p$. Also, notice that for paradoxical p, we have $\neg p \equiv p$. The translation of $\neg p$ into S5 is $\Diamond\neg p$. Thus, for a paradoxical p, both players have a winning strategy in the game in S5.

Now, let $\varphi = \neg\psi$. Suppose that only Astrolabe has a winning strategy. By the game rules of $\mathsf{GTS}^{\mathrm{LP}}$, Astrolabe has a winning strategy for ψ as well as the negation of a paradoxical formula is also paradoxical. Now, we will use the co-inductive part of the argument. By the induction hypothesis for the same result for K3, Abelard and Heloise have winning strategies in the translated game in S5 for $\mathsf{Tr}_{K3}(\psi)$. Taking one step back, with their roles switched, both Abelard and Heloise have winning strategies in a game for $\neg\mathsf{Tr}_{K3}(\psi)$, too. Then, by the translation, they have winning strategies for $\mathsf{Tr}_{LP}(\neg\psi)$, which is $\mathsf{Tr}_{LP}(\varphi)$ in S5. A symmetric argument for the K3-S5 is straight forward.

The cases for the binary connectives are straight forward, hence skipped.

Proof (Proof of Theorem 6). In [10], while constructing the LP model based on a given S5 model, the authors associate the propositions that are true *everywhere* with the LP truth value T, the propositions that are true *nowhere* with F, and the propositions that are true *somewhere* with P. They also show that the given S5 model and the LP model obtained in this fashion are Tr_{LP}-equivalent [10].

Based on these observation, then, if Heloise has a winning strategy for $\Gamma_{\mathrm{S5}}(M, \varphi)$ at all points in M, then φ has a truth value T in LP. By Theorem 1, Heloise has a winning strategy in $\Gamma_{\mathrm{LP}}(M', \varphi)$. Similarly, if Abelard has a winning strategy for $\Gamma_{\mathrm{S5}}(M, \varphi)$ at all points in M, then φ has a truth value F in LP. Again, by Theorem 1, Abelard has a winning strategy in $\Gamma_{\mathrm{LP}}(M', \varphi)$. Finally, if Astrolabe has a winning strategy for $\Gamma_{\mathrm{S5}}(M, \varphi)$ at some points in M, then φ has a truth value P in LP. By Theorem 1, Astrolabe has a winning strategy in $\Gamma_{\mathrm{LP}}(M', \varphi)$.

Generalized Ultraproduct
and Kirman-Sondermann Correspondence
for Vote Abstention*

Geghard Bedrosian[1], Alessandra Palmigiano[2,3], and Zhiguang Zhao[2]

[1] Faculty of Mathematics, Bielefeld University, Germany
gbedrosi@math.uni-bielefeld.de
[2] Faculty of Technology, Policy and Management, Delft University of Technology,
The Netherlands
{a.palmigiano,z.zhao-3}@tudelft.nl
[3] Department of Pure and Applied Mathematics, University of Johannesburg,
South Africa

Abstract. The present paper refines Herzberg and Eckert's model-theoretic approach to aggregation. The proposed refinement is aimed at naturally accounting for vote abstention, and is technically based on a more general notion of ultraproduct than the one standardly occurring in model theory textbooks. Unlike the standard ultraproduct construction, which yields the empty model as soon as any one single coordinate features the empty model, this generalized ultraproduct construction faithfully reflects the indication of 'large sets'. Thus, our proposed refinement naturally accounts for those situations in which e.g. a voting round is non-null if and only if a 'large set' of voters actually participate in the vote. In the present setting, Arrow's impossibility theorem also covers 'elections with only two candidates'.

Keywords: Social choice, model theory, ultrafilter, generalized ultraproduct, Kirman-Sonderman correspondence, vote abstention.
Mathematics Subject Classification (2010): 03C20, 03C98, 91B14.
Journal of Economic Literature Classification: D71.

1 Introduction

The present paper pertains to a line of research in social choice theory aimed at understanding the logical underpinning of Arrow's impossibility theorem [1] and also exploring its scope, as well as extending Arrow-type results to infinite electorates.

* The first author gratefully acknowledges financial support by the German Research Foundation (DFG) through the International Graduate College (IGK) *Stochastics and Real World Models* (Bielefeld–Beijing). The research of the second and third author has been made possible by the NWO Vidi grant 016.138.314, the NWO Aspasia grant 015.008.054, and a Delft Technology Fellowship awarded in 2013. The authors thank Frederik Herzberg for his very valuable comments which helped to improve the paper.

This line of research originates in the work of Kirman and Sondermann [11], which characterizes the so-called Arrow-rational social welfare functions by establishing a bijective correspondence between them and the collection of ultrafilters over the set of individuals. The Kirman-Sonderman correspondence hinges on the fact that the decisive coalitions associated with any Arrow-rational social welfare function form an ultrafilter over the set of individuals. Herzberg and Eckert [8] gave a very elegant generalization of the Kirman-Sondermann correspondence in a model-theoretic setting by characterizing Arrow-rational social welfare functions as exactly those defined in terms of an ultraproduct construction parametrized by the ultrafilter of their associated decisive coalitions.

In the literature on social choice, there are several ways to treat abstention[1]. The first approach is to ignore any voters that abstain, and thus working in a variable domain model (see Pivato [14]). The second approach is to treat abstention as if the voters ranked all candidates equally. The third approach is to treat abstention as a separate type of input that may be elicited from a voter. This means that there are two types of inputs that voters may submit: ranking of candidates or abstention. Our method belongs to the third approach.

In the present paper, the results in [8] are extended to a setting in which the assumption that every individual votes/expresses a judgment is dropped. Allowing the empty model to occur in profiles is a natural way to formalize the vote abstention of the corresponding individual. However, the standard model-theoretic notion of ultraproduct is not amenable to support this natural formalization of vote abstention, given that it is enough for a coordinate to be empty for the standard ultraproduct construction to yield the empty set/model. This would correspond to situations in which the abstention of one voter would be enough to declare the voting round null. While this is true in some situations, there are many settings (e.g. referenda) in which the voting round is declared null unless a certain *quorum* of voters is met. Technically, the contribution of the present paper is based on replacing the standard model-theoretic ultraproduct construction with a generalized one, introduced by Makkai [13] in a category-theoretic setting. The main advantage of Makkai's ultraproduct is that it yields the empty model unless nonempty models occur in each coordinate belonging to some member of its associated ultrafilter. In this respect, Makkai's ultraproduct reflects more faithfully than the standard one the indications of the 'large sets' of the ultrafilter.

We observe that, in the extended setting accounting for vote abstention, Arrow's impossibility theorem strengthens. Indeed, the usual assumption, also required in [8], on the existence of three non-isomorphic models of the theory is dropped, and replaced by the weaker requirement on the existence of two non-

---

[1] Notice that we use the term "abstention" in a way which is different from how it is typically used in the social choice literature. In particular, abstention does not mean being indifferent between two options (this would correspond, in our setting, to allowing the model associated with any voter to be a partial but not necessarily linear order). By abstention, we mean that voters do not take part in the voting process altogether.

isomorphic models. This allows us to extend e.g. Arrow's impossibility theorem [1] to a setting of elections with only two candidates (cf. discussions at the end of Section 2).

Finally, from a more methodological perspective, besides allowing for the extension of the results in [8] to a setting accounting for vote abstention, Makkai's ultraproduct construction lends itself to connecting the model-theoretic approach to judgment aggregation to the algebraic and category-theoretic approaches in [4], [7], [9] and [10]. Establishing these systematic connections is the focus of ongoing research.

Structure of the Paper. In Section 2, preliminaries are collected about Arrow-rational aggregators, the first leg of the generalized Kirman-Sondermann correspondence is introduced, and the Arrow's impossibility theorem for vote abstention is briefly discussed. In Section 3, the generalized ultraproduct construction is introduced as a specialization of Makkai's general definition to the present model-theoretic setting. In Section 4, relevant properties are collected of the generalized ultraproduct construction. In Section 5, the second leg of the generalized Kirman-Sondermann correspondence is introduced, and the proof of the Kirman-Sondermann isomorphism is given. In Section 6, the case study of preference aggregation in the setting of vote abstention is discussed.

2   Arrow-Rational Aggregators

Fix a first-order language $\mathcal{L}$, consisting of identity $\approx$, constant symbols c for each element in a given non-empty set A and of relation symbols R each of which of finite arity $k = k(R)$. Let $\mathcal{S}$ denote the set of atomic $\mathcal{L}$-formulas, and $\mathcal{I}$ the Boolean closure of $\mathcal{S}$. Fix a consistent set T of universal $\mathcal{L}$-sentences, let Ω be the class of models M of T the domain of which coincides with the subset A^M of the interpretations in M of the constant symbols in $\mathcal{L}$. In what follows, we will always consider models up to isomorphism. Hence, models in Ω can be thought of as equivalence classes of isomorphic models. We let $|\Omega|$ denote the cardinality of Ω modulo isomorphism. We will denote $\mathcal{L}$-structures by $\mathcal{B}$, and elements in Ω by M, N, possibly with subscripts or superscripts. Sometimes, abusing notation, we will use M, N for elements in $\Omega \cup \{\varnothing\}$. We let $R, \ldots, c, \ldots$ denote the symbols in the language $\mathcal{L}$ and let $R^{\mathcal{B}}, \ldots, c^{\mathcal{B}}, \ldots$ denote the corresponding semantic object in the $\mathcal{L}$-structure $\mathcal{B}$. For each $\mathcal{L}$-structure $\mathcal{B}$, let $A^{\mathcal{B}} := \{c^{\mathcal{B}} \mid c \text{ constant symbol in } \mathcal{L}\}$. We let $|M|$ denote the domain of M. As usual, for any model M and formula λ, we write $M \models \lambda$ to indicate that λ is true of M.

The extra assumption that the universe of each model M in Ω is the set $A^M = \{c^M \mid c \text{ is a constant symbol in } \mathcal{L}\}$ guarantees the following

Fact 1. *Any two models $M_1, M_2 \in \Omega \cup \{\varnothing\}$ such that $M_1 \models \lambda$ iff $M_2 \models \lambda$ for any $\lambda \in \mathcal{I}$ are isomorphic.*

Proof. The claim trivially holds both when M_1 and M_2 coincide with the empty set, and when only one of the two coincides with the empty set (in the latter case the assumptions do not hold: indeed, the sentence $c \approx c$ for any constant symbol c holds of the nonempty model and does not hold of the empty model). If M_1 and M_2 are both nonempty, then their domains bijectively correspond: indeed, $|M_1| = A^{M_1} \cong A^{M_2} = |M_2|$. By definition, this bijective correspondence identifies the interpretations of all constant symbols. Since by assumption $M_1 \models R(c_1, \ldots c_k)$ iff $M_2 \models R(c_1, \ldots c_k)$ for any relation symbol R and all constant symbols $c_1, \ldots c_k$, it is a straightforward verification that this correspondence identifies also the interpretations of each relation symbol.

Fix a non-empty set I, which we will think of as the set of *individuals*. The subsets of I will be referred to as *coalitions*. Elements $\underline{M} \in (\Omega \cup \{\varnothing\})^I$ are the *profiles*. For any such profile, and any $\lambda \in \mathcal{L}$, the *coalition supporting λ given $\underline{M}$* is the set $C(\underline{M}, \lambda) := \{i \in I \mid M_i \models \lambda\}$.

An *aggregator* is a partial map $f : (\Omega \cup \{\varnothing\})^I \to \Omega \cup \{\varnothing\}$. The domain of f is denoted $dom(f)$.

Definition 1. *(cf. [8], definition before Remark 3.3) An aggregator f is **Arrow-rational** if it satisfies the following conditions:*

(A1) **Universal Domain:** *$dom(f) = (\Omega \cup \{\varnothing\})^I$.*
(A2) **Generalized Pareto Principle:** *for any $\underline{M} \in dom(f)$ and any $\lambda \in \mathcal{L}$,*

$$\text{if } f(\underline{M}) \models \lambda, \text{ then } C(\underline{M}, \lambda) \neq \varnothing.$$

(A3) **Generalized Systematicity:** *for all $\underline{M}, \underline{N} \in dom(f)$ and all $\lambda, \mu \in \mathcal{L}$,*

$$\text{if } C(\underline{M}, \lambda) = C(\underline{N}, \mu), \text{ then } f(\underline{M}) \models \lambda \text{ iff } f(\underline{N}) \models \mu.$$

The collection of Arrow-rational aggregators is denoted by $\mathcal{AR}$.

Definition 2. (Decisive Coalition) *For any aggregator f, a coalition $C \subseteq I$ is f-decisive if, for any $\lambda \in \mathcal{L}$ and any $\underline{M} \in dom(f)$,*

$$\text{if } C = C(\underline{M}, \lambda), \text{ then } f(\underline{M}) \models \lambda.$$

Let $\mathcal{D}_f$ denote the set of the f-decisive coalitions.

The following lemma is an immediate consequence of the definitions involved:

Lemma 1. *For any aggregator f satisfying (A3), any $\underline{M} \in dom(f)$ and $\lambda \in \mathcal{L}$,*

$$C(\underline{M}, \lambda) \in \mathcal{D}_f \quad \text{iff} \quad f(\underline{M}) \models \lambda.$$

The following lemma shows that the assignment $f \mapsto \mathcal{D}_f$ defines a map $\Lambda : \mathcal{AR} \to \beta I$, where βI denotes the set of ultrafilters over I. The map Λ provides one direction of the generalized Kirman-Sondermann correspondence we aim at obtaining. The following lemma is a variant of Lemma 5.3 in [8], which assumes the aggregator to be weakly Arrow-rational[2] instead of Arrow-rational, as is done

---

[2] An aggregator is *weakly Arrow-rational* if it satisfies conditions of (A2), (A3) of Definition 1 and the following condition (A1'): there exist models $M_1, M_2, M_3 \in \Omega$ s.t. $\{M_1, M_2, M_3\}^I \subseteq dom(f)$, and M_1, M_2, M_3 respectively are models of three pairwise inconsistent $\mathcal{L}$-sentences.

here. Another perhaps more interesting difference is that here we assume that there are at least two non-isomorphic models in Ω, whereas Lemma 5.3 in [8] assumes the existence of at least three non-isomorphic models in Ω. The proof of this lemma can be found in an expanded version of the present paper [2].

Lemma 2. *For any $f \in \mathcal{AR}$, the collection $\mathcal{D}_f$ is an ultrafilter over I.[3]*

Notice that there are significant cases in which the lemma above is not implied by Lemma 5.3 in [8]. The reason is that, in significant cases, Arrow-rationality does not imply weak Arrow-rationality. Indeed, it was shown in [8, Remark 3.2] that condition (A1) implies condition (A1') if $\mu, \nu \in \mathcal{S}$ exist such that $\mu \wedge \nu, \mu \wedge \neg \nu$ and $\neg \mu \wedge \nu$ are each consistent with T. In this case, three pairwise different models M_1, M_2, M_3 exist in Ω such that $M_1 \models \mu \wedge \nu, M_2 \models \mu \wedge \neg \nu$ and $M_3 \models \neg \mu \wedge \nu$, which then makes (A1) sufficient for (A1'). However, let us provide a significant example in which such μ and ν do not exist, and Arrow-rationality does not imply weak Arrow-rationality. Indeed, let $\mathcal{L}$ consist of one binary relation symbol $<$ and two constant symbols a and b. Let T be the $\mathcal{L}$-theory that says that $<$ is a strict linear order and that there are exactly two alternatives a and b (this example models elections with only two candidates). Then, up to isomorphism, there are exactly two models for T. Hence, in this case, condition (A1) does not imply condition (A1'). Moreover, the assumptions of the lemma above are satisfied by this example, whereas those of Lemma 5.3 in [8] are not.

2.1   Arrow-type Impossibility for Vote Abstention

Definition 3. *An aggregator $f : (\Omega \cup \{\varnothing\})^I \rightarrow \Omega \cup \{\varnothing\}$ is dictatorial if there exists some $i \in I$ such that $f(\underline{M}) = M_i$ for any profile $\underline{M}$.*

Lemma 3. *Any aggregator $f : (\Omega \cup \{\varnothing\})^I \rightarrow \Omega \cup \{\varnothing\}$ satisfying (A3) and such that $\mathcal{D}_f$ is a principal ultrafilter is dictatorial.*

Proof. Let $i_0 \in I$ be the generator of $\mathcal{D}_f$. It is enough to show that $f(\underline{M})$ is isomorphic to M_{i_0} for any profile $\underline{M}$. By Fact 1, it is enough to show that $f(\underline{M}) \models \lambda$ iff $M_{i_0} \models \lambda$ for any $\lambda \in \mathcal{I}$. Indeed, by Lemma 1,

$$f(\underline{M}) \models \lambda \text{ iff } C(\underline{M}, \lambda) \in \mathcal{D}_f \text{ iff } M_{i_0} \models \lambda.$$

As an immediate consequence of the lemmas above we obtain:

---

[3] Recall that, for every non-empty set I, a *filter* $\mathcal{D}$ over I is a collection of subsets of I which is closed under supersets and intersection of finitely many members. A filter $\mathcal{D}$ is *proper* if $\varnothing \notin \mathcal{D}$. An *ultrafilter* over I is a maximal proper filter. Maximality can be equivalently characterized by the following conditions: (a) for any $X \subseteq I$, if $X \notin \mathcal{D}$ then $I \setminus X \in \mathcal{D}$; (b) for all $X, Y \subseteq \mathcal{D}$, if $X \cup Y \in \mathcal{D}$, then either $X \in \mathcal{D}$ or $Y \in \mathcal{D}$. An ultrafilter $\mathcal{D}$ over I is *principal* if it is of the form $\{X \subseteq I \mid i_0 \in X\}$ for some $i_0 \in I$, and is *nonprincipal* otherwise. An immediate consequence of that is, if I is finite, all ultrafilters over I are principal.

Corollary 1. *If T is a universal $\mathcal{L}$-theory such that $|\Omega| \geq 2$, then any Arrow-rational aggregator $f : (\Omega \cup \{\varnothing\})^I \to \Omega \cup \{\varnothing\}$ such that the ultrafilter $\mathcal{D}_f$ is principal is dictatorial.*

The assumption $|\Omega| \geq 2$ in the statement of the corollary above is needed in order to apply Lemma 2. As is well known, in the standard setting of Arrow's theorem, the analogous corollary fails for $|\Omega| = 2$, the majority rule being a counterexample. However, notice that, in the present setting in which aggregators are maps $f : (\Omega \cup \{\varnothing\})^I \to \Omega \cup \{\varnothing\}$, the majority rule is not guaranteed anymore to define an aggregator. Indeed, let $I = \{i_1, i_2, i_3\}$ and $A = \{a, b\}$. Let T be the universal theory of two-element linear orders (cf. Section 6).

Then Ω consists, up to isomorphism, of the models M_a (the one in which a is preferred to b, that is, in which Rab is true), and M_b (the one in which b is preferred to a, that is, in which Rba is true). No universal aggregator $f : (\Omega \cup \{\varnothing\})^I \to \Omega \cup \{\varnothing\}$ satisfies the following condition:

$$f(\underline{M}) \models \lambda \quad \text{iff} \quad |\{i \mid M_i \models \lambda\}| > |\{i \mid M_i \models \neg\lambda\}|. \tag{2.1}$$

Indeed, consider the input $\underline{M} = (M_a, M_b, \varnothing)$ and the sentences Rab, Rba and $a \equiv a$. Clearly, $\{i \mid M_i \models Rab\} = \{i_1\}$, $\{i \mid M_i \models Rba\} = \{i_2\}$ and $\{i \mid M_i \models a \equiv a\} = \{i_1, i_2\}$. If f satisfies (2.1), this implies that $f(\underline{M}) \models \neg Rab$, $f(\underline{M}) \models \neg Rba$ and $f(\underline{M}) \models a \equiv a$. However, none of $M_a, M_b, \varnothing$ satisfy the three sentences simultaneously, therefore f cannot be well-defined at $\underline{M} = (M_a, M_b, \varnothing)$, and thus f cannot be universal.

3   Generalized Ultraproduct Construction

The remainder of the paper is aimed at providing a setting which incorporates the Arrow-type impossibility result for vote abstention as a special case. Towards this aim, in the present section a construction is introduced which, for each (ultra)filter $\mathcal{D}$ over I and each profile $\underline{M} \in (\Omega \cup \{\varnothing\})^I$, yields an $\mathcal{L}$-model $U(\underline{M}, \mathcal{D})$. This construction amounts to the specialization of Makkai's ultraproduct construction (cf. [13, Section 1.3]) from a more general category-theoretic setting to the model-theoretic setting of interest here. In the remainder of this subsection we fix a set I and an (ultra)filter $\mathcal{D}$ over I.

We find it useful to make use of the following auxiliary definition: for any I-indexed family of sets $\underline{S} = \{S_i \mid i \in I\}$, let the *generalized union product of $\underline{S}$* be defined as follows:

$$GUP_\mathcal{D}(\underline{S}) := \coprod_{J \in \mathcal{D}} \prod_{j \in J} S_j = \bigcup \{\{(s_i)_{i \in J} \mid s_i \in S_i\} \mid J \in \mathcal{D}\}.$$

Notice that we are not excluding S_i to be empty for some $i \in I$. This definition naturally applies also to I-indexed families $\underline{R} = \{R_i \mid i \in I\}$ where R_i is a k-ary relation (for a fixed $k \geq 1$) on a given set S_i for each $i \in I$:[4]

$$GUP_\mathcal{D}(\underline{R}) := \coprod_{J \in \mathcal{D}} \prod_{j \in J} R_j = \bigcup \{\{(\overline{s}_i)_{i \in J} \mid \overline{s}_i \in R_i\} \mid J \in \mathcal{D}\}.$$

---

[4] In this case, we will say that $\underline{R}$ is a family of k-ary relations over $\underline{S}$.

The definition above also applies when $k = 0$, if we regard any element $c_i \in S_i$ as a 0-ary relation R_i on S_i.[5] Under this stipulation, I-indexed families $\underline{R}$ of 0-ary relations can be identified with I-indexed sequences $\underline{c} = (x_i)_{i \in I}$ such that for every $i \in I$

$$x_i = \begin{cases} c_i & \text{if } c_i \in S_i \\ * & \text{if } S_i = \varnothing, \end{cases}$$

where $* \notin \bigcup_{i \in I} S_i$. Then, for every $J \in \mathcal{D}$, the product set $\prod_{j \in J} R_j$ reduces to the sequence $(x_j)_{j \in J}$, and hence

$$GUP_{\mathcal{D}}(\underline{c}) := \coprod_{J \in \mathcal{D}} \prod_{j \in J} R_j = \bigcup \{(x_j)_{j \in J} \mid J \in \mathcal{D}\}.$$

For the sake of readability, we will drop the subscripted $\mathcal{D}$ when this causes no confusion. Clearly, $GUP(\underline{c}) \cap GUP(\underline{S}) \neq \varnothing$ iff some $J \in \mathcal{D}$ exists such that $c_j \in S_j$ for every $i \in J$.

Notice that if $\underline{R}$ is an I-indexed family of k-ary relations over $\underline{S}$, then $GUP(\underline{R})$ is not a k-ary relation on $GUP(\underline{S})$. Fortunately, this situation can be remedied as follows. For any set S and $k \geq 1$, let S^k denote the k-ary universal relation on S. The following isomorphism holds for any $J \in \mathcal{D}$ and any $k \geq 1$:

$$\sigma_J : \prod_{j \in J} (S_j)^k \longrightarrow (\prod_{j \in J} S_j)^k$$

which maps the J-indexed array $(\bar{s}_j)_{j \in J}$ of k-tuples $\bar{s}_j = (s_1^j, \ldots, s_k^j) \in (S_j)^k$ to the k-tuple of J-indexed arrays $((s_1^j)_{j \in J}, \ldots, (s_k^j)_{j \in J})$. Since $\prod_{j \in J} R_j \subseteq \prod_{j \in J} (S_j)^k$, the σ_J-direct image of $\prod_{j \in J} R_j$ is a k-ary relation:

$$\sigma_J[\prod_{j \in J} R_j] \subseteq (\prod_{j \in J} S_j)^k.$$

Hence, $GUP(\underline{R})$ induces the k-ary relation

$$GUP'(\underline{R}) := \bigcup \{\sigma_J[\prod_{j \in J} R_j] \mid J \in \mathcal{D}\} \subseteq (GUP(\underline{S}))^k.$$

Consider the equivalence relation on $GUP(\underline{S})^6$ defined as follows:

$$(s_j)_{j \in J} \equiv_{\underline{S}}^{\mathcal{D}} (t_h)_{h \in H} \quad \text{iff} \quad \{i \in J \cap H \mid s_i = t_i\} \in \mathcal{D}.$$

Definition 4. *(cf. [13], Section 1.3) For any profile $\underline{M} \in (\Omega \cup \{\varnothing\})^I$, the **generalized ultraproduct** of $\underline{M}$ over $\mathcal{D}$ is the $\mathcal{L}$-model $U = U(\underline{M}, \mathcal{D})$ specified as follows:*

---

[5] Regarding elements $c \in S$ as 0-ary relations on S departs from the usual convention in model theory, according to which 0-ary relations are truth-values.

[6] For ease of notation, we will often drop the subscript in $\equiv_{\underline{S}}^{\mathcal{D}}$ and rely on the context for its correct interpretation.

– the universe $|U(\underline{M}, \mathcal{D})|$ of $U(\underline{M}, \mathcal{D})$ is

$$U(\underline{S}, \mathcal{D}) := GUP(\underline{S})/\equiv^{\mathcal{D}}_{\underline{S}},$$

where $\underline{S} = \{|M_i| \mid i \in I\}$;
– for any constant symbol c,

$$c^U = c^{U(\underline{M}, \mathcal{D})} := [(x_j)_{j \in J}]_{\equiv^{\mathcal{D}}_{\underline{S}}};$$

where $(x_j)_{j \in J} \in GUP(\underline{c}) \cap GUP(\underline{S})$, and $\underline{c} = (x_i)_{i \in I}$ such that for every $i \in I$

$$x_i = \begin{cases} c^{M_i} & \text{if } M_i \neq \varnothing \\ * & \text{otherwise} ; \end{cases}$$

– for any k-ary relation symbol R $(k \geq 1)$, the k-ary relation $R^U = R^{U(\underline{M}, \mathcal{D})}$ on U is defined as follows:

$$([(s_1^j)_{j \in J_1}]_{\equiv^{\mathcal{D}}_{\underline{S}}}, \dots, [(s_k^j)_{j \in J_k}]_{\equiv^{\mathcal{D}}_{\underline{S}}}) \in R^U \quad \text{iff} \quad ((t_1^j)_{j \in J}, \dots (t_k^j)_{j \in J}) \in GUP'(\underline{R})$$

for some $J \in \mathcal{D}$ and some $(t_1^j)_{j \in J}, \dots, (t_k^j)_{j \in J}$ such that, for every $1 \leq \ell \leq k$,

$$(t_\ell^j)_{j \in J} \equiv^{\mathcal{D}}_{\underline{S}} (s_\ell^j)_{j \in J_\ell}.$$

Notice that the elements of $GUP(\underline{c}) \cap GUP(\underline{S})$ are all identified by $\equiv^{\mathcal{D}}_{\underline{S}}$, so c^U is well-defined. Notice also that c^U is defined only if $GUP(\underline{c}) \cap GUP(\underline{S}) \neq \varnothing$, and as discussed early on, this is the case iff some $J \in \mathcal{D}$ exists such that $M_j \neq \varnothing$ for every $j \in J$. On the other hand, as we will discuss next (cf. Fact 2), this condition also characterizes the non-emptiness of $U(\underline{M}, \mathcal{D})$.

4   Properties of the Generalized Ultraproduct Construction

Let $\underline{S}$ be an I-indexed family of sets. For any ultrafilter $\mathcal{D}$ over I and any $J \in \mathcal{D}$, if $S_i = \varnothing$ for some $i \in J$, then $\prod_{i \in J} S_i = \varnothing$. Hence:

Fact 2. *For every I-indexed family of sets $\underline{S}$ and any ultrafilter $\mathcal{D}$ over I,*

$$GUP(\underline{S}) \neq \varnothing \quad \text{iff} \quad \text{some } J \in \mathcal{D} \text{ exists s.t. } S_i \neq \varnothing \text{ for all } i \in J.$$

Recall that if $\mathcal{D}$ is a principal ultrafilter, $\mathcal{D}$ is generated by the singleton $\{i_0\}$ for some individual $i_0 \in I$, which can be identified with the *dictator*. The following fact is an immediate consequence of the fact above:

Fact 3. *For every profile $\underline{M}$ and any principal ultrafilter $\mathcal{D}$ over I,*

$$U(\underline{M}, \mathcal{D}) = \varnothing \quad \text{iff} \quad M_{i_0} = \varnothing. \tag{4.1}$$

Definition 4 generalizes the following

Definition 5. *For any (ultra)filter $\mathcal{D}$ on I, and any profile $\underline{M} \in \Omega^I$, the **standard ultraproduct** of $\underline{M}$ over $\mathcal{D}$ is the $\mathcal{L}$-model $U' = U'(\underline{M}, \mathcal{D})$ specified as follows:*

– *the universe $|U'(\underline{M}, \mathcal{D})|$ of $U'(\underline{M}, \mathcal{D})$ is*

$$\prod_{i \in I} M_i / \sim_{\mathcal{D}},$$

where for any $(s_i)_{i \in I}, (t_i)_{i \in I} \in \prod_{i \in I} M_i$,

$$(s_i)_{i \in I} \sim_{\mathcal{D}} (t_i)_{i \in I} \quad \text{iff} \quad \{i \in I \mid s_i = t_i\} \in \mathcal{D};$$

– *for any constant symbol c,*

$$c^{U'} := [\underline{c}]_{\sim_{\mathcal{D}}}$$

where $\underline{c} = (c^{M_i})_{i \in I}$;

– *for any k-ary relation symbol R ($k \geq 1$), the k-ary relation $R^{U'} = R^{U'(\underline{M}, \mathcal{D})}$ on U' is defined as follows:*

$$([(s_1^i)_{i \in I}]_{\sim_{\mathcal{D}}}, \ldots, [(s_k^i)_{i \in I}]_{\sim_{\mathcal{D}}}) \in R^{U'} \quad \text{iff} \quad \{i \in I \mid (s_1^i, \ldots, s_k^i) \in R^{M_i}\} \in \mathcal{D}.$$

The definition above is in general different from Definition 4. Indeed, if $M_i = \varnothing$ for some $i \in I$, then $U'(\underline{M}, \mathcal{D}) = \varnothing$, while $U(\underline{M}, \mathcal{D})$ does not need to be empty (cf. Fact 2). However, if $M_i \neq \varnothing$ for any $i \in I$, then the two constructions can be identified, as shown in the following.

Fact 4. *For any (ultra)filter $\mathcal{D}$ on I, and any profile $\underline{M} \in (\Omega \cup \{\varnothing\})^I$, if $M_i \neq \varnothing$ for every $i \in I$ then $U(\underline{M}, \mathcal{D})$ and $U'(\underline{M}, \mathcal{D})$ are isomorphic.*

Proof. Clearly, for all $(y_i)_{i \in I}$ and $(y_i')_{i \in I}$,

$$(y_i)_{i \in I} \equiv_{\mathcal{D}} (y_i')_{i \in I} \quad \text{iff} \quad \{i \in I \mid y_i = y_i'\} \in \mathcal{D} \quad \text{iff} \quad (y_i)_{i \in I} \sim_{\mathcal{D}} (y_i')_{i \in I}.$$

Moreover, for every $J \in \mathcal{D}$ and for every $(t_j)_{j \in J}$ there exists some $(y_i)_{i \in I}$ s.t. $(t_j)_{j \in J} \equiv_{\mathcal{D}} (y_i)_{i \in I}$: indeed, the assumption that $M_i \neq \varnothing$ for every $i \in I$ guarantees that there exists at least one I-indexed array defined as follows:

$$y_i = \begin{cases} t_i & \text{if } i \in K \\ \text{any } y \in M_i \neq \varnothing & \text{otherwise.} \end{cases}$$

By construction, $\{i \in I \cap J = J \mid y_i = t_i\} = J \in \mathcal{D}$, and hence $(t_j)_{j \in J} \equiv_{\mathcal{D}} (y_i)_{i \in I}$. From the facts above, it follows that the map $\varphi : |U(\underline{M}, \mathcal{D})| \to |U'(\underline{M}, \mathcal{D})|$ defined by the assignment $[(t_j)_{j \in J}]_{\equiv_{\mathcal{D}}} \mapsto [(y_i)_{i \in I}]_{\sim_{\mathcal{D}}}$ is well defined and has an inverse $\psi : |U'(\underline{M}, \mathcal{D})| \to |U(\underline{M}, \mathcal{D})|$ defined by the assignment $[(y_i)_{i \in I}]_{\sim_{\mathcal{D}}} \mapsto [(y_i)_{i \in I}]_{\equiv_{\mathcal{D}}}$. Moreover, these assignments identify $c^{U'}$ and c^U for every constant symbol c, and also identify R^U and $R^{U'}$ for every k-ary relation symbol R. Indeed, it can be easily verified that $\varphi(c^U) = c^{U'}$ and that

$$([(s_1^j)_{j \in J_1}]_{\equiv_{\mathcal{D}}}, \ldots, [(s_k^j)_{j \in J_k}]_{\equiv_{\mathcal{D}}}) \in R^U \quad \text{iff} \quad (\varphi([(s_1^j)_{j \in J_1}]_{\equiv_{\mathcal{D}}}), \ldots, \varphi([(s_k^j)_{j \in J_k}]_{\equiv_{\mathcal{D}}})) \in R^{U'}.$$

The following is a restatement of [13, Theorem 1.3.1] specialized to the model-theoretic setting of our interest. The proof of this theorem appears in the extended version of the present paper (cf. [2]).

Theorem 5. *(Generalized Łoś's Theorem). The following are equivalent for any formula* $\lambda(x_1, \ldots, x_n)$ *with n free variables and any profile* $\underline{M} \in (\Omega \cup \{\varnothing\})^I$:

- $U(\underline{M}, \mathcal{D}) \models \lambda\left([\underline{s_1}^{J_1}]_{\equiv_\mathcal{D}}, \ldots, [\underline{s_n}^{J_n}]_{\equiv_\mathcal{D}}\right)$;
- $\{i \in J_1 \cap \ldots \cap J_n \mid M_i \models \lambda(s_{1,i}, \ldots s_{n,i})\} \in \mathcal{D}$.

5 Generalized Kirman-Sondermann Correspondence

The present section is aimed at introducing the second half of the generalized Kirman-Sondermann correspondence (the first half was discussed at the end of Section 2, before Lemma 2), and characterizing Arrow-rational aggregators in terms of the generalized ultraproduct construction introduced in the previous subsection. Recall that, for any $\mathcal{L}$-structure $\mathcal{B}$ with domain B and any $C \subseteq B$ such that $A^\mathcal{B} \subseteq C$, the *restriction* of $\mathcal{B}$ to C is the $\mathcal{L}$-structure the universe of which is C, which is obtained by restricting the interpretation of all relation symbols to C. For every $M \in \Omega$, let $res_A M$ denote the restriction of M to A^M. In what follows, we find it convenient to define $res_A M$ also when M is the empty model. If $M = \varnothing$, then we stipulate that $res_A M = \varnothing$.

Lemma 4. *For all* $\lambda \in \mathcal{I}$,

$$res_A U(\underline{M}, \mathcal{D}) \models \lambda \qquad iff \qquad C(\underline{M}, \lambda) \in \mathcal{D}.$$

Proof. By the generalized Łoś's theorem, $C(\underline{M}, \lambda) = \{i \in I \mid M_i \models \lambda\} \in \mathcal{D}$ iff $U(\underline{M}, \mathcal{D}) \models \lambda$. Since by assumption λ is quantifier-free, the latter condition is equivalent to $res_A U(\underline{M}, \mathcal{D}) \models \lambda$.

Definition 6. *For every ultrafilter* $\mathcal{D}$ *over I, let* $f_\mathcal{D} : (\Omega \cup \{\varnothing\})^I \to \Omega \cup \{\varnothing\}$ *be defined by the assignment*

$$\underline{M} \mapsto res_A U(\underline{M}, \mathcal{D}).$$

By Łoś's theorem, $U(\underline{M}, \mathcal{D}) \models T$ for every profile $\underline{M}$. Since T is a universal theory, this implies that $res_A U(\underline{M}, \mathcal{D}) \models T$, which shows that $f_\mathcal{D}$ is well defined. The following proposition shows that the assignment $\mathcal{D} \mapsto f_\mathcal{D}$ defines a map $\Phi : \beta I \to \mathcal{AR}$.

Proposition 1. *For every ultrafilter* $\mathcal{D}$ *over I, the aggregator* $f_\mathcal{D}$ *is Arrow-rational.*

Proof. Condition (A1) is verified by construction. As to (A2), fix a profile $\underline{M}$ and $\lambda \in \mathcal{I}$, and assume that $f_\mathcal{D}(\underline{M}) \models \lambda$, that is, $res_A U(\underline{M}, \mathcal{D}) \models \lambda$. Then Lemma 4 implies that $C(\underline{M}, \lambda) \in \mathcal{D}$. Hence $C(\underline{M}, \lambda)$ must be nonempty, since $\mathcal{D}$ is an ultrafilter, and hence is proper. As to (A3), let $C(\underline{M}, \lambda) = C(\underline{N}, \mu)$ for some $\underline{M}, \underline{N}$ and $\lambda, \mu \in \mathcal{I}$. Hence, by Lemma 4,

$$f_\mathcal{D}(\underline{M}) \models \lambda \ \text{ iff } \ C(\underline{M}, \lambda) \in \mathcal{D} \ \text{ iff } \ C(\underline{N}, \mu) \in \mathcal{D} \ \text{ iff } \ f_\mathcal{D}(\underline{N}) \models \mu.$$

Next, we are going to show that the maps Λ and Φ defining the Kirman-Sondermann correspondence (cf. discussions before Lemma 2 and before Proposition 1) are inverse to one another.

Proposition 2. *For every $f \in \mathcal{AR}$, $f_{\mathcal{D}_f}$ and f can be identified up to isomorphism.*

Proof. By Lemmas 4 and 1,

$$\mathrm{res}_A U(\underline{M}, \mathcal{D}_f) \models \lambda \text{ iff } C(\underline{M}, \lambda) \in \mathcal{D} \text{ iff } f(\underline{M}) \models \lambda$$

for any profile $\underline{M}$ and any $\lambda \in \mathcal{I}$. Then the statement follows from Fact 1.

In the proof of the next proposition, we make crucial use of the assumption that at least two non-isomorphic models exist in Ω.

Proposition 3. *For every $\mathcal{D} \in \beta I$, $\mathcal{D}_{f_{\mathcal{D}}} = \mathcal{D}$.*

Proof. Fix $X \subseteq I$, and let us show that $X \in \mathcal{D}_{f_{\mathcal{D}}}$ iff $X \in \mathcal{D}$. By assumption, two non-isomorphic models M, N exist in $\Omega \cup \{\varnothing\}$. As shown in the proof of Proposition 2, this implies that $M \models \lambda$ and $N \not\models \lambda$ for some $\lambda \in \mathcal{I}$. Let us define the profile $\underline{M} \in (\Omega \cup \{\varnothing\})^I$ as follows: for any $i \in I$, let

$$M_i = \begin{cases} M & \text{if } i \in X \\ N & \text{if } i \notin X. \end{cases}$$

By construction, $C(\underline{M}, \lambda) = X$, and hence the required equivalence can be proved as follows:

$$
\begin{aligned}
C(\underline{M}, \lambda) \in \mathcal{D}_{f_{\mathcal{D}}} \text{ iff } & f_{\mathcal{D}}(\underline{M}) \models \lambda && \text{(Lemma 1)} \\
\text{iff } & \mathrm{res}_A U(\underline{M}, \mathcal{D}) \models \lambda && \text{(Definition 6)} \\
\text{iff } & C(\underline{M}, \lambda) \in \mathcal{D}. && \text{(Lemma 4)}
\end{aligned}
$$

The following is an immediate consequence of Propositions 2 and 3:

Theorem 6. (Kirman-Sondermann Correspondence for Vote Abstention). *For any language $\mathcal{L}$, any universal $\mathcal{L}$-theory T with at least two non-isomorphic models, and any set I of individuals, the set $\mathcal{AR}$ of Arrow-rational aggregators (cf. Definition 1) and the set βI of the ultrafilters over I bijectively correspond via the map $\Lambda : \mathcal{AR} \to \beta I$ defined by the assignment $f \mapsto \mathcal{D}_f$. The inverse of Λ is the map $\Phi : \beta I \to \mathcal{AR}$, defined by the assignment $\mathcal{D} \mapsto f_{\mathcal{D}}$.*

6   Arrow-Type Impossibility Theorem for Vote Abstention

By taking concrete universal theories T, the treatment developed so far specializes to concrete settings in social choice. As an example, in the present section, we capture and discuss the theory of preference aggregation in settings in which individuals might abstain from voting.

The case of preference aggregation over n candidates is modelled, as is done in [9], by taking $\mathcal{L}$ to be a language with n constant symbols $a_1, \ldots, a_n$ and one binary relation symbol R. Consider the following theory T_n:

- $\forall x(\neg Rxx)$ (irreflexivity);
- $\forall x \forall y \forall z(Rxy \wedge Ryz \rightarrow Rxz)$ (transitivity);
- $\forall x \forall y(Rxy \vee Ryx \vee x \approx y)$ (completeness);
- $\forall x(\neg x \approx x) \vee \forall x(x \approx a_1 \vee \ldots \vee x \approx a_n)$;
- $\forall x \forall y(x \approx a_j \wedge y \approx a_k \rightarrow \neg x \approx y)$ for $j \neq k$;

The first three sentences state that each model M_i of T is a linear order given by the individual i, and the last two items state that the domain of each model is either empty or consists of n pairwise distinct elements $a_1^M, \ldots, a_n^M$. Therefore, the aggregator $f : (\Omega \cup \{\varnothing\})^I \rightarrow \Omega \cup \{\varnothing\}$ aggregates a collection of linear orders (or empty order, corresponding to the voter abstention case) into a single linear order (or empty order).

When $n \geq 2$, it is easy to see that $|\Omega| \geq 2$; therefore Corollary 1 applies, yielding:

Theorem 7. *(Generalized Arrow impossibility theorem for preference aggregation). For T_n given above ($n \geq 2$) and for any finite I, any Arrow-rational aggregator $f : (\Omega \cup \{\varnothing\})^I \rightarrow \Omega \cup \{\varnothing\}$ is dictatorial.*

The present setting for vote-abstention allows to prove a strengthened version of Arrow's impossibility theorem in preference aggregation which, unlike the standard one, holds e.g. also for 2-candidate elections. The technical reason for this is to be traced in the proof of Lemma 2, omitted in the present paper but available in [2], which is a variant of Lemma 5.3 in [8]. Indeed, given two non-isomorphic models, the empty model plays the role of the third one. As discussed after Corollary 1, the features of the present set up are such that the counterexamples to the analogous strengthening in the standard setting are not definable anymore.

References

1. Arrow, K.: Social choice and individual values, 2nd edn. Cowles Foundation Monograph, vol. 12. John Wiley & Sons (1963)
2. Bedrosian, G., Palmigiano, A., Zhao, Z.: Generalized ultraproduct and Kirman-Sondermann correspondence for vote abstention, working paper (2015), http://www.appliedlogictudelft.nl/publications/
3. Bell, J.L., Slomson, A.B.: Models and ultraproducts: an introduction, North-Holland, Amsterdam (1974)
4. Esteban, M., Palmigiano, A., Zhao, Z.: An Abstract Algebraic Logic view on Judgment Aggregation. In: Proceedings of Logic, Rationality and Interaction, 5th International Workshop, LORI 2015 (2015)
5. Fey, M.: Mays theorem with an infinite population. Social Choice and Welfare 23(2), 275–293 (2004)
6. Fishburn, P.: Arrow's impossibility theorem: concise proof and infinite voters. Journal of Economic Theory 2, 103–106 (1970)
7. Herzberg, F.: Universal algebra for general aggregation theory: many-valued propositional-attitude aggregators as MV-homomorphisms. Journal of Logic and Computation (2013)

8. Herzberg, F., Eckert, D.: The model-theoretic approach to aggregation: impossibility results for finite and infinite electorates. Journal of Mathematical Social Science 64, 41–47 (2012)

9. Eckert, D., Herzberg, F.: The problem of judgment aggregation in the framework of Boolean-valued models. In: Bulling, N., van der Torre, L., Villata, S., Jamroga, W., Vasconcelos, W. (eds.) CLIMA 2014. LNCS, vol. 8624, pp. 138–147. Springer, Heidelberg (2014)

10. Keiding, H.: The categorical approach to social choice theory. Journal of Mathematical Social Sciences 1, 177–191 (1981)

11. Kirman, A., Sondermann, D.: Arrow's theorem, many agents, and invisible dictators. Journal of Economic Theory 5(2), 267–277 (1972)

12. Lauwers, L., Van Liedekerke, L.: Ultraproducts and aggregation. Journal of Mathematical Economics 24(3), 217–237 (1995)

13. Makkai, M., Ultraproducts and categorical logic. Lecture Notes in Math., vol. 1130, pp. 222–309. Springer (1985)

14. Pivato, M.: Variable-population voting rules. Journal of Mathematical Economics 49(3), 210–221 (2013)

Learning Actions Models: Qualitative Approach

Thomas Bolander[1] and Nina Gierasimczuk[2]

[1] DTU Compute, Technology University of Denmark, Copenhagen, Denmark
tobo@dtu.dk
[2] ILLC, University of Amsterdam, Amsterdam, The Netherlands
nina.gierasimczuk@gmail.com

Abstract. In dynamic epistemic logic, actions are described using
action models. In this paper we introduce a framework for studying learnability of action models from observations. We present first results concerning propositional action models. First we check two basic learnability
criteria: finite identifiability (conclusively inferring the appropriate action
model in finite time) and identifiability in the limit (inconclusive convergence to the right action model). We show that deterministic actions are
finitely identifiable, while non-deterministic actions require more learning
power—they are identifiable in the limit. We then move on to a particular
learning method, which proceeds via restriction of a space of events within
a learning-specific action model. This way of learning closely resembles the
well-known update method from dynamic epistemic logic. We introduce
several different learning methods suited for finite identifiability of particular types of deterministic actions.

Dynamic epistemic logic (DEL) allows analyzing knowledge change in a systematic way. The static component of a situation is represented by an epistemic
model, while the structure of the dynamic component is encoded in an *action
model*. An action model can be applied to the epistemic model via so-called
product update operation, resulting in a new up-to-date epistemic model of the
situation after the action has been executed. A language, interpreted on epistemic models, allows expressing conditions under which an action takes effect
(so-called preconditions), and the effects of such actions (so-called postconditions). This setting is particularly useful for modeling the process of epistemic
planning (see [7,1]): one can ask which sequence of actions should be executed
in order for a given epistemic formula to hold in the epistemic model after the
actions are executed.

The purpose of this paper is to investigate possible learning mechanisms involved in discovering the 'internal structure' of actions on the basis of their
executions. In other words, we are concerned with qualitative learning of action
models on the basis of observations of pairs of the form (initial state, resulting
state). We analyze learnability of action models in the context of two learning conditions: finite identifiability (conclusively inferring the appropriate action
model in finite time) and identifiability in the limit (inconclusive convergence to
the right action model). The paper draws on the results from formal learning
theory applied to DEL (see [11,13,12]).

© Springer-Verlag Berlin Heidelberg 2015
W. van der Hoek et al. (Eds.): LORI 2015, LNCS 9394, pp. 40–52, 2015.
DOI: 10.1007/978-3-662-48561-3_4

Learning of action models is highly relevant in the context of epistemic planning. A planning agent might not initially know the effects of her actions, so she will initially not be able to plan to achieve any goals. However, if she can learn the relevant action models through observing the effect of the actions (either by executing the actions herself, or by observing other agents), she will eventually learn how to plan. Our ultimate goal is to integrate learning of actions into (epistemic) planning agents. In this paper, we seek to lay the foundations for this goal by studying learnability of action models from streams of observations.

The structure of the paper is as follows. In Section 1 we recall the basic concepts and notation concerning action models and action types in DEL. In Section 2 we specify our learning framework and provide general learnability results. In Section 3 we study particular learning functions, which proceed via updating action models with new information. In the end we briefly discuss related and further work. The full version of the paper is available at http://arxiv.org/abs/1507.04285.

1 Languages and Action Types

Let us first present the basic notions required for the rest of the article (see [6,8] for more details). Following the conventions of automated planning, we take the set of atomic propositions and the set of actions to be finite. Given a finite set P of atomic propositions, we define the (single-agent) *epistemic language* over P, $\mathcal{L}_{epis}(P)$, by the following BNF: $\phi ::= p \mid \neg\phi \mid \phi \wedge \phi \mid K\phi$, where $p \in P$. The language $\mathcal{L}_{prop}(P)$ is the propositional sublanguage without the $K\phi$ clause. When P is clear from the context, we write $\mathcal{L}_{epis}$ and $\mathcal{L}_{prop}$ instead of $\mathcal{L}_{epis}(P)$ and $\mathcal{L}_{prop}(P)$, respectively. By means of the standard abbreviations we introduce the additional symbols $\rightarrow$, $\vee$, $\leftrightarrow$, $\perp$, and $\top$.

Definition 1 (Epistemic Models and States). *An* epistemic model *over a set of atomic propositions P is $\mathcal{M} = (W, R, V)$, where W is a finite set of* worlds, *$R \subseteq W \times W$ is an equivalence relation, called the* indistinguishability relation, *and $V : P \rightarrow \mathcal{P}(W)$ is a valuation function. An* epistemic state *is a pointed epistemic model $(\mathcal{M}, w)$ consisting of an epistemic model $\mathcal{M} = (W, R, V)$ and a distinguished world $w \in W$ called the* actual world.

A *propositional state* (or simply *state*) over P is a subset of P (or, equivalently, a propositional valuation $\nu : P \rightarrow \{0, 1\}$). We identify propositional states and singleton epistemic models via the following canonical isomorphism. A propositional state $s \subseteq P$ is isomorphic to the epistemic model $\mathcal{M} = (\{w\}, \{(w, w)\}, V)$ where $V(p) = \{w\}$ if $p \in s$ and $V(p) = \emptyset$ otherwise. Truth in epistemic states $(\mathcal{M}, w)$ with $\mathcal{M} = (W, R, V)$ (and hence propositional states) is defined as usual and hence omitted.

Dynamic epistemic logic (DEL) introduces the concept of an action model for modelling the changes to states brought about by the execution of actions [6]. We here use a variant that includes postconditions [8,7], which means that actions can have both epistemic effects (changing the beliefs of agents) and ontic effects (changing the factual states of affairs).

Definition 2 (Action Models). *An* action model *over a set of atomic propositions* P *is* $\mathcal{A} = (E, Q, pre, post)$, *where* E *is a finite set of* events; $Q \subseteq E \times E$ *is an equivalence relation called the* indistinguishability relation; $pre : E \to \mathcal{L}_{epis}(P)$ *assigns to each event a* precondition; $post : E \to \mathcal{L}_{prop}(P)$ *assigns to each event a* postcondition. *Postconditions are conjunctions of literals (atomic propositions and their negations) or* $\top$.[1] $dom(\mathcal{A}) = E$ *denotes the domain of* $\mathcal{A}$. *The set of all action models over* P *is denoted* Actions(P).

Intuitively, events correspond to the ways in which an action changes the epistemic state, and the indistinguishability relation codes (an agent's) ability to recognize the difference between those different ways. In an event e, $pre(e)$ specifies what conditions have to be satisfied for it to take effect, and $post(e)$ specifies its outcome.

Example 1. Consider the action of tossing a coin. It can be represented by the following action model (h means that the coin is facing heads up):

$$\mathcal{A} = \quad \overset{\bullet}{e_1 : \langle \top, h \rangle} \qquad \overset{\bullet}{e_2 : \langle \top, \neg h \rangle}$$

We label each event by a pair whose first argument is the event's precondition while the second is its postcondition. Hence, formally we have $\mathcal{A} = (E, Q, pre, post)$ with $E = \{e_1, e_2\}$, Q is the identity on E, $pre(e_1) = pre(e_2) = \top$, $post(e_1) = h$ and $post(e_2) = \neg h$. The action model encodes that tossing the coin will either make h true (e_1) or h false (e_2).

Definition 3 (Product Update). *Let* $\mathcal{M} = (W, R, V)$ *and* $\mathcal{A} = (E, Q, pre, post)$ *be an epistemic model and action model (over a set of atomic propositions P), respectively. The* product update *of* $\mathcal{M}$ *with* $\mathcal{A}$ *is the epistemic model* $\mathcal{M} \otimes \mathcal{A} = (W', R', V')$, *where* $W' = \{(w, e) \in W \times E \mid (\mathcal{M}, w) \models pre(e)\}$; $R' = \{((w, e), (v, f)) \in W' \times W' \mid wRv \text{ and } eQf\}$; $V'(p) = \{(w, e) \in W' \mid post(e) \models p \text{ or } ((\mathcal{M}, w) \models p \text{ and } post(e) \not\models \neg p)\}$. *For* $e \in dom(\mathcal{A})$, *we define* $\mathcal{M} \otimes e = \mathcal{M} \otimes (\mathcal{A} \restriction \{e\})$.

The product update $\mathcal{M} \otimes \mathcal{A}$ represents the result of executing the action $\mathcal{A}$ in the state(s) represented by $\mathcal{M}$.

Example 2. Continuing Example 1, consider a situation of an agent seeing a coin lying heads-up, i.e., the singleton epistemic state $\mathcal{M} = (\{w\}, \{w, w\}, V)$ with $V(h) = \{w\}$. Let us now calculate the result of executing the coin toss in this model.

$$\mathcal{M} \otimes \mathcal{A} = \quad \overset{\bullet}{(w_1, e_1) : h} \qquad \overset{\bullet}{(w_1, e_2) :}$$

Here each world is labelled by the propositions being true at the world.

---

[1] We are here using the postcondition conventions from [7], which are slightly non-standard. Any action model with standard postconditions can be turned into one of our type, but it might become exponentially larger in the process [8,7].

We say that two action models $\mathcal{A}_1$ and $\mathcal{A}_2$ are *equivalent*, written $\mathcal{A}_1 \equiv \mathcal{A}_2$, if for any epistemic model $\mathcal{M}$, $\mathcal{M} \otimes \mathcal{A}_1 \underline{\leftrightarrow} \mathcal{M} \otimes \mathcal{A}_2$, where $\underline{\leftrightarrow}$ denotes standard bisimulation on epistemic models [17].

Definition 4 (Action Types). *An action model* $\mathcal{A} = (E, Q, pre, post)$ *is:*

- atomic *if* $|E| = 1$.
- deterministic *if all preconditions are mutually inconsistent, that is,* $\models pre(e) \wedge pre(f) \to \bot$ *for all distinct* $e, f \in E$.
- fully observable *if* Q *is the identity relation on* E. *Otherwise it is* partially observable.
- precondition-free *if* $pre(e) = \top$ *for all* $e \in E$.
- propositional *if* $pre(e) \in \mathcal{L}_{prop}$ *for all* $e \in E$.
- universally applicable *if* $\models \bigvee_{e \in E} pre(e)$.
- normal *if for all propositional literals* l *and all* $e \in E$, $pre(e) \models l$ *implies* $post(e) \not\models l$.
- with basic preconditions *if all* $pre(e)$ *are conjunctions of literals (propositional atoms and their negations).*
- with maximal preconditions *if all* $pre(e)$ *are maximally consistent conjunctions of literals (i.e., preconditions are conjunctions of literals in which each atomic proposition* p *occurs exactly once, either as* p *or as* $\neg p$*).*

Some of the notions defined above are known from existing literature [7,8,16]. The newly introduced notions are precondition-free, universally applicable, and normal actions, as well as actions with basic preconditions. Note that action types interact with each other, atomic actions are automatically both deterministic and fully observable, and precondition-free actions can only be deterministic if atomic.[2]

In the remainder of this section we set a uniform representation of action models that we will later on use in learning methods. We also specify and justify the restrictions we impose on action models. In this paper we are concerned with product updates of *propositional* states with *propositional* action models. Let s denote a propositional state over P, and let $\mathcal{A} = (E, Q, pre, post)$ be any propositional action model. Using the definition above and the canonical isomorphism between propositional states and singleton epistemic states, we get that $s \otimes \mathcal{A}$ is isomorphic to the epistemic model (W', R', V'), where $W' = \{e \in E \mid s \models pre(e)\}$, $R' = \{(e, f) \in W' \times W' \mid eQf\}$, $V'(p) = \{e \in W' \mid post(e) \models p$ or $(s \models p$ and $post(e) \not\models \neg p)\}$. If $\mathcal{A}$ is fully observable, then the indistinguishability of $s \otimes \mathcal{A}$ is the identity relation. This means that we can think of $s \otimes \mathcal{A}$ as a set of propositional states (via the canonical isomorphism between singleton epistemic models and propositional states). In this case we write $s' \in s \otimes \mathcal{A}$ to mean that s' is one of the propositional states in $s \otimes \mathcal{A}$. When $\mathcal{A}$ is atomic we have $s \otimes a = s'$ for some propositional state s' (using again the canonical isomorphism).

---

[2] The actions considered in propositional STRIPS planning (called *set-theoretic planning* in [9]) correspond to epistemic actions that are atomic and have basic postconditions.

Example 3. Consider the action model $\mathcal{A}$ of Example 1 (the coin toss). It is a precondition-free, fully observable, non-deterministic action. Consider an initial propositional state $s = \{h\}$. Then $s \otimes \mathcal{A}$ is the epistemic model of Example 2. It has two worlds, one in which h is true, and another in which h is false. So we have $\emptyset, \{h\} \in s \otimes \mathcal{A}$, i.e., the outcome of tossing the coin is either the propositional state where h is false ($\emptyset$) or the one where h is true ($\{h\}$).

Proposition 1. *Any propositional action model is equivalent to a normal action model with basic preconditions.*

The condition for being universally applicable intuitively means that the action specifies an outcome no matter what state it is applied to. In this paper we will only be concerned with universally applicable action models.

2   Learning Action Models

In the following we will use the expressions *action* and *action model* interchangeably. Below we will first present general results on learnability of various types of action models, and then, in Section 3, we study particular learning methods and exemplify them.

We are concerned with learning fully observable actions (action models). Partially observable actions are generally not learnable in the strict sense to be defined below. Consider for instance an agent trying to learn an action that controls the truth value of a proposition p, but where the agent cannot observe p (events making p true and events making p false are indistinguishable). Then clearly there is no way for that agent to learn exactly how the action works. The case of fully observable actions is much simpler. If initially the agent has no uncertainty, her "belief state" can be represented by a propositional state. Executing any sequence of fully observable actions will then again lead to a propositional state. So in the case of fully observable actions, we can assume actions to make transitions between propositional states.

For the rest of this section, except in examples, we fix a set P of atomic propositions.

Definition 5. *A stream $\mathcal{E}$ is an infinite sequence of pairs (s, s') of propositional states over P, i.e., $\mathcal{E} \in (\mathcal{P}(P) \times \mathcal{P}(P))^\omega$. The elements (s, s') of $\mathcal{E}$ are called observations. Let $\mathbb{N} := \mathbb{N}^+ \cup \{0\}$, let $\mathcal{E}$ be a stream over P, and let $s, t \in \mathcal{P}(P)$. $\mathcal{E}_n$ stands for the n-th observation in $\mathcal{E}$. $\mathcal{E}[n]$ stands for the the initial segment of $\mathcal{E}$ of length n, i.e., $\mathcal{E}_0, \dots, \mathcal{E}_{n-1}$. $\mathrm{set}(\mathcal{E}) := \{(x, y) \mid (x, y) \text{ is an element of } \mathcal{E}\}$ stands for the set of all observations in $\mathcal{E}$; we similarly define $\mathrm{set}(\mathcal{E}[n])$ for initial segments of streams.*

Definition 6. *Let $\mathcal{E}$ be a stream over P and $\mathcal{A}$ a fully observable action model over P. The stream $\mathcal{E}$ is sound with respect to $\mathcal{A}$ if for all $(s, s') \in \mathrm{set}(\mathcal{E})$, $s' \in s \otimes \mathcal{A}$. The stream $\mathcal{E}$ is complete with respect to $\mathcal{A}$ if for all $s \subseteq P$ and all $s' \in s \otimes \mathcal{A}$, $(s, s') \in \mathrm{set}(\mathcal{E})$. In this paper we always assume the streams to be sound and complete. For brevity, if $\mathcal{E}$ is sound and complete wrt $\mathcal{A}$, we will write: '$\mathcal{E}$ is for $\mathcal{A}$'.*

A *learning function* is a computable $L : (\mathcal{P}(P) \times \mathcal{P}(P))^* \to \text{Actions}(P) \cup \{\uparrow\}$. In other words, a learning function takes a finite sequence of observations (pairs of propositional states) and outputs an action model or a symbol corresponding to 'undecided'.

We will study two types of learning: finite identifiability and identifiability in the limit. First let us focus on *finite identifiability*. Intuitively, finite identifiability corresponds to conclusive learning: upon observing some finite amount of action executions the learning function outputs, with certainty, a correct model for the action in question (up to equivalence). This certainty can be expressed in terms of the function being once-defined: it is allowed to output an action model only once, there is no chance of correction later on. Formally, we say that a learning function L is *(at most) once defined* if for any stream $\mathcal{E}$ for an action over P and $n, k \in \mathbb{N}$ such that $n \neq k$, we have that $L(\mathcal{E}[n]){=}\uparrow$ or $L(\mathcal{E}[k]){=}\uparrow$.

Definition 7. *Let $\mathcal{X}$ be a class of action models and $\mathcal{A} \in \mathcal{X}$, L be a learning function, and $\mathcal{E}$ be a stream. We say that:*

1. *L finitely identifies $\mathcal{A}$ on $\mathcal{E}$ if L is once-defined and there is an $n \in \mathbb{N}$ s.t. $L(\mathcal{E}[n]) \equiv \mathcal{A}$.*
2. *L finitely identifies $\mathcal{A}$ if L finitely identifies $\mathcal{A}$ on every stream for $\mathcal{A}$.*
3. *L finitely identifies $\mathcal{X}$ if L finitely identifies every $\mathcal{A} \in \mathcal{X}$.*
4. *$\mathcal{X}$ is finitely identifiable if there is a function L which finitely identifies $\mathcal{X}$.*

The following definition and theorem are adapted from [15,14,13].

Definition 8. *Let $\mathcal{X} \subseteq \text{Actions}(P)$. A set $D_{\mathcal{A}} \subseteq \mathcal{P}(P) \times \mathcal{P}(P)$ is a definite finite tell-tale set (DFTT) for $\mathcal{A}$ in $\mathcal{X}$ if*

1. *$D_{\mathcal{A}}$ is sound for $\mathcal{A}$ (i.e., for all $(s, s') \in D_{\mathcal{A}}$, $s' \in s \otimes \mathcal{A}$),*
2. *$D_{\mathcal{A}}$ is finite, and*
3. *for any $\mathcal{A}' \in \mathcal{X}$, if $D_{\mathcal{A}}$ is sound for $\mathcal{A}'$, then $\mathcal{A} \equiv \mathcal{A}'$.*

Lemma 1. *$\mathcal{X}$ is finitely identifiable iff there is an effective procedure $\mathsf{D} : \mathcal{X} \to \mathcal{P}(\mathcal{P}(P) \times \mathcal{P}(P))$, given by $\mathcal{A} \mapsto D_{\mathcal{A}}$, that on input $\mathcal{A}$ produces a definite finite tell-tale of $\mathcal{A}$.*

In other words, the finite set of observations $D_{\mathcal{A}}$ is consistent with only one action $\mathcal{A}$ in the class (up to equivalence of actions). D is a computable function that gives a $D_{\mathcal{A}}$ for any action $\mathcal{A}$.

Theorem 1. *For any finite set of propositions P the set of (fully observable) deterministic propositional actions over P is finitely identifiable.*

Example 4. Theorem 1 shows that deterministic actions are finitely identifiable. We will now show that this does not carry over to non-deterministic actions, that is, non-deterministic actions are in general not finitely identifiable. Consider the action of tossing a coin, given by the action model $\mathcal{A}$ in Example 1. If in fact the coin is fake and it will always land tails (so it only consists of the event e_2), in no finite amount of tosses the agent can exclude that the coin is fair, and that heads will start appearing in the long run (that e_1 will eventually occur). So the agent will never be able to say "stop" and declare the action model to only consist of e_2. This argument can be generalised, leading to the theorem below.

Theorem 2. *For any finite set of propositions P the set of arbitrary (including non-deterministic) fully observable propositional actions over P is not finitely identifiable.*

A weaker condition of learnability, *identifiability in the limit*, allows widening the scope of learnable actions, to cover also the case of arbitrary actions. Identifiability in the limit requires that the learning function after observing some finite amount of action executions outputs a correct model (up to equivalence) for the action in question and then forever keeps to this answer (up to equivalence) in all the outputs to follow. This type of learning can be called 'inconclusive', because certainty cannot be achieved in finite time.

Definition 9. *Let $\mathcal{X}$ be a class of action models and $\mathcal{A} \in \mathcal{X}$, L be a learning function, and $\mathcal{E}$ be a stream. We say that:*

1. *L identifies $\mathcal{A}$ on $\mathcal{E}$ in the limit if there is $k \in \mathbb{N}$ such that for all $n \geq k$, $L(\mathcal{E}[n]) \equiv \mathcal{A}$.*
2. *L identifies $\mathcal{A}$ in the limit if L identifies $\mathcal{A}$ in the limit on every $\mathcal{E}$ for $\mathcal{A}$.*
3. *L identifies $\mathcal{X}$ in the limit if L identifies in the limit every $\mathcal{A} \in \mathcal{X}$.*
4. *$\mathcal{X}$ is identifiable in the limit if there is an L which identifies $\mathcal{X}$ in the limit.*

The following theorem is adapted from [2].

Theorem 3. *For any finite set of propositions P the set of (fully observable) propositional actions over P is identifiable in the limit.*

Having established the general facts about finite identifiability and identifiability in the limit of propositional fully-observable actions, we will now turn to studying particular learning methods suited for such learning conditions.

3   Learning Actions via Update

Standard DEL, and in particular public announcement logic, deals with learning within epistemic models. If an agent is in a state described by an epistemic model $\mathcal{M}$ and learns from a reliable source, that ϕ is true, her state will be updated by eliminating all the worlds where ϕ is false. That is, the model $\mathcal{M}$ will be restricted to the worlds where ϕ is true. This can also be expressed in terms of action models, where the learning of ϕ corresponds to taking the product update of $\mathcal{M}$ with the event model $\langle \phi, \top \rangle$ (public announcement of ϕ).

Now we turn to learning *actions* rather than learning *facts*. Actions are represented by action models, so to learn an action means to infer the action model that describes it. Consider again the action model $\mathcal{A}$ of Example 1. The coin toss is non-deterministic and fully observable: either h or $\neg h$ will nondeterministically be made true and the agent is able to distinguish these two outcomes (no edge between e_1 and e_2). However, we can also think of $\mathcal{A}$ as the hypothesis space of a *deterministic* action, that is, the action $\mathcal{A}$ is in fact deterministically making h true or false, but the agent is currently uncertain about which one it is. Given the prior knowledge that the action in question must be deterministic, learning the action could proceed in a way analogous to that of update in the usual DEL setting.

It could for instance be that the agent knows that the coin is fake and always lands on the same side, but the agent initially does not know which. After the agent has executed the action once, she will know. She will observe either h becoming false or h becoming true, and can hence discard either e_1 or e_2 from her hypothesis space. She has now *learned* the correct deterministic action model for tossing the fake coin. Note the nice symmetry to learning of facts: here, learning of facts means eliminating worlds in epistemic models, learning of actions means eliminating events in action models.

In the rest of this section, all action models are silently assumed to be: fully observable, propositional, and universally applicable. Furthermore, we can assume them to be normal and have basic preconditions, due to Proposition 1.

3.1 Learning Precondition-Free Atomic Actions

We will first propose and study an update learning method especially geared towards learning the simplest possible type of ontic actions: precondition-free atomic actions.

Definition 10. *For any deterministic action model $\mathcal{A}$ and any pair of propositional states (s, s'), the update of $\mathcal{A}$ with (s, s') is defined by $\mathcal{A} \mid (s, s') := \mathcal{A} \upharpoonright \{e \in E \mid if\, pre(e) \models s\, then\, s \otimes e = s'\}$. For a set S of pairs of propositional states, we define: $\mathcal{A} \mid S := \mathcal{A} \upharpoonright \{e \in E \mid for\, all\, (s, s') \in S,\, if\, pre(e) \models s\, then\, s \otimes e = s'\}$.*

The update $\mathcal{A} \mid (s, s')$ restricts the action model $\mathcal{A}$ to the events that are consistent with observing s' as the result of executing the action in question in the state s.

Definition 11. *The update learning function for precondition-free atomic actions over P is the learning function L_1 defined by $L_1(\mathcal{E}[n]) = \mathcal{A}^1_{init} \mid set(\mathcal{E}[n])$ where $\mathcal{A}^1_{init} = (E, Q, pre, post)$ with $E = \{\psi \mid \psi\, is\, a\, consistent\, conjunction\, of\, literals\, over\, P\}$; Q is the identity relation on $E \times E$; $pre(e) = \top$ for all $e \in E$; $post(\psi) = \psi$.*

In Figure 1 we show a generic example of such update learning for $P = \{p, q\}$.

Theorem 4. *The class of precondition-free atomic actions is finitely identifiable by the update learning function L_1^{update}, defined in the following way:*

$$L_1^{update}(\mathcal{E}[n]) = \begin{cases} L_1(\mathcal{E}[n]) & if\, card(dom(L_1(\mathcal{E}[n]))) = 1 \\ & and\, for\, all\, k < n,\, L_1^{update}(\mathcal{E}[k]) = \uparrow; \\ \uparrow & otherwise. \end{cases}$$

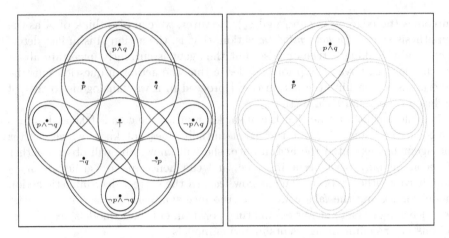

Fig. 1. On the left hand side $\mathcal{A}^1_{init}$ for $P = \{p, q\}$, together with sets corresponding to possible observations. We have labelled each event e by $post(e)$. On the right hand side the state of learning after observing $\mathcal{E}_0 = (\{q\}, \{p, q\})$.

3.2   Learning Deterministic Actions with Preconditions

We now turn to learning of action models with preconditions. First we only treat the case of maximal preconditions, then afterwards we generalise to arbitrary (not necessarily maximal) preconditions.

Definition 12. *The* update learning function for deterministic action models with maximal preconditions *over P is the learning function L_2 defined by $L_2(\mathcal{E}[n]) = \mathcal{A}^2_{init} \mid set(\mathcal{E}[n])$ where $\mathcal{A}^2_{init} = (E, Q, pre, post)$ with $E = \{(\phi, \psi) \mid \phi$ is a maximally consistent conjunction of literals over P and ψ is a conjunction of literals over P not containing any of the conjuncts of ϕ $\}$; Q is the identity on $E \times E$; $pre((\phi, \psi)) = \phi$; $post((\phi, \psi)) = \psi$.*

Theorem 5. *The class of deterministic action models with maximal preconditions is finitely identifiable by the following update learning function L_2^{update}.*

$$L_2^{update}(\mathcal{E}[n]) = \begin{cases} L_2(\mathcal{E}[n]) & \text{if for all } e, e' \in dom(L_2(\mathcal{E}[n])) \\ & \text{if } e \neq e', \text{ then } pre(e) \neq pre(e') \\ & \text{and for all } k < n, \ L_2^{update}(\mathcal{E}[k]) = \uparrow; \\ \uparrow & \text{otherwise.} \end{cases}$$

Example 5. Consider a simple scenario with a pushbutton and a light bulb. Assume there is only one proposition p: 'the light is on', and only one action: pushing the button. We assume an agent wants to learn the functioning of the pushbutton. There are 4 distinct possibilities: 1) the button does not affect the light (i.e., the truth value of p); 2) it is an *on button*: it turns on the light unconditionally (makes p true); 3) it is an *off button*: it turns off the light unconditionally (makes p false); 4) it is an *on/off button* (flips the truth value of p). If the agent is learning by update, it starts with the action model $\mathcal{A}^2_{init}$ containing the following events: $\langle p, \top \rangle$, $\langle \neg p, \top \rangle$, $\langle p, \neg p \rangle$, and $\langle \neg p, p \rangle$. Note that by definition $\mathcal{A}^2_{init}$ does not contain the events $\langle p, p \rangle$ and $\langle \neg p, \neg p \rangle$, since they both have a postcondition conjunct which is also a precondition conjunct. Assume the first two observations the learner receives (the first elements of a stream $\mathcal{E}$) are $(\emptyset, \{p\})$ and $(\{p\}, \emptyset)$. Since the agent uses learning by update, she revises her model as follows (cf. Definition 12):

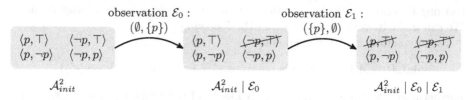

Now the agent has reached a deterministic action model $\mathcal{A}^2_{init} \mid \mathrm{set}(\mathcal{E}[2])$, and can report this to be the correct model of the action, cf. Theorem 5. Note that the two observations correspond to first pushing the button when the light is off ($\mathcal{E}_0$), and afterwards pushing the button again after the light has come on ($\mathcal{E}_1$). These two observations are sufficient to learn that the pushbutton is of the on/off type (it has one event that makes p true if p is currently false, and another event making p true if currently false).

Consider now another stream $\mathcal{E}'$ where the first two elements are $(\emptyset, \{p\})$ and $(\{p\}, \{p\})$. Update learning will now instead reduce the initial action model $\mathcal{A}^2_{init}$ to the action model only containing $\langle p, \top \rangle$ and $\langle \neg p, p \rangle$. This time the learner identifies the button to be an on button, again after only two observations. It is not hard to show that in a setting with only one propositional symbol p, any deterministic action will be identified after having received the first two distinct observations.

Example 6. Consider learning the functioning of an n-bit binary counter, where the action to be learned is the increment operation. For $i = 1, \ldots, n$, we use the proposition c_i to denote that the ith least significant bit is 1. Consider first the case $n = 2$. A possible stream for the increment operation is the following:

$$(\emptyset, \{c_1\}), \qquad (\{c_1\}, \{c_2\}), \qquad (\{c_2\}, \{c_2, c_1\}), \qquad (\{c_2, c_1\}, \{\emptyset\}), \qquad \cdots$$

| 0 0 → 0 1 | 0 1 → 1 0 | 1 0 → 1 1 | 1 1 → 0 0 | $\cdots$ |
| $c_2\ c_1\quad c_2\ c_1$ | $c_2\ c_1\quad c_2\ c_1$ | $c_2\ c_1\quad c_2\ c_1$ | $c_2\ c_1\quad c_2\ c_1$ | |

Using the update learning method on this stream, it is easy to show that the learner will after the first 4 observations be able to report the correct action model containing the following events: $\langle \neg c_2 \wedge \neg c_1, c_1 \rangle, \langle \neg c_2 \wedge c_1, c_2 \wedge \neg c_1 \rangle, \langle c_2 \wedge \neg c_1, c_1 \rangle, \langle c_2 \wedge c_1, \neg c_2 \wedge \neg c_1 \rangle$. Note that since $\mathcal{A}^2_{init}$ has maximal preconditions, the action model learned for an n-bit counter will necessarily contain 2^n events: one for each possible configuration of the n bits. If we did not insist on maximal preconditions, we would only need $n + 1$ events to describe the n-bit counter: $\langle \neg c_i \wedge c_{i-1} \wedge c_{i-2} \wedge \cdots \wedge c_1, c_i \wedge \neg c_{i-1} \wedge \neg c_{i-2} \wedge \cdots \wedge \neg c_1 \rangle$ for all $i = 2, \ldots, n$, $\langle \neg c_1, c_1 \rangle$ and $\langle c_n \wedge \cdots \wedge c_1, \neg c_n \wedge \cdots \wedge \neg c_1 \rangle$. This means that there is room for improvement in our learning method.

To allow learning of deterministic action models where preconditions are not required to be maximal we need a different learning condition. Consider learning an action on $P = \{p\}$ that sets p true unconditionally. With non-maximal preconditions, all of the following events would be consistent with any stream for the action: $\langle \top, p \rangle, \langle \neg p, p \rangle, \langle p, \top \rangle$. To get to a deterministic action model, the learning function would have to delete either the first or the two latter events. We can make it work as described in the following.

For any action model $\mathcal{A}$ we define

$$\min(\mathcal{A}) = \mathcal{A} \upharpoonright \{e \mid \text{there is no event } e' \neq e \text{ with } pre(e) \models pre(e')\}.$$

Furthermore, we define L_3 to be exactly like L_2 of Definition 12 except in the definition of E, ϕ can be any conjunction of literals, not only maximally consistent ones.

Theorem 6. *The class of deterministic action models is finitely identifiable by the following update learning function* L_3^{update}.

$$L_3^{update}(\mathcal{E}[n]) = \begin{cases} \min(L_3(\mathcal{E}[n])) & \textit{if for all } s \in \mathcal{P}(P) \textit{ there exists an } s' \textit{ s.t.} \\ & (s, s') \in \text{set}(\mathcal{E}[n]) \textit{ and for all } k < n, \\ & L_3^{update}(\mathcal{E}[k]) = \uparrow; \\ \uparrow & \textit{otherwise.} \end{cases}$$

The theorem can be seen as a generalisation of Theorem 5 in that it allows the learner to learn more compact action models in which maximal consistency of preconditions is not enforced (on the contrary, by the way the min operator is defined above, the learner will learn an action model with *minimal* preconditions). For instance, in the case of the n-bit counter considered in Example 5, it can be shown that the learner will learn the action model with $n + 1$ events instead of the one with 2^n events.

4   Conclusions and Related Work

This paper is the first to study the problem of learnability of action models in dynamic epistemic logic (DEL). We provided an original learnability framework and several early results concerning fully observable propositional action models with respect to conclusive (finite identifiability) and inconclusive (identifiability in the limit) learnability. Apart from those general results, we proposed various learning functions which code particular learning algorithms. Here, by implementing the update method (commonly used in DEL), we demonstrated how the learning of action models can be seen as transitioning from nondeterministic to deterministic actions.

Related Work. A similar qualitative approach to learning actions has been addressed by [18] within the STRIPS planning formalism. The STRIPS setting is more general than ours in that it uses atoms of first-order predicate logic for pre- and postconditions. It is however less general in neglecting various aspects of actions which we have successfully treated in this paper: negative preconditions (i.e., negative literals as precondition conjuncts), negative postconditions, conditional effects (which we achieve through non-atomic action models). We believe that the ideas introduced here can be applied to generalize the results of [18] to richer planning frameworks allowing such action types. It is also worth mentioning here that there has been quite substantial amount of work in relating DEL and learning theory (see [11,12] for overviews), which concerns a different setting: treating update and upgrade revision policies as long term learning methods, where learning can be seen as convergence to certain types of knowledge (see [3,4,5]). A study of abstract properties of finite identifiability in a setting similar to ours, including various efficiency considerations, can be found in [13].

Further Directions. In this short paper we only considered fully observable actions applied in fully observable states, and hence did not use the full expressive power of the DEL formalism. The latter still remains adequate, since action models provide a very well-structured and principled way of describing actions in a logical setting, and since its use opens ways to various extensions. The next steps are to cover more DEL action models: those with arbitrary pre- and postconditions, and those with partial observability and multiple agents. As described earlier, partially observable actions are not learnable in the strict sense considered above, but we can still investigate agents learning "as much as possible" given their limitations in observability. The multi-agent case is particularly interesting due to the possibility of agents with varied limitations on observability, and the possibility of communication within the learning process.

We plan to study the computational complexity of learning proposed in this paper, but also to investigate other more space-efficient learning algorithms. We are also interested algorithms that produce minimal action models. Furthermore, we here considered only what we call *reactive learning*: the learner has no influence over observations. We would also like to study the case of *proactive learning*, where the learner gets to choose which actions to execute, and hence observe their effects. This is probably the most relevant type of learning for a general learning-and-planning agent. In this context, we also plan to focus on *consecutive streams*: streams corresponding to executing sequences of actions rather than observing arbitrary state transitions. Our ultimate aim is to relate learning and planning within the framework of DEL. Those two cognitive capabilities are now investigated mostly in separation—our goal is to bridge them.

Acknowledgements. Nina Gierasimczuk is funded by NWO Veni grant 275-20-043. We are grateful to Martin Holm Jensen, Mikko Berggren Ettienne and the anonymous reviewers for valuable ideas and feedback.

References

1. Andersen, M.B., Bolander, T., Jensen, M.H.: Conditional epistemic planning. In: del Cerro, L.F., Herzig, A., Mengin, J. (eds.) JELIA 2012. LNCS, vol. 7519, pp. 94–106. Springer, Heidelberg (2012)
2. Angluin, D.: Inductive inference of formal languages from positive data. Information and Control 45(2), 117–135 (1980)
3. Baltag, A., Gierasimczuk, N., Smets, S.: Belief revision as a truth-tracking process. In: Apt, K. (ed.) TARK 2011: Proceedings of the 13th Conference on Theoretical Aspects of Rationality and Knowledge, pp. 187–190. ACM (2011)
4. Baltag, A., Gierasimczuk, N., Smets, S.: Truth tracking by belief revision. ILLC Prepublication Series PP-2014-20 (to appear in Studia Logica 2015) (2014)
5. Baltag, A., Gierasimczuk, N., Smets, S.: On the solvability of inductive problems: A study in epistemic topology. ILLC Prepublication Series PP-2015-13 (to appear in Proceedings of TARK 2015) (2015)
6. Baltag, A., Moss, L.S., Solecki, S.: The logic of public announcements and common knowledge and private suspicions. In: Gilboa, I. (ed.) TARK 1998: Proceedings of the 7th Conference on Theoretical Aspects of Rationality and Knowledge, pp. 43–56. Morgan Kaufmann (1998)
7. Bolander, T., Andersen, M.B.: Epistemic planning for single- and multi-agent systems. Journal of Applied Non-Classical Logics 21, 9–34 (2011)
8. van Ditmarsch, H., Kooi, B.: Semantic results for ontic and epistemic change. In: Bonanno, G., van der Hoek, W., Wooldridge, M. (eds.) LOFT 7: Logic and the Foundation of Game and Decision Theory. Texts in Logic and Games, vol. 3, pp. 87–117. Amsterdam University Press (2008)
9. Ghallab, M., Nau, D.S., Traverso, P.: Automated Planning: Theory and Practice. Morgan Kaufmann (2004)
10. Gierasimczuk, N.: Learning by erasing in dynamic epistemic logic. In: Dediu, A.H., Ionescu, A.M., Martín-Vide, C. (eds.) LATA 2009. LNCS, vol. 5457, pp. 362–373. Springer, Heidelberg (2009)
11. Gierasimczuk, N.: Knowing One's Limits. Logical Analysis of Inductive Inference. Ph.D. thesis, Universiteit van Amsterdam, The Netherlands (2010)
12. Gierasimczuk, N., de Jongh, D., Hendricks, V.F.: Logic and learning. In: Baltag, A., Smets, S. (eds.) Johan van Benthem on Logical and Informational Dynamics. Springer (2014)
13. Gierasimczuk, N., de Jongh, D.: On the complexity of conclusive update. The Computer Journal 56(3), 365–377 (2013)
14. Lange, S., Zeugmann, T.: Types of monotonic language learning and their characterization. In: COLT 1992: Proceedings of the 5th Annual ACM Conference on Computational Learning Theory, pp. 377–390. ACM (1992)
15. Mukouchi, Y.: Characterization of finite identification. In: Jantke, K.P. (ed.) AII 1992. LNCS, vol. 642, pp. 260–267. Springer, Heidelberg (1992)
16. Sadzik, T.: Exploring the Iterated Update Universe, ILLC Prepublications PP-2006-26 (2006)
17. Sietsma, F., van Eijck, J.: Action emulation between canonical models. Journal of Philosophical Logic 42(6), 905–925 (2013)
18. Walsh, T.J., Littman, M.L.: Efficient learning of action schemas and web-service descriptions. In: AAAI 2008: Proceedings of the 23rd National Conference on Artificial Intelligence, vol. 2, pp. 714–719. AAAI Press (2008)

Great Expectations

Eddy Keming Chen and Daniel Rubio

Rutgers University

Abstract. Standard expected utility theory faces familiar problems with
the infinities. We propose a new theory based on surreal numbers and suggest that it solves many of those problems, including Pascal's Wager and
the Pasadena Game.

1 Introduction

The standard expected utility maximization account of rationality has been very
successful in the finite domain. But the introduction of infinities leads to puzzle
and paradox. There are two ways for the infinite to make its entrance. In some
problems, a state or proposition has an infinite utility attached to it. In other
problems, there are infinitely many states or propositions with finite utilities
attached, leading to infinite expected utilities.

There is a classic puzzle associated with each. Pascal's wager, nearly as old
as decision theory itself, is the best-known problem involving infinite utilities
associated with a single state. The St. Petersburg Game is the best known associated with an infinite state space leading to an infinite expected utility, but
it has given rise to an even more fearsome cousin: the Pasadena Game.

We propose to conservatively extend classic expected utility theory with non-
standard analysis; namely, Conway's surreal numbers. In §2, we introduce surreal
mathematics and prove a representation theorem. In §3, we analyze pascal's
wager, showing that we can deal with infinite utilities arising from a single state
(and that there is nothing malformed about such problems. In §4, we analyze the
Pasadena game, and show that our theory has the resources to deal even with
very hard problems involving infinite state spaces (problems that more resist
analysis by more standard mathematical methods).

2 A Surreal Solution

Transfinite decision theory requires the ability to perform arithmetic operations on finite and infinite numbers, with commutativity and non-absorption
(and other standard desirable properties of addition) intact, and where every
number—finite and transfinite—has an additive inverse (so that, for example,
$\omega - \omega$, is defined). In short: we need a totally ordered field including both the
finite and transfinite numbers. More precisely, we require:

1. an ordered-field including all reals and ordinals;

© Springer-Verlag Berlin Heidelberg 2015
W. van der Hoek et al. (Eds.): LORI 2015, LNCS 9394, pp. 53–63, 2015.
DOI: 10.1007/978-3-662-48561-3_5

2. addition in that field that is commutative, non-absorptive, and such that each element has an additive inverse;

3. multiplication in that field that is commutative, non-absorptive, and such that each non-zero element has a multiplicative inverse.

In short: we need a number system and accompanying operations that allow us to treat finite and transfinite numbers in similar and familiar ways.

2.1   Surreal Numbers: The Basics

Fortunately, John Conway discovered (or invented, depending on your philosophy of mathematics) such a field, and began its exploration in his *On Numbers and Games* (1974). Conway called the objects he discovered *surreal* numbers. For those familiar with Dedekind's construction of the reals out of the rationals, it may be helpful to note that Conway's construction is quite similar to Dedekind's. Except, rather than using the rationals, Conway uses the ordinals. Nevertheless, we can think of Surreal numbers as being something analogous to performing "Dedekind cuts" on ordinals. They are defined recursively as follows:[1]

Definition 1. If L and R are sets of numbers, and no $x \in L \geq$ any $y \in R$, then $\{L|R\}$ is a number.

Convention 1. If x=$\{L|R\}$, we will write x^L as a convention for the typical member of L, and x^R for the typical member of R

Definition 2. $x \geq y$ iff no $x^R \leq y$ and no $y^L \geq x$

Other familiar ordering relations are defined in the usual way.[2]

Definition 1 looks circular. Fortunately, the null set is trivially a set of numbers, and so our first surreal number is $\{\varnothing|\varnothing\} = 0$.[3] From 0, we gain two new numbers: $\{0|\varnothing\} = 1$ and $\{\varnothing|0\} = -1$. From these numbers, we can find yet more numbers. In order to avoid tedious iterations, we can see the structure of the surreals laid out in figure 1. We use **No** to denote the class of numbers created by repeated application of definition 1, and the iteration of definition 1 on which n is found its 'birthday.'[4]

With a hearty stock of numbers, we can now set about defining arithmetic operations.

Definition 3: x + y = $\{x^L + y, x + y^L | x^R + y, y^R + x\}$

Definition 4: -x = $\{-x^L|-x^R\}$

---

[1] Conway [1974]
[2] $x \ngeq y$ iff not $x \geq y, x > y$ iff $x \geq y$ and $y \ngeq x, x = y$ iff $x \geq y$ and $y \geq x$
[3] A similar trick saves definition 2 from circularity, for it allows us to prove that $0 \geq 0$.
[4] So the birthday of 0 is day 0 the birthday of $1, -1$ is day 1, etc.

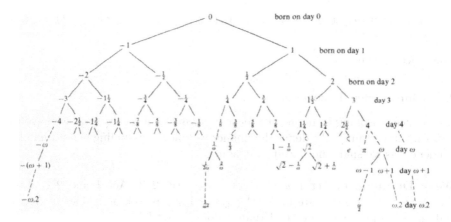

Fig. 1. The Surreal Tree

Definition 5: $x \times y = \{x^L \times y + y^L \times x - x^L \times y^L, x^R \times y + y^R \times x - x^R \times y^R | x^L \times y + y^R \times x - x^L \times y^R, x^R \times y + y^L \times x - x^R \times y^L\}$

These definitions make **No** an ordered field including all reals and all ordinals (in fact, Conway proves that it is a universally embedding field). We refer the interested reader to Conway for the proofs and further details.[5]

2.2   Representation Theorem

The first step in any decision theory is a representation theorem. Here, we state a von Neumann and Morgenstern-style representation theorem for surreal utilities. See Appendix for the proof.

Notation 1: Let $\star$ denote an embedding from the standard universe into the surreal universe. Let **No** denote a surreal model.

Theorem 1 (Surreal Von Neumann-Morgenstern Theorem). Let X be a space of lotteries, and let $\preceq$ be a binary relation $\subseteq X \times X$. There exists an affine function $U : X \to \mathbf{No}$ such that $\forall x, y \in X$

$$U(x) \leq U(y) \Leftrightarrow x \preceq y$$

if and only if $\preceq$ satisfies all of the following:

1. Completeness: $\forall x, y \in X$, either $x \preceq y$ or $y \preceq x$.
2. Transitivity: $\forall x, y, z \in X$, if $x \preceq y$ and $y \preceq z$, then $x \preceq z$.
3. Continuity$\star$: $\forall x, y, z \in X$, if $x \prec y \prec z$, then there exist surreals $p, q \in \star(0, 1)$ such that $px + (1 - p)z \prec y \prec qx + (1 - q)z$.

---

[5] Conway [1974], 15-44

4. Independence⋆: $\forall x, y, z \in X, \forall p \in \star(0, 1]$, $x \preceq y$ if and only if $px + (1 - p)z \preceq px + (1 - p)z$.

Proof: See Appendix.

2.3 Dominance Theorems

Dominance reasoning is among the most venerable and secure ways we have to deal with uncertainty. We define dominance as follows (assuming act-state independence, as we shall and have throughout):

Weak Dominance. Act 1 *weakly dominates* Act 2 iff Act 1 and 2 contain the same states, every state in Act 1 pays at least as well as it does in Act 2, and one state pays more in Act 1 than it does in Act 2.

Strict Dominance. Act 1 *strictly dominates* Act 2 iff Act 1 and 2 contain the same states, and every state in Act 1 pays better than it does in Act 2.

It is known that in the finite case, expected utility maximization respects dominance.[6] But it is also known that in the infinite case, expected utility maximization need not respect dominance.[7] These results show that some revision of the EU axioms is required in order to put transfinite decision theory on a even footing with its finite cousin, and may represent the most serious challenge emerging from the Pasadena Game literature.

Traditionally, utility functions are real-valued. This property is enforced by the Archimedean Axiom (or by the choice of Continuity Axiom):

Archimedean. If $G_1 \prec G_2 \prec G_3$, then there exists some probability $\epsilon \in (0, 1)$ s.t. $(1 - \epsilon G_1) + \epsilon G_3 \prec G \prec (1 - \epsilon G_3) + \epsilon G_1$.

But because **No** is a non-archimedean field, our proposal and results require that we give it up (as astute readers of §4.3 will note, even though we replace it with a "weaker" axiom which allows surreal probabilities). Happily, it is the Archimedean Axiom that is the source of our trouble with Dominance in the first place (as Fine, Hajek and Nover note). In fact, in the framework of surreal numbers, we can easily prove that both Weak Dominance and Strict Dominance hold up to countable sums.

Theorem 2 (Surreal Strict Dominance Theorem.) Let $\{a_i\}_{i \in \mathbb{N}}$ and $\{b_i\}_{i \in \mathbb{N}}$ be two surreal-valued expected payoff series. If $\{a_i\}_{i \in \mathbb{N}}$ strictly point-wise dominates $\{b_i\}_{i \in \mathbb{N}}$, then $\sum_{i=1}^{\infty} a_i > \sum_{i=1}^{\infty} b_i$.

Proof: See Appendix.

---

[6] See Easwaran [Ms.].

[7] Fine [2008], see Hajek and Nover [2008] for further discussion.

Theorem 3 (Surreal Weak Dominance Theorem.) Let $\{a_i\}_{i\in\mathbb{N}}$ and $\{b_i\}_{i\in\mathbb{N}}$ be two surreal-valued expected payoff series. If $\{a_i\}_{i\in\mathbb{N}}$ strictly dominates $\{b_i\}_{i\in\mathbb{N}}$ at least in one term and weakly dominates $\{b_i\}_{i\in\mathbb{N}}$ in all other terms, then

$$\sum_{i=1}^{\infty} a_i > \sum_{i=1}^{\infty} b_i.$$

Proof: We leave this as an exercise for the reader. $\square$

3   Pascal's Wager

With the technical work out of the way, we turn to our opening puzzles to showcase our theory. First up: Pascal's wager. Many criticisms have been leveled against this argument.[8] Two of these in particular display the difficulties involved in infinitely valued states. Hajek's 'mixed strategies' objection reveals the odd feature that (in cardinal arithmetic), a 1 in 1000 shot at an infinite utility is as valuable as a 999 in 1000 shot. Diderot's 'many gods' objection requires us to add and subtract infinities, resulting in sums that typically go undefined.

3.1   Mixed Strategies

Typically, Pascal's wager has been set up as a decision problem with two options.

Table 1. Pascal's Wager, Classical Presentation

	God	No God
Christian	∞	10
Non-Christian	5	10
Expected Payoff	∞	10

But decision theorists know better. Whenever we have gambles, we can adopt mixtures of those gambles. We can think of mixtures heuristically as using coin flips to decide which gamble to take. So someone presented with the decision in Table 1 might make her choice by flipping a fair coin. In that case, the expected utility of the flip strategy = the expected utility of simply picking "Christian." In fact, the coin can be arbitrarily biased against Christian, and still the mixed strategy has the same expected utility as the pure "Christian" option. This is counterintuitive because gambles with arbitrary biases will have the same expected utility and the agent ought to be indifferent among the different gambles.

This is because ∞, at least in the extended reals and more familiar Cantorian realms, has the absorption property.[9] By standard lights, a chance at ∞ is as good as the genuine artifact. But in surreal arithmetic, this is not true. The

---

[8] See Jordan [2007] for a review.

[9] In Hajek's terms: *reflexive under multiplication.*

surreal ω is strictly greater than the surreal $.5\omega$. Thus, our proposal correctly predicts that the pure "Christian" strategy beats all mixed strategies.[10]

Hajek noted the potential for surreal valued utilities to escape his objection.[11] Instead, he argues that surreal infinite numbers do not have the same properties as the infinity Pascal seems to be talking about. Hajek's Pascal sees salvation as the greatest good, and thus possessing the absorption property for addition. We do not dispute Hajek's reading of Pascal (although we express some skepticism about the theology underlying a view according to which salvation is the greatest good, or indeed the rationality of a view whereby salvation and no apple is as preferable as salvation plus an apple),[12] but we are less interested in giving a faithful representation of Pascal's original argument than we are in applying our more general proposal to this problem in transfinite decision theory.

3.2 Many Gods

A common objection to Pascal is that his decision problem is too simple, and as a result, the use of infinite utilities looks less problematic than it is.[13] For there are a great many purported gods, many of which treat their followers well, and their doubters cruelly. Moreover, there are any number of other potential eschatological situations. Perhaps there is a god, but god is a universalist, so that everyone ends well. Perhaps there is a god, but god is a rationalist, and so anyone who makes epistemic decisions (like belief in a god) for pragmatic reasons ends poorly. The objection goes that once we see all these situations, and their accompanying infinite utilities and disutilities in the decision problem, we conclude that there's nothing interesting to say, and so problems of this sort aren't sensibly posed.[14]

But our proposal allows us to formulate and analyze this objection precisely. Let $E_1...E_n...$ be a (potentially infinite) partition over states in an expanded Pascalian decision problem. With each E_i, we associate some surreal number n, corresponding to $u(E_i)$ in the agent's utility function. Let $cr(E_i)$ be value of the agent's credence function over E_i. We can then give the EU of each of the E_i's.

For example, suppose our agent thinks there are three live divine candidates: Zeus, Apollo and Athena. She then has four religious options: Zeusianism, Athenianism, Apollinism, and Atheism. Zeusianism is an exclusivist religion. Zuesians

---

[10] We note that our proposal is not the only one to do this. See Bartha [2007] and Herzberg [2011] for alternate proposals that make the same prediction. We do note that all of these proposals make use of non-Archimedean utilities, of which we shall say more soon.

[11] Hajek [2003].

[12] But see Herzberg [2011] for discussion.

[13] The objection is as old as Diderot [1746], but has received a more rigorous formulatin in e.g. Cargile [1966]. Herzberg [2011] briefly discusses this possibility but focuses instead on modeling agents who do not countenance it.

[14] Rescher [1985] presses this line of reasoning, although not explicitly in connection with the many gods objection.

get infnite utiity, but everyone else is damned. But according to Athenian theology, Athena is a universalist who will give everyone infinite utility. According
to Apollinism, Apollo rewards atheists and damns everyone else.[15] We may represent the problem as a table, representing the finite utility of a regular life as 100:

Table 2. Pascal's Wager With Three Gods

	Zeus	Athena	Apollo	Atheism
Zeusian	ω	ω	$-\omega$	100
Athenian	$-\omega$	ω	$-\omega$	100
Apollinist	$-\omega$	ω	$-\omega$	100
Atheist	$-\omega$	ω	ω	100

Already, using surreal values allows her to assign a sensible ranking of her
options. Which religion is best will depend on what her credence function is
like. If, for instance, Cr(Zeus) = .5, Cr(Athena) = .3, Cr(Apollo) = .1, and
Cr(Atheism) = .1, then EU(Zeusian) = $.7\omega - 10$ > EU(Atheist) = $-.1\omega + 10$ >
EU(Athenian) = EU(Apollinist) = $-.3\omega + 10$; with the credence favoring Zeus,
Zeus is the best option. On the other hand, if Cr(Zeus) = .1, Cr(Athena) =
.2, Cr(Apollo) = .2, and Cr(Atheism) = .5, then EU(Atheist) = $.3\omega + 50$ >
EU(Zeusian) = $.1\omega + 50$ > EU(Athenian) = EU(Apollinist) = $-.1\omega + 50$; with
the credence favoring atheism, Atheist is the best option.[16]

We can do this for arbitrarily complicated decision problems of this sort. So
there is nothing incoherent or problematic about the use of infinite utilities. We
leave evaluation of the argument's ultimate success or failure to future work.

4   Infinite State Spaces

In this section, we will discuss the problems of infinite state spaces, specifically
the Pasadena Game. We start by giving an overview of the problem, which stems
from results about countable sums and convergence. Then we show how our theory is able to make sense of the problem. Our solution here is less straightforward
than the analysis of Pascal's wager, but still leads to a way for consistently valu-
ing games like it.

---

[15] Such Apollo is thus an example of the sort of god no one believes, but is regularly
trotted out by philosophers objecting to Wager-style arguments. We propose Silly
Theism as a technical name for this type of religion.

[16] As we have set things up, the exclusivist and atheist options each dominate the
universalist and silly options, but other combinations (such as scenarios with multiple
exclusivist gods in play, or mildly inclusivist options where some gods favor some
infidels over others) can bring out the benefits of those.

4.1   The Pasadena Game

Hajek and Nover's Pasadena game is played by flipping a coin until a coin lands heads, paying on the nth flip $\$(-1)^{n-1}2^n/n$.[17]

Table 3. Pasadena Game

Heads on nth Flip	1	2	3	4	...
Probability	1/2	1/4	1/8	1/16	...
Payoff	\$2	\$-2	\$8/3	\$-4	...
Expected Payoff	\$1	\$-1/2	\$1/3	\$-1/4	...

The expected payoffs of this game match the alternating harmonic series, assuming utility linear in dollars. This series is conditionally convergent. But the Riemann Rearrangement Theorem tells us that in a conditionally convergent series, infinite summation is non-commutative. For any such conditionally convergent series $\sum_{i=1}^{\infty} a_i$, we can make it to converge to a different number simply by changing the order of the summation. In fact, for any real number n, we can rearrange the terms in some way such that it converges to that number. The theorem also says that $\sum_{i=1}^{\infty} a_i$ can be rearranged such that it diverges to $+\infty$ or $-\infty$.

This mathematical result, therefore, leaves us no way to consistently assign a expected value to the game; however we add up the expected payoffs, for every real number (or positive/negative infinity) there is a rearrangement of the series which sums to it.

It is initially tempting to argue that there is something wrong with the Pasadena Game itself. After all, asking for the sum of the members of the set {1, -1/2, 1/3, -1/4...} will likely draw a bemused look and an explanation of conditional convergence from a mathematician, in the same way asking for the largest natural number will draw a bemused look and an explanation of infinity. But Hajek and Nover give us several compelling reasons against thinking that the game is somehow flawed, incoherent, or ill-stated.[18] In doing so, they introduce the Altadena Game, which is exactly like the Pasadena game, but worth $1 more in every state. And with it, a new constraint on solutions: any solution to the Pasadena Game must predict that the Altadena Game (which strictly dominates it) is more valuable. Standard EU theory cannot satisfy this contstraint.[19]

The key to evading pasadena-style problems is to evade conditional convergence. By relying on the equivalence of affine transformations of the utility function in faithfully representing preferences, we can always produce a 'stable' way to value a game.

---

[17] Hajek and Nover [2004].

[18] Hajek & Nover [2004], [2006] and [2008]. See Colyvan [2006] and [2008] for an opposing view.

[19] Fine [2008], Hajek and Nover [2008].

Before we proceed, a few definitions and another result:

cc-vulnerability. Let u be a utility function and p the probability function which together with u represents some agent's preferences. We will say that u is cc-vulnerable with respect to p iff there is some $X \subseteq \{x_i : x$ is the product of p_i and u_i $\}$ such that the members of X can be all and only the members of a conditionally convergent series.

cc-invulnerability. Let u be a utility function and p be a probability function. We will say that u is cc-invulnerable with respect to p iff u is not CC-Vulnerable with respect to p.

cc-invulnerable Transformation. Let p be a probability function, u be a utility function and u' be a positive affine transformation of u. We say that u' is a cc-invulerable transformation of u with respect to p iff u is cc-vulnerable with respect to p and u' is cc-invulnerable with respect to p.

Corollary 6.1. Let u be a surreal-valued utility function. There exists a utility function u' that is a cc-invulnerable transformation of u.

We will fix the probability function p and drop the "with respect to p" in the following discussion. By Corollary 6.1,[20] every utility function with codomain **No** (which is to say, surreal-valued) has a cc-invulnerable transformation (in fact, infinitely many). We note that this is not true of utility functions with codomain $\mathbb{R}$, and thus surreal utilities are vital to our proposed resolution.

Because any representable preference structure can be adequately represented with a cc-invulnerable utility function, we contend that cc-vulnerability in the utility function is inessential to adequately representing a preference structure. Because cc-vulnerable utility functions give rise to problematic gambles like the Pasadena Game, we contend that cc-vulnerability is a representational defect. We propose that cc-vulnerable representations be rejected in favor of their cc-invulnerable transformations. This proposal does not allow the problematic payoff series to arise; but so long as we have a representation theorem, it ensures that any preference structure—and, *a fortiori*, any preference structure which can be represented by a utility function that is linear in dollars—can still be represented.

With the surreal representation theorem in hand, we can deploy the Surreal Dominance Theorems (theorems 2-3) and Corollary 6.1 to give our solution to the Pasadena problems. We suggest the following procedure for an agent whose utility is linear in dollars: first, take an agent's entire preference structure; next, locate a cc-invulnerable surreal representation of it (whose existence is guaranteed by the Surreal Representation Theorem and Corollary 6.1). When this has been done, the original Pasadena game will have a fixed utility; call

---

[20] Proof in Appendix.

it x.[21] Furthermore, a dollar will also have a fixed utility; call it y. For the Altadena Game and others lie, the utility well be will be $x + ny$, where n is the number of 'steps' distant (dollars added) it is from the original game . This gives the value of the game in utils. If a dollar value is desired, we may simply look at the relation between dollars and utils in the fixed representation. A similar but more complicated procedure will work agents whose utility is not linear in dollars. Since the surreals (codomain of the utility function) form a totally ordered field, we have given a systematic solution that (1) consistently values the Pasadena games; (2) respects dominance reasoning; and (3) relative to an adequate representation, ensures a fixed interval between 'steps' on the ladder of the Pasadena sequence.

A   Proofs of Key Theorems

See http://www.danielkfrubio.com/papers/.

References

1. Bartha, P.: Taking Stock of Infinite Value: Pascal's Wager and Relative Utilities. Synthese 154, 5–52 (2007)
2. Cargile, J.: Pascal's Wager. Philosophy 35, 250–257 (1966)
3. Colyvan, M.: No Expectations. Mind 115, 695–702 (2006)
4. Colyvan, M.: Relative Expectation Theory. Journal of Philosophy 105(1), 37–44 (2008)
5. Conway, J.: On Numbers and Games. A.K. Peters/CRC Press, Natick (1974)
6. Diderot, D.: Pensees Philosophiques, reprinted Whitefish. Kessinger Publishing, MN (2009)
7. Easwaran, K.: Dr. Truthlove, or How I Learned to Stop Worrying and Love Bayesian Probabilities. Nous (forthcoming)
8. Ehrlich, P.: The Absolute Arithmetic Continuum and The Unification of All Numbers Great and Small. The Bulletin of Symbolic Logic 18(1), 1–45 (2012)
9. Fine, T.: Evaluating the Pasadena, Altadena, and St. Petersburg Gambles. Mind 117, 613–632 (2008)
10. Hajek, A.: Waging War on Pascal's Wager. Mind 113, 27–56 (2003)
11. Hajek, A.: Unexpected Expectations. Mind 123(490), 533–567 (2014)

---

[21] We note that there are open problems in the neighborhood. For example, what is an infinite sum over surreals? Usually in real analysis, we define infinite sums over reals by taking the limit to infinity over partial sums. But this is not feasible for surreals since the limits in many cases do not exist (partly because **No** is not Cauchy-complete). So how can we relate the infinite sums to the algebraic operations defined in surreal analysis? Conway himself suggests global approaches with much departure from real analysis. It is also possible to define it in a piece-wise fashion. Since surreal analysis is a new and open field, and we remain optimistic that future work will offer more natural solutions. Another approach is to take the simplest element of the convex subclass that results from adding surreal numbers, an option that we hope to explore in future work.

12. Hajek, A., Nover, H.: Vexing Expectations. Mind 113, 237–249 (2004)
13. Hajek, A., Nover, H.: Perplexing Expectations. Mind 115, 703–720 (2006)
14. Hajek, A., Nover, H.: Complex Expectations. Mind 117, 643–664 (2008)
15. Herzberg, F.: Hyperreal Utilities and Pascal's Wager. Logique Et Analyse 213, 69–108 (2011)
16. Jordan, J.: Pascal's Wager and James's Will To Believe. In: Oxford Handbook of Philosophy of Religion. OUP, Oxford (2007)
17. Keisler, H.J.: Foundations of Infinitesimal Calculus. Prindle, Weber & Schmidt, Boston (1976)
18. Meacham, C.J.G., Weisberg, J.: Representation Theorems and the Foundations of Decision Theory. Australasian Journal of Philosophy 89(4), 641–663 (2011)
19. Pascal, B.: Pensées, translated by W. F. Trotter. Dent, London (1910)
20. Rescher, N.: Pascal's Wager. Notre Dame University Press, South Bend (1985)

Probabilistic Epistemic Updates on Algebras

Willem Conradie[1], Sabine Frittella[2],
Alessandra Palmigiano[1,3], and Apostolos Tzimoulis[3]

[1] Department of Pure and Applied Mathematics, University of Johannesburg
wconradie@uj.ac.za
[2] LIF, Aix-Marseille Université, CNRS UMR 7279, France
sabine.frittella@lif.univ-mrs.fr
[3] Faculty of Technology, Policy and Management, Delft University of Technology
{a.palmigiano,a.tzimoulis-1}@tudelft.nl

Abstract. The present paper contributes to the development of the mathematical theory of epistemic updates using the tools of duality theory. Here we focus on Probabilistic Dynamic Epistemic Logic (PDEL). We dually characterize the product update construction of PDEL-models as a certain construction transforming the complex algebras associated with the given model into the complex algebra associated with the updated model. Thanks to this construction, an interpretation of the language of PDEL can be defined on algebraic models based on Heyting algebras. This justifies our proposal for the axiomatization of the intuitionistic counterpart of PDEL.

Keywords: intuitionistic probabilistic dynamic epistemic logic, duality, intuitionistic modal logic, algebraic models, pointfree semantics.
Math. Subject Class.: 03B42, 06D20, 06D50, 06E15.

1 Introduction

The contributions of the present paper pertain to the research program, started in [MPS14,KP13] and continued in [GKP13,FGK$^+$14b,Riv14,BR15,FGK$^+$14a] [FGK$^+$14c,FGKP14], which is aimed at developing the mathematical theory of epistemic updates with the tools of duality theory.

The present paper lays the semantic ground for the introduction of a logical framework generalizing probabilistic dynamic epistemic logic (PDEL) [Koo03], [vBGK09]. The generalization concerns the following respects:
(a) weakening the underlying reasoning machinery from classical propositional logic to nonclassical formalisms (e.g. intuitionistic logic);
(b) generalizing the formal treatment of agents' epistemics by relaxing the requirement of normality for the epistemic modal operators;
(c) considering intuitionistic probability theory as the background framework for probabilistic reasoning.

A major motivation for (c) is the need to account for situations in which the probability of a certain proposition p is interpreted as an agent's propensity to bet on p given some evidence for or against p. If there is little or no evidence for

© Springer-Verlag Berlin Heidelberg 2015
W. van der Hoek et al. (Eds.): LORI 2015, LNCS 9394, pp. 64–76, 2015.
DOI: 10.1007/978-3-662-48561-3_6

or against p, it should be reasonable to attribute low probability values to both p and $\neg p$, which is forbidden by classical probability theory (cf. [Wea03]).

A major motivations for (a) is the need to account for situations in which truth emerges as the outcome of a complex procedure (rather than e.g. being ascertained instantaneously). Examples of these situations are ubiquitous in social science. For instance, consider the case of the assessment of the authenticity of works of art. Turner's painting *The Beacon Light* is a case in point: after doubts had been cast on its being a genuine Turner, recent investigations into the materials and painting techniques have established its authenticity[1]. A fully fledged formalisation of such cases will be reported on in an extended version of the present paper [CFP+]. By its main features, intuitionistic logic is particularly suited to account for situations like the one mentioned above, where truth is ascertained by means of a procedure (a 'proof'). Moreover, the intuitionistic environment allows for a finer-grained analysis when serving as a base for more expressive formalisms such as modal and dynamic logics. Indeed, the fact that the box-type and the diamond-type modalities are no longer interdefinable makes several mutually independent choices possible which cannot be disentangled in the classical setting. It should be remarked at this point that of course it is possible in principle to use formalisms based on classical propositional logic to analyse situations in which truth emerges as a social construct (e.g. the outcome of a procedure), and that an 'automatic' and powerful way of generating such a formalism is via Gödel-type encodings. However, the resulting treatment is significantly more cumbersome and ad hoc, and from a technical point of view such an encoding might destroy nice properties enjoyed by the original intuitionistic framework (see e.g. discussion at the end of [CGP14, Section 36.9]). Insisting on a Boolean propositional base could have been motivated by the need to rely on a well developed and solid mathematical environment. However, recent developments (cf. e.g. [CPS,CGP14,CP15,CFPS15,CC15,PSZ15a,PSZ15b,GMP+15]) have made available a mathematical environment for non-classical logics[2] that is as advanced and solid as the classical one, and on which it is now possible to capitalise. Finally, these mathematical developments appear in tandem with interesting analyses on the philosophical side of formal logic (e.g. [AP14]), exploring epistemic logic in an evidentialist key, which is congenial with the kind of social situations targeted by our research programme.

Our methodology follows [MPS14,KP13], and is based on the dual characterization of the product update construction for standard PDEL-models as a certain construction transforming the complex algebras associated with a given model into the complex algebra associated with the updated model. This dual characterization naturally generalizes to much wider classes of algebras, which

---

[1] cf. e.g. Darren Devine, *End to doubts over museum's Turner paintings as all found to be genuine.* Wales Online, 23 September 2012. Retrieved from http://www.walesonline.co.uk/news/wales-news/end-doubts-over-museums-turner-2024586 .

[2] By non-classical logics we mean logics the propositional base of which is weaker than classical propositional logic.

include, but are not limited to, arbitrary BAOs and arbitrary modal expansions of Heyting algebras (HAOs). Thanks to this construction, the benefits and the wider scope of applications given by a point-free, nonclassical theory of epistemic updates are made available: for instance, this construction makes it possible to derive the definition of product updates on topological spaces by means of an effective computation. As an application of this dual characterization, we present the axiomatization for the intuitionistic analogue of PDEL which arises semantically from this construction.

Structure of the Paper: In Section 2, we give an alternative, two-step treatment of the PDEL-update on relational models. In Section 3, we expand on the methodology underlying the application of the duality toolkit. Section 4 is the main section, in which the construction of the PDEL-updates on Heyting algebras is introduced. In Section 5, we very briefly describe how the updates on algebras can be used to define the intuitionistic version of PDEL.

2 PDEL Language and Updates

In the present section, we report on the language of PDEL, and give an alternative, two-step account of the product update construction on PDEL-models. This account is similar to the treatment of epistemic updates in [MPS14,KP13], and as explained in Section 3, it lays the ground to the dualization procedure which motivates the construction introduced in Section 4. The specific PDEL framework we report on shares common features with those of [BCHS13] and [vBGK09].

2.1 PDEL-Formulas, Event Structures, and PES-Models

In the remainder of the paper, we fix a countable set AtProp of proposition letters p, q and a set Ag of agents i. We let $\alpha_1, ..., \alpha_n, \beta$ denote rational numbers.

Definition 1. *The set $\mathcal{L}$ of PDEL-formulas φ and the class $\mathsf{PEM}_{\mathcal{L}}$ of probabilistic event structures $\mathcal{E}$ over $\mathcal{L}$ are built by simultaneous recursion as follows:*

$$\varphi ::= p \mid \bot \mid \varphi \wedge \varphi \mid \varphi \vee \varphi \mid \varphi \rightarrow \varphi \mid \Diamond_i\varphi \mid \Box_i\varphi \mid \langle \mathcal{E}, e\rangle\varphi \mid [\mathcal{E}, e]\varphi \mid (\sum_{k=1}^{n} \alpha_k\mu_i(\varphi)) \geq \beta.$$

The connectives $\top$, $\neg$, and $\leftrightarrow$ are defined by the usual abbreviations. A probabilistic event structure over $\mathcal{L}$ is a tuple $\mathcal{E} = (E, (\sim_i)_{i\in\mathsf{Ag}}, (P_i)_{i\in\mathsf{Ag}}, \Phi, \mathsf{pre})$, such that E is a non-empty finite set, each $\sim_i$ is an equivalence relation on E, each $P_i : E \rightarrow [0,1]$ assigns a probability distribution on each $\sim_i$-equivalence class (i.e., $\sum\{P_i(e') : e' \sim_i e\} = 1$), Φ is a finite set of pairwise inconsistent $\mathcal{L}$-formulas, and pre assigns a probability distribution $\mathsf{pre}(\bullet|\phi)$ over E for every $\phi \in \Phi$.

Informally, elements of E encode possible events, the relations $\sim_i$ encode as usual the epistemic uncertainty of the agent i, who assigns probability $P_i(e)$ to

e being the actually occurring event, formulas in Φ are intended as the preconditions of the event, and $\mathsf{pre}(e|\phi)$ expresses the prior probability that the event $e \in E$ might occur in a(ny) state satisfying precondition ϕ. In what follows, we will refer to the structures $\mathcal{E}$ defined above as *event structures over* $\mathcal{L}$.

Definition 2. *A* probabilistic epistemic state model (PES-model) *is a structure* $\mathbb{M} = (S, (\sim_i)_{i \in \mathsf{Ag}}, (P_i)_{i \in \mathsf{Ag}}, [\![\cdot]\!])$ *such that S is a non-empty set, each $\sim_i$ is an equivalence relation on S, each $P_i : S \to [0,1]$ assigns a probability distribution on each $\sim_i$-equivalence class, (i.e., $\sum\{P_i(s') : s' \sim_i s\} = 1$), and $[\![\cdot]\!] : \mathsf{AtProp} \to \mathcal{P}S$ is a valuation map.*

As usual, the map $[\![\cdot]\!]$ will be identified with its unique extension to $\mathcal{L}$, so that we will be able to write $[\![\varphi]\!]$ for every $\varphi \in \mathcal{L}$.

Notation 1. *For any probabilistic epistemic model $\mathbb{M}$, any probabilistic event structure $\mathcal{E}$, any $s \in S$ and $e \in E$ we let $\mathsf{pre}(e \mid s)$ denote the value $\mathsf{pre}(e \mid \phi)$, for the unique $\phi \in \Phi$ such that $\mathbb{M}, s \Vdash \phi$ (recall that the formulas in Φ are pairwise inconsistent). If no such ϕ exists then we let $\mathsf{pre}(e \mid s) = 0$.*

2.2 Epistemic Updates

Throughout the present subsection, we fix a PES-model $\mathbb{M}$ and a probabilistic event structure $\mathcal{E}$ over $\mathcal{L}$. The updated model is given in two steps, the first of which is detailed in the following

Definition 3. *Let the* intermediate structure *of $\mathbb{M}$ and $\mathcal{E}$ be the tuple*

$$\coprod\nolimits_{\mathcal{E}} \mathbb{M} := (\coprod\nolimits_{|E|} S, (\sim_i^{\coprod})_{i \in \mathsf{Ag}}, (P_i^{\coprod})_{i \in \mathsf{Ag}}, [\![\cdot]\!]_{\coprod})$$

where $\coprod_{|E|} S \cong S \times E$ is the $|E|$-fold coproduct of S, each binary relation $\sim_i^{\coprod}$ on $\coprod_{|E|} S$ is defined as follows:

$$(s, e) \sim_i^{\coprod} (s', e') \quad \textit{iff } s \sim_i s' \textit{ and } e \sim_i e';$$

each map $P_i^{\coprod} : \coprod_{|E|} S \to [0,1]$ is defined by $(s, e) \mapsto P_i(s) \cdot P_i(e) \cdot \mathsf{pre}(e \mid s)$ and $[\![p]\!]_{\coprod} := \{(s, e) \mid s \in [\![p]\!]_{\mathbb{M}}\} = [\![p]\!]_{\mathbb{M}} \times E$ for every $p \in \mathsf{AtProp}$.

Remark 1. In general $P_i^{\coprod}$ does not induce probability distributions over the $\sim_i^{\coprod}$-equivalence classes. Hence, $\coprod_{\mathcal{E}} \mathbb{M}$ is not a PES-model.[3] However, the second step of the construction will yield a PES-model.

Finally, in order to define the updated model, observe that the map pre in $\mathcal{E}$ induces the map $\mathit{pre} : E \to \mathcal{L}$ defined by $\quad e \mapsto \bigvee\{\phi \in \Phi \mid \mathsf{pre}(e \mid \phi) \neq 0\}$.

---

[3] Indeed, Definition 9 will be introduced in Section 4 precisely with the purpose of capturing the dual of $P_i^{\coprod}$.

Definition 4. *For any PES-model* $\mathbb{M}$ *and any probabilistic event structure* $\mathcal{E}$ *over* $\mathcal{L}$, *let*

$$\mathbb{M}^{\mathcal{E}} := (S^{\mathcal{E}}, (\sim_i^{\mathcal{E}})_{i \in \mathsf{Ag}}, (P_i^{\mathcal{E}})_{i \in \mathsf{Ag}}, [\![\cdot]\!]_{\mathbb{M}^{\mathcal{E}}})$$

with

1. $S^{\mathcal{E}} := \{(s,e) \in \coprod_{|E|} S \mid \mathbb{M}, s \Vdash pre(e)\}$;
2. $[\![p]\!]_{\mathbb{M}^{\mathcal{E}}} := [\![p]\!]_{\coprod} \cap S^{\mathcal{E}}$;
3. $\sim_i^{\mathcal{E}} = \sim_i^{\coprod} \cap (S^{\mathcal{E}} \times S^{\mathcal{E}})$ *for any* $i \in \mathsf{Ag}$;
4. *each map* $P_i^{\mathcal{E}} : S^{\mathcal{E}} \to [0,1]$ *is defined by the assignment*

$$(s,e) \mapsto \frac{P_i^{\coprod}(s,e)}{\sum \{P_i^{\coprod}(s',e') \mid (s,e) \sim_i (s',e')\}}.$$

3 Methodology

In the present section, we expand on the methodology of the paper. In the previous section, we gave a two-step account of the *product update* construction which, for any PES-model $\mathbb{M}$ and any event model $\mathcal{E}$ over $\mathcal{L}$, yields the updated model $\mathbb{M}^{\mathcal{E}}$ as a certain *submodel* of a certain *intermediate model* $\coprod_{\mathcal{E}} \mathbb{M}$. This account is analogous to those given in [MPS14,KP13] of the product updates of models of PAL and Baltag-Moss-Solecki's dynamic epistemic logic EAK. In each instance, the original product update construction can be illustrated by the following diagram (which uses the notation introduced in the instance treated in the previous section):

$$\mathbb{M} \hookrightarrow \coprod_{\mathcal{E}} \mathbb{M} \hookleftarrow \mathbb{M}^{\mathcal{E}}.$$

As is well known (cf. e.g. [DP02]) in duality theory, coproducts can be dually characterized as products, and subobjects as quotients. In the light of this fact, the construction of *product update*, regarded as a "subobject after coproduct" concatenation, can be dually characterized on the algebras dual to the relational structures of PES-models by means of a "quotient after product" concatenation, as illustrated in the following diagram:

$$\mathbb{A} \twoheadleftarrow \prod_{\mathcal{E}} \mathbb{A} \rightarrow \mathbb{A}^{\mathcal{E}},$$

resulting in the following two-step process. First, the coproduct $\coprod_{\mathcal{E}} M$ is dually characterized as a certain *product* $\prod_{\mathcal{E}} \mathbb{A}$, indexed as well by the states of $\mathcal{E}$, and such that $\mathbb{A}$ is the algebraic dual of $\mathbb{M}$; second, an appropriate *quotient* of $\prod_{\mathcal{E}} \mathbb{A}$ is then taken, which dually characterizes the submodel step. On which algebras are we going to apply the "quotient after product" construction? The prime candidates are the algebras associated with the PES-models via standard Stone-type duality:

Definition 5. *For any PES-model* $\mathbb{M}$, *its complex algebra is the tuple*

$$\mathbb{M}^+ := \langle \mathcal{P}S, (\Diamond_i)_{i \in \mathsf{Ag}}, (\Box_i)_{i \in \mathsf{Ag}}, (P_i^+)_{i \in \mathsf{Ag}} \rangle$$

where for each $i \in \mathsf{Ag}$ *and* $X \in \mathcal{P}S$,

$$\Diamond_i X = \{s \in S \mid \exists x(s \sim_i x \text{ and } x \in X)\},$$
$$\Box_i X = \{s \in S \mid \forall x(s \sim_i x \implies x \in X)\},$$
$$\mathsf{dom}(P_i^+) = \{X \in \mathcal{P}S \mid \exists y \forall x(x \in X \implies x \sim_i y)\}^4$$
$$P_i^+ X = \sum_{x \in X} P_i(x)$$

In this setting, the "quotient after product" construction behaves exactly in the desired way, in the sense that one can check *a posteriori* that the following holds:[5]

Proposition 1. *For every PES-model* $\mathbb{M}$ *and any event structure* $\mathcal{E}$ *over* $\mathcal{L}$, *the algebraic structures* $(\mathbb{M}^+)^{\mathcal{E}}$ *and* $(\mathbb{M}^{\mathcal{E}})^+$ *can be identified.*

Moreover, the "quotient after product" construction holds in much greater generality than the class of complex algebras of PES-models, which is exactly its added value over the update on relational structures. In the following section, we are going to define it in detail in the setting of epistemic Heyting algebras.

4 Probabilistic Dynamic Epistemic Updates on Heyting Algebras

The present section aims at introducing the algebraic counterpart of the event update construction presented in Section 2.

For the sake of enforcing a neat separation between syntax and semantics, throughout the present section we will disregard the logical language $\mathcal{L}$, and work on *algebraic probabilistic epistemic structures* (APE-structures, cf. Definition 10) rather than on APE-models (i.e. APE-structures endowed with valuations). To be able to define the update construction, we will need to base our treatment on the following, modified definition of event structure over an algebra, rather than over $\mathcal{L}$:

Definition 6. *For any epistemic Heyting algebra* $\mathbb{A}$ *(cf. Definition 7), a* probabilistic event structure over $\mathbb{A}$ *is a tuple* $\mathbb{E} = (E, (\sim_i)_{i \in \mathsf{Ag}}, (P_i)_{i \in \mathsf{Ag}}, \Phi, \mathsf{pre})$ *such that* E, $\sim_i$, P_i *are as in Definition 1;* Φ *is a finite subset of* $\mathbb{A}$ *such that* $a_j \wedge a_k = \bot$ *for all* $a_i, a_j \in \Phi$ *such that* $a_i \neq a_j$; pre *assigns a probability distribution* $\mathsf{pre}(\bullet|a)$ *over* E *for every* $a \in \Phi$.

In what follows, we will typically refer to the structures defined above as *event structures*. In the next subsection, we introduce APE-structures based on epistemic Heyting algebras. In Subsection 4.2 we introduce the first step of the two-step update, namely, the 'product' construction. In Subsection 4.3, we introduce the second and final step, the 'quotient' construction.

---

[4] i.e. the domain of P_i^+ consists of all the subsets of the equivalence classes of $\sim_i$.

[5] Caveat: we are abusing notation here. Proposition 1 should be formulated using Definition 13 and Fact 2.

4.1   Algebraic Probabilistic Epistemic Structures

Definition 7. *An epistemic Heyting algebra is a tuple* $\mathbb{A} := \langle \mathbb{L}, (\Diamond_i)_{i\in\mathsf{Ag}}, (\Box_i)_{i\in\mathsf{Ag}} \rangle$ *such that* $\mathbb{L}$ *is a Heyting algebra, and each* $\Diamond_i$ *and* $\Box_i$ *is a monotone unary operation on* $\mathbb{L}$ *such that for all* $a, b \in \mathbb{L}$,

$$\Diamond_i(a \to b) \leq \Box_i a \to \Diamond_i b \qquad \Diamond_i a \to \Box_i b \leq \Box_i(a \to b)$$
$$\Diamond_i a \wedge b \leq \Diamond_i(a \wedge \Diamond_i b) \qquad \Box_i(a \vee \Box_i b) \leq \Box_i a \vee b$$
$$a \leq \Diamond_i a \qquad\qquad\qquad \Box_i a \leq a$$
$$\Diamond_i \Diamond_i a \leq \Diamond_i a \qquad\qquad \Box_i a \leq \Box_i \Box_i a.$$

In what follows, $\mathbb{A}$ will denote an epistemic Heyting algebra.

Definition 8. *An element* $a \in \mathbb{A}$ *is* i-*minimal if*

1. $a \neq \bot$,
2. $\Diamond_i a = a$ *and*
3. *if* $b \in \mathbb{A}$, $b < a$, *and* $\Diamond_i b = b$, *then* $b = \bot$.

Let $\mathsf{Min}_i(\mathbb{A})$ *denote the set of the* i-*minimal elements of* $\mathbb{A}$.

Notice that for any $b \in \mathbb{A} \setminus \{\bot\}$ there exists at most one $a \in \mathsf{Min}_i(\mathbb{A})$ such that $b \leq a$. Indeed every such a must coincide with $\Diamond_i b$. The next definition uses insights from [Wea03].

Definition 9. *A partial function* $\mu : \mathbb{A} \to \mathbb{R}^+$ *is an* i-*premeasure on* $\mathbb{A}$ *if* $\mathrm{dom}(\mu) = \mathsf{Min}_i(\mathbb{A})\!\downarrow$, *and* μ *is order-preserving,* $\mu(\bot) = 0$ *if* $\mathrm{dom}(\mu) \neq \varnothing$ *and for every* $a \in \mathsf{Min}_i(\mathbb{A})$ *and all* $b, c \in a\!\downarrow$ *it holds that* $\mu(b \vee c) = \mu(b) + \mu(c) - \mu(b \wedge c)$. *An* i-*premeasure on* $\mathbb{A}$ *is an* i-*measure if* $\mu(a) = 1$ *for every* $a \in \mathsf{Min}_i(\mathbb{A})$.

Definition 10. *An algebraic pre-probabilistic epistemic structure (ApPE-structure) is a tuple* $\mathcal{F} := \langle \mathbb{A}, (\mu_i)_{i\in\mathsf{Ag}} \rangle$ *such that* $\mathbb{A}$ *is an epistemic Heyting algebra (cf. Definition 8), and each* μ_i *is an* i-*premeasure on* $\mathbb{A}$. *An ApPE-structure* $\mathcal{F}$ *is an* algebraic probabilistic epistemic structure (APE-structure) *if each* μ_i *is a* i-*measure on* $\mathbb{A}$. *We refer to* $\mathbb{A}$ *as the* support *of* $\mathcal{F}$.

Lemma 1. *For any PES-model* $\mathbb{M}$, *the* i-*minimal elements of its complex algebra* $\mathbb{M}^+$ *are exactly the equivalence classes of* $\sim_i$.

Proposition 2. *For any PES-model* $\mathbb{M}$, *the complex algebra* $\mathbb{M}^+$ *(cf. Definition 5) is an APE-structure.*

4.2   The Intermediate (Pre-)Probabilistic Epistemic Structure

In the present subsection, we define the intermediate ApPE-structure $\prod_\mathbb{E} \mathcal{F}$ associated with any APE-structure $\mathcal{F}$ and any event structure $\mathbb{E}$ over the support of $\mathcal{F}$ (cf. Definition 10 for the definition of support):

$$\prod_\mathbb{E} \mathcal{F} := \langle \prod_{|E|} \mathbb{A}, (\Diamond_i')_{i\in\mathsf{Ag}}, (\Box_i')_{i\in\mathsf{Ag}}, (\mu_i')_{i\in\mathsf{Ag}} \rangle \tag{4.1}$$

Let us start by defining the algebra which will become the support of the intermediate APE-structure above:

Definition 11. *For every epistemic Heyting algebra* $\mathbb{A} = (\mathbb{L}, (\Diamond_i)_{i \in \mathsf{Ag}}, (\Box_i)_{i \in \mathsf{Ag}})$
and every event structure $\mathbb{E}$ *over* $\mathbb{A}$*, let*

$$\prod_{\mathbb{E}} \mathbb{A} := (\prod_{|E|} \mathbb{L}, (\Diamond'_i)_{i \in \mathsf{Ag}}, (\Box'_i)_{i \in \mathsf{Ag}}),$$

where

1. $\prod_{|E|} \mathbb{L}$ *is the* $|E|$*-fold power of* $\mathbb{L}$*, the elements of which can be seen either as* $|E|$*-tuples of elements in* $\mathbb{A}$*, or as maps* $f : E \to \mathbb{A}$*.*
2. *For any* $f : E \to \mathbb{A}$*, the map* $\Diamond'_i(f) : E \to \mathbb{A}$ *is defined by the assignment* $e \mapsto \bigvee\{\Diamond_i f(e') \mid e' \sim_i e\};$
3. *For any* $f : E \to \mathbb{A}$*, the map* $\Box'_i(f) : E \to \mathbb{A}$ *is defined by the assignment* $e \mapsto \bigwedge\{\Box_i f(e') \mid e' \sim_i e\}.$

Below, the algebra $\prod_{\mathbb{E}} \mathbb{A}$ *will be sometimes abbreviated as* $\mathbb{A}'$*.*

We refer to [KP13, Section 3.1] for an extensive justification of the definition of the operations $\Diamond'_i$ and $\Box'_i$.

Proposition 3. *For every epistemic Heyting algebra* $\mathbb{A}$ *and every event structure* $\mathbb{E}$ *over* $\mathbb{A}$*, the algebra* $\mathbb{A}'$ *is an epistemic Heyting algebra.*

Proposition 4. *For every* $\mathbb{A}$ *and* i*,* $\mathsf{Min}_i(\mathbb{A}') = \{f_{e,a} \mid e \in E \text{ and } a \in \mathsf{Min}_i(\mathbb{A})\}$*, where for any* $e \in E$ *and* $a \in \mathsf{Min}_i(\mathbb{A})$*, the map* $f_{e,a} : E \to \mathbb{A}$ *is defined by the following assignment:*

$$e' \mapsto \begin{cases} a & \text{if } e' \sim_i e \\ \bot & \text{otherwise.} \end{cases}$$

Definition 12. *For any APE-structure* $\mathcal{F}$ *and any event structure* $\mathbb{E}$ *over the support of* $\mathcal{F}$*, let*

$$\prod_{\mathbb{E}} \mathcal{F} := \langle \prod_{\mathbb{E}} \mathbb{A}, (\mu'_i)_{i \in \mathsf{Ag}} \rangle$$

where

1. $\prod_{\mathbb{E}} \mathbb{A} = \mathbb{A}'$ *is defined as in Definition 11;*
2. *each* $\mu'_i : \mathbb{A}' \to [0,1]$ *is defined as follows:*

$$\mathsf{dom}(\mu'_i) = \mathsf{Min}_i(\mathbb{A}') \!\downarrow$$
$$\mu'_i(f) = \sum_{e \in E} \sum_{a \in \Phi} P_i(e) \cdot \mu_i(f(e) \wedge a) \cdot \mathsf{pre}(e \mid a).$$

Proposition 5. *For every APE-structure* $\mathcal{F}$ *and every event structure* $\mathbb{E}$ *over the support of* $\mathcal{F}$*, the intermediate structure* $\prod_{\mathbb{E}} \mathcal{F}$ *is an ApPE-structure (cf. Definition 10).*

Proof. The proof that $\prod_{\mathbb{E}} \mathbb{A}$ is an epistemic Heyting algebra is entirely analogous to the proof of [KP13, Proposition 8], and is omitted. Let us assume that the domain of μ'_i is non-empty. By definition, μ'_i is order-preserving and $\mu'_i(\bot) = 0$. Finally, by Proposition 4, i-minimal elements of $\mathbb{A}'$ are of the form $f_{e,a} : E \to \mathbb{A}$ for some $e \in E$ and some i-minimal element $a \in \mathbb{A}$. Fix one such element, and let $g, h : E \to \mathbb{A}$ such that $g \vee h \leq f_{e,a}$. By definition, $f \leq f_{e,a}$ can be rewritten as $f(e') \leq f_{e,a}(e')$ for any $e' \in E$. Since $f_{e,a}(e') = \bot$ for any $e' \nsim_i e$, we can deduce that $g(e') = h(e') = \bot$ for any $e' \nsim_i e$. Hence,

$$\mu_i'(g \vee h)$$

$$= \sum_{e' \in E} \sum_{a \in \Phi} P_i(e') \cdot \mu_i((g(e') \vee h(e')) \wedge a) \cdot \mathsf{pre}(e' \mid a) \qquad \text{(by definition)}$$

$$= \sum_{e' \sim_i e} \sum_{a \in \Phi} P_i(e') \cdot \mu_i((g(e') \vee h(e')) \wedge a) \cdot \mathsf{pre}(e' \mid a)$$

$$\qquad\qquad (g(e') = h(e') = \bot \text{ for any } e' \nsim_i e \text{ and } \mu_i(\bot) = 0)$$

$$= \sum_{e' \sim_i e} \sum_{a \in \Phi} P_i(e') \cdot \mu_i((g(e') \wedge a) \vee (h(e') \wedge a)) \cdot \mathsf{pre}(e' \mid a) \qquad \text{(distributivity)}$$

$$= \sum_{e' \sim_i e} \sum_{a \in \Phi} P_i(e') \cdot (\mu_i(g(e') \wedge a) + \mu_i(h(e') \wedge a) - \mu_i(g(e') \wedge h(e') \wedge a)) \cdot \mathsf{pre}(e' \mid a)$$

$$= \sum_{e \in E} \sum_{a \in \Phi} P_i(e) \cdot (\mu_i(g(e) \wedge a) + \mu_i(h(e) \wedge a) - \mu_i(g(e) \wedge h(e) \wedge a)) \cdot \mathsf{pre}(e \mid a)$$

$$\qquad\qquad (\mu_i(\bot) = 0 \text{ by Definition 16 and } g(e') = h(e') = \bot \text{ for any } e' \nsim_i e)$$

$$= \mu_i'(g) + \mu_i'(h) - \mu_i'(g \wedge h) \qquad \text{(by definition)}$$

Definition 13. *For any PES-model* $\mathbb{M}$ *and any event structure* $\mathcal{E} = (E, (\sim_i)_{i \in \mathsf{Ag}}, (P_i)_{i \in \mathsf{Ag}}, \Phi, \mathsf{pre})$ *over* $\mathcal{L}$, *let* $\mathbb{E}_\mathcal{E} := (E, (\sim_i)_{i \in \mathsf{Ag}}, (P_i)_{i \in \mathsf{Ag}}, \Phi_\mathbb{M}, \mathsf{pre}_\mathbb{M})$, *where* $\Phi_\mathbb{M} := \{[\![\phi]\!]_\mathbb{M} \mid \phi \in \Phi\}$, *and* $\mathsf{pre}_\mathbb{M}$ *assigns a probability distribution* $\mathsf{pre}(\bullet|a)$ *over* E *for every* $a \in \Phi_\mathbb{M}$.

Fact 2. *For any PES-model* $\mathbb{M}$ *and any event structure* $\mathcal{E}$ *over* $\mathcal{L}$, *the tuple* $\mathbb{E}_\mathcal{E}$ *is an event structure over the epistemic Heyting algebra underlying* $\mathbb{M}^+$.

Proposition 6. *For every PES-model* $\mathbb{M}$ *and any event structure* $\mathcal{E}$ *over* $\mathcal{L}$,

$$\left(\coprod_\mathcal{E} \mathbb{M}\right)^+ \cong \prod_{\mathbb{E}_\mathcal{E}} \mathbb{M}^+.$$

4.3   The Pseudo-Quotient and the Updated APE-Structure

In the present subsection, we define the APE-structure $\mathcal{F}^\mathbb{E}$, resulting from the update of the APE-structure $\mathcal{F}$ with and the event structure $\mathbb{E}$ over the support of $\mathcal{F}$, by taking a suitable pseudo-quotient of the intermediate APE-structure $\prod_\mathbb{E} \mathcal{F}$. Some of the results which are relevant for the ensuing treatment (such as the characterization of the i-minimal elements in the pseudo-quotient) are independent of the fact that we will be working with the intermediate algebra. Therefore, in what follows, we will discuss them in the more general setting of arbitrary epistemic Heyting algebras $\mathbb{A}$:

Definition 14. (cf. [MPS14, Sections 3.2, 3.3]) For any $\mathbb{A}$ and any $a \in A$, let $\mathbb{A}^a := (\mathbb{L}/\cong_a, (\Diamond_i^a)_{i \in \mathsf{Ag}}, (\Box_i^a)_{i \in \mathsf{Ag}})$, where $\cong_a$ is defined as follows: $b \cong_a c$ iff $b \wedge a = c \wedge a$ for all $b, c \in \mathbb{L}$, each operation $\Diamond_i^a$ is defined by the assignment $\Diamond_i^a[b] := [\Diamond_i(b \wedge a)]$ and each operation $\Box_i^a$ is defined by the assignment $\Box_i^a[b] := [\Box_i(a \to b)]$, where $[c]$ denotes the $\cong_a$-equivalence class of any given $c \in \mathbb{L}$.

Proposition 7. (cf. [MPS14, Fact 12]) The algebra $\mathbb{A}^a$ of Definition 14 is an epistemic Heyting algebra.

Proposition 8. The following are equivalent for any $\mathbb{A}$ and any $a \in A$:

1. $[b] \in \mathsf{Min}_i(\mathbb{A}^a)$;
2. $[b] = [b']$ for a unique $b' \in \mathsf{Min}_i(\mathbb{A})$ such that $b' \wedge a \neq \bot$.

Hence, in what follows, whenever $[b] \in \mathsf{Min}_i(\mathbb{A}^a)$, we will assume w.l.o.g. that $b \in \mathsf{Min}_i(\mathbb{A})$ is the "canonical" (in the sense of Proposition 8) representant of $[b]$.

For any APE-structure $\mathcal{F}$ and any event structure $\mathbb{E}$ over the support $\mathbb{A}$ of $\mathcal{F}$, the map pre in $\mathbb{E}$ induces the map $\overline{pre} : E \to A$ defined by $e \mapsto \bigvee_{\substack{a \in \Phi \\ \mathsf{pre}(e|a) \neq 0}} a$.

It immediately follows from Propositions 4 and 8 that the i-minimal elements of $\mathbb{A}^{\mathbb{E}}$ are exactly the elements $[f_{e,a}]$ for $e \in E$ and $a \in \mathsf{Min}_i(\mathbb{A})$ such that $a \wedge \overline{pre}(e') \neq \bot$ for some $e' \sim_i e$.

Definition 15. For any APE-structure $\mathcal{F}$ and any event structure $\mathbb{E}$ over the support of $\mathcal{F}$, the updated APE-structure is the tuple $\mathcal{F}^{\mathbb{E}} := (\mathbb{A}^{\mathbb{E}}, (\mu_i^{\mathbb{E}})_{i \in \mathsf{Ag}})$, s.t.:

1. $\mathbb{A}^{\mathbb{E}} := (\prod_{\mathbb{E}} \mathbb{A})^{\overline{pre}}$, i.e. $\mathbb{A}^{\mathbb{E}}$ is obtained by instantiating Definition 14 to $\prod_{\mathbb{E}} \mathbb{A}$ and $\overline{pre} \in \prod_{\mathbb{E}} \mathbb{A}$;
2. $\mathsf{dom}(\mu_i^{\mathbb{E}}) = \mathsf{Min}_i(\mathbb{A}^{\mathbb{E}})\!\downarrow$ for each partial map $\mu_i^{\mathbb{E}} : \mathbb{A}^{\mathbb{E}} \to [0, 1]$ and $\mu_i^{\mathbb{E}}([g]) := \frac{\mu_i'(g)}{\mu_i'(f)}$ for every $[g] \in \mathsf{dom}(\mu_i^{\mathbb{E}})$ where $[g] \leq [f]$ for some $[f] \in \mathsf{Min}_i(\mathbb{A}^{\mathbb{E}})$.

Notice that if $[g] \neq \bot$ then $[f]$ is unique (cf. discussion after Definition 8). If $[g] = \bot$ then $\mu'(g) = 0$. Hence the above is well-defined.

Proposition 9. For any APE-structure $\mathcal{F}$ and any event structure $\mathbb{E}$ over the support of $\mathcal{F}$, the tuple $\mathcal{F}^{\mathbb{E}}$ is an APE-structure.

Proof. By Proposition 7, $\mathbb{A}^{\mathbb{E}}$ is an epistemic Heyting algebra. Let us assume that the domain of $\mu_i^{\mathbb{E}}$ is non-empty. To finish the proof, it remains to be shown that each partial map $\mu_i^{\mathbb{E}}$ satisfies the conditions of Definition 9. Clearly, $\mu_i^{\mathbb{E}}(\bot) = 0$ and $\mu_i^{\mathbb{E}}([f]) = 1$ for all $[f] \in \mathsf{Min}_i(\mathbb{A}^{\mathbb{E}})$.

To argue that $\mu_i^{\mathbb{E}}$ is monotone, observe preliminarily that $\mu_i'(g) = \mu_i'(g \wedge \overline{pre})$. This follows by the definition of μ_i' and the fact that if $\mathsf{pre}(e \mid a) \neq 0$ then $a \leq \overline{pre}(e)$. Assume that $[g_1] \leq [g_2] \leq [f_{e,a}]$. This means that $g_1 \wedge \overline{pre} \leq g_2 \wedge \overline{pre}$. Since μ_i' is monotone, $\mu_i'(g_1) = \mu_i'(g_1 \wedge \overline{pre}) \leq \mu_i'(g_2 \wedge \overline{pre}) = \mu_i(g_2)$. This implies that

$$\frac{\mu_i'(g_1)}{\mu_i'(f_{e,a})} \leq \frac{\mu_i'(g_2)}{\mu_i'(f_{e,a})}$$

that is, $\mu_i^{\mathbb{E}}([g_1]) \leq \mu_i^{\mathbb{E}}([g_2])$.

As for the last condition, let $[g_1]$ and $[g_2]$ in $\mathcal{F}^{\mathbb{E}}$ such that $[g_1] \leq [f_{e,a}]$ and $[g_2] \leq [f_{e,a}]$. We have:

$$
\begin{aligned}
\mu_i^{\mathbb{E}}([g_1] \vee [g_2]) &= \frac{\mu_i'((g_1 \wedge \overline{pre}) \vee (g_2 \wedge \overline{pre}))}{\mu_i'(f_{e,a})} \\
&= \frac{\mu_i'(g_1 \wedge \overline{pre}) + \mu_i'(g_2 \wedge \overline{pre}) - \mu_i'((g_1 \wedge g_2) \wedge \overline{pre})}{\mu_i'(f_{e,a})} \\
&= \frac{\mu_i'(g_1 \wedge \overline{pre})}{\mu_i'(f_{e,a})} + \frac{\mu_i'(g_2 \wedge \overline{pre})}{\mu_i'(f_{e,a})} - \frac{\mu_i'((g_1 \wedge g_2) \wedge \overline{pre})}{\mu_i'(f_{e,a})} \\
&= \frac{\mu_i'(g_1)}{\mu_i'(f_{e,a})} + \frac{\mu_i'(g_2)}{\mu_i'(f_{e,a})} - \frac{\mu_i'(g_1 \wedge g_2)}{\mu_i'(f_{e,a})} \\
&= \mu_i^{\mathbb{E}}([g_1]) + \mu_i^{\mathbb{E}}([g_2]) - \mu_i^{\mathbb{E}}([g_1 \wedge g_2]).
\end{aligned}
$$

Lemma 2. *For any PES-model $\mathbb{M}$ and any event structure $\mathcal{E}$ over $\mathcal{L}$,*

$$(P_i^+)^{\mathbb{E}_{\mathcal{E}}} = (P_i^{\mathcal{E}})^+.$$

Proposition 1 follows from the above lemma and [KP13, Proposition 3.6].

5   PDEL, Intuitionistically

In the present section, we apply the update construction on algebras introduced in the previous section to the definition of the intuitionistic counterpart of PDEL.

Definition 16. *Algebraic probabilistic epistemic models (APE-models) are tuples $\mathcal{M} = \langle \mathcal{F}, v \rangle$ s.t. $\mathcal{F} = \langle \mathbb{A}, (\mu_i)_{i \in \mathsf{Ag}} \rangle$ is an APE-structure, and $v : \mathsf{AtProp} \to \mathbb{A}$.*

The update construction of Section 4 extends from APE-structures to APE-models. Indeed, for any APE-model $\mathcal{M}$ and any event structure $\mathcal{E}$ over $\mathcal{L}$ (cf. Definition 1), the following tuple is an event structure over $\mathbb{A}$:

$$\mathbb{E}_{\mathcal{E}} := (E, (\sim_i)_{i \in \mathsf{Ag}}, (P_i)_{i \in \mathsf{Ag}}, \Phi_{\mathcal{M}}, \mathsf{pre}_{\mathcal{M}}),$$

where $\Phi_{\mathcal{M}} := \{[\![\phi]\!]_{\mathcal{M}} \mid \phi \in \Phi\}^6$, and $\mathsf{pre}_{\mathcal{M}}$ assigns a probability distribution $\mathsf{pre}(\bullet | a)$ over E for every $a \in \Phi_{\mathcal{M}}$. Then,

$$\mathcal{M}^{\mathcal{E}} := \langle \mathcal{F}^{\mathcal{E}}, v^{\mathcal{E}} \rangle,$$

where $\mathcal{F}^{\mathcal{E}} := \mathcal{F}^{\mathbb{E}_{\mathcal{E}}}$ as in Definition 15, and $v^{\mathcal{E}}(p) = [v^{\Pi}(p)]$ for every $p \in \mathsf{AtProp}$, where $v^{\Pi}(p) : E \to \mathbb{A}$ is defined by the assignment $e \mapsto v(p)$. For every $e \in E$, let $\pi_e : \prod_{\mathbb{E}_{\mathcal{E}}} \mathbb{A} \to \mathbb{A}$ be the eth projection; also, let $\pi : \prod_{\mathbb{E}_{\mathcal{E}}} \mathbb{A} \to \mathbb{A}^{\mathbb{E}_{\mathcal{E}}}$ be the quotient map. As explained in [MPS14, Section 3.2], the map $\iota : \mathbb{A}^{\mathbb{E}_{\mathcal{E}}} \to \prod_{\mathbb{E}_{\mathcal{E}}} \mathbb{A}$ defined by the assignment $[g] \mapsto g \wedge \overline{pre}$ is well defined.

---

[6] Caveat: the definition of $\mathbb{E}_{\mathcal{E}}$ should more appropriately be given by simultaneous induction together with the interpretation of formulas.

Definition 17. *The interpretation of $\mathcal{L}$-formulas on any APE-model $\mathcal{M}$ is defined recursively as follows:*

$$[\![p]\!]_{\mathcal{M}} = v(p) \qquad\qquad [\![\varphi \to \psi]\!]_{\mathcal{M}} = [\![\varphi]\!]_{\mathcal{M}} \to^{\mathbb{A}} [\![\psi]\!]_{\mathcal{M}}$$

$$[\![\bot]\!]_{\mathcal{M}} = \bot^{\mathbb{A}} \qquad\qquad [\![\top]\!]_{\mathcal{M}} = \top^{\mathbb{A}}$$

$$[\![\varphi \wedge \psi]\!]_{\mathcal{M}} = [\![\varphi]\!]_{\mathcal{M}} \wedge^{\mathbb{A}} [\![\psi]\!]_{\mathcal{M}} \qquad\qquad [\![\varphi \vee \psi]\!]_{\mathcal{M}} = [\![\varphi]\!]_{\mathcal{M}} \vee^{\mathbb{A}} [\![\psi]\!]_{\mathcal{M}}$$

$$[\![\Diamond_i \varphi]\!]_{\mathcal{M}} = \Diamond_i [\![\varphi]\!]_{\mathcal{M}} \qquad\qquad [\![\Box_i \varphi]\!]_{\mathcal{M}} = \Box_i [\![\varphi]\!]_{\mathcal{M}}$$

$$[\![\langle \mathcal{E}, e \rangle \varphi]\!]_{\mathcal{M}} = [\![\overline{pre}(e)]\!]_{\mathcal{M}} \wedge^{\mathbb{A}} \pi_e \circ \iota([\![\varphi]\!]_{\mathcal{M}^{\mathbb{E}_\mathcal{E}}}) \qquad [\![[\mathcal{E}, e] \varphi]\!]_{\mathcal{M}} = [\![\overline{pre}(e)]\!]_{\mathcal{M}} \to^{\mathbb{A}} \pi_e \circ \iota([\![\varphi]\!]_{\mathcal{M}^{\mathbb{E}_\mathcal{E}}})$$

$$[\![(\textstyle\sum_{k=1}^{n} \alpha_k \mu_i(\varphi_k)) \geq \beta]\!]_{\mathcal{M}} = \bigvee\{a \in \mathbb{A} \mid a \in \mathrm{Min}_i(\mathbb{A}) \text{ and } (\textstyle\sum_{k=1}^{n} \alpha_k \mu_i([\![\varphi_k]\!]_{\mathcal{M}} \wedge a)) \geq \beta\}$$

The following axioms are sound on APE-models under the interpretation above:

$$\langle \mathcal{E}, e \rangle (\sum_{k=1}^{n} \alpha_k \mu_i(\varphi_k) \geq \beta) \;\leftrightarrow\; Pre(e) \wedge \Big(\sum_{k=1}^{n} \sum_{\substack{e' \sim_i e \\ \phi \in \Phi}} \alpha_k \cdot P_i(e') \cdot pre(e' \mid \phi) \mu_i(\phi \wedge \langle \mathcal{E}, e' \rangle \varphi_k)$$

$$+ \sum_{\substack{e' \sim_i e \\ \phi \in \Phi}} -\beta \cdot P_i(e') \cdot pre(e' \mid \phi) \mu_i(\phi) \geq 0 \Big)$$

$$[\mathcal{E}, e] (\sum_{k=1}^{n} \alpha_k \mu_i(\varphi_k) \geq \beta) \;\leftrightarrow\; Pre(e) \to \Big(\sum_{k=1}^{n} \sum_{\substack{e' \sim_i e \\ \phi \in \Phi}} \alpha_k \cdot P_i(e') \cdot pre(e' \mid \phi) \mu_i(\phi \wedge [\mathcal{E}, e'] \varphi_k)$$

$$+ \sum_{\substack{e' \sim_i e \\ \phi \in \Phi}} -\beta \cdot P_i(e') \cdot pre(e' \mid \phi) \mu_i(\phi) \geq 0 \Big)$$

References

AP14.   Artemov, S., Protopopescu, T.: Intuitionistic epistemic logic. preprint arXiv:1406.1582 (2014)

BCHS13.   Baltag, A., Christo, Z., Hansen, J.U., Smets, S.: Logical models of informational cascades. Studies in Logic. College Publications (2013)

BR15.   Bakhtiari, Z., Rivieccio, U.: Epistemic updates on bilattices (submitted) (2015)

CC15.   Conradie, W., Craig, A.: Canonicity results for mu-calculi: An algorithmic approach. Journal of Logic and Computation (forthcoming) (2015)

CFP+.   Conradie, W., Frittella, S., Palmigiano, A., Tzimoulis, A., Wijnberg, N.: Probabilistic epistemic updates on algebras (in preparation)

CFPS15.   Conradie, W., Fomatati, Y., Palmigiano, A., Sourabh, S.: Algorithmic correspondence for intuitionistic modal mu-calculus. Theoretical Computer Science 564, 30–62 (2015)

CGP14.   Conradie, W., Ghilardi, S., Palmigiano, A.: Unified correspondence. In: Baltag, A., Smets, S. (eds.) Johan F.A.K. van Benthem on Logical and Informational Dynamics. Outstanding Contributions to Logic. Springer (2014) (in print)

CP15.   Conradie, W., Palmigiano, A.: Algorithmic correspondence and canonicity for non-distributive logics. Journal of Logic and Computation (forthcoming) (2015)

CPS.   Conradie, W., Palmigiano, A., Sourabh, S.: Algebraic modal correspondence: Sahlqvist and beyond (submitted) (2015)

DP02.   Davey, B.A., Priestley, H.A.: Lattices and Order. Cambridge Univerity Press (2002)

FGK⁺14a.   Frittella, S., Greco, G., Kurz, A., Palmigiano, A., Sikimić, V.: A multi-type display calculus for dynamic epistemic logic. Journal of Logic and Computation, Special Issue on Substructural Logic and Information Dynamics (2014)

FGK⁺14b.   Frittella, S., Greco, G., Kurz, A., Palmigiano, A., Sikimić, V.: Multi-type sequent calculi. In: Zawidzki, M., Indrzejczak, A., Kaczmarek, J. (eds.) Trends in Logic XIII, pp. 81–93. Lodź University Press (2014)

FGK⁺14c.   Frittella, S., Greco, G., Kurz, A., Palmigiano, A., Sikimić, V.: A proof-theoretic semantic analysis of dynamic epistemic logic. Journal of Logic and Computation (2014)

FGKP14.   Frittella, S., Greco, G., Kurz, A., Palmigiano, A.: Multi-type display calculus for propositional dynamic logic. Journal of Logic and Computation, Special Issue on Substructural Logic and Information Dynamics (2014)

GKP13.   Greco, G., Kurz, A., Palmigiano, A.: Dynamic epistemic logic displayed. In: Grossi, D., Roy, O., Huang, H. (eds.) LORI 2014. LNCS, vol. 8196, pp. 135–148. Springer, Heidelberg (2013)

GMP⁺15.   Greco, G., Ma, M., Palmigiano, A., Tzimoulis, A., Zhao, Z.: Unified correspondence as a proof-theoretic tool (sumbitted) (2015)

Koo03.   Kooi, B.P.: Probabilistic dynamic epistemic logic. Journal of Logic, Language and Information 12(4), 381–408 (2003)

KP13.   Kurz, A., Palmigiano, A.: Epistemic updates on algebras. Logical Methods in Computer Science, abs/1307.0417 (2013)

MPS14.   Ma, M., Palmigiano, A., Sadrzadeh, M.: Algebraic semantics and model completeness for intuitionistic public announcement logic. Annals of Pure and Applied Logic 165(4), 963–995 (2014)

PSZ15a.   Palmigiano, A., Sourabh, S., Zhao, Z.: Jónsson-style canonicity for ALBA-inequalities. Journal of Logic and Computation (forthcoming) (2015)

PSZ15b.   Palmigiano, A., Sourabh, S., Zhao, Z.: Sahlvist theory for impossible worlds. Journal of Logic and Computation (forthcoming) (2015)

Riv14.   Rivieccio, U.: Algebraic semantics for bilattice public announcement logic. In: Studia Logica, Proc. Trends in Logic XIII. Springer (2014)

vBGK09.   van Benthem, J., Gerbrandy, J., Kooi, B.P.: Dynamic update with probabilities. Studia Logica 93(1), 67–96 (2009)

Wea03.   Weatherson, B.: From classical to intuitionistic probability. Notre Dame Journal of Formal Logic 44(2), 111–123 (2003)

An Abstract Algebraic Logic View on Judgment Aggregation*

María Esteban[1], Alessandra Palmigiano[2,3], and Zhiguang Zhao[2]

[1] Departament de Lògica, Història i Filosofia de la Ciència, Facultat de Filosofia,
Universitat de Barcelona, Spain
mariaesteban.edu@gmail.com
[2] Faculty of Technology, Policy and Management, Delft University of Technology,
The Netherlands
{a.palmigiano,z.zhao-3}@tudelft.nl
[3] Department of Pure and Applied Mathematics, University of Johannesburg,
South Africa

Abstract. In the present paper, we propose Abstract Algebraic Logic (AAL) as a general logical framework for Judgment Aggregation. Our main contribution is a generalization of Herzberg's algebraic approach to characterization results on judgment aggregation and propositional-attitude aggregation, characterizing certain Arrovian classes of aggregators as Boolean algebra and MV-algebra homomorphisms, respectively. The characterization result of the present paper applies to agendas of formulas of an arbitrary *selfextensional* logic. This notion comes from AAL, and encompasses a vast class of logics, of which classical, intuitionistic, modal, many-valued and relevance logics are special cases. To each selfextensional logic $\mathcal{S}$, a unique class of algebras $\mathbb{Alg}\mathcal{S}$ is canonically associated by the general theory of AAL. We show that for any selfextensional logic $\mathcal{S}$ such that $\mathbb{Alg}\mathcal{S}$ is closed under direct products, any algebra in $\mathbb{Alg}\mathcal{S}$ can be taken as the set of truth values on which an aggregation problem can be formulated. In this way, judgment aggregation on agendas formalized in classical, intuitionistic, modal, many-valued and relevance logic can be uniformly captured as special cases. This paves the way to the systematic study of a wide array of "realistic agendas" made up of complex formulas, the propositional connectives of which are interpreted in ways which depart from their classical interpretation. This is particularly interesting given that, as observed by Dietrich, nonclassical (subjunctive) interpretation of logical connectives can provide a strategy for escaping impossibility results.

Keywords: Judgment aggregation, Systematicity, Impossibility theorems, Abstract Algebraic Logic, Logical filter, Algebra homomorphism. *Math. Subject Class.* 91B14; 03G27.

* The research of the second and third author has been made possible by the NWO Vidi grant 016.138.314, by the NWO Aspasia grant 015.008.054, and by a Delft Technology Fellowship awarded in 2013.

W. van der Hoek et al. (Eds.): LORI 2015, LNCS 9394, pp. 77–89, 2015.
DOI: 10.1007/978-3-662-48561-3_7

1   Introduction

Social Choice and Judgment Aggregation. The theory of *social choice* is
the formal study of mechanisms for collective decision making, and investigates
issues of philosophical, economic, and political significance, stemming from the
classical Arrovian problem of how the preferences of the members of a group can
be "fairly" aggregated into one outcome.

In the last decades, many results appeared generalizing the original Arrovian
problem, which gave rise to a research area called *judgment aggregation* (JA) [25].
While the original work of Arrow [1] focuses on preference aggregation, this can
be recognized as a special instance of the aggregation of consistent judgments,
expressed by each member of a group of individuals over a given set of logically
interconnected propositions (the *agenda*): each proposition in the agenda is either
accepted or rejected by each group member, so as to satisfy certain requirements
of logical consistency. Within the JA framework, the Arrovian-type *impossibility
results* (axiomatically providing sufficient conditions for aggregator functions to
turn into degenerate rules, such as dictatorship) are obtained as consequences of
characterization theorems [26], which provide necessary and sufficient conditions
for agendas to have aggregator functions on them satisfying given axiomatic
conditions.

In the same logical vein, in [24], *attitude aggregation theory* was introduced;
this direction has been further pursued in [19], where a characterization theorem has been given for certain many-valued propositional-attitude aggregators
as MV-algebra homomorphisms.

The Ultrafilter Argument and its Generalizations. Methodologically, the
ultrafilter argument is the tool underlying the generalizations and unifications
mentioned above. It can be sketched as follows: to prove impossibility theorems
for finite electorates, one shows that the axiomatic conditions on the aggregation
function force the set of all decisive coalitions to be an (ultra)filter on the powerset of the electorate. If the electorate is finite, this implies that all the decisive
coalitions must contain one and the same (singleton) coalition: the oligarchs (the
dictator). Employed in [11] and [23] for a proof of Arrow's theorem alternative to
the original one[1], this argument was applied to obtain elegant and concise proofs
of impossibility theorems also in judgment aggregation [7]. More recently, it gave
rise to characterization theorems, e.g. establishing a bijective correspondence between Arrovian aggregation rules and ultrafilters on the set of individuals [20].
Moreover, the ultrafilter argument has been generalized by Herzberg and Eckert [20] to obtain a generalized Kirman-Sondermann correspondence as a consequence of which Arrow-rational aggregators can be identified with those arising
as ultraproducts of profiles (see also [2], in which the results in [20] have been generalized to a setting accounting for vote abstention), and—using the well-known
correspondence between ultrafilters and Boolean homomorphisms—similar cor-

---

[1] See also [16] for further information about the genesis and application of the technique.

respondences have been established between Arrovian judgment aggregators and Boolean algebra homomorphisms [18].

Escaping Impossibility via Nonclassical Logics. While much research in this area explored the limits of the applicability of Arrow-type results, at the same time the question of how to 'escape impossibility' started attracting increasing interest. In [5], Dietrich provides a unified model of judgment aggregation which applies to predicate logic as well as to modal logic and fuzzy logics. In [6], Dietrich argues that impossibility results do not apply to a wide class of realistic agendas once propositions of the form 'if *a* then *b*' are modelled as *subjunctive* implications rather than material implications. Besides their theoretical value, these results are of practical interest, given that subjunctive implication models the meaning of if-then statements in natural language more accurately than material implication. In [27] and [28], Porello discusses judgment aggregation in the setting of intuitionistic, linear and substructural logics. In particular, in [28], it is shown that linear logic is a viable way to circumvent impossibility theorems in judgment aggregation.

Aim. A natural question arising in the light of these results is how to uniformly account for the role played by the different logics (understood both as formal language and deductive machinery) underlying the given agenda in characterization theorems for JA.

The present paper focuses on *Abstract Algebraic Logic* as a natural theoretical setting for Herzberg's results [17, 19], and the theory of *(fully) selfextensional logics* as the appropriate logical framework for a nonclassical interpretation of logical connectives, in line with the approach of [6].

Abstract Algebraic Logic and Selfextensional Logics. Abstract Algebraic Logic (AAL) [14] is a forty-year old research field in mathematical logic. It was conceived as the framework for an algebraic approach to the investigation of classes of logics. Its main goal was establishing a notion of *canonical algebraic semantics* uniformly holding for classes of logics, and using it to systematically investigate (metalogical) properties of logics in connection with properties of their algebraic counterparts.

Selfextensionality is the metalogical property holding of those logical systems whose associated relation of logical equivalence on formulas is a congruence of the term algebra. Wójcicki [29] characterized selfextensional logics as the logics which admit a so-called *referential semantics* (which is a general version of the well known possible-world semantics of modal and intuitionistic logics), and in [22], a characterization was given of the particularly well behaved subclass of the fully selfextensional logics in general duality-theoretic terms. This subclass includes many well-known logics, such as classical, intuitionistic, modal, many-valued and relevance logic. These and other results in this line of research (cf. e.g. [8, 9, 15, 21]) establish a systematic connection between possible world semantics and the logical account of intensionality.

Contributions. In the present paper, we generalize and refine Herzberg's characterization result in [19] from the MV-algebra setting to any class of algebras canonically associated with some selfextensional logic. This generalization simultaneously accounts for agendas expressed in the language of such logics as modal, intuitionistic, relevance, substructural and many-valued logics. Besides having introduced the connection between AAL and Judgment Aggregation, the added value of this approach is that it is parametric in the logical system $\mathcal{S}$. In particular, the properties of agendas are formulated independently of a specific logical signature and are slightly different than those in Herzberg's setting. In contrast with Herzberg's characterization result, which consisted of two slightly asymmetric parts, the two propositions which yield the characterization result in the present paper (cf. Propositions 1 and 2) are symmetric. Aggregation of propositional attitudes modeled in classical, intuitionistic, modal, Łukasiewicz and relevance logic can be uniformly captured as special cases of the present result. This makes it possible to fine-tune the expressive and deductive power of the formal language of the agenda, so as to capture e.g. intensional or vague statements.

Structure of the Paper. In Section 2, relevant preliminaries are collected on Abstract Algebraic Logic. In Section 3, Herzberg's algebraic framework for aggregation theory is generalized from MV-algebras to $\mathcal{S}$-algebras, where $\mathcal{S}$ is an arbitrary selfextensional logic. In Section 4, the main characterization result is stated. In Section 5, the impossibility theorem for judgment aggregation is deduced as a corollary of the main result, and one well known setting accounting for the subjunctive reading of implication is discussed.

2   Preliminaries on Abstract Algebraic Logic

The present section collects the basic concepts of Abstract Algebraic Logic that we will use in the paper. For a general view of AAL the reader is addressed to [13] and the references therein.

2.1   General Approach.

As mentioned in the introduction, in AAL, logics are not studied in isolation, and in particular, investigation focuses on classes of logics and their identifying *metalogical properties*. Moreover, the notion of *consequence* rather than the notion of *theoremhood* is taken as basic: consequently, *sentential logics*, the primitive objects studied in AAL, are defined as tuples $\mathcal{S} = \langle \mathbf{Fm}, \vdash_{\mathcal{S}} \rangle$ where $\mathbf{Fm}$ is the algebra of formulas of type $\mathcal{L}_{\mathcal{S}}$ over a denumerable set of propositional variables Var, and $\vdash_{\mathcal{S}}$ is a *consequence relation* on (the carrier of) $\mathbf{Fm}$ (cf. Subsection 2.3).

This notion encompasses logics that are defined by any sort of proof-theoretic calculus (Gentzen-style, Hilbert-style, tableaux, etc.), as well as logics arising from some classes of (set-theoretic, order-theoretic, topological, algebraic, etc.) semantic structures, and in fact it allows to treat logics independently of the way in which they have been originally introduced. Another perhaps more common approach in logic takes the notion of theoremhood as basic and consequently sees logics as sets of formulas (possibly closed under some rules of inference).

This approach is easily recaptured by the notion of sentential logic adopted in AAL: Every sentential logic $\mathcal{S}$ is uniquely associated with the set $Thm(\mathcal{S}) = \{\varphi \in Fm \mid \emptyset \vdash_{\mathcal{S}} \varphi\}$ of its *theorems*.

2.2   Consequence Operations

For any set A, a *consequence operation* (or closure operator) on A is a map $C : \mathcal{P}(A) \to \mathcal{P}(A)$ such that for every $X, Y \subseteq A$: (1) $X \subseteq C(X)$, (2) if $X \subseteq Y$, then $C(X) \subseteq C(Y)$ and (3) $C(C(X)) = C(X)$. The closure operator C is *finitary* if in addition satisfies (4) $C(X) = \bigcup\{C(Z) : Z \subseteq X, Z \text{ finite}\}$. For any consequence operation C on A, a set $X \subseteq A$ is *C-closed* if $C(X) = X$. Let $\mathcal{C}_C$ be the collection of C-closed subsets of A.

For any set A, a *closure system* on A is a collection $\mathcal{C} \subseteq \mathcal{P}(A)$ such that $A \in \mathcal{C}$, and $\mathcal{C}$ is closed under intersections of arbitrary non-empty families. A closure system is *algebraic* if it is closed under unions of up-directed[2] families.

For any closure operator C on A, the collection $\mathcal{C}_C$ of the C-closed subsets of A is a closure system on A. If C is finitary, then $\mathcal{C}_C$ is algebraic. Any closure system $\mathcal{C}$ on A defines a consequence operation $C_{\mathcal{C}}$ on A by setting $C_{\mathcal{C}}(X) = \bigcap\{Y \in \mathcal{C} : X \subseteq Y\}$ for every $X \subseteq A$. The $C_{\mathcal{C}}$-closed sets are exactly the elements of $\mathcal{C}$. Moreover, $\mathcal{C}$ is algebraic if and only if $C_{\mathcal{C}}$ is finitary.

2.3   Logics

Let $\mathcal{L}$ be a propositional language type (i.e. a set of connectives and their arities, which we will also regard as a set of function symbols) and let $\mathbf{Fm}_{\mathcal{L}}$ denote the algebra of formulas (or term algebra) of $\mathcal{L}$ over a denumerable set V of propositional variables. Let $Fm_{\mathcal{L}}$ be the carrier of the algebra $\mathbf{Fm}_{\mathcal{L}}$. A *logic* (or deductive system) of type $\mathcal{L}$ is a pair $\mathcal{S} = \langle \mathbf{Fm}_{\mathcal{L}}, \vdash_{\mathcal{S}} \rangle$ such that $\vdash_{\mathcal{S}} \subseteq \mathcal{P}(Fm_{\mathcal{L}}) \times Fm_{\mathcal{L}}$ such that the operator $C_{\vdash_{\mathcal{S}}} : \mathcal{P}(Fm_{\mathcal{L}}) \to \mathcal{P}(Fm_{\mathcal{L}})$ defined by

$$\varphi \in C_{\vdash_{\mathcal{S}}}(\Gamma) \quad \text{iff} \quad \Gamma \vdash_{\mathcal{S}} \varphi$$

is a consequence operation with the property of *invariance under substitutions*; this means that for every substitution σ (i.e. for every $\mathcal{L}$-homomorphism $\sigma : \mathbf{Fm}_{\mathcal{L}} \to \mathbf{Fm}_{\mathcal{L}}$) and for every $\Gamma \subseteq Fm_{\mathcal{L}}$,

$$\sigma[C_{\vdash_{\mathcal{S}}}(\Gamma)] \subseteq C_{\vdash_{\mathcal{S}}}(\sigma[\Gamma]).$$

For every $\mathcal{S}$, the relation $\vdash_{\mathcal{S}}$ is the *consequence* or *entailment* relation of $\mathcal{S}$. A logic is *finitary* if the consequence operation $C_{\vdash_{\mathcal{S}}}$ is finitary. Sometimes we will use the symbol $\mathcal{L}_{\mathcal{S}}$ to refer to the propositional language of a logic $\mathcal{S}$.

The *interderivability relation* of a logic $\mathcal{S}$ is the relation $\equiv_{\mathcal{S}}$ defined by

$$\varphi \equiv_{\mathcal{S}} \psi \quad \text{iff} \quad \varphi \vdash_{\mathcal{S}} \psi \text{ and } \psi \vdash_{\mathcal{S}} \varphi.$$

$\mathcal{S}$ satisfies the *congruence property* if $\equiv_{\mathcal{S}}$ is a congruence of $\mathbf{Fm}_{\mathcal{L}}$.

---

[2] For $\langle P, \leq \rangle$ a poset, $U \subseteq P$ is *up-directed* when for any $a, b \in U$ there exists $c \in U$ such that $a, b \leq c$.

2.4 Logical Filters

Let S be a logic of type $\mathcal{L}$ and let $\boldsymbol{A}$ be an $\mathcal{L}$-algebra (from now on, we will drop reference to the type $\mathcal{L}$, and when we refer to an algebra or class of algebras in relation with S, we will always assume that the algebra and the algebras in the class are of type $\mathcal{L}$).

A subset $F \subseteq A$ is an S-*filter* of $\boldsymbol{A}$ if for every $\Gamma \cup \{\varphi\} \subseteq Fm$ and every $h \in \mathrm{Hom}(\mathbf{Fm}_{\mathcal{L}}, \boldsymbol{A})$,

$$\text{if } \Gamma \vdash_S \varphi \text{ and } h[\Gamma] \subseteq F, \text{ then } h(\varphi) \in F.$$

The collection $\mathrm{Fi}_S(\boldsymbol{A})$ of the S-filters of $\boldsymbol{A}$ is a closure system. Moreover, $\mathrm{Fi}_S(\boldsymbol{A})$ is an algebraic closure system if S is finitary. The consequence operation associated with $\mathrm{Fi}_S(\boldsymbol{A})$ is denoted by $C_S^{\boldsymbol{A}}$. For every $X \subseteq A$, the closed set $C_S^{\boldsymbol{A}}(X)$ is the S-filter of $\boldsymbol{A}$ generated by X. If S is finitary, then $C_S^{\boldsymbol{A}}$ is finitary for every algebra $\boldsymbol{A}$.

On the algebra of formulas $\mathbf{Fm}$, the closure operator $C_S^{\mathbf{Fm}}$ coincides with $C_{\vdash_S}$ and the $C_S^{\mathbf{Fm}}$-closed sets are exactly the S-*theories*; that is, the sets of formulas which are closed under the relation $\vdash_S$.

2.5 S-algebras and Selfextensional Logics

One of the basic topics of AAL is how to associate in a uniform way a class of algebras with an arbitrary logic S. According to contemporary AAL [13], the canonical algebraic counterpart of S is the class $\mathbb{Alg}S$, whose elements are called S-*algebras*. This class can be defined via the notion of Tarski congruence.

For any algebra $\mathbf{A}$ (of the same type as S) and any closure system $\mathcal{C}$ on $\mathbf{A}$, the *Tarski congruence of* $\mathcal{C}$ *relative to* $\mathbf{A}$, denoted by $\tilde{\boldsymbol{\Omega}}_{\mathbf{A}}(\mathcal{C})$, is the greatest congruence which is compatible with all $F \in \mathcal{C}$, that is, which does not relate elements of F with elements which do not belong to F. The Tarski congruence of the closure system consisting of all S-theories relative to $\mathbf{Fm}$ is denoted by $\tilde{\boldsymbol{\Omega}}(S)$. The quotient algebra $\mathbf{Fm}/\tilde{\boldsymbol{\Omega}}(S)$ is called the *Lindenbaum-Tarski algebra* of S.

For any algebra $\mathbf{A}$, we say that $\mathbf{A}$ is an S-*algebra* (cf. [13, Definition 2.16]) if the Tarski congruence of $\mathrm{Fi}_S(\boldsymbol{A})$ relative to $\mathbf{A}$ is the identity. It is well-known (cf. [13, Theorem 2.23] and ensuing discussion) that $\mathbb{Alg}S$ is closed under direct products. Moreover, for any logic S, the Lindenbaum-Tarski algebra is an S-algebra (see page 36 in [13]).

A logic S is *selfextensional* (cf. [29]) when the relation of *logical equivalence* between formulas

$$\varphi \equiv_S \psi \qquad \text{iff} \qquad \varphi \vdash_S \psi \quad \text{and} \quad \psi \vdash_S \varphi$$

is a congruence relation of the formula algebra $\mathbf{Fm}$. An equivalent definition of selfextensionality (see page 48 in [13]) is given as follows: S is selfextensional iff the Tarski congruence $\tilde{\boldsymbol{\Omega}}(S)$ and the relation of logical equivalence $\equiv_S$ coincide. In such case the Lindenbaum-Tarski algebra reduces to $\mathbf{Fm}/\equiv_S$. Examples of selfextensional logics besides classical propositional logic are intuitionistic logic, positive modal logic [4], the $\{\wedge, \vee\}$-fragment of classical propositional logic, Belnap's four-valued logic [3], the local consequence relation on modal formulas arising from Kripke frames, the (order-induced) consequence relation associated with MV-algebras and defined by "preserving degrees of truth" (cf. [12]), and the order-induced consequence relation of linear logic. Examples of non-selfextensional logics include the $\top$-induced consequence relation of linear logic, the ($\top$-induced) consequence relation associated with MV-algebras and defined by "preserving absolute truth" (cf. [12]), and the global consequence relation on modal formulas arising from Kripke frames.

From now on we assume that $\mathcal{S}$ is a selfextensional logic and $\mathbf{B} \in \mathbb{A}\mathrm{lg}\mathcal{S}$. For any formula $\varphi \in Fm$, we say that φ is *provably equivalent* to a propositional variable iff there exist a propositional variable x such that $\varphi \equiv x$.

3   Formal Framework

In the present section, we generalize Herzberg's algebraic framework for aggregation theory from MV-propositional attitudes to $\mathcal{S}$-propositional attitudes, where $\mathcal{S}$ is an arbitrary selfextensional logic. Our conventional notation is similar to [19]. Let $\mathcal{L}$ be a logical language which contains countably many connectives, each of which has arity at most n, and let Fm be the collection of $\mathcal{L}$-formulas.

3.1   The Agenda

The *agenda* will be given by a set of formulas $X \subseteq \mathbf{Fm}$. Let $\bar{X}$ denote the closure of X under the connectives of the language, i.e. the smallest set containing all formulas in X and the 0-ary connectives in $\mathcal{L}$, and closed under the connectives in the language. Notice that for any constant $c \in \mathcal{L}$, we have $c \in \bar{X}$.

We want the agenda to contain a sufficiently rich collection of formulas. In the classical case, it is customary to assume that the agenda contains at least two propositional variables. In our general framework, this translates in the requirement that the agenda contains at least n formulas that 'behave' like propositional variables, in the sense that their interpretation is not constrained by the interpretation of any other formula in the agenda.

We could just assume that the agenda contains at least n different propositional variables, but we will deal with a slightly more general situation, namely, we assume that the agenda is n-pseudo-rich:

Definition 1. *An agenda is n-pseudo-rich if it contains at least n formulas $\{\delta_1, \ldots, \delta_n\}$ such that each δ_i is provably equivalent to x_i for some set $\{x_1, \ldots, x_n\}$ of pairwise different propositional variables.*

3.2   Attitude Functions, Profiles and Attitude Aggregators

An *attitude function* is a function $A \in \mathbf{B}^X$ which assigns an element of the algebra $\mathbf{B}$ to each formula in the agenda.

The *electorate* will be given by some (finite or infinite) set N. Each $i \in N$ is called an *individual*.

An *attitude profile* is an N-sequence of attitude functions, i.e. $\boldsymbol{A} \in (\mathbf{B}^X)^N$. For each $\varphi \in X$, we denote the N-sequence $\{A_i(\varphi)\}_{i \in N} \in \mathbf{B}^N$ by $\boldsymbol{A}(\varphi)$.

An attitude aggregator is a function which maps each profile of individual attitude functions in some domain to a collective attitude function, interpreted as the set of preferences of the electorate as a whole. Formally, an *attitude aggregator* is a partial map $F : (\mathbf{B}^X)^N \nrightarrow \mathbf{B}^X$.

3.3 Rationality

Let the agenda contain formulas $\varphi_1, \ldots, \varphi_m, g(\varphi_1, \ldots, \varphi_m) \in X$, where $g \in \mathcal{L}$ is an m-ary connective of the language and $m \leq n$. Among all attitude functions $A \in \mathbf{B}^X$, those for which it holds that $A(g(\varphi_1, \ldots, \varphi_n)) = g^{\mathbf{B}}(A(\varphi_1), \ldots, A(\varphi_n))$ are of special interest. In general, we will focus on attitude functions which are 'consistent' with the logic $\mathcal{S}$ in the following sense.

We say that an attitude function $A \in \mathbf{B}^X$ is *rational* if it can be extended to a homomorphism $\bar{A} : \mathbf{Fm}_{/\equiv} \longrightarrow \mathbf{B}$ of $\mathcal{S}$-algebras. In particular, if A is rational, then it can be uniquely extended to $\bar{X}$, and we will implicitly use this fact in what follows.

We say that a profile $\boldsymbol{A} \in (\mathbf{B}^X)^N$ is *rational* if A_i is a rational attitude function for each $i \in N$.

We say that an attitude aggregator $F : (\mathbf{B}^X)^N \nrightarrow \mathbf{B}^X$ is *rational* if for all rational profiles $\boldsymbol{A} \in dom(F)$ in its domain, $F(\boldsymbol{A})$ is a rational attitude function. Moreover, we say that F is *universal* if $\boldsymbol{A} \in dom(F)$ for any rational profile $\boldsymbol{A}$. In other words, an aggregator is universal whenever its domain contains all rational profiles, and it is rational whenever it gives a rational output provided a rational input.

3.4 Decision Criteria and Systematicity

A *decision criterion* for F is a partial map $f : \mathbf{B}^N \nrightarrow \mathbf{B}$ such that for all $\boldsymbol{A} \in dom(F)$ and all $\varphi \in X$,

$$F(\boldsymbol{A})(\varphi) = f(\boldsymbol{A}(\varphi)).\tag{3.1}$$

As observed by Herzberg [19], an aggregator is independent if the aggregate attitude towards any proposition φ does not depend on the individuals attitudes towards propositions other than φ:

An aggregator F is *independent* if there exists some map $g : \mathbf{B}^N \times X \nrightarrow \mathbf{B}$ such that for all $\boldsymbol{A} \in dom(F)$, the following diagram commutes (whenever the partial maps are defined):

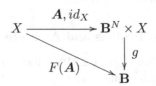

An aggregator F is *systematic* if there exists some *decision criterion* f for F, i.e. there exists some map $f : \mathbf{B}^N \nrightarrow \mathbf{B}$ such that for all $\boldsymbol{A} \in dom(F)$, the following diagram commutes (whenever the partial maps are defined):

Systematic aggregation is a special case of independent aggregation, in which the output of g does not depend on the input in the second coordinate. Thus, g is reduced to a decision criterion $f : \mathbf{B}^N \nrightarrow \mathbf{B}$.

An aggregator F is *strongly systematic* if there exists some decision criterion f for F, such that for all $\boldsymbol{A} \in dom(F)$, the following diagram commutes (whenever the partial maps are defined):

Notice that the diagram above differs from the previous one in that the agenda X is now replaced by its closure $\bar{X}$ under the connectives of the language. If X is closed under the operations in $\mathcal{L_S}$, then systematicity and strong systematicity coincide.

A formula $\varphi \in \mathbf{Fm}$ is *strictly contingent* if for all $a \in \mathbf{B}$ there exists some homomorphism $v : \mathbf{Fm} \to \mathbf{B}$ such that $v(\varphi) = a$. Notice that for any $n \geq 1$, any n-pseudo rich agenda (cf. Definition 1) always contains a strictly contingent formula. Moreover, if the agenda contains some strictly contingent formula φ, then any universal systematic attitude aggregator F has a unique decision criterion (cf. [19, Remark 3.5]).

Before moving on to the main section, we mention four definitions which appear in Herzberg's paper, namely that of *Paretian* attitude aggregator (cf. [19, Definition 3.7]), *complex* and *rich* agendas (cf. [19, Definition 3.8]), and *strongly systematizable* aggregators (cf. [19, Definition 3.9]). Unlike the previous ones, these definitions rely on the specific MV-signature, and thus do not have a natural counterpart in the present, vastly more general setting. However, as we will see, our main result can be formulated independently of these definitions. Moreover, a generalization of the Pareto condition follows from the assumptions of F being universal, rational and strongly systematic, as then it holds that for any constant $c \in \mathcal{L_S}$, and any $\varphi \in \mathbf{Fm}$, if $A_i(\varphi) = c$ for all $i \in N$, then $F(\boldsymbol{A})(\varphi) = c$.

4   Characterization Results

In the present section, the main results of the paper are presented. In what follows, we fix a language type $\mathcal{L}$ and a selfextensional logic $\mathcal{S}$. Recall that $\equiv$ indicates the interderivability relation associated with $\mathcal{S}$ and $\mathbf{B}$ is an arbitrary algebra in $\mathbb{A}\mathrm{lg}\mathcal{S}$.

Lemma 1. *Let X be an n-pseudo-rich agenda, $m \leq n$, $g \in \mathcal{L}$ be an m-ary connective and $a_1, \ldots, a_m \in \mathbf{B}$. Then there exist formulas $\delta_1, \ldots, \delta_m \in X$ in the agenda and a rational attitude function $A : X \longrightarrow \mathbf{B}$ such that $A(\delta_j) = a_j$ for each $j \in \{1, \ldots, m\}$.*

Proof. As the agenda is n-pseudo-rich, there are formulas $\delta_1, \ldots, \delta_m \in X$ each of which is provably equivalent to a different propositional variable x_i. Notice that this implies that the formulas $\delta_1, \ldots, \delta_m$ are not pairwise interderivable. So the $\equiv$-equivalence cells $[\delta_1], \ldots, [\delta_m]$ are pairwise different, and moreover there exists a valuation $v : \mathbf{Fm}/\equiv \longrightarrow \mathbf{B}$ such that $v(\delta_i) = a_i$ for all $i \in \{1, \ldots, m\}$. Let $A := v \circ \pi_{\restriction X}$, where $\pi_{\restriction X} : X \to \mathbf{Fm}/\equiv$ is the restriction of the canonical projection $\pi : \mathbf{Fm} \to \mathbf{Fm}/\equiv$ to X. Then clearly $A : X \to \mathbf{B}$ is the required rational attitude function.

Lemma 2. *Let X be an n-pseudo-rich agenda, $m \leq n$, $g \in \mathcal{L}$ be an m-ary connective and $\boldsymbol{a}_1, \ldots, \boldsymbol{a}_m \in \mathbf{B}^N$. Then there exist formulas $\delta_1, \ldots, \delta_m \in X$ in the agenda and a rational attitude profile $\boldsymbol{A} : X \longrightarrow \mathbf{B}^N$ such that $\boldsymbol{A}(\delta_j) = \boldsymbol{a}_j$ for each $j \in \{1, \ldots, m\}$.*

Proof. As the agenda is n-pseudo-rich, there are formulas $\delta_1, \ldots, \delta_m \in X$ each of which is provably equivalent to a different propositional variable x_i. By the previous lemma, for each $i \in N$, there exists a rational attitude function $A_i : X \longrightarrow \mathbf{B}$ such that $A_i(\delta_j) = \boldsymbol{a}_j(i)$ for each $j \in \{1, \ldots, m\}$. Thus it is easy to check that the sequence of attitudes $\boldsymbol{A} := \{A_i\}_{i \in N}$ is a rational profile such that $\boldsymbol{A}(\delta_j) = \boldsymbol{a}_j$ for each $j \in \{1, \ldots, m\}$.

Recall that given that X is n-pseudo rich, there exists a unique decision criterion for any strongly systematic attitude aggregator F (cf. page 85). We omit the proofs of the following propositions, which can be found in an extended version of the present paper (cf. [10]):

Proposition 1. *Let F be a rational, universal and strongly systematic attitude aggregator. Then the decision criterion of F is a homomorphism of $\mathcal{S}$-algebras.*

Proposition 2. *Let $f : \mathbf{B}^N \twoheadrightarrow \mathbf{B}$ be a homomorphism of $\mathcal{S}$-algebras. Then the function $F : (\mathbf{B}^X)^N \twoheadrightarrow \mathbf{B}^X$, defined for any rational profile $\boldsymbol{A}$ and any $\varphi \in X$ by the following assignment:*

$$F(\boldsymbol{A})(\varphi) = f(\boldsymbol{A}(\varphi)),$$

is a rational, universal and strongly systematic attitude aggregator.

Finally, the conclusion of the following corollary expresses a property which is a generalization of the Pareto condition (cf. [19, Definition 3.7]).

Corollary 1. *If F is universal, rational and strongly systematic, then for any constant $c \in \mathcal{L}_\mathcal{S}$ and $\varphi \in \mathbf{Fm}$, if $A_i(\varphi) = c^{\mathbf{B}}$ for all $i \in N$, then $F(\mathbf{A})(\varphi) = c^{\mathbf{B}}$.*

Proof. Let $c \in \mathcal{L}_\mathcal{S}$ and $\varphi \in \mathbf{Fm}$. Notice that by definition of the product algebra, the sequence $\{c^{\mathbf{B}}\}_{i \in N}$ is precisely $c^{\mathbf{B}^N}$. If $A_i(\varphi) = c^{\mathbf{B}}$ for all $i \in N$, i.e. $\mathbf{A}(\varphi) = c^{\mathbf{B}^N}$, then by Proposition 1, $F(\mathbf{A})(\varphi) = f(\mathbf{A}(\varphi)) = f(c^{\mathbf{B}^N}) = c^{\mathbf{B}}$, as required.

5 Applications

In the present section, we show how the setting in the present paper relates to existing settings in the literature.

5.1 Arrow-Type Impossibility Theorem for Judgment Aggregation

Let $\mathcal{S}$ be the classical propositional logic. Its algebraic counterpart $\mathrm{Alg}\mathcal{S} = \mathbb{BA}$ is the variety of Boolean algebras. Let $\mathcal{L} = \{\neg, \vee\}$ be its language (the connectives $\wedge, \rightarrow, \leftrightarrow$ are definable from the primitive ones). Let $\mathbf{B} = \mathbf{2}$ be the two-element Boolean algebra. Let $X \subseteq Fm_\mathcal{L}$ be a 2-pseudo-rich agenda.

By Propositions 1 and 2, for every electorate N, there exists a bijection between rational, universal and strongly systematic attitude aggregators $F : (\mathbf{2}^X)^N \longrightarrow \mathbf{2}^X$ [3] and Boolean homomorphisms $f : \mathbf{2}^N \longrightarrow \mathbf{2}$.

Recall that there is a bijective correspondence between Boolean homomorphisms $f : \mathbf{2}^N \longrightarrow \mathbf{2}$ and ultrafilters of $\mathbf{2}^N$. Moreover, if N is finite, every ultrafilter of $\mathbf{2}^N$ is principal. In this case, a decision criterion corresponds to an ultrafilter exactly when it is dictatorial.

5.2 A Mathematical Environment for the Subjunctive Interpretation of 'if – then' Statements

In [6], Dietrich argues that, in order to reflect the meaning of connection rules (i.e. formulas of the form $p \rightarrow q$ or $p \leftrightarrow q$ such that p and q are conjunctions of atomic propositions or negated atomic propositions) as they are understood and used in natural language, the connective $\rightarrow$ should be interpreted subjunctively. That is, the formula $p \rightarrow q$ should not be understood as a statement about the actual world, but about whether q holds in hypothetical world(s) where p holds, depends on q's truth value in possibly non-actual worlds. Dietrich proposes that, in the context of connection rules, any such implication should satisfy the following conditions:

(a) for any atomic propositions p and q, $p \rightarrow q$ is inconsistent with $\{p, \neg q\}$ but consistent with each of $\{p, q\}$ $\{\neg p, q\}$ $\{\neg p, \neg q\}$;

---

[3] Note that in this case an alternative presentation of F is $F : \mathcal{P}(X)^N \longrightarrow \mathcal{P}(X)$, which is the standard one.

(b) for any atomic propositions p and q, $\neg(p \to q)$ is consistent with each of $\{p, \neg q\}$, $\{p, q\}$, $\{\neg p, q\}$ and $\{\neg p, \neg q\}$.

Clearly, the classical interpretation of $p \to q$ as $\neg p \vee q$ satisfies only condition (a) but not (b). The subjunctive interpretation of $\to$ has been formalised in various settings based on possible-worlds semantics. One such setting, which is different from the one adopted by Dietrich's, is given by Boolean algebras with operators (BAOs). These are Boolean algebras endowed with an additional unary operation $\square$ satisfying the identities $\square 1 = 1$ and $\square(x \wedge y) = \square x \wedge \square y$. Let us further restrict ourselves to the class of BAOs such that the inequality $\square x \leq x$ is valid. This class coincides with $\mathbb{A}\text{lg}\mathcal{S}$, where $\mathcal{S}$ is the normal modal logic **T** with the so-called local consequence relation. It is well known that **T** is selfextensional and is complete w.r.t. the class of reflexive Kripke frames. In this setting, let us stipulate that $p \to q$ is interpreted as $\square(\neg p \vee q)$.

It is easy to see that this interpretation satisfies both conditions (a) and (b). To show that $p \to q$ is inconsistent with $\{p, \neg q\}$, observe that $\square(\neg p \vee q) \wedge p \wedge \neg q \leq (\neg p \vee q) \wedge p \wedge \neg q = (\neg p \wedge (p \wedge \neg q)) \vee (q \wedge (p \wedge \neg q)) = \bot \vee \bot = \bot$.

To show that $p \to q$ is consistent with $\{p, q\}$, consider the two-element BAO s.t. $\square 1 = 1$ and $\square 0 = 0$. The assignment mapping p and q to 1 witnesses the required consistency statement. The remaining part of the proof is similar and hence is omitted.

Clearly, the characterization theorem given by Propositions 1 and 2 applies also to this setting. However, the main interest of this setting is given by the possibility theorems. It would be a worthwile future research direction to explore the interplay and the scope of these results.

References

1. Arrow, K.J.: Social choice and individual values, 2nd edn., vol. 12. John Wiley, New York (1963)
2. Bedrosian, G., Palmigiano, A., Zhao, Z.: Generalized ultraproduct and Kirman-Sondermann correspondence for vote abstention. In: Proceedings of Logic, Rationality and Interaction, 5th International Workshop, LORI 2015 (2015)
3. Belnap Jr., N.D.: A useful four-valued logic. In: Modern uses of multiple-valued logic, pp. 5–37. Springer (1977)
4. Celani, S., Jansana, R.: Priestley duality, a Sahlqvist theorem and a Goldblatt-Thomason theorem for positive modal logic. Logic Journal of IGPL 7(6), 683–715 (1999)
5. Dietrich, F.: A generalised model of judgment aggregation. Social Choice and Welfare 28(4), 529–565 (2007)
6. Dietrich, F.: The possibility of judgment aggregation on agendas with subjunctive implications. Journal of Economic Theory 145(2), 603–638 (2010)
7. Dietrich, F., Mongin, P.: The premiss-based approach to judgment aggregation. Journal of Economic Theory 145(2), 562–582 (2010)
8. Esteban, M.: Duality theory and Abstract Algebraic Logic. PhD thesis, Universitat de Barcelona (November 2013), http://www.tdx.cat/handle/10803/125336
9. Esteban, M., Jansana, R.: Priestley style duality for filter distributive congruential logics (2015)

10. Esteban, M., Palmigiano, A., Zhao, Z.: An Abstract Algebraic Logic view on Judgment Aggregation. Working paper (2015),
    http://www.appliedlogictudelft.nl/publications/
11. Fishburn, P.: Arrow's impossibility theorem: concise proof and infinite voters. Journal of Economic Theory 2(1), 103–106 (1970)
12. Font, J.M.: An Abstract Algebraic Logic view of some multiple-valued logics, pp. 25–57. Physica-Verlag GmbH, Heidelberg (2003)
13. Font, J.M., Jansana, R.: A general algebraic semantics for sentential logics. In: The Association for Symbolic Logic, 2nd edn., Ithaca, N.Y. Lectures Notes in Logic, vol. 7 (2009)
14. Font, J.M., Jansana, R., Pigozzi, D.: A survey of Abstract Algebraic Logic. Studia Logica 74(1/2), 13–97 (2003)
15. Gehrke, M., Jansana, R., Palmigiano, A.: Canonical extensions for congruential logics with the deduction theorem. Annals of Pure and Applied Logic 161(12), 1502–1519 (2010)
16. Grossi, D., Pigozzi, G.: Judgment aggregation: a primer. Synthesis Lectures on Artificial Intelligence and Machine Learning 8(2), 1–151 (2014)
17. Herzberg, F.: Judgment aggregation functions and ultraproducts. Institute of Mathematical Economics, University of Bielefeld (2008)
18. Herzberg, F.: Judgment aggregators and Boolean algebra homomorphisms. Journal of Mathematical Economics 46(1), 132–140 (2010)
19. Herzberg, F.: Universal algebra for general aggregation theory: many-valued propositional-attitude aggregators as MV-homomorphisms. Journal of Logic and Computation (2013)
20. Herzberg, F., Eckert, D.: Impossibility results for infinite-electorate abstract aggregation rules. Journal of Philosophical Logic 41, 273–286 (2012)
21. Jansana, R.: Selfextensional logics with implication. In: Logica Universalis, pp. 65–88. Birkhäuser Basel (2005)
22. Jansana, R., Palmigiano, A.: Referential semantics: duality and applications. Reports on Mathematical Logic 41, 63–93 (2006)
23. Kirman, A.P., Sondermann, D.: Arrow's theorem, many agents, and invisible dictators. Journal of Economic Theory 5(2), 267–277 (1972)
24. List, C., Dietrich, F.: The aggregation of propositional attitudes: towards a general theory. In: Oxford Studies in Epistemology, vol. 3, pp. 215–234. Oxford University Press (2010)
25. List, C., Polak, B.: Introduction to judgment aggregation. Journal of Economic Theory 145(2), 441–466 (2010)
26. Nehring, K., Puppe, C.: Strategy-proof social choice on single-peaked domains: possibility, impossibility and the space between. University of California at Davis (2002)
27. Porello, D.: A proof-theoretical view of collective rationality. In: Proceedings of the Twenty-Third International Joint Conference on Artificial Intelligence, pp. 317–323. AAAI Press (2013)
28. Porello, D.: Logics for collective reasoning. In: Proceedings of the European Conference on Social Intelligence (ECSI 2014) (2014)
29. Wójcicki, R.: Referential matrix semantics for propositional calculi. Bulletin of the Section of Logic 8(4), 170–176 (1979)

Context-Dependent Utilities
A Solution to the Problem of Constant Acts in Savage

Haim Gaifman and Yang Liu

Department of Philosophy, Columbia University
New York, NY 10027, USA
{hg17,y.liu}@columbia.edu

Abstract. Savage's framework of subjective preference among acts provides a paradigmatic derivation of rational subjective probabilities within a more general theory of rational decisions. The system is based on a set of possible states of the world, and on acts, which are functions that assign to each state a consequence. The representation theorem states that the given preference between acts is determined by their expected utilities, based on uniquely determined probabilities (assigned to sets of states), and numeric utilities assigned to consequences. Savage's derivation, however, is based on a highly problematic well-known assumption not included among his postulates: for any consequence of an act in some state, there is a "constant act" which has that consequence in all states. This ability to transfer consequences from state to state is, in many cases, miraculous – including simple scenarios suggested by Savage as natural cases for applying his theory. We propose a simplification of the system, which yields the representation theorem without the constant act assumption. We need only postulates P1-P6. This is done at the cost of reducing the set of acts included in the setup. The reduction excludes certain theoretical infinitary scenarios, but includes the scenarios that should be handled by a system that models human decisions.

Keywords: subjective expected utility, Savage's postulates, constant acts, context-dependent decision making.

1 Introduction

In his classic *The Foundations of Statistics*[1] Savage sets up a foundational system within which he derives both subjective probabilities and utilities from the preferences of a rational agent, provided that the preferences satisfy certain plausible postulates. The upshot is that the expected utilities come out as a measure that defines the agent's given preferences. The derivation relies however on additional implicit assumptions, one of which, the CAA discussed below, is quite problematic. Let us first recall the basic structure of the Savage system. It is based on the following four components:

---

[1] The first edition [4] of Savage's book was published in 1954, all citations made in this paper refer to the second and revised edition [5] published in 1972.

© Springer-Verlag Berlin Heidelberg 2015
W. van der Hoek et al. (Eds.): LORI 2015, LNCS 9394, pp. 90–101, 2015.
DOI: 10.1007/978-3-662-48561-3_8

1. A set S of *states* (or states of the world),
2. A set C of *consequences*, which are the consequences of the agent's acts,
3. A set $\mathcal{A}$ of *acts*, where each act is a function, f, which associates with every state, s, the consequence $f(s)$ of performing f in a world that is in state s,
4. The (rational) agent's *preference relation*, $\succcurlyeq$, defined over acts, which is a total preorder. Here, as is customary in current mathematics, "preorder" means a reflexive and transitive relation. A preorder is *total* or *complete* if for any f, g either $f \succcurlyeq g$ or $g \succcurlyeq f$.

The intended meaning of $f \succcurlyeq g$ is: f is weakly preferable to g, i.e., is at least as good as g; it is also written $g \preccurlyeq f$. If both $f \succcurlyeq g$ and $g \succcurlyeq f$, then we denote it by $f \equiv g$. Obviously this is an equivalence relation, it means that f and g are equi-preferable: the agent considers them to be equally good. We define: $f \succ g =_{\mathrm{Df}} f \succcurlyeq g$ and $g \not\succcurlyeq f$. This means that f is strictly preferred to g. Note that our notation and terminology differ from Savage's and this can be more than a technicality. For instance, after defining "constant acts" he does not use this term and one has to infer that certain acts are constant only from the notation; that notation, however, is sometimes ambiguous.[2]

Other elements are introduced in Savage's presentation at later stages, as the system is being developed in the book. Thus, there are *events*, which are sets of states that form, under the usual set-theoretic operations, a Boolean algebra, $\mathcal{B}$, in which S is the universal set. And there is the notion of *conditional preference*, that is: $f \succcurlyeq g$ *given* E where E is an event, which is defined using P2 (the sure-thing postulate) and which is supposed to express what the agent prefers under the assumption that $s \in E$. Furthermore, for any $f, g \in \mathcal{A}$, the *combination* of f and g with respect to an event E, in symbols $f|E + g|\overline{E}$, is defined as: $f(s)$ if $s \in E$, $g(s)$ if $s \in \overline{E}$, where $\overline{E} = S - E$ is the compliment of E with respect to S.[3] We sometimes refer to this operation as "cut-and-paste". This notation can be easily generalized to define combinations of n many acts: $f_1|P_1 + \cdots + f_n|P_n$ is the act h such that $h(s) = f_i(s)$ for $s \in P_i$ ($i = 1, \ldots, n$), and this is used under the assumption that $P_1, \ldots, P_n$ is a partition of the set of all states.

1.1   The Problem of the Constant-act Assumption

One crucial element of the system is the notion of *constant acts* or, in Savage's phrasing, "acts that are constant" (p.25). The idea is that a constant act has the same consequence in all states. To be precise, being a constant act is not a

---

[2]   Savage's "simple ordering" is, in our terminology, a total preorder. He uses 'F' for the set of consequences and he characterizes total preorders as "simple orderings". In particular, he uses boldface letters **f**, **g**, ... for acts and italics f, g, ... for values of "acts that are constant", writing $\mathbf{f} \equiv g$ when $\mathbf{f}(s) = g$ for all states s. He also uses 'f' for constant act whose value is f. Furthermore, he sometimes switches to italicized notation even when the function is not constant, as he does in the statement of P4 on p.31, where he writes $f_A(s)$ instead of $\mathbf{f}_A(s)$, or in Theorem 1 on page 70, where he writes $f(s) = f_i$ instead of $\mathbf{f}(s) = f_i$ as he should.

[3]   Some writers use '$f \oplus_E g$' or 'fEg' or '$[f$ on E, g on $\overline{E}]$' for combined acts.

property of a single act, but is subject to an axiom that applies to a bunch of acts: the preference between two constant acts, given some event, does not depend on the event. The fifth postulate (P5) posits the existence of two non-equivalent constant acts.

Savage's representation theorem claims that a preference relation that satisfies the postulates determines a unique (finitely additive) probability on $\mathcal{B}$ and a utility function (unique up to a linear transformation) which assigns numeric utilities to consequences, such that $f \succcurlyeq g$ iff the expected utility of f is greater or equal to that of g. The derivation of a probability and a utility is carried out in two stages. In the first stage a finitely additive probability is derived from a preference relation, which satisfies the postulates P1–P6. As far as constant acts are concerned, this derivation does not require more than P5 (the existence of two non-equivalent constant acts is sufficient). But in the second stage—the derivation of a utility in chapter 5—Savage tacitly assumes the following:

CAA (Constant-acts Assumption). For every consequence $a \in C$ there exists a constant act $\mathfrak{c}_a$, such that $\mathfrak{c}_a(s) = a$, for all $s \in S$.

Note that after introducing "acts that are constant" Savage hardly uses the term anymore and one has to infer that such and such acts are constant only from the notation, which is not always consistent (see Footnote 2). Fishburn ([2]) who observed that CAA is required for the proof of the representation theorem, has also pointed out the problematic nature of CAA (cf. Footnote 4 below). Among others who have also emphasized the need for CAA in Savage's system are [3,6,7]. This assumption, we shall argue, does not sit well with certain simple scenarios of decision making, which Savage considers as the kind of situations that his system is supposed to handle.

The difficulty is the fact that the very possibility of some consequence may depend on the world being in a certain state: *the consequence could not exist in a different state of the world.* At the beginning of his book ([5, p.14]) Savage proposes the following omelet-making problem to illustrate the way his system works. The agent, call him John (in the book it is 'you'), has to finish making an omelet, which was begun by his wife. She broke into a bowl five good eggs and John finds a sixth egg, which can be added to the bowl or thrown away (we assume that there is no option of keeping it for future use). John does not know if the egg is good or rotten and has to decide between three acts: (1) Break it into the bowl (2) break it into a saucer to see if it is good or rotten (3) throw it away. There are two possible states of the world *good* and *rotten*, which are determined by the state of the sixth egg. The consequences of each act are given in Table 1, as it appears in the book.

John's ranking of the acts (that is, his preference relation, $\succcurlyeq$) reflects both his probabilistic estimates regarding the likeliness of each state, as well as the utility values of the consequences; for example, if he is sufficiently confident that the egg is good and if washing the saucer is, for him, of considerable nuisance, he will prefer "break into bowl" to "break into saucer". His preferences for these three acts cannot, of course, determine the probabilities and utilities, but if the set of acts over which the preference relation is defined is sufficiently rich (where

Table 1. Savage omelet example.

Act	State	
	Good	Rotten
break into bowl	six-egg omelet	no omelet and all five eggs destroyed
break into saucer	six-egg omelet and a saucer to wash	five-egg omelet and a saucer to wash
throw away	five-egg omelet and one good egg wasted	five-egg omelet

"sufficiently rich" is determined by the postulates), then we get probabilities and utilities. Obviously the consequence "six-egg omelet" means an omelet made of the six eggs of the story, in the case where the sixth egg is good. Yet CAA requires that there should be a constant act that yields that consequence also in the state in which the sixth egg is rotten. It would involve a miraculous production of a good six-egg omelet out of five good eggs and a rotten one.[4]

The problem arises also in the second scenario, which Savage proposes for the very purpose of clarifying what is implied by a constant act (*ibid.* p.25). A person, call her Jane, plans to go with friends on a picnic, and she has to choose between buying a tennis racquet and buying a bathing suit (assume that buying both is ruled out for financial reasons). The bathing suit would be handier if the picnic is held near water where one can swim; the racquet would be better, if the picnic is not held near water but near a tennis court. One might consider the possession of a bathing suit and the possession of a tennis racquet as constant, state-independent consequences. But Savage makes it clear that this would not do, since the preference order of possessing a racquet and possessing a bathing suit depends on the state of the world, where the state of the world includes the picnic-location. Savage argues that the payoffs should be entities such as: "a refreshing swim with friends, or sitting on a shadeless beach twiddling a brand-new tennis racquet while one's friends swim". That, however, does not make the constant-acts problem easier. To get a constant act, we have to appeal to the theoretical possibility that while Jane sits on a shadeless beach twiddling a brand new tennis racket, she has somehow the enjoyment of a refreshing swim with her friends.

Perhaps the constant-acts problem is not so difficult if we consider getting sums of money, or some other quantitative goods, as being of equivalent value to

---

[4] In passing, Fishburn ([2, p.166-7]) also voiced this unsatisfactory feature of CAA. He pointed out that, for any states $s, s' \in S$, if $W(s)$ and $W(s')$ are respectively the sets of consequences that may occur under s and s', then it might well be that $W(s) \neq W(s')$ (or even that $W(s) \cap W(s') = \emptyset$), in which case the CAA fails. He remarked that he is not aware of any axiomatic system that does not make the assumption that $W(s) = W(s') = C$ for all $s, s' \in S$, and he left this line of research as an open question (see also [1, p.162]).

the consequences in question. In the omelet scenario, John may consider getting $k as being equivalent to a six-egg omelet and this can serve also as a payoff in the state "rotten". But it is not clear what the equivalence of $k with a six-egg omelet means in the given context where John has to finish making the omelet. We may consider replacing Table 1 by the following table, in which the entries are dollar amounts; this would turn the problem into a problem of choosing between gambles. (Obviously, k is assumed to be the largest payoff, l is the

Act	State	
	Good	Rotten
Gamble 1	$k	$l
Gamble 2	$m	$n
Gamble 3	$p	$q

smallest, $m > n$ and $q > n$.) And we may consider offering John the choice of not completing the task – throwing out all eggs – and getting in return to choose a gamble from the table above. But this artificial dubious device undermines the big attraction of Savage's system: its ability to evaluate consequences that do not consist in winning or loosing sums of money or goods. If all consequences are to be replaced by dollar sums before the system is applied, the main point of the system is lost.

One objective of this paper is to show that CAA is not required for applying Savage's system to any finitistic problem, that is to say, *a problem that is stated in terms of finitely many evants, finitely many acts and finitely many possible consequences.* All that we need is the existence of two distinguished constant acts.

1.2   The Significance of the Set of Acts and the Boolean Algebra

The weaker the postulates and the presuppositions which are needed to get the representation theorem, the stronger the theorem is. The basic presupposition of Savages system is that the preference relation is defined over some very rich set of acts. In some places Savage even considers every function from states to consequences to be an act, in situations in which the set of states, as well as the set of consequences, has the cardinality of the continuum. This is exorbitant. Of course the set of acts should be sufficient for handling the kind of problems that the system is designed for. As a rule, these problems are stated in terms of finitely many simple acts, where a simple act is an act, f, which has finitely many values, such that, $f^{-1}(x)$ is an event (a member of the Boolean algebra $\mathcal{B}$) for each consequence x that is a value of f. Such acts are called by Savage gambles. It is easily seen that a simple act, f, can be written in the form $f = f|P_1 + \ldots + f|P_n$, where $P_1, \ldots, P_n$ is a partition of S, $P_i = f^{-1}(x_i)$ and the x_i are consequences.

In the initial scenario the agent is supposed to decide between given options that belong to some finite set of simple acts. P6 implies however that the preference is to be defined over richer sets that involve more refined events (cf. Theorem 2.3 below). But, as we shall show, we never need more than simple acts. (In Section 3, we comment on how our model can be generalized to treat certain infinitary cases.)

Now the richness of the set of acts is also determined by the richness of the Boolean algebra $\mathcal{B}$ of events, namely the collection of subsets that constitute events. As noted, Savage considers possibilities in which this Boolean algebra consists of all subsets of real numbers. But his proof of the representation theorem requires only that it be a σ-algebra, that is, closed under unions of countable many sets. Our results can be now stated as follows:

i. While we assume that the Boolean algebra is a σ-algebra, we can derive the representation theorem if we consider only a preference defined over simple acts, which include two non-equivalent constant ones.
ii. Moreover, we can also give up the assumption that the algebra is a σ-algebra and get the representation theorem, nonetheless. In fact, we need only a countable Boolean algebra so that the simple acts defined over it satisfy P6.

(i) is proved by using Savage's derivation of probabilities from two constant acts. We deviate from him in the derivation of expected utilities for simple acts (where the set of consequences is arbitrary). In the next section, we lay out the basic ideas behind our construction, the full technical details will be left to the full paper. (ii) is a more difficult result that is based on a more difficult derivation of probabilities. We do not have the space for getting into it here.

2   Context-Dependent Decision-Making

2.1   Subjective Probability

To derive subjective probability from preferences, Savage uses P1-P6. The construction starts with a derivation of qualitative probabilities.

Definition 2.1 *For any events E, F, say that E is weakly more probable than F, written $E \succeq F$, if, for any constant acts $\mathfrak{c}_a$ and $\mathfrak{c}_b$ such that $\mathfrak{c}_a \succ \mathfrak{c}_b$,*

$$\mathfrak{c}_a | E + \mathfrak{c}_b | \overline{E} \succeq \mathfrak{c}_a | F + \mathfrak{c}_b | \overline{F}. \tag{2.1}$$

Savage's P4 guarantees that (2.1) does not depend on the choice of the pair of constant acts. It is also not difficult to show that $\succeq$ is a qualitative probability. The task is to show that this qualitative probability admits a numerical *representation*: there exists a real-valued probability measure μ defined on an algebra of events satisfying:

$$E \succeq F \iff \mu(E) \geq \mu(F). \tag{2.2}$$

Savage's proof of the existence of a quantitative probability that satisfies (2.2) requires the assumption that the algebra of events is closed under countable unions, i.e., it is a σ-algebra. (That one can do without this assumption is, as noted above, the content of our second result.) So far only two non-equivalent constant acts are required.[5]

Theorem 2.2 (Savage) *Let $\succeq$ be a preference relation among acts. Suppose that $\succeq$ satisfies P1-6 and that the Boolean algebra $\mathcal{B}$ of events is a σ-algebra, then there exists a unique (finitely additive) probability measure μ for which (2.2) holds.*

The proof of the theorem establishes also the following theorem, which holds under the assumption that the algebra of events is a σ-algebra.

Theorem 2.3 *Given the probability measure μ obtained above, for any event E and any $0 \le \rho \le 1$, there exists some $F \subseteq E$ such that $\mu(F) = \rho\mu(E)$.*

Note that, unlike Theorem 2.2, Theorem 2.3 fails if the assumption that the Boolean algebra is a σ-algebra is omitted. A weaker version of it holds: The set of all ρ for which the equality holds is dense in $(0, 1)$.

2.2   Utility for All Acts

The following are some simple properties of the two distinguished constant acts, which are immediate from the definitions above and Theorem 2.2.

Lemma 2.4 *For any events E, F,*

1. *$\mu(E) > \mu(F)$ iff $c_1|E + c_0|\overline{E} \succ c_1|F + c_0|\overline{F}$,*
2. *$\mu(E) = \mu(F)$ iff $c_1|E + c_0|\overline{E} \equiv c_1|F + c_0|\overline{F}$.*

We show that, under P1-6 and the assumption that there exist two constant acts c_0 and c_1, the agent's preferences can be represented by a utility function in Savage's system without appealing to CAA. To this end, we first observe that to each act $f \in \mathcal{A}$ satisfying $c_1 \succeq f \succeq c_0$ there corresponds a combined act using the two distinguished constant acts which is indifferent to f under $\succeq$.

Lemma 2.5 *For and $f \in \mathcal{A}$, if $c_1 \succeq f \succeq c_0$, there exists an event E_f such that*

$$c_1|E_f + c_0|\overline{E_f} \equiv f. \tag{2.3}$$

In proving this lemma, we make full use of the derived personal probability μ from Theorem 2.2, the proof given here is somewhat standard in utility theory. Figure 1 provides an illustration of the general method involved in the proof, where $c_1|E_f + c_0|\overline{E_f}$ is the act that yields c_1 if E_f occurs, status quo otherwise. The aim is to find the appropriate E_f so that the given event f is indifferent to this combined act.

---

[5] This observation is also noted in [1, p.161] where the author remarked that "[as far as obtaining a unique probability measure is concerned] Savage's $\mathscr{C}$ [i.e., the set of consequences] can contain as few as two consequences." See [2, §14.1-3] for a clean exposition of Savage's proof of (2.2), and see especially §14.3 for an illustration of the role of P1-6 played in deriving numerical probability.

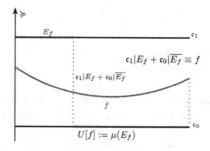

Fig. 1. The case where $c_1 \succcurlyeq f \succcurlyeq c_0$

Proof of Lemma 2.5. Let us consider the following two sets of events.

$$B := \left\{ E \mid c_1 | E + c_0 | \overline{E} \succcurlyeq f \right\};$$
$$C := \left\{ E \mid c_1 | E + c_0 | \overline{E} \preccurlyeq f \right\}. \tag{2.4}$$

It is easily seen that B and C are nonempty, for at least we have $S \in B$ and $\emptyset \in C$. Let μ be the probability measure derived from Theorem 2.2, Next, consider the following sets defined in terms of B, C and μ:

$$B_\mu := \left\{ \mu(E) \mid E \in B \right\};$$
$$C_\mu := \left\{ \mu(E) \mid E \in C \right\}. \tag{2.5}$$

Let $\alpha_* = \inf B_\mu$ and $\alpha^* = \sup C_\mu$. Note that, for any $a > \alpha_*$, there must exist some $a' \in B_\mu$ such that $a > a' \geq \alpha_*$ (for, otherwise, a is a lower bound of B_μ strictly greater than α_*, which contradicts the assumption $\alpha_* = \inf B_\mu$). Since $a' \in B_\mu$ then, by the definition of B_μ in (2.5), there is some event, say, $F' \in B$ such that $\mu(F') = a'$. Further, let F be an event such that $\mu(F) = a$ (the existence of F is guaranteed by Theorem 2.3). Then, by Lemma 2.4, $\mu(F) = a > \mu(F') = a' \geq \alpha_*$ implies $c_1 | F + c_0 | \overline{F} \succ c_1 | F' + c_0 | \overline{F'} \succcurlyeq f$. It follows, via P1, that, for any F,

$$\mu(F) > \alpha_* \implies F \notin C. \tag{2.6}$$

The contrapositive of (2.6) says that, for any F, $F \in C$ implies that $\mu(F) \leq \alpha_*$. In other words, α_* is an upper bound of C_μ, and hence $\alpha^* = \sup C_\mu \leq \alpha_*$. Using a symmetric argument one can show that $\alpha^* \geq \alpha_*$. Hence $\alpha^* = \alpha_*$.

Next, let E_f be such that $\mu(E_f) = \alpha^* = \alpha_*$ (again, the existence of E_f is guaranteed by Theorem 2.3). The proof is completed if we can show that $E_f \in B \cap C$. Suppose, to the contrary, $E_f \notin B$, then, by P1, $f \succ c_1 | E_f + c_0 | \overline{E_f}$. The latter implies, via P6, there exists a partition $\{P_i\}_{i=1}^n$ such that,

$$c_1 \Big| P_i + \left(c_1 | E_f + c_0 | \overline{E_f} \right) \Big| \overline{P_i} \quad \text{for all } i = 1, \dots, n, \tag{2.7}$$

that is,

$$f \succ c_1 \Big| E_f \cup P_i + c_0 \Big| \overline{E_f \cup P_i} \quad \text{for all } i = 1, \dots, n. \tag{2.8}$$

Then, it follows that $E_f \cup P_i \in C$ for *all* $i = 1, \ldots, n$. On the other hand, note that P_i's form a partition of S, we consider two cases:

(1) If for *some* P_j in the partition we have $\mu(E_f \cup P_j) > \mu(E_f) = \alpha_*$, then, by (2.6), $E_f \cup P_j \notin C$, a contradiction.
(2) If $\mu(E_f \cup P_j) \le \mu(E_f) = \alpha_*$ for *all* $j = 1, \ldots, n$, then it is easily seen that $\mu(E_f) = 1$. By Lemma 2.4(2), it follows that $c_1 | E_f + c_0 | \overline{E_f} \equiv c_1 | S + c_0 | \overline{S} \equiv c_1$, and hence $E_f \in B$, but this contradicts the hypothesis $E_f \notin B$.

Hence, E_f must be in B. Similarly, it can be shown that $E_f \in C$. Then we have $E_f \in B \cap C$. This completes the proof of the lemma.   □

Remark 1. 1. In light of the lemma, for any $f \in \mathcal{A}$ satisfying $c_1 \succcurlyeq f \succcurlyeq c_0$, let E_f be such that (2.3) holds, we define the *utility* of f to be

$$U[f] := \mu(E_f), \tag{2.9}$$

where μ is obtained through Theorem 2.2 and E_f is from (2.3).
2. Notice that, if there exists another event E_f' for which (2.3) holds, then we have $c_1 | E_f + c_0 | \overline{E_f} \equiv c_1 | E_f' + c_0 | \overline{E_f'}$. It follows, via Lemma 2.4(2), that $\mu(E_f') = \mu(E_f)$, hence $U[f]$ is well defined.
3. For the two distinguished constant acts c_1 and c_0, trivially we have $E_{c_1} = S$ and $E_{c_0} = \emptyset$, then (2.9) yields that $U[c_1] = 1$ and $U[c_0] = 0$.
4. It is plain that U does not need to be uniquely defined by (2.9): if h is any monotonically increasing function on the reals (or any order preserving function), then U can also be defined by $h \circ \mu$.
5. If $f \succ c_1$ (or $c_0 \succ f$), it is easy to see that Lemma 2.5 can be adjusted to show that there exists some E_f such that $f | E_f + c_0 | \overline{E_f} \equiv c_1$ (or $c_1 | E_f + f | \overline{E_f} \equiv c_0$), in which case U can be defined standardly as in (2.11) below.

Theorem 2.6 *Let $\succcurlyeq$ be a preference relation over acts, if $\succcurlyeq$ satisfies P1-6, then there exists a real-valued function U on $\mathcal{A}$ satisfying, for all $f, g \in \mathcal{A}$,*

$$f \succcurlyeq g \iff U[f] \ge U[g], \tag{2.10}$$

where

$$U[f] := \begin{cases} \frac{1}{\mu(E_f)} & \text{if } f \succ c_1, \\ \mu(E_f) & \text{if } c_1 \succcurlyeq f \succcurlyeq c_0, \\ \frac{\mu(E_f)}{\mu(E_f) - 1} & \text{if } c_0 \succ f. \end{cases} \tag{2.11}$$

2.3   Context-Dependent Expected Utility for Simple Acts

We now proceed to show that, assuming P1-6, the utility of a simple act can be further expressed as its expected utility of its consequences. Let us denote the set of all simple acts by $\mathcal{A}_0$. Recall that a simple act $f \in \mathcal{A}_0$ is one that has a finite number of consequences, say, $x_1, \ldots, x_n$, and let $P_1, \ldots, P_n$ be the corresponding sets of states under which they obtain. It is easily seen that $\{P_i\}_{i=1}^{n}$ forms a partition of S:

$$P_i = f^{-1}(x_i) \quad (i = 1, \ldots, n),$$

$$P_i \cap P_j = \emptyset \quad (i \neq j) \quad \text{and} \quad \bigcup_{i=1}^{n} P_i = S. \tag{2.12}$$

We seek to define a *context-dependent* utility function u over consequences such that the utility of a simple act $U[f]$ can be represented by its expected utility:

$$U[f] = \sum_{i=1}^{n} \mu(P_i)u(P_i, x_i), \tag{2.13}$$

where $u(P_i, x_i)$ is the utility of consequences x_i *given* P_i. As it will be shortly shown, in all cases in which $\mu(P_i) > 0$ this value depends only on the consequence x_i. And this value is the same across different acts. We thus can speak of context-dependent utilities. We can assign utilities to consequences, but these utilities can be used for the purpose of calculating expected utilities as long as the consequence is obtained as a value of states that constitute a set of probability greater than 0.

We adopt the following notation:

$$\mathfrak{c}_x^*(s) := \begin{cases} x & \text{if } s \in E, \\ 0 & \text{if } s \notin E, \end{cases} \quad \text{for some } E \in \mathcal{B}. \tag{2.14}$$

We refer to $\mathfrak{c}_x^*$ as a *locally constant act* which yields x in all states in E, 0 (status quo) otherwise. It is obvious that $\mathfrak{c}_x^*$ is a generalization of Savage's notion of constant act. Now with (2.14), a simple act f satisfying (2.12) can be expressed by the combination of a series of locally constant acts as follows

$$f = \mathfrak{c}_{x_1}^*|P_1 + \cdots + \mathfrak{c}_{x_n}^*|P_n. \tag{2.15}$$

The goal is to represent simple acts in the form of (2.15) by expected utilities.[6] Observe that, if $\mu(P_i) = 0$ for some P_i, then the term $\mu(P_i)u(P_i, x_i)$ in (2.13) is 0, in which case consequence x_i can be seen as having no contribution to the total utility calculation. As a rule, one can assign in this situation an arbitrary

---

[6]   Savage ([5, p.71]) uses $\sum_i \rho_i f_i$ to denote the class of simple acts for which, to use his notations, there exist partitions B_i of S such that $P(B_i) = \rho_i$ and $f(s) = f_i$ for $s \in B_i$. He further remarks that if a simple act $\mathbf{f}$ is such that "the consequences f_i will befall the person in case B_i occurs, then the value of $\mathbf{f}$ is independent of how the partition B_i is chosen." In other words, his utility function, once derived, is *state-independent*. We, on the other hand, take that the value of a consequence depends on the states under which it obtains. Thus, we allow that for two simple acts f, g with different partitions $\{P_i\}_{i=1}^{n}$ and $\{Q_i\}_{i=1}^{n}$ for which $\mu(P_i) = \mu(Q_i)$ and $f(s) = g(t)$ for $s \in P_i$ and $t \in Q_i$ $(i = 1, \ldots, n)$, $f \not\equiv g$. That is, we allow Theorem 1 ([5, p.70]) to fail in our decision model where utilities are *context-dependent*.

finite value to the consequence $f(s)$ where $s \in P_i$. If, on the other hand, $\mu(P_i) \neq 0$, consider act $c_{x_i}^*|P_i + c_0|\overline{P_i}$. Then in light of Theorem 2.6, define a *context-dependent* utility of x_i in P_i in terms of the utility of $c_{x_i}^*|P_i + c_0|\overline{P_i}$ as follows

$$u(P_i, x_i) := \begin{cases} c & \text{if } \mu(P_i) = 0, \\ \dfrac{U\left[c_{x_i}^*\big|P_i + c_0\big|\overline{P_i}\right]}{\mu(P_i)} & \text{if } \mu(P_i) \neq 0, \end{cases} \qquad (2.16)$$

where c can be any number in $[0, 1]$. Finally, it remains to verify that $\succcurlyeq$ among simple acts indeed admits an expected utility representation using the probability measure μ and utility function u given above. We put this claim in the form of the following theorem. The rather straightforward proof is omitted.

Theorem 2.7 *Let $\succcurlyeq$ be a preference relation over acts, if $\succcurlyeq$ satisfies P1-6, then there exist a probability measure μ on events and a utility function u on the consequences such that, for any $f, g, \in \mathcal{A}_0$,*

$$f \succcurlyeq g \iff \sum_{x \in f(S)} \mu[f(s) = x] u(f^{-1}(x), x) \geq \sum_{x \in g(S)} \mu[g(s) = x] u(g^{-1}(x), x).$$

3   Infinitary Cases

Our method can be generalized to treat certain infinitary case. There are acts, f, in which there are countably many consequences, say $x_1, x_2, \ldots, x_n, \ldots$ such that $f^{-1}(x_n)$ is a non-null set for every n. In other words, we allow the number of cells of the partition in (2.12) to be unbounded. Then (2.16) and Theorem 2.7 also apply to this case, where the expected utility of f can be defined by

$$\sum_{i=1}^{\infty} \mu[f(s) = x_i] u(f^{-1}(x_i), x_i) \qquad (3.1)$$

provided that $\sum_{i=1}^{\infty} \mu[f(s) = x_i] \cdot |u(f^{-1}(x_i), x_i)|$ converges. It is defined as the sum of the positive values minus the sum of the negative ones. Note that μ does not need to be countably additive. The expectation in that case is defined for discrete random variables, for which the sum absolutely converges.

Finally, we point out that Savage needed the CAA because he wanted to extend the expectation to continuous random variables, that is, he wanted to define the integral:

$$\int X(s) \, d\mu(s) \qquad (3.2)$$

where X is a measurable function, which is interpreted in his system as a general act with potentially uncountably many consequences, and μ is a finitely additive probability. Mathematically this is interesting. But we do not think that it is required for applying his system to decision scenarios which a rational human agent is expected to face.

Acknowledgments. Thanks are due to three anonymous reviewers for helpful comments.

References

1. Fishburn, P.: Subjective expected utility: A review of normative theories. Theory and Decision 13(2), 139–199 (1981)
2. Fishburn, P.C.: Utility Theory for Decision Making. Wiley, New York (1970)
3. Pratt, J.W.: Some comments on some axioms for decision making under uncertainty. In: Balch, M., McFadden, D., Wu, S. (eds.) Essays on Economic Behavior Under Uncertainty, pp. 82–92. North-Holland Pub. Co., American Elsevier Pub. Co., Amsterdam, New York (1974)
4. Savage, L.J.: The Foundations of Statistics. John Wiley & Sons, Inc. (1954)
5. Savage, L.J.: The Foundations of Statistics. Second revised edn. Dover Publications, Inc. (1972)
6. Seidenfeld, T., Schervish, M.: A conflict between finite additivity and avoiding dutch book. Philosophy of Science, 398–412 (1983)
7. Shafer, G.: Savage revisited. Statistical Science 1(4), 463–485 (1986)

Graph-Based Belief Merging

Konstantinos Georgatos*

Department of Mathematics and Computer Science, John Jay College
City University of New York, 524 West 59th Street
New York, New York 10019, USA
kgeorgatos@jjay.cuny.edu

Abstract. Graphs are employed to define a variety of distance-based binary merging operators. We provide logical characterization results for each class of merging operators introduced and discuss the extension of this approach to the merging of sequences and multisets.

1   Introduction

Belief Merging ([1,2]), also called arbitration ([3,4]), belief fusion ([5]), and non-prioritized belief revision ([6]), is the process of combining two or more, possibly inconsistent, propositions into a single consistent proposition. The purpose of such a combination is to model, depending on the application, the process of agreement between possibly disagreeing parties, of group decision among multiple possibly conflicting courses of action, or, simply, of making sense out of a number of possibly conflicting sources of information.

Consider the following example: let Analyst1 and Analyst2 be two stock market analysts, and Stock1 and Stock2 be the stocks of two companies. Let p the proposition "Stock1's price will rise," and q the proposition "Stock2's price will rise." Now suppose Analyst1 believes that $A = p \wedge q$ and Analyst2 believes that $B = \neg p \wedge q$. That is, Analyst1 believes that Stock1 and Stock2 will rise while Analyst2 believes that Stock1 will fall and Stock2 will rise. If one had to merge the opinions of Analyst1 and Analyst2, then most likely one would keep the belief that Stock2 will rise and believe nothing about Stock1, where the two analysts disagree. If we denote the operation of merge with $\otimes$ then $A \otimes B = q$
. Observe that the conjunction $A \wedge B$ cannot model agreement because it is a contradiction. Similarly, if instead Analyst1 believes that $A = p \vee q$ and Analyst2 believes that $B = \neg p \vee q$, then disjunction $A \vee B$ cannot model agreement either because it is a tautology.

The problem of merging has been extensively studied in different fields. For example, in distributed computing and theory of networks ([7]), multiple agents need to negotiate a common agreement or adopt a common view. In databases ([8]), we need to merge databases that may have inconsistent information. In social settings, voting and or belief aggregation schemes are nothing but

---

* Support for this project was provided by a PSC-CUNY Award, jointly funded by The Professional Staff Congress and The City University of New York.

© Springer-Verlag Berlin Heidelberg 2015
W. van der Hoek et al. (Eds.): LORI 2015, LNCS 9394, pp. 102–115, 2015.
DOI: 10.1007/978-3-662-48561-3_9

algorithms that merge different views or preferences (votes) into a single one (elected outcome).

There have been several approaches to belief merging. We have model based operators ([3]), syntax sensitive operators ([2]), and default based operators ([9]) among others (for a survey see [10]) as well as merging operators at the higher level of preference relations ([11]). Further, there have been significant steps towards an understanding of the theoretical underpinnings of merging. A form of binary merging called arbitration has been logically characterized with postulates similar to AGM postulates in [4]. Arbitration based on distance has been characterized in [12]. Other representation results have been obtained for multiset merging ([13]) and multiset merging with integrity constraints ([14]). However, in the binary case, merging operator systems do not correspond uniquely to the distance spaces from which they were generated (as shown in [12]). In the multiset case, a characterization result with respect to distance spaces is missing entirely ([10]). The purpose of this paper is to introduce and characterize three classes of distance-based binary merging operators that are modeled by graphs.

We will proceed as follows: in the next section we will introduce the basic idea of geodesic reasoning, how it applies to merging, and present the first formal definitions. In Section 3, we define three binary merging operators through sets of logical postulates and show that these are characterized by corresponding graph theoretic operators. In Section 4, we explore the ways how those binary operators can be generalized to sequences and multisets of propositions thus modeling belief merging for several agents. In Section 5, we place the present framework among previous significant results in the area and conclude.

2   Geodesic Reasoning

We will define and characterize three merging operators that can be defined on graphs using the metric defined on graphs, called *geodesic*. The use of geodesic metric rests on a novel view of similarity as a derived concept. Traditionally, similarity has been conceived as a primitive concept usually represented by distance; that is, the following identification is made:

$$\text{similarity} = \text{distance}$$

Our idea ([15]) is that similarity is not primitive but it can be generated by a relation of indistinguishability. This idea can be summarized by the following maxim: two objects are similar when there is a context within which they are indistinguishable. Therefore, similarity can be *measured* with degrees of indistinguishability.

For example, although two similar houses might appear different in various details when we stand in front of them, they will appear identical if we observe them from a larger distance x. Thus, similarity implies indistinguishability at a certain distance x. The smaller the distance, the more similar the objects are.

A simple representation of indistinguishability by a reflexive symmetric non-transitive relation goes back to [16]. Such relations have been studied together with a set under various names such as tolerance spaces ([17]), proximity spaces, and others, but the best way to describe a set of worlds with an indistinguishability relation is simply a graph. Similarity now will be the distance map defined on the graph defined by the shortest path. Given a relation R the distance from y to x is the least number of times we need to apply R in order to reach y from x. Traditionally, this kind of relation has been called *geodesic*. We have

$$\text{similarity} = \text{geodesic distance (of a graph)}$$

Using graphs with their geodesic metric generalizes several popular formalisms such as threshold and integer metrics as well as, hamming distance (see [18]).

When we perform merging of two beliefs we choose the models of the two beliefs that are the most similar and therefore the closest with respect to the geodesic distance.

Example 1. We illustrate the process with the following example (edges represent the reflexive symmetric tolerance relation).

Fig. 1. Non-commutative revision

In Figure 1, let $A = \{a, b\}$ and $B = \{d, e\}$. Then the merging, denoted by $A \otimes B$, of A with B equals the subset $\{b, d\}$ containing the elements of A and B whose distance is the least among the elements of the two sets: the distance of b from d is 2 while the distance of a from d and e from b is 3. This form of merging corresponds to arbitration of [4], and is a special case of the distance-based merging operator of [12].

The merging operator defined has an important property, namely, it implies disjunction:

$$\phi \otimes \psi \vdash \phi \vee \psi.$$

There are are cases, however, where this is not possible or not desirable.

Example 2. Suppose that we count the pennies saved in a jar. An initial count finds 112 pennies. A second count finds 114 pennies. It seems plausible that the merge of these two counts is the set $\{112, 113, 114\}$ as one or both counts could have been wrong. Using the propositions of the previous example, we would like that the extension of $A \otimes B$ is the set $\{b, c, d\}$ (see Figure 2). We will call this form of merging *convex merging*.

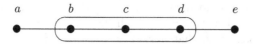

Fig. 2. Convex merging

Example 3. Suppose now that we need to classify submitted articles into three groups *accept, reject* and *borderline*. For a given paper, we receive two reviews from the referees. One thinks it belongs to the accept group and the other to the reject group. It seems to me that the merge of those two opinions is borderline. Obviously this notion of merging seems more appropriate when beliefs have different sources such as the case of voting. Using the above example, we would like that $A \otimes B$ is modeled by the set $\{c\}$ (see Figure 3). We will call this form of merging *barycentric merging*.

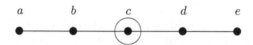

Fig. 3. Barycentric merging

Although the above examples describe binary merging operators, we believe that similar ideas can be applied to multiset merging. Geodesic semantics have been successfully developed for a variety of belief change operators such as revision, update, conditionalization, and contraction ([18,19,20]) and this paper is an effort towards extending geodesic semantics to belief merging.

2.1   Tolerance Spaces and Their Geodesic

We will use a reflexive and symmetric relation to model indistinguishability. A set equipped with such a relation is frequently called a tolerance space. In addition, we will assume that the space is connected:

Definition 1. *Let X be a set and $R \subseteq X \times X$ a relation on X. Then (X, R) is called a* (connected) tolerance space *when R is reflexive, symmetric, and $(X$ is) connected, i.e., for all $x, y \in X$ there is a non negative integer n such that $xR^n y$.*

In the above definition, we assume $R^0 = \mathrm{id}_X$, $R^n = R^{n-1} \circ R$ for $n > 0$.

Given a tolerance space (X, R) we can define a metric called *geodesic* with a map d from $X \times X$ to Z^+ (the set of non-negative integers) where

$$d(x, y) = \min\{n \mid xR^n y\}.$$

Note that a geodesic metric is not any integer metric. The values of the geodesic metric are determined by adjacency. The results of this paper depend heavily on this property which can be described with: for all $x, y \in X$ such that $d(x, y) = n$ with $1 < n < \infty$ there is $z \in V$ with $z \neq x, y$ such that $d(x, y) = d(x, z) + d(z, y)$. In particular we can choose z so that $d(x, z) = 1$. Note here that a geodesic metric is a topological metric, that is, it satisfies identity, symmetry and triangle inequality.

The geodesic distance extends to distance between non-empty subsets with

$$d(A, B) = \min\{d(x, y) \mid x \in A, y \in B\}. \tag{1}$$

We shall also write $d(x, A)$ for $d(\{x\}, A)$. Similarly for $d(A, x)$. We will write A^c for the complement of A and A^n for the set $\{x \in X : d(A, x) \leq n\}$ (where $n = 0, 1, \ldots$). The proof of the following is straightforward

Lemma 1. *If A and A^c are non-empty, we have $d(A, A^c) = 1$.*

3   Merging Based on a Geodesic

We will define and characterize three different notions of belief merging. We will use a propositional language $\mathcal{L}$ with a finite set of atomic propositions. We will also assume classic propositional calculus and write $\phi \vdash \psi$ if ϕ implies ψ and $\phi \equiv \psi$ if they are equivalent. An interpretation w is a function from atomic propositions to $\{T, F\}$. An interpretation extends to a map from $\mathcal{L}$ to $\{T, F\}$ and will be called a *model* of ϕ if it maps ϕ to T. We write M for the set of all models. If A is a set of formulas then we write $v(A)$ to denote the set of all models of A. If X is a set of models then ϕ_X denotes a formula whose set of models is X. We say ϕ is complete, if for any propositional formula ψ, ϕ implies ψ or ϕ implies $\neg\psi$. ϕ_w is complete for all w and if ϕ is complete then there is a model w such that $\phi_w \equiv \phi$ (because our language is finite).

3.1   Non-prioritized Revision Merging

This notion of merging picks the "closest" models of the propositions to be merged. It has been introduced first in [3] and its logical properties have been studied in [4]. Schlechta gave a characterization of this merging where closeness is defined by a general notion of distance between models ([12]) (see Example 1). In this section we will characterize this form of merging using the geodesic distance of a graph. Note that both Revesz and Schlechta's characterizations are based on distance notions more general than ours so their postulates are also valid in our framework. Our postulated refer explicitly neither to points (complete theories) nor to distance as in Schlechta's characterization. In other words we give a purely logical characterization of the underlying graph.

We will now characterize the revision merge of a geodesic space. To this end call a merging operator *revision geodesic* if it satisfies the rules of Table 1.

A few words about the rules appearing in Table 1: Rule 1 guarantees that merging returns a consistent formula if one of the formulas is consistent and Rule 2 merging with an inconsistent has no effect. Using Rule 3 if two formulas

Table 1. Geodesic merging rules

1. If ϕ is consistent then $\phi \wedge (\phi \otimes \psi)$ is consistent
2. If ϕ is inconsistent then $\phi \otimes \psi \equiv \psi$
3. If $\phi \wedge \psi$ is consistent, then $\phi \otimes \psi \equiv \phi \wedge \psi$
4. If $\phi_1 \equiv \phi_2$ then $\phi_1 \otimes \psi \equiv \phi_2 \otimes \psi$
5. $\phi \otimes \psi \equiv \psi \otimes \phi$
6. If $\phi \vdash \neg\psi$ then $(\psi \otimes \neg\psi) \wedge \phi \vdash (\phi \otimes \neg\phi) \wedge \phi$
7. If $\phi \vdash \neg\psi$ and $\phi \otimes \neg\phi \vdash \neg\psi$ then $\psi \otimes \neg\psi \vdash \neg\phi$
8. If $\phi \vdash \neg\psi$ then $(\phi \otimes \psi) \wedge \psi \equiv ((\phi \otimes \neg\phi) \otimes \psi) \wedge \psi$
9. $\phi \otimes \psi \vdash \phi \vee \psi$

are consistent together then their merging becomes their conjunction. Rule 4 postulates substitution of logically equivalent formulas and Rule 5 commutativity. Rule 6 is a form of monotonicity for the second argument. Rule 7 implies symmetry for the underlying relation of indistinguishability. Rule 8 is an induction axiom that allows us to define merging from less distant formulas. Finally, Rule 9 postulates arbitration. Rule 9 implies the following

$$\phi \otimes \psi \equiv ((\phi \otimes \psi) \wedge \phi) \vee ((\phi \otimes \psi) \wedge \psi). \tag{2}$$

We will characterize the class of geodesic merging operators using a merging operation on subsets based on geodesic distance, thus the use of the term "geodesic". The distance d is the geodesic distance of the tolerance space (M, R), where M is the set of models and the indistinguishability relation R is defined from the merging operator $\otimes$ with

$$(x, y) \in R \quad \text{if and only if} \quad \phi_y \vdash \phi_x \otimes \neg\phi_x. \tag{3}$$

The distance d between models lifts to a distance between subsets of models using (1). We can now define a merging operator on subsets with

$$A \otimes B = \begin{cases} \{x \in A, y \in B : d(x, y) = d(A, B)\} & \text{if } A, B \neq \emptyset \\ A \cup B & \text{otherwise.} \end{cases}$$

(We use the same symbol for the merging operators between formulas and subsets.)

Observe that the above definition is equivalent to the following:

$$A \otimes B = \begin{cases} (A^{d(A,B)} \cap B) \cup (B^{d(A,B)} \cap A) & \text{if } A, B \neq \emptyset \\ A \cup B & \text{otherwise.} \end{cases}$$

Now the following characterization theorem holds:

Theorem 1. *Let $\otimes$ be a geodesic merging operator. Then there exists a binary relation R (defined by (3)) such that (M, R) is a tolerance space, where M is the set of models, and the following holds*

$$v(\phi \otimes \psi) = v(\phi) \otimes v(\psi). \tag{4}$$

Conversely, if an operator satisfies (4) then it is a geodesic merging operator.

The above proposition shows that the set of rules of Table 1 characterizes the class of geodesic metrics.

Now the question that arises is in what sense this is a non-prioritized revision. The answer is that the definition of merge can be based on the definition of revision:

Definition 2. *A merge operator* $\otimes$ *will be called* non-prioritized revision operator *if there exists a revision operator* $*$ *such that*

$$\phi \otimes \psi \equiv (\phi * \psi) \vee (\psi * \phi).$$

It is not hard to show that a revision geodesic merge operator is a non-prioritized revision. It suffices to define the revision operator. Simply let (as in [14])

$$\phi * \psi = (\phi \otimes \psi) \wedge \psi \tag{5}$$

We now have the following.

Proposition 1. *Let* $*$ *be the operator defined as above then*

$$\phi \otimes \psi \equiv (\phi * \psi) \vee (\psi * \phi).$$

A characterization of the revision operator can be given in terms of distance. Let

$$A * B = \begin{cases} \{y \in B : d(A, y) = d(v(A), v(B))\} & \text{if } A, B \neq \emptyset \\ B & \text{otherwise} \end{cases}$$

or, equivalently,

$$A * B = \begin{cases} A^{d(A,B)} \cap B & \text{if } A, B \neq \emptyset \\ B & \text{otherwise.} \end{cases}$$

Corollary 1. *If* $*$ *is defined by (5) then*

$$v(\phi * \psi) = v(\phi) * v(\psi). \tag{6}$$

The operator $*$ is a revision operator because it is defined through distance minimization as in [21,18]. Nevertheless it is useful to know what properties exactly this revision operator satisfies. This question has been answered in [18]. Call a revision operator *geodesic* if it satisfies the properties of Table 2.

The following has been proved in [18].

Proposition 2. *If* $*$ *is an operator that satisfies (6), then* $*$ *is a geodesic revision operator. Conversely, given a geodesic revision operator* $*$*, then there exists a binary relation* R *such that* (M, R) *is a tolerance space, where* M *is the set of models, and* $*$ *satisfies (6).*

Table 2. Geodesic revision rules

1. $\phi * \psi \vdash \psi$
2. If ψ is consistent, then $\phi * \psi$ is consistent
3. If ϕ is inconsistent, then $\phi * \psi \equiv \psi$
4. If $\phi \wedge \psi$ is consistent, then $\phi * \psi \equiv \phi \wedge \psi$
5. If $\psi_1 \equiv \psi_2$ and $\phi_1 \equiv \phi_2$ then $\phi_1 * \psi_1 \equiv \phi_2 * \psi_2$
6. If $\psi \vdash \neg\phi$ then $\phi * \psi \equiv (\phi * \neg\phi) * \psi$
7. If $\psi \vdash \neg\phi$ then $\phi * \psi \equiv (\neg\phi * \phi) * \psi$
8. If $\phi * \psi \equiv \chi * \psi$ then $\phi * \psi \equiv (\phi \vee \chi) * \psi$
9. If $\psi \vdash \neg\phi$ then $\phi * \psi \vdash \neg\psi * \psi$
10. If $\phi \vdash \neg\psi$ then $\phi * \neg\phi \vdash \neg\psi$ iff $\psi * \neg\psi \vdash \neg\phi$

Table 3. Convex merging rules

1. If ϕ is consistent then $\phi \wedge (\phi \otimes_c \psi)$ is consistent
2. If ϕ is inconsistent then $\phi \otimes_c \psi \equiv \psi$
3. If $\phi \wedge \psi$ is consistent, then $\phi \otimes_c \psi \equiv \phi \wedge \psi$
4. If $\phi_1 \equiv \phi_2$ then $\phi_1 \otimes_c \psi \equiv \phi_2 \otimes_c \psi$
5. $\phi \otimes_c \psi \equiv \psi \otimes_c \phi$
6. If $\phi \vdash \neg\psi$ then $\psi \otimes_c \neg\psi \wedge \phi \vdash \phi \otimes_c \neg\phi \wedge \phi$
7. If $\phi \vdash \neg\psi$ and $\phi \otimes_c \neg\phi \vdash \neg\psi$ then $\psi \otimes_c \neg\psi \vdash \neg\phi$
8. If $\phi \vdash \neg\psi$ then $\phi \otimes_c \psi \equiv ((\phi \otimes_c \neg\phi) \otimes_c \psi) \vee ((\psi \otimes_c \neg\psi) \otimes_c \phi)$

3.2 Convex Merging

The second binary operator we will consider is convex merging. This form of merging is illustrated by Example 2. The idea is not only picking the closest worlds modeling the propositions to be merged, but also including in-between worlds not necessarily belonging to the propositions to be merged.

Call an operator (geodesic) *convex merging* if it satisfies the rules of Table 3. Notice that the only changes are the omission of the arbitration rule and the replacement of the induction rule with the more appropriate Rule 8.

Now define an operator on subsets with

$$A \otimes_c B = \begin{cases} \{x : d(A, x) + d(B, x) = d(A, B)\} & \text{if } A, B \neq \emptyset \\ A \cup B & \text{otherwise.} \end{cases}$$

The following characterization theorem holds:

Theorem 2. *Let $\otimes_c$ be a geodesic convex merging operator. Then there exists a binary relation R (defined by (3)) such that (M, R) is a tolerance space and the following holds*

$$v(\phi \otimes_c \psi) = v(\phi) \otimes_c v(\psi). \tag{7}$$

Table 4. Barycentric merging rules

1. If ϕ is inconsistent, then $\phi \odot \psi \equiv \psi$
2. If $\phi \wedge \psi$ is consistent, then $\phi \odot \psi \equiv \phi \wedge \psi$
3. If $\phi_1 \equiv \phi_2$ then $\phi_1 \odot \psi \equiv \phi_2 \odot \psi$
4. $\phi \odot \psi \equiv \psi \odot \phi$
5. If $\phi \vdash \neg\psi$ then $(\psi \odot \neg\psi) \wedge \phi \vdash (\phi \odot \neg\phi) \wedge \phi$
6. If $\phi \vdash \neg\psi$ and $\phi \odot \neg\phi \vdash \neg\psi$ then $\psi \odot \neg\psi \vdash \neg\phi$
7. If $\phi \vdash \neg\psi$ and $(\phi \odot \neg\phi) \wedge \psi$ is consistent,
   then $\phi \odot \psi \equiv ((\phi \odot \neg\phi) \odot \psi) \vee ((\psi \odot \neg\psi) \odot \phi)$
8. If $\phi \vee (\phi \odot \neg\phi) \vdash \neg\psi$,
   then $\phi \odot \psi \equiv ((\phi \odot \neg\phi) \odot \psi) \odot ((\psi \odot \neg\psi) \odot \phi)$

Conversely, if an operator satisfies (7) *then it is a convex merging operator.*

As a contraction operator can generate a revision operator using the Levi identity, a convex merging operator generates a revision merging operator. To see that notice that we have

$$A \otimes B = (A \otimes_c B) \cap (A \cup B).$$

3.3   Barycentric Merging

In this section, we will characterize the third notion of binary merging called barycentric. This notion of merging corresponds to Example 3. The elements of the barycentric merge fall between and are spaced equally from the merged subsets. This notion of merging has been early identified by Revesz (See Example 3.1)

Definition 3. *Call an operator (geodesic) barycentric merging if it satisfies the rules of Table 4.*

Observe that Rule 9 of Table 1 does not hold in barycentric merging because the equidistant elements might not belong in the merged subsets.
Define the following operator on subsets:

$$A \odot B = \begin{cases} \{x \in \mathrm{mid}(A, B) : d(A, x) + d(B, x) = d(A, B)\} & \text{if } A, B \neq \emptyset \\ A \cup B & \text{otherwise.} \end{cases}$$

where

$$\mathrm{mid}(A, B) = A^k \cap B^k, \quad k = \min\{l : A^l \cap B^l \neq \emptyset\}.$$

The barycentric merging arises by selecting those elements of the convex merging that are "midway" between the merged subsets.
The need for barycentric merging has been early identified.

Example 4. This a modified version of Example 3.1 in [3]. Suppose two students are tutored in programming using examples in Datalog (D), SQL (S) or Query-by-example (Q). One student prefers examples in all three languages ($D \wedge S \wedge Q$) while the other prefers examples only in Query-by-example ($\neg D \wedge \neg S \wedge Q$). If we need to merge the preferences of the students, that is

$$D \wedge S \wedge Q \otimes \neg D \wedge \neg S \wedge Q$$

then geodesic merging would pick their disjunction $(D \wedge S \wedge Q) \vee (\neg D \wedge \neg S \wedge Q)$. Barycentric merging, in contrast, would pick $(\neg D \wedge S \wedge Q) \vee (D \wedge \neg S \wedge Q)$. Revesz points out that in this example barycentric merging should be chosen over geodesic merging, as the former satisfies both students (albeit not completely), while the latter, should the tutor chooses to teach all three, might result to one student dropping out of the tutoring sessions.

Now the following characterization theorem holds:

Theorem 3. *Let $\odot$ be a geodesic barycentric merging operator. Then there exists a binary relation R (defined by (3)) such that (M, R), where M is the set of models, is a tolerance space and the following holds*

$$v(\phi \odot \psi) = v(\phi) \odot v(\psi). \tag{8}$$

Conversely, if an operator satisfies (8) then it is a barycentric merging operator.

4   Further Work

Extending the previous results to multisets is perhaps the most important future direction of this work. However, it is not immediately clear how one may proceed.

A straightforward definition of multisets of formulas, or rather sequences of formulas, is to use binary merging to define merging of sequences of formulas. Let

$$\otimes(\phi_1, \phi_2, \ldots, \phi_n)$$

denote the merging of the sequence $\phi_1, \phi_2 \ldots, \phi_n$. Using a binary merging operator we have

$$\otimes(\phi_1, \ldots, \phi_n) = ((\ldots (\phi_1 \otimes \phi_2) \cdots) \otimes \phi_n).$$

Therefore there can be defined three different operators depending on the basic binary operator used. Unfortunately, the order of operations need to be specified as none of the merging operators introduced is associative. A counterexample for barycentric merging appears in Figure 4: notice that

$$\{a\} \odot (\{b\} \odot \{c\}) = \{a\} \odot \{a, d_3\} = \{a\}$$

whereas

$$(\{a\} \odot (\{b\})) \odot \{c\} = \{d_1\} \odot \{c\} = \{a, d_5\}.$$

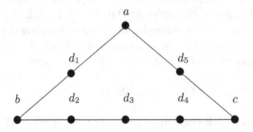

Fig. 4. Barycentric revision is not associative

Nevertheless, fixing a specific order of evaluation to sequences can be of interest. For example we can choose to favor latter received information and this feature can be part of a belief change strategy (as in belief revision).

Another more principled approach will be to employ indistinguishability once more to define a tolerance space among multisets. Let (X, R) be a tolerance space. Then, let (X^n, R_n) be the product tolerance space, where R_n is defined by

$$(x_1, \ldots, x_n) R_n (y_1, \ldots, y_n) \quad \text{iff} \quad x_i R y_i \text{ for } i = 1, \ldots, n.$$

The product space is not commutative but by regarding X^n as the multiset space rather than the product space—we keep the same notation for multiset and vectors—the relation of indistinguishability can be adjusted as follows: the multiset $(x_1, \ldots, x_n)$ is indistinguishable from $(y_1, \ldots, y_n)$ if there exists a permutation π such that $x_i R y_{\pi(i)}$ for $i = 1, \ldots, n$. Now that we have an indistinguishability relation among multisets we can define a tolerance space with a geodesic metric. The geodesic metric is defined as the shortest path (of R_n) between the two multisets but one may show that it can be reduced to the geodesic metric of the original space X:

$$d((x_1, \ldots, x_n), (y_1, \ldots, y_n)) = \min_{\pi}\{\max_i\{d(x_i, y_{\pi(i)})\}\}.$$

The geodesic metric lifts to a metric between subsets of multisets and therefore a geodesic revision operator can be defined the same way as in the previous section.

In case of multiple agents, a subset of multisets is not arbitrary but rather it has the following form

$$(A_1, \ldots, A_n) = \{(x_1, \ldots, x_n) \mid x_i \in A_i\}.$$

Each subset in the multiset contains all the worlds that the agent considers possible and represents the view of an agent. Merging translates to agreement or common view therefore we need to look for elements that are common to all subsets. So if $\bigcap A_i$ is not empty then merging should result to the following set

$$(A_1, \ldots, A_n) \cap D_X^n,$$

where $D_X^n = \{(x,\ldots,x) \mid x \in X\} \subseteq X^n$. If $\bigcap A_i$ is empty then the merging process should pick the closest elements to $(A_1,\ldots,A_n)$ from the diagonal set D_X^n. Therefore the merging of $(A_1,\ldots,A_n)$ is the revision

$$(A_1,\ldots,A_n) * D_X^n$$

The above revision operator is the one induced by the geodesic metric on the tolerance space of the multisets of elements defined earlier. This notion of multiset merging readily reduces to a binary revision operator.

The above options for extending merging to a richer framework reveal the variety of meanings merging can have, as well as, the ways it can be applied. We believe that geodesic semantics lends a useful tool, and the study of the above definitions and their characterization within the geodesic framework is currently under investigation. There is no characterization of distance based merging operators on multisets. We know many properties that such operators satisfy but we do not know the properties that characterize them ([10]). We believe that the employment of geodesic semantics is a right step towards representation results of this kind.

5   Comparison with Other Work and Conclusion

Geodesic merging characterized in Section 3.1 is a special case of arbitration (see [3,4]). In particular, the rules of Table 1 imply all postulates of arbitration operators in Section 3 of [4]. Postulates A_1 to A_5 of [4] correspond to our Rules 2 to 5 while A_7 and A_8 are translations of Rules 9 and 1, respectively, of Table1. Postulate A_6 holds true for all merging operators based on a notion of a global distance (see Remark 6.9 in [14]). A simpler way to show that geodesic merging is a special case of arbitration is to show that every graph gives rise to a set of parametrized orderings of complete interpretations (worlds) (the models of [4]) using the following definition

$$w \leq_\phi w' \quad \text{iff} \quad d(\phi,w) \leq d(\phi,w'),$$

where $\phi \in \mathcal{L}$ and w,w' are worlds.

Similarly, geodesic merging is easily seen to be a special case of Schlechta's non prioritized revision operator ([12]). The axioms of Definition 2.2 of [12] are implied by our rules of Table 1 when restricted to finite language and assuming a Katsuno-Mendelzon presentation like ours (for example Condition ($\uparrow$ 2) on page 47 of [12] corresponds to Rule 3 of Table 1). Semantically, we minimize over a geodesic metric which is a special case of minimization over the semi-metric used in [12].

Finally, one can see that our rules for geodesic merging satisfy the IC merging postulates of [14] in the binary case when merging is constrained by the disjunction of the formulas to be merged. This follows from the results of Section 6 of [14] where it is shown that arbitration operators can be seen as a special case of IC merging operators. However, it is not immediately obvious how one can induce a convex or a barycentric operator from an IC merging operator. It can be shown using techniques from the previous section that convex and barycentric operators can also be thought of as special cases of IC merging operators. Therefore, all of the classes of merging operators introduced herein do not deviate from the IC merging framework.

Now, it is clear where our merging operators are situated. They are all special cases of well studied classes of merging operators. The advantage of our approach is that we have shown that our classes of operators are characterized by a global metric on a graph. Other merging operators have been shown to correspond to either families of orderings on worlds (that may represent closeness), or generalized distance relations (where the triangle inequality does not necessarily hold). In both cases, the framework is too general. On one hand, families of orderings are parametrized by epistemic states which means that we need a relation for each formula in our language. In contrast, our approach is a single global relation (a graph). Generalized distance relations, on the other hand, do not necessarily satisfy a triangle inequality which forms the basis of most examples appearing in studies of merging operators. Hamming distance examples are ubiquitous and geodesic metric is a (qualitative) generalization of hamming distance. Moreover, geodesic metrics also encompass the metrics generated by threshold (see [19]), so we believe that such merging operators will be especially useful to applications where continuous metrics are mapped to integer ones (usually because of rounding). In other words, our framework is at an appropriate level of abstraction.

References

1. Borgida, A., Imielinski, T.: Decision making in commitees - a framework for dealing with inconsistency and non-monotonicity. In: NMR, pp. 21–32 (1984)
2. Baral, C., Kraus, S., Minker, J.: Combining multiple knowledge bases. IEEE Transactions on Knowledge and Data Engineering 3, 208–220 (1991), doi:10.1109/69.88001
3. Revesz, P.Z.: On the semantics of theory change: Arbitration between old and new information. In: Proceedings of the Twelfth ACM SIGACT-SIGMOD-SIGART Symposium on Principles of Databases, pp. 71–82 (1993)
4. Liberatore, P., Schaerf, M.: Arbitration (or how to merge knowledge bases). IEEE Trans. Knowl. Data Eng. 10(1), 76–90 (1998)
5. Smets, P.: The combination of evidence in the transferable belief model. IEEE Transactions on Pattern Analysis and Machine Intelligence 12, 447–458 (1990), doi:10.1109/34.55104
6. Rabinowicz, W.: Global belief revision based on similarities between worlds. In: Hansson, S.O., Rabinowicz, W. (eds.) Logic for a Change: Essays Dedicated to Sten Lindström on the Occasion of His Fiftieth Birthday. Uppsala prints and preprints in philosophy, no. 9, pp. 80–105. Department of Philosophy, Uppsala University (1995)
7. Hall, D., Martin Liggins, I., Chong, C., Linas, J.: Distributed Data Fusion for Network-Centric Operations. CRC PressINC (2012)
8. Elmagarmid, A., Rusinkiewicz, M., Sheth, A.: Management of heterogeneous and autonomous database systems. Morgan Kaufmann Publishers Inc., San Francisco (1999)

9. Delgrande, J.P., Schaub, T.: A consistency-based approach for belief change. Artif. Intell. 151(1-2), 1–41 (2003)
10. Konieczny, S., Pérez, R.P.: Logic based merging. J. Philosophical Logic 40(2), 239–270 (2011)
11. Andréka, H., Ryan, M., Schobbens, P.: Operators and laws for combining preference relations. Journal of Logic and Computation 12(1), 13–53 (2002)
12. Schlechta, K.: Non-prioritized belief revision based on distances between models. Theoria 63(1-2), 34–53 (1997)
13. Konieczny, S., Pérez, R.P.: Merging with integrity constraints. In: Hunter, A., Parsons, S. (eds.) ECSQARU 1999. LNCS (LNAI), vol. 1638, pp. 233–244. Springer, Heidelberg (1999)
14. Konieczny, S., Pérez, R.P.: Merging information under constraints: A logical framework. Journal of Logic and Computation 12(5), 773–808 (2002)
15. Georgatos, K.: On indistinguishability and prototypes. Logic Journal of the IGPL 11(5), 531–545 (2003)
16. Poincaré, H.: La Valeur de la Science. Flammarion, Paris (1905)
17. Zeeman, E.C.: The topology of the brain and visual perception. In: Fort, M.K. (ed.) The Topology of 3-Manifolds, pp. 240–256. Prentice Hall, Englewood Cliffs (1962)
18. Georgatos, K.: Geodesic revision. Journal of Logic and Computation 19(3), 447–459 (2009), doi:10.1093/logcom
19. Georgatos, K.: Conditioning by minimizing accessibility. In: Bonanno, G., Löwe, B., van der Hoek, W. (eds.) LOFT 2008. LNCS (LNAI), vol. 6006, pp. 20–33. Springer, Heidelberg (2010)
20. Georgatos, K.: Iterated contraction based on indistinguishability. In: Artemov, S., Nerode, A. (eds.) LFCS 2013. LNCS, vol. 7734, pp. 194–205. Springer, Heidelberg (2013)
21. Lehmann, D.J., Magidor, M., Schlechta, K.: Distance semantics for belief revision. J. Symb. Log. 66(1), 295–317 (2001)

Human Strategic Reasoning in Dynamic Games: Experiments, Logics, Cognitive Models

Sujata Ghosh[1], Tamoghna Halder[2], Khyati Sharma[3] and Rineke Verbrugge[4]

[1] Indian Statistical Institute, Chennai
sujata@isichennai.res.in
[2] 'University of California, Davis
thaldera@gmail.com
[3] Dunia Finance LLC, Dubai
khyati.sharma27@gmail.com
[4] University of Groningen, The Netherlands
L.C.Verbrugge@rug.nl

Abstract. This article provides a three-way interaction between experiments, logic and cognitive modelling so as to bring out a shared perspective among these diverse areas, aiming towards better understanding and better modelling of human strategic reasoning in dynamic games.

1 Introduction

How suitable are idealized formal models of social reasoning processes with respect to the nuances of the real world? In particular, do these formal methods represent human strategic reasoning satisfactorily or should we instead concentrate on empirical studies and models based on those empirical data? Ghosh, Meijering and Verbrugge [6] made an effort to bridge the gap between logical and cognitive treatments of strategic reasoning in dynamic games. They proposed to combine empirical studies, formal modeling and cognitive modeling to study human strategic reasoning. In their words, "rather than thinking about logic and cognitive modeling as completely separate ways of modeling, we consider them to be complementary and investigate how they can aid one another to bring about a more meaningful model of real-life scenarios". In the current article, we apply this combination of methods to the question to what extent people use backward induction or forward induction in dynamic games.

Backward and Forward Induction Reasoning. Backward Induction (BI) is the textbook approach for solving extensive-form games with perfect information. In generic games without payoff ties, BI yields the unique subgame perfect equilibrium. The assumptions underpinning BI are that all players commonly believe in everybody's future rationality, no matter how irrational players' past behaviour has already proven. See [15,18] for more details.

In Forward Induction (FI) reasoning, on the other hand, a player tries to rationalize the opponent's past behaviour in order to assess his future moves. Thus, in a subgame where no strategy of the opponent is consistent with common knowledge of rationality *and* his past behaviour, the player may still rationalize

© Springer-Verlag Berlin Heidelberg 2015
W. van der Hoek et al. (Eds.): LORI 2015, LNCS 9394, pp. 116–128, 2015.
DOI: 10.1007/978-3-662-48561-3_10

the opponent's past behaviour by attributing to him a strategy which is optimal against a presumed *suboptimal* strategy of hers, or by attributing to him a strategy which is optimal vis-a-vis a *rational* strategy of hers, which is only optimal against a suboptimal strategy of *his*. If the player pursues this rationalizing reasoning to the highest extent possible [2] and reacts accordingly, she ends up choosing what is called an *Extensive-Form Rationalizable* (EFR) strategy [17] (see also [18,16,5]). Thus EFR strategies are based on FI reasoning, and in the following we use the terms EFR and FI synonymously.

There have been extensive debates among game theorists and logicians about the merits of backward induction. Experimental economists and psychologists have shown that human subjects do not always follow the backward induction strategy in large centipede games [10,14]. Recently, based on an eye-tracking study and complexity considerations, it turned out that even when human subjects produce the outwardly correct 'backward induction answer' in smaller games, they may use a different internal reasoning strategy to achieve it [13,3]. To investigate human reasoning strategies, Ghosh, Meijering and Verbrugge [6] presented a formal language to represent strategies on a finer-grained level than was possible before. The language and its semantics helped to precisely distinguish different cognitive reasoning strategies, that can then be tested on the basis of computational cognitive models and experiments with human subjects. The syntactic framework of the formal system provided a generic way of constructing computational cognitive models of the participants of a 'marble drop' game.

Aims of This Article. Ghosh, Heifetz and Verbrugge [5] conducted a game-theoretic experiment that involves a participant's expectations about the opponent's reasoning strategies, that may in turn depend on expectations about the participant's reasoning. It deals with the following question: In a dynamic game of perfect information, are people inclined to do forward induction reasoning (i.e. show EFR behaviour)? In the current work, we extend our aim of bridging formal and empirical studies to this question from behavioural game theory, utilizing the experimental findings from [5]. The main new elements of this work with respect to [6,5,8] are as follows:

- We study robustness of the findings of [5], to alleviate concern that different participants might follow a variety of reasoning patterns. Thus, more grounding is given to the outcomes, which is used for formal modelling.
- Unlike the eye-tracking studies used in [13,6], the experiment which forms the backbone of this paper includes participants' verbal comments regarding the reasoning they applied to perform their actions (see [8]), which made it possible to introduce agents' beliefs about their opponents' moves and beliefs in the logical language. We conjecture that this language is more succinct than the language proposed in [6] in describing strategic reasoning, which in turn may lead to a more efficient modelling.

In what follows, we briefly recall Ghosh and colleagues' recent experiment on forward induction [5,8], report a robustness study of the findings of the experiment, and extend the language introduced in [6] to describe players' reasoning strategies, adding a belief operator to reflect players' expectations. Finally, we

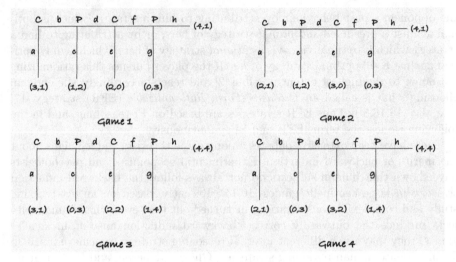

Fig. 1. Collection of the main games used in the experiment. The ordered pairs at the leaves represent pay-offs for the computer (C) and the participant (P), respectively.

sketch how strategy-formulas in this extended language can be turned into computational cognitive models that help to distinguish what is going on in people's minds when they play dynamic games of perfect information.

2   An Experimental Study: Do People Use FI?

We provide a brief summary of the experimental games and the experimental procedure underlying the current work. The experiment (previously reported in [5]) was designed to tackle the question whether people are inclined to use forward induction (FI) reasoning when they play dynamic perfect information games. The main interest was to examine participants' behaviour following a deviation from BI behaviour by their opponent right at the beginning of the game; for details, see [5,8].

The games that were used in the experiment are given in Figures 1 and 2. In these two-player games, the players play alternately. Let C denote the computer and P the participant. In the first four games (Figure 1), the computer plays first, followed by the participant. The players control two decision nodes each. In the last two games (Figure 2), which are truncated versions of two of the games of Figure 1, the participant moves first.

To explicate the difference between BI and EFR behaviour consider game 1, one of the experimental games (cf. Figure 1). Here, the unique Backward Induction (BI) strategies for player C and player P are $a; e$ and $c; g$, respectively, which indicate that the game will end at the first node, going down. In contrast, EFR would proceed as follows, starting from the scenario in which the game reaches the first decision node of P. Among the two strategies of player C that are compatible with this event, namely $b; e$ and $b; f$, only the latter is rational for player C. This is because of the fact that $b; e$ is dominated by $a; e$, while $b; f$ is optimal for player C if

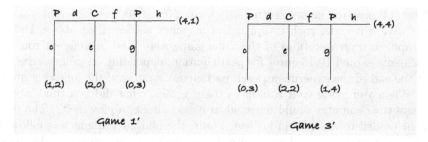

Fig. 2. Truncated versions of Game 1 and Game 3. The ordered pairs at the leaves represent pay-offs for C and P, respectively.

she believes that player P will play d; h with a high enough probability. Attributing to player C the strategy b; f is thus player P's best way to rationalize player C's choice of b, and in reply, d; g is player P's best response to b; f. Thus, the unique Extensive-Form Rationalizable (EFR, [17]) strategy (an *FI* strategy) of player P is d; g, which is distinct from his BI strategy c; g. For a detailed discussion on BI and EFR strategies in games $2, 3, 4, 1', 3'$, see [5].

The experiment was conducted at the Institute of Artificial Intelligence at the University of Groningen, the Netherlands. A group of 50 Bachelor's and Master's students from different disciplines took part. They had little or no knowledge of game theory, so as to ensure that neither backward induction nor forward induction was already known to them.[1] The participants played the finite perfect-information games in a graphical interface on the computer screen (cf. Figure 3). In each case, the opponent was the computer which had been programmed to play according to plans that were best responses to some plan of the participant, and this was told to the participants.

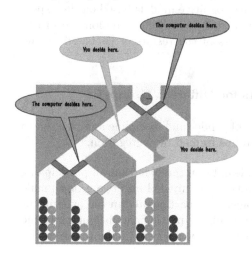

Fig. 3. Graphical interface for the participants. The computer controls the blue trapdoors and acquires blue marbles (represented as dark grey in a black and white print) as pay-offs, while the participant controls the orange trapdoors and acquires orange marbles (light grey in a black and white print) as pay-offs.

---

[1] The candidate participants were asked about their educational details. Two students who had followed a course on game theory were excluded.

After 14 practice games, each participant played 48 experimental games. There were 8 rounds, each comprised of 6 games as described above. Different graphical representations of the same game were used in different rounds. Participants earned 10-15 euros for participation, depending on points earned.

At the end of the experiment, each participant was asked the following question: 'When you made your choices in these games, what did you think about the ways the computer would move when it was about to play next?' The participant needed to describe in his own words, the plan he thought was followed by the computer on its next move after the participant's initial choice. We used these answers to classify various strategic reasoning processes applied by the participants while playing the experimental games.

To analyse whether participants P played FI strategies in the games described in figures 1 and 2, we can formulate the following *hypothesis* (see [5] for an explanation) concerning the participant's choice in his first decision node (if reached in games 1, 2, 3, 4, and in all rounds of games 1′ and 3′):

> "d will be played most often in game 3, less so in game 1, even less in games 3′ and 4, least often in games 1′ and 2", which we henceforth abbreviate as "$d : 3 > 1 > 3′, 4 > 1′, 2$."

In games 1 and 3, d is the only EFR move; in games 1′ and 2, d is neither a BI nor an EFR move; and in games 3′ and 4, both c and d are EFR moves. Moreover, in game 3, reaching the first decision node is compatible with common knowledge of rationality.

Ghosh et al. [5] found that in the aggregate, participants were indeed more likely to make decisions in accordance with their best-rationalization EFR conjecture, i.e., consistent with FI reasoning. For a detailed study and a discussion of some alternative explanations of the results, see [5,8]. Our main concern in the current paper is how we can construct cognitive models based on the experimental findings and how logic can play a role in such construction. To justify our aim, we first investigate the robustness of the results of [5] based on the available group-divisions.

2.1 Robustness: Different Results for Different Groups?

We segregated the participants in terms of gender and discipline and went on to test the hypothesis over the different groups formed by segregation.[2]

Segregation by Gender. The available data on the behaviour of participants at their first decision node in the six games were divided into two groups: male and female. Overall, 40 men and 10 women had participated in the experiment reported in [5,8]. We studied the choices made by participants belonging to the two groups.[3] For the hypothesis, we have the following, very similar to the results reported in [5]:

---

[2] Because of little variance among participants, we did not segregate by age.

[3] The results are based on one sample and two sample proportion tests.

- **Male** $d : 3, 3' > 4 > 1' > 1 > 2$
- **Female** $d : 3, 3' > 4 > 1', 1 > 2$

As to individual games, the tests revealed the following behaviour. We use the notations $i \sim j$ to denote that options i and j are chosen equally often, and $i > j$ to denote that i is chosen more often than j. The null hypothesis was that c and d were chosen equally often at the first decision node:

- Game 1: $c > d$ (male, female).
- Game 2: $c > d$ (male, female).
- Game 3: $d > c$ (male, female).
- Game 4: $d \sim c$ (male), $d > c$ (female).
- Game $1'$: $c > d$ (male, female).
- Game $3'$: $d > c$ (male, female).

Segregation by Discipline. For this study, the data on 50 participants was separated into three broad groups based on the nature of the study fields of the participants:

Artificial Intelligence (AI): artificial intelligence and human-machine communication (27 students);

Behavioural and Social Sciences (BSS): accountancy, economics and business economics, human resource management, international relations, law and business economics, and psychology (10 students);

Exact Sciences (ES): biology, biomedical sciences, drug innovation, computer science, mathematics, and physics (13 students).

Similar statistical analysis was done over the choices made by the participants belonging to the three groups. We summarize the results for the hypothesis:

- **AI** $d : 3, 3' > 4 > 1', 1 > 2$
- **BSS** $d : 3, 3' > 4 > 1', 1 > 2$
- **ES** $d : 3, 3' > 4 > 1', 1 > 2$

For the hypotheses on the individual games:

- Game 1: $c > d$ (AI, BSS), $d \sim c$ (ES).
- Game 2: $c > d$ (AI, BSS, ES).
- Game 3: $d > c$ (AI, BSS, ES).
- Game 4: $d \sim c$ (BSS, ES), $d > c$ (AI).
- Game $1'$: $c > d$ (AI, BSS), $d \sim c$ (ES).
- Game $3'$: $d > c$ (AI, BSS, ES).

The statistical analyses based on gender and discipline suggest that the results mentioned in Section 2 about participants' behaviour at their first decision node are robust. We only found minor variations corresponding to certain groups.

3   A Language for Strategies

In the line of [6], we propose a logical language specifying strategies of players. Our motivation for introducing this logical framework is to build a pathway from empirical to cognitive modelling studies. A detailed formal study of this framework regarding its expressive power and axiomatics is left for future work.

This framework uses empirical studies to provide insights into cognitive models of human strategic reasoning as performed during the experiment discussed in Section 2. The main idea is to use the logical syntax to express the different reasoning procedures as performed and conveyed by the participants and use these formulas to systematically build up reasoning rules of computational cognitive models of strategic reasoning.

A novel part of the proposed language is that we add an explicit notion of belief to the language proposed in [6] in order to describe participants' expectations regarding future moves of the computer. This belief operator is parametrized by both players and nodes of the game tree so that the possible expectations of players at each of their nodes can be expressed within the language itself. The whole point is to explicate the human reasoning process, therefore the participants' beliefs and expectations need to come to the fore. Such expectations formed an essential part of the current experimental study. We first build a syntax for game trees (cf. [19,7]). Let N denote a finite set of players and let Σ denote a countable set of actions.

Syntax for Extensive form Game Trees. Let *Nodes* be a countable set. The syntax for specifying finite extensive form game trees is given by:

$$\mathbb{G}(\textit{Nodes}) := (i, x) \mid \Sigma_{a_m \in J}((i, x), a_m, t_{a_m})$$

where $i \in N$, $x \in \textit{Nodes}$, $J(\text{finite}) \subseteq \Sigma$, and $t_{a_m} \in \mathbb{G}(\textit{Nodes})$.

Given $h \in \mathbb{G}(\textit{Nodes})$, we define the tree T_h generated by h inductively as follows (see Figure 4 for an example):

- $h = (i, x)$: $T_h = (S_h, \Rightarrow_h, \widehat{\lambda}_h, s_x)$ where $S_h = \{s_x\}$, $\widehat{\lambda}_h(s_x) = i$.
- $h = ((i, x), a_1, t_{a_1}) + \cdots + ((i, x), a_k, t_{a_k})$: Inductively we have trees $T_1, \ldots T_k$ where for $j : 1 \leqslant j \leqslant k$, $T_j = (S_j, \Rightarrow_j, \widehat{\lambda}_j, s_{j,0})$.
  Define $T_h = (S_h, \Rightarrow_h, \widehat{\lambda}_h, s_x)$ where
  - $S_h = \{s_x\} \cup S_{T_1} \cup \ldots \cup S_{T_k}$;
  - $\widehat{\lambda}_h(s_x) = i$ and for all j, for all $s \in S_{T_j}$, $\widehat{\lambda}_h(s) = \widehat{\lambda}_j(s)$;
  - $\Rightarrow_h = \bigcup_{j:1\leqslant j\leqslant k}(\{(s_x, a_j, s_{j,0})\} \cup \Rightarrow_j)$.

Given $h \in \mathbb{G}(\textit{Nodes})$, let $\textit{Nodes}(h)$ denote the set of distinct pairs (i, x) that occur in the expression of h.

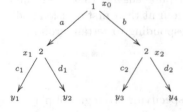

Fig. 4. Extensive form game tree. The nodes are labelled with turns of players and the edges with the actions. The syntactic representation of this tree can be given by:

$h = ((1, x_0), a, t_1) + ((1, x_0), b, t_2)$, where
$t_1 = ((2, x_1), c_1, (2, y_1)) + ((2, x_1), d_1, (2, y_2))$;
$t_2 = ((2, x_2), c_2, (2, y_3)) + ((2, x_2), d_2, (2, y_4))$.

3.1   Strategy Specifications

A syntax for specifying partial strategies and their compositions in a structural manner involving simultaneous recursion has been used in [6] to describe empirical reasoning of participants involved in a game experiment in a dynamic game called 'marble drop' [12,11], as demonstrated by an eye-tracking study [13]. The main case specifies, for a player, which conditions she tests before making a move. In what follows, the pre-condition for a move depends on observables that hold at the current game position, some belief conditions, as well as some simple finite past-time conditions and some finite look-ahead that each player can perform in terms of the structure of the game tree. Both the past-time and future conditions may involve some strategies that were or could be enforced by the players. These pre-conditions are given by the syntax defined below.

For any countable set X, let $BPF(X)$ (the boolean, past and future combinations of the members of X) be sets of formulas given by the following syntax:

$$BPF(X) := x \in X \mid \neg\psi \mid \psi_1 \vee \psi_2 \mid \langle a^+ \rangle \psi \mid \langle a^- \rangle \psi,$$

where $a \in \Sigma$, a countable set of actions.

Formulas in $BPF(X)$ can be read as usual in a dynamic logic framework and are interpreted at game positions. The formula $\langle a^+ \rangle \psi$ (respectively, $\langle a^- \rangle \psi$) refers to one step in the future (respectively, past). It asserts the existence of an a edge after (respectively, before) which ψ holds. Note that future (past) time assertions up to any bounded depth can be coded by iteration of the corresponding constructs. The 'time free' fragment of $BPF(X)$ is formed by the boolean formulas over X. We denote this fragment by $Bool(X)$.

For each $h \in \mathbb{G}(Nodes)$ and $(i,x) \in Nodes(h)$, we now add a new operator $\mathbb{B}_h^{(i,x)}$ to the syntax of $BPF(X)$ to form the set of formulas $BPF_b(X)$. The formula $\mathbb{B}_h^{(i,x)}\psi$ can be read as "in the game tree h, player i *believes* at node x that ψ holds". One might feel that it is not elegant that the belief operator is parametrized by the nodes of the tree, however, our main aim is not to propose a logic for the sake of its nice properties, but to have a logical language that can be used suitably for constructing computational cognitive models corresponding to participants' strategic reasoning.

Syntax. Let $P^i = \{p_0^i, p_1^i, \ldots\}$ be a countable set of observables for $i \in N$ and $P = \bigcup_{i \in N} P^i$. To this set of observables we add two kinds of propositional variables $(u_i = q_i)$ to denote 'player i's utility (or payoff) is q_i' and $(r \leqslant q)$ to denote that 'the rational number r is less than or equal to the rational number q'[4] The syntax of strategy specifications is given by:

$$Strat^i(P^i) := [\psi \mapsto a]^i \mid \eta_1 + \eta_2 \mid \eta_1 \cdot \eta_2,$$

where $\psi \in BPF_b(P^i)$. For a detailed explanation see [6]. The basic idea is to use the above constructs to specify properties of strategies as well as to combine them to describe a play of the game. For instance, the interpretation of a player

---

[4] as in [6] and inspired by [4].

i's specification $[p \mapsto a]^i$ where $p \in P^i$, is to choose move a at every game position belonging to player i where p holds. At positions where p does not hold, the strategy is allowed to choose any enabled move. The strategy specification $\eta_1 + \eta_2$ says that the strategy of player i conforms to the specification η_1 or η_2. The construct $\eta_1 \cdot \eta_2$ says that the strategy conforms to specifications η_1 and η_2.

Semantics. We consider perfect information games with belief structures as models. The idea is very similar to that of temporal belief revision frame presented in [4]. Let $M = (T, \{\overset{x}{\longrightarrow}_i\}, V)$ with $T = (S, \Rightarrow, s_0, \widehat{\lambda}, \mathcal{U})$, where $(S, \Rightarrow, s_0, \widehat{\lambda})$ is an extensive form game tree, $\mathcal{U} : frontier(T) \times N \rightarrow \mathbb{Q}$ is a utility function. Here, $frontier(T)$ denotes the leaf nodes of the tree T. For each $s_x \in S$ with $\widehat{\lambda}(s_x) = i$, we have a binary relation $\overset{x}{\longrightarrow}_i$ over S (cf. the connection between h and T_h presented above). Finally, $V : S \rightarrow 2^P$ is a valuation function. The truth value of a formula $\psi \in BPF_b(P)$ at the state s, denoted $M, s \models \psi$, is defined as follows:

- $M, s \models p$ iff $p \in V(s)$.
- $M, s \models \neg\psi$ iff $M, s \not\models \psi$.
- $M, s \models \psi_1 \vee \psi_2$ iff $M, s \models \psi_1$ or $M, s \models \psi_2$.
- $M, s \models \langle a^+ \rangle\psi$ iff there exists an s' such that $s \overset{a}{\Rightarrow} s'$ and $M, s' \models \psi$.
- $M, s \models \langle a^- \rangle\psi$ iff there exists an s' such that $s' \overset{a}{\Rightarrow} s$ and $M, s' \models \psi$.
- $M, s \models \mathbb{B}_h^{(i,x)}\psi$ iff the underlying game tree of T_M is the same as T_h and for all s' such that $s \overset{x}{\longrightarrow}_i s'$, $s' \models \psi$.

The truth definitions for the new propositions are as follows:

- $M, s \models (u_i = q_i)$ iff $\mathcal{U}(s, i) = q_i$.
- $M, s \models (r \leqslant q)$ iff $r \leqslant q$, where r, q are rational numbers.

Strategy specifications are interpreted on strategy trees of T. We also assume the presence of two special propositions **turn**$_1$ and **turn**$_2$ that specify which player's turn it is to move, i.e. the valuation function satisfies the property

- for all $i \in N$, **turn**$_i \in V(s)$ iff $\widehat{\lambda}(s) = i$.

One more special proposition **root** is assumed to indicate the root of the game tree, that is the starting node of the game. The valuation function satisfies the property

- **root** $\in V(s)$ iff $s = s_0$.

We recall that a strategy for player i is a function μ^i which specifies a move at every game position of the player, i.e. $\mu^i : S^i \rightarrow \Sigma$. A strategy μ can also be viewed as a subtree of T where for each node belonging to the opponent player i, there is a unique outgoing edge and for nodes belonging to player $\bar{i}$, every enabled move is included. A partial strategy for player i is a partial function σ^i which specifies a move at some (but not necessarily all) game positions of the player, i.e. $\sigma^i : S^i \rightharpoonup \Sigma$. A partial strategy can be viewed as a set of total strategies of the player [6].

The semantics of the strategy specifications are given as follows. Given a model M and a partial strategy specification $\eta \in Strat^i(P^i)$, we define a semantic function $[\![\cdot]\!]_M : Strat^i(P^i) \rightarrow 2^{\Omega^i(T_M)}$, where each partial strategy specification is associated with a set of total strategy trees and $\Omega^i(T)$ denotes the set of all player i strategies in the game tree T.

For any $\eta \in Strat^i(P^i)$, the semantic function $[\![\eta]\!]_M$ is defined inductively:

- $[\![[\psi \mapsto a]^i]\!]_M = \Upsilon \in 2^{\Omega^i(T_M)}$ satisfying: $\mu \in \Upsilon$ iff μ satisfies the condition that, if $s \in S_\mu$ is a player i node then $M, s \models \psi$ implies $out_\mu(s) = a$.
- $[\![\eta_1 + \eta_2]\!]_M = [\![\eta_1]\!]_M \cup [\![\eta_2]\!]_M$
- $[\![\eta_1 \cdot \eta_2]\!]_M = [\![\eta_1]\!]_M \cap [\![\eta_2]\!]_M$

Above, $out_\mu(s)$ is the unique outgoing edge in μ at s. Recall that s is a player i node and therefore by definition of a strategy for player i, there is a unique outgoing edge at s.

Before describing specific strategies found in the empirical study, we would like to focus on the new operator of belief, $\mathbb{B}_h^{(i,x)}$ proposed above. Note that this operator is considered for each node in each game. The idea is that the same player might have different beliefs at different nodes of the game. We had to introduce the syntax of the extensive form game trees to make this definition sound, otherwise we would have had to restrict our discussion to single game trees. The semantics given to the operator is entangled in both the syntax and semantics, which might create problems in finding an appropriate axiom system. A possible solution would be to introduce some generic classes of games similar to the idea of generic game boards [20], using the notion of enabled game trees [7]. This is left for future work, as well as a comparison of the expressiveness of the current language with those of existing logics of belief and strategies.

3.2   Describing Specific Strategies in the Experimental Games

Let us now express some actual reasoning processes that participants displayed during the experiment. Some participants described how they reasoned in their answers to the final question. Example 1 of such reasoning: "If the game reaches my first decision node and if the payoffs are such that I believe that the computer would not play e if its second decision node is reached, then I play d at my current decision node". This kind of strategic reasoning can be expressed using the following formal notions.

Let us assume that actions are part of the observables, that is, $\Sigma \subseteq P$. The semantics for the actions can be defined appropriately. Let $n_1, \ldots, n_4$ denote the four decision nodes of game 1, with C playing at n_1 and n_3, and P playing at the remaining two nodes n_2 and n_4. We have four belief operators for this game - two for each player. We abbreviate some formulas which describe the payoff structure of the game:

$$\langle d \rangle \langle f \rangle \langle h \rangle ((u_C = p_C) \wedge (u_P = p_P)) = \alpha$$
$$\langle d \rangle \langle f \rangle \langle g \rangle ((u_C = q_C) \wedge (u_P = q_P)) = \beta$$
$$\langle d \rangle \langle e \rangle ((u_C = r_C) \wedge (u_P = r_P)) = \gamma$$
$$\langle c \rangle ((u_C = s_C) \wedge (u_P = s_P)) = \delta$$
$$\langle b^- \rangle \langle a \rangle ((u_C = t_C) \wedge (u_P = t_P)) = \chi$$
$$\varphi := \alpha \wedge \beta \wedge \gamma \wedge \delta \wedge \chi$$

Let ψ_i denote the conjunction of all the order relations of the rational payoffs for player i given in game 1. A strategy specification describing the strategic reasoning of Example 1 (at the node n_2) is:

$$\eta_P^1 : [(\varphi \wedge \psi_P \wedge \psi_C \wedge \langle b^- \rangle \mathbf{root} \wedge \mathbb{B}_{g1}^{n_2,P}\langle d \rangle \neg e \wedge \mathbb{B}_{g1}^{n_2,P}\langle d \rangle\langle f \rangle g) \mapsto d]^P$$

A *BI* reasoning at the same node can be formulated as follows:

$$\eta_P^2 : [(\varphi \wedge \psi_P \wedge \psi_C \wedge \langle b^- \rangle \mathbf{root} \wedge \mathbb{B}_{g1}^{n_2,P}\langle d \rangle e \wedge \mathbb{B}_{g1}^{n_2,P}\langle d \rangle\langle f \rangle g) \mapsto c]^P$$

The example above shows how strategic reasoning of participants can be formulated in the proposed framework (which could then be converted to appropriate reasoning rules to build up computational cognitive models). Note that our representations have become quite succinct using the belief operator, compared to the representations we had in [6], because expressions for response strategies are not needed anymore. We leave the details for future work.

4   Modelling in ACT-R

We now provide a brief description of the cognitive architecture at the basis of our computational cognitive model. ACT-R is an integrated theory of cognition as well as a cognitive architecture that many cognitive scientists use [1]. ACT-R consists of modules that link with cognitive functions, for example, vision, motor processing, and declarative processing. Each module maps onto a specific brain region. Furthermore, each module is associated with a buffer and the modules communicate among themselves via these buffers.

The computational cognitive models that we propose are inspired by [6]. We consider a class of models, where each model is based on a set of strategy specifications that can be generated using the logical framework we presented in Section 3. The specifications can represent both backward induction reasoning or forward induction reasoning (in particular, *EFR* reasoning), among others.

Each of the specifications defined in Subsection 3.2 comprises comparisons between relevant payoffs for both the players. For each comparison, a cognitive model has a set of production rules that specify what the model should do. To compare player C's payoffs, say at two leaf nodes, the model first has to find, attend, and encode them in the so-called problem state buffer [1]. For each subsequent payoff, the model performs the following procedure (cf. Figure 5):
 – request the visual module to find the payoffs' visual locations;
 – direct visual attention to that location; and
 – update the problem state (buffer).

The specifications η_P^1 and η_P^2 (see Subsection 3.2) specify what the model should do after encoding the payoffs in the problem state. First, the payoffs need to be compared and the comparison needs to stored. Then the belief operators are dealt with as follows (cf. Figure 5):
 – attend visual location of the node depicted by the belief operator; and
 – encode the actions and beliefs at the problem state (buffer).

The decisions are made corresponding to the recorded payoffs and the resulting beliefs. An example production rule could be as follows; the model will select and fire this production rule to generate a response:

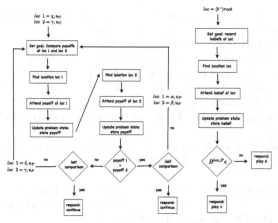

Fig. 5. Flowcharts for reasoning processes as described in Example 1 and BI

IF

Goal is to record Player P's belief at node n,	If the current goal is to record Player P's beliefs at node n,
Problem State represents Player P's actions at n, c and d	and the problem state has stored the actions,
$\mathbb{B}^{(P,n)}f$	and belief is f will be played (by C),

THEN

Decision is play d	then request the manual (or motor) module to produce a key press (i.e., play d).

5 Conclusion

In this paper we have continued the line of work started in [6] and proposed another logical language to aid in the construction of computational cognitive models based on the findings of a game-theoretic experiment. We have shown that logic can play a major role in Marr's computational and algorithmic levels of inquiry for cognitive sciences [9]. In future we aim to implement various sets of specifications in separate models, and to simulate repeated game play to study possible learning effects. An advantage of constructing ACT-R models, not only logical formulas, is that quantitative predictions are generated, for example, concerning decision times, locus of attention and activity of brain regions, which can then be tested in further experiments.

References

1. Anderson, J.: How Can Human Mind Occur in the Physical Universe? OUP (2007)
2. Battigalli, P.: Strategic rationality orderings and the best rationalizability principle. Games and Economic Behavior 13, 178–200 (1996)

3. Bergwerff, G., Meijering, B., Szymanik, J., Verbrugge, R., Wierda, S.: Computational and algorithmic models of strategies in turn-based games. In: Proceedings of the 36th Annual Conference of the Cognitive Science Society, pp. 1778–1783 (2014)

4. Bonanno, G.: Axiomatic characterization of the AGM theory of belief revision in a temporal logic. Artificial Intelligence 171(2-3), 144–160 (2007)

5. Ghosh, S., Heifetz, A., Verbrugge, R.: Do players reason by forward induction in dynamic perfect information games? In: Ramanujam, R. (ed.) Proceedings of the 15th Conference on Theoretical Aspects of Rationality and Knowledge, pp. 121–130 (2015)

6. Ghosh, S., Meijering, B., Verbrugge, R.: Strategic reasoning: Building cognitive models from logical formulas. Journal of Logic, Language and Information 23(1), 1–29 (2014)

7. Ghosh, S., Ramanujam, R.: Strategies in games: A logic-automata study. In: Bezhanishvili, N., Goranko, V. (eds.) ESSLLI 2010/2011, Lectures. LNCS, vol. 7388, pp. 110–159. Springer, Heidelberg (2012)

8. Halder, T., Sharma, K., Ghosh, S., Verbrugge, R.: How do adults reason about their opponent? Typologies of players in turn-taking games. In: Proceedings of the 37th Annual Conference of the Cognitive Science Society, pp. 854–859 (2015)

9. Marr, D.: Vision. Freeman and Company, New York (1982)

10. McKelvey, R.D., Palfrey, T.R.: An experimental study of the centipede game. Econometrica: Journal of the Econometric Society, 803–836 (1992)

11. Meijering, B., Taatgen, N.A., van Rijn, H., Verbrugge, R.: Modeling inference of mental states: As simple as possible, as complex as necessary. Interaction Studies 15(3), 455–477 (2014)

12. Meijering, B., Van Rijn, H., Taatgen, N.A., Verbrugge, R.: I do know what you think I think: Second-order theory of mind in strategic games is not that difficult. In: Proc. 33rd Annual Conf. Cognitive Science Society, pp. 2486–2491 (2011)

13. Meijering, B., Van Rijn, H., Taatgen, N.A., Verbrugge, R.: What eye movements can tell about theory of mind in a strategic game. PloS One 7(9), e45961 (2012)

14. Nagel, R., Tang, F.F.: Experimental results on the centipede game in normal form: An investigation on learning. Journal of Mathematical Psychology 42(2), 356–384 (1998)

15. Osborne, M.J., Rubinstein, A.: A Course in Game Theory. MIT Press (1994)

16. Pacuit, E.: Dynamic models of rational deliberation in games. In: van Benthem, J., Ghosh, S., Verbrugge, R. (eds.) Models of Strategic Reasoning: Logics, Games and Communities. LNCS-FoLLI, vol. 8972, Springer, New York (2015)

17. Pearce, D.: Rationalizable strategic behaviour and the problem of perfection. Econometrica 52, 1029–1050 (1984)

18. Perea, A.: Epistemic Game Theory: Reasoning and Choice. CUP (2012)

19. Ramanujam, R., Simon, S.: A logical structure for strategies. In: Logic and the Foundations of Game and Decision Theory (LOFT 7). Texts in Logic and Games, vol. 3, pp. 183–208. Amsterdam University Press (2008)

20. van Benthem, J., Ghosh, S., Liu, F.: Modelling simultaneous games with dynamic logic. Knowledge, Rationality and Action 165, 247–268 (2008)

A Note on Reliability-Based Preference Dynamics

Sujata Ghosh[1] and Fernando R. Velázquez-Quesada[2]

[1] Indian Statistical Institute, Chennai, India
sujata@isichennai.res.in
[2] Grupo de Lógica, Lenguaje e Información, Universidad de Sevilla, Spain
FRVelazquezQuesada@us.es

Abstract. This paper continues a line of work that studies individual preference upgrades in order to model situations akin to a process of public deliberation in collective decision making. It proposes a general upgrade policy, presenting its semantic definition and a corresponding modality for describing its effects as well as a complete axiom system.

1 Introduction

Deliberation and aggregation are essential and complementary components of any democratic decision making process. While the well-studied process of aggregation focuses on accumulating individual preferences without discussing their origin [1], deliberation can be seen as a conversation through which individuals justify their preferences, a process that might lead to changes in their opinions as they are influenced by one another. Even if deliberation does not lead to unanimity, the discussion can lead to some 'preference uniformity' (see how deliberation can help in bypassing social choice theory's impossibility results in [2]), which might facilitate their eventual aggregation. In addition, the combination of both processes provides a more realistic model for decision making scenarios.

In [3], the authors presented a framework where agents have both preferences over a set of objects and reliability over the agents themselves. The main focus was to study how the public announcement of the individual preferences affects the preferences themselves. The paper proposed several preference upgrade policies based on the agents' reliability orderings, and then introduced a general lexicographic upgrade operation subsuming all of them. Decision procedures were provided to decide whether, under such upgrade policies, the iterative and public announcement of individual preferences can eventually lead to preference unanimity/stability.

But not every 'reasonable' policy for upgrading individual preferences falls under the scope of the general lexicographic upgrade (see page 132 for a discussion) - this paper presents a more general upgrade policy, viz. the general layered upgrade. As we see in the example discussed later (cf. Example 1), the general definition provided in this paper captures intuitive upgrades which could not be formalised by policies discussed in [3]. Moreover, this short and technical note constitutes a necessary step towards formalizing reasonable deliberation

W. van der Hoek et al. (Eds.): LORI 2015, LNCS 9394, pp. 129–142, 2015.
DOI: 10.1007/978-3-662-48561-3_11

processes which would facilitate their combination with aggregation processes in decision making. We leave this combination/reconciliation part for future work.

2   Recalling the Framework

This section briefly recalls (and, in some cases, extends) the definitions of the *PR* framework; further details can be found in [3]. Throughout this paper, let *Ag* be a *finite non-empty* set of agents, with $|Ag| = n$.

Definition 1 (*PR* Frame). *A preference and reliability (PR) frame F is a tuple* $\langle W, \{\leq_i, \preccurlyeq_i\}_{i \in Ag} \rangle$ *where* (I) *W is a finite non-empty set of worlds,* (II) $\leq_i \subseteq (W \times W)$ *is a total preorder (a total, reflexive and transitive relation), agent i's preference relation over worlds in W ($u \leq_i v$ is read as "world v is at least as preferable as world u for agent i");* (III) $\preccurlyeq_i \subseteq (Ag \times Ag)$ *is a total order (a total, reflexive, transitive and antisymmetric relation), agent i's reliability relation over agents in Ag ($j \preccurlyeq_i j'$ is read as "agent j' is at least as reliable as agent j for agent i").*

The motivations for the restrictions on the preference and the reliability relations are discussed in [3]. For now, here are further useful definitions.

Definition 2. *Let* $F = \langle W, \{\leq_i, \preccurlyeq_i\}_{i \in Ag} \rangle$ *be a frame.*

- $mr(i) = j$ *(j is agent i's most reliable agent) iff*$_{def}$ $j' \preccurlyeq_i j$ *for every* $j' \in Ag$;
- $Max_{\leq_i}(U)$, *the set containing agent i's most preferred worlds among those in* $U \subseteq W$, *is formally defined as* $\{v \in U \mid u \leq_i v \text{ for every } u \in U\}$.

2.1   A Formal Language

Throughout this paper, let *At* be a countable set of atomic propositions.

Definition 3 (Language). *Formulas* φ, ψ *and relational expressions* π, σ *of the language* $\mathcal{L}^{PR}$ *are given by*

$$\varphi, \psi ::= \top \mid p \mid j \sqsubseteq_i j' \mid \neg\varphi \mid \varphi \vee \psi \mid \langle \pi \rangle \varphi$$
$$\pi, \sigma ::= 1 \mid \leq_i \mid \geq_i \mid ?(\varphi, \psi) \mid -\pi \mid \pi \cup \sigma \mid \pi \cap \sigma$$

with $p \in At$ *and* $i, j, j' \in Ag$. *Standard abbreviations as the* converse operator $^{-1}$ *over relational expressions*[1] *will facilitate the writing of formulas.*[2]

The set of formulas of $\mathcal{L}^{PR}$ contains atomic propositions (p) and formulas describing the agents' reliability relations ($j \sqsubseteq_i j'$), and it is closed under negation ($\neg$), disjunction ($\vee$) and modal operators of the form $\langle \pi \rangle$ with π a relational expression. The set of relational expressions contains the constant 1 (the global

---

[1] Such operator is given by $1^{-1} := 1$, $(\leq_i)^{-1} := \geq_i$, $(\geq_i)^{-1} := \leq_i$, $(?(\varphi, \psi))^{-1} := ?(\psi, \varphi)$, $(-\pi)^{-1} := -(\pi^{-1})$, $(\pi \cup \sigma)^{-1} := \pi^{-1} \cup \sigma^{-1}$ and $(\pi \cap \sigma)^{-1} := \pi^{-1} \cap \sigma^{-1}$.

[2] Additionally, $\langle <_i \rangle \varphi := \langle \leq_i \cap -\geq_i \rangle \varphi$ and $\langle >_i \rangle \varphi := \langle -\leq_i \cap \geq_i \rangle \varphi$.

relation), the preference relations ($\leq_i$), their respective converse ($\geq_i$; [4,5]) and an additional construction of the form $?(\varphi, \psi)$ with φ and ψ formulas of the language, and it is closed under Boolean operations over relations (the so called *boolean modal logic*; [6]).

The following two definitions establish what a model is and how formulas of $\mathcal{L}^{PR}$ are interpreted over such structures.

Definition 4 (*PR* Model). *A PR model M is a tuple $\langle F, V \rangle$ where F is a PR frame and $V : At \to \wp(W)$ is a valuation function.*

Definition 5 (Semantic Interpretation). *Let $M = \langle W, \{\leq_i, \preccurlyeq_i\}_{i \in Ag}, V \rangle$ be a PR model. The function $\llbracket \cdot \rrbracket^M$ from formulas in $\mathcal{L}^{PR}$ to subsets of W and the function $\ulcorner \cdot \urcorner^M$ from relational expressions in $\mathcal{L}^{PR}$ to binary relations over W are defined simultaneously in the following way.*

$$\llbracket \top \rrbracket^M := W \qquad \llbracket p \rrbracket^M := V(p) \qquad \llbracket j \sqsubseteq_i j' \rrbracket^M := \begin{cases} W & \text{if } j \preccurlyeq_i j' \\ \varnothing & \text{otherwise} \end{cases}$$

$$\llbracket \neg\varphi \rrbracket^M := W \setminus \llbracket \varphi \rrbracket^M \qquad \llbracket \varphi \vee \psi \rrbracket^M := \llbracket \varphi \rrbracket^M \cup \llbracket \psi \rrbracket^M$$

$$\llbracket \langle \pi \rangle \varphi \rrbracket^M := \{ w \in W \mid \text{there is } u \in \llbracket \varphi \rrbracket^M \text{ with } (w, u) \in \ulcorner \pi \urcorner^M \}$$

and

$$\ulcorner 1 \urcorner^M := W \times W \qquad\qquad \ulcorner -\pi \urcorner^M := (W \times W) \setminus \ulcorner \pi \urcorner^M$$

$$\ulcorner \leq_i \urcorner^M := \leq_i \qquad\qquad \ulcorner \pi \cup \sigma \urcorner^M := \ulcorner \pi \urcorner^M \cup \ulcorner \sigma \urcorner^M$$

$$\ulcorner \geq_i \urcorner^M := \{ (v, u) \in (W \times W) \mid u \leq_i v \} \qquad \ulcorner \pi \cap \sigma \urcorner^M := \ulcorner \pi \urcorner^M \cap \ulcorner \sigma \urcorner^M$$

$$\ulcorner ?(\varphi, \psi) \urcorner^M := \llbracket \varphi \rrbracket^M \times \llbracket \psi \rrbracket^M$$

Note, in particular, how $\ulcorner ?(\varphi, \psi) \urcorner^M$ is the set of those pairs $(u, v) \in (W \times W)$ such that u satisfies φ and v satisfies ψ.[3] A formula φ is true at world $w \in W$ in model M when $w \in \llbracket \varphi \rrbracket^M$. A formula is valid when it is true at every world of every model, as usual.

The operator $?(\varphi, \psi)$, useful for providing the axiom system for the general upgrade operation to be introduced in Subsection 3.1, is the only construction in $\mathcal{L}^{PR}$ that does not appear in [3]. Thus, an axiom system characterising formulas in $\mathcal{L}^{PR}$ valid on *PR* models is given by the axioms and rules in Table 1 of [3] plus the formula $\langle ?(\psi_1, \psi_2) \rangle \varphi \leftrightarrow (\psi_1 \wedge \langle 1 \rangle (\psi_2 \wedge \varphi))$, which characterises the extra operator.

3   Individual Preference Upgrades

Intuitively, a public announcement of the agents' individual preferences might induce an agent i to adjust her own preferences according to what has been

---

[3] The relation $\ulcorner ?(\varphi, \psi) \urcorner^M$ is a natural generalisation of the relation $\ulcorner ?\varphi \urcorner^M := \{ (u, u) \in (W \times W) \mid u \in \llbracket \varphi \rrbracket^M \}$ for the traditional *PDL* test operation $?\varphi$ [7].

announced and the reliability ordering she assigns to the set of agents.[4] For example, an agent might adopt the preferences of the agent on whom she relies the most, or might use such preference for 'breaking ties' among her equally-preferred zones.

In [3] the authors introduced the *general lexicographic upgrade* operation, which creates a preference ordering following a priority list of orderings.

Definition 6 (General Lexicographic Upgrade). *A lexicographic list $\mathcal{R}$ over W is a finite non-empty list whose elements are indexes of preference orderings over W, with $|\mathcal{R}|$ the list's length and $\mathcal{R}[k]$ its kth element ($1 \leq k \leq |\mathcal{R}|$). Intuitively, $\mathcal{R}$ is a priority list of preference orderings, with $\leq_{\mathcal{R}[1]}$ having the highest priority. Given $\mathcal{R}$, the preference ordering $\leq_{\mathcal{R}} \subseteq (W \times W)$ is defined as*

$$u \leq_{\mathcal{R}} v \quad \textit{iff}_{def} \quad \underbrace{\left(u \leq_{\mathcal{R}[|\mathcal{R}|]} v \; \wedge \; \bigwedge_{k=1}^{|\mathcal{R}|-1} u \simeq_{\mathcal{R}[k]} v \right)}_{1} \vee \underbrace{\bigvee_{k=1}^{|\mathcal{R}|-1} \left(u <_{\mathcal{R}[k]} v \; \wedge \; \bigwedge_{l=1}^{k-1} u \simeq_{\mathcal{R}[l]} v \right)}_{2}$$

Thus, $u \leq_{\mathcal{R}} v$ holds if this agrees with the least prioritised ordering ($\leq_{\mathcal{R}[|\mathcal{R}|]}$) and for the rest of them u and v are equally preferred (part 1), or if there is an ordering $\leq_{\mathcal{R}[k]}$ with a strict preference for v over u and all orderings with higher priority see u and v as equally preferred (part 2).

This operation allows an agent i to upgrade her preferences by taking $\leq_i' := \leq_{\mathcal{R}}$, with $\mathcal{R}$ a lexicographic list containing the ordered indexes of the agents whose preferences will be used. It subsumes not only the natural instance in which $\mathcal{R}$ is given directly by the agent's reliability ordering, but also other possibilities as, e.g., one in which the agent adopts 'as is' the preferences of the agent on whom she relies the most. A sound and complete axiom system for a modality representing the operation can be found in [3].

Even though the general lexicographic upgrade covers many natural upgrades, there are also 'reasonable' policies that fall outside its scope. Sometimes we are not interested in considering the complete order among the choices of the most reliable agent, but only her most preferred choices. For example, consider a girl planning to take a boy out for a movie of his choice, and let w_i denote the world where 'movie i is the most preferred' ($i = 1, 2, 3, 4$). Rather than considering the complete preference ordering it makes more sense to consider the most preferred movies of the boy and among that what she would like to watch most as well. In any case they can take note of their choices among all the options, as they may not know which movie ticket they would get. We will get back to this example in a moment. For the following definition, recall that $\text{Max}_{\leq_i}(W)$ denotes agent i's most preferred worlds among those in W.

Definition 7 (Conservative Upgrade). *Agent j put her most reliable agent's most preferred worlds above the rest, using her old ordering to break ties in both zones.[5] More precisely, with $U := \text{Max}_{\leq_{\text{mr}(j)}}(W)$,*

---

[4] Note that this work does not focus on the formal representation of such announcement, but rather on the formal representation of its effects.

[5] This upgrade is called *lexicographic* in [8] and [9].

$$u \leq'_j v \quad \textit{iff}_{def} \ (\{u,v\} \cap U = \{u,v\} \ \wedge \ u \leq_j v) \ \vee \ (\{u,v\} \cap U = \{v\}) \ \vee$$
$$(\{u,v\} \cap U = \varnothing \ \wedge \ u \leq_j v)$$

The conservative upgrade is not an instance of the general lexicographic upgrade, as there are cases in which the output of the former cannot be reproduced by any instance of the latter.

Example 1. Suppose agent a is agent b's most reliable agent and their individual preferences are as below (reflexive and transitive arrows omitted).

A conservative upgrade on b's preferences will create two zones, the upper one with a's most preferred worlds (w_3 and w_4), and the lower one with the remaining worlds (w_1 and w_2). Within each zone, b's old preferences will apply, thus producing $w_3 <'_b w_4$ and $w_1 <'_b w_2$. The final result is then

$$b: \quad w_1 \longrightarrow w_2 \longrightarrow w_3 \longrightarrow w_4$$

Observe how no lexicographic list can produce this outcome. First, no singleton list does the job, as $\leq'_b$ is different from both $\leq_a$ and $\leq_b$. The list $\langle\langle a \ ; b \rangle\rangle$ (with the leftmost ordering having the highest priority) also fails, as it would give $\leq_a$ the highest priority, thus producing an ordering with w_2 strictly below w_1, different from what $\leq'_b$ states. Finally, $\langle\langle b \ ; a \rangle\rangle$ fails too, as it will give priority to $\leq_b$, thus putting w_4 strictly below w_1, again different from what $\leq'_b$ establishes.

Now agents a and b can be considered as the boy and girl respectively in the earlier example, with their preference orders about movies given as above. After a conservative preference upgrade, the first choice movie for the girl is movie 4.

3.1 The General Layered Upgrade

The conservative upgrade does not create a preference ordering following a priority list of orderings. Instead, it puts a set of elements of the domain at the topmost layer of the ordering (in Definition 7, the set $\text{Max}_{\leq_{\text{mr}(j)}}(W)$), then using a 'default' ordering (in Definition 7, $\leq_j$) to sort both this layer and those worlds that do not appear in it. This observation leads to the following definition.

Definition 8 (General Layered Upgrade). *A layered list $\mathcal{S}$ over W is a finite (possibly empty) list of pairwise disjoint subsets of W together with a default preference ordering over W. The list's length is denoted by $|\mathcal{S}|$, its kth element is denoted by $\mathcal{S}[k]$ (with $1 \leq k \leq |\mathcal{S}|$), and $\leq^{\mathcal{S}}_{\text{def}}$ is its default preference ordering. Intuitively, $\mathcal{S}$ defines layers of elements of W in the new preference ordering $\leq_{\mathcal{S}}$, with $\mathcal{S}[1]$ the set of worlds that will be in the topmost layer and $\leq^{\mathcal{S}}_{\text{def}}$ the preference ordering that will be applied to each individual set and to those worlds not in $\bigcup_{k=1}^{|\mathcal{S}|} \mathcal{S}[k]$. Formally, given $\mathcal{S}$, the ordering $\leq_{\mathcal{S}} \subseteq (W \times W)$ is defined as*

$$u \leq_{\mathcal{S}} v \quad iff_{def} \quad \underbrace{\left(u \leq_{def}^{\mathcal{S}} v \ \wedge \ \left(\{u,v\} \cap \bigcup_{k=1}^{|\mathcal{S}|} \mathcal{S}[k] = \varnothing \vee \bigvee_{k=1}^{|\mathcal{S}|} \{u,v\} \subseteq \mathcal{S}[k]\right)\right)}_{1}$$

$$\vee \quad \underbrace{\bigvee_{k=1}^{|\mathcal{S}|} \left(v \in \mathcal{S}[k] \ \wedge \ u \notin \bigcup_{l=1}^{k} \mathcal{S}[l]\right)}_{2}$$

Thus, $u \leq_{\mathcal{S}} v$ holds if this agrees with the default ordering $\leq_{def}^{\mathcal{S}}$ and either neither u nor v are in any of the specified sets in $\mathcal{S}$ or else both are in the same set (part 1), or if there is a set $\mathcal{S}[k]$ in which v appears and u appears neither in the same set (a case already covered in part 1) nor in one with higher priority (part 2).

Here are two useful observations. First, if $|\mathcal{S}| = 0$, then while both the whole part 2 and the right-hand side of the rightmost disjunct in part 1 collapse to $\bot$, the left-hand side of the rightmost disjunct in part 1 collapses to $\top$. Thus,

$$u \leq_{\mathcal{S}} v \quad iff \quad u \leq_{def}^{\mathcal{S}} v$$

On the other hand, if $\mathcal{S}$'s sets form a partition of W (i.e., the sets are not only mutually exclusive but also collectively exhaustive), then $\bigcup_{k=1}^{|\mathcal{S}|} \mathcal{S}[k] = W$ so the left-hand side of the rightmost disjunct in part 1 collapses to $\bot$. Then,

$$u \leq_{\mathcal{S}} v \quad iff \quad \underbrace{\left(u \leq_{def}^{\mathcal{S}} v \wedge \bigvee_{k=1}^{|\mathcal{S}|} \{u,v\} \subseteq \mathcal{S}[k]\right)}_{1} \vee \underbrace{\bigvee_{k=1}^{|\mathcal{S}|} \left(v \in \mathcal{S}[k] \ \wedge \ u \notin \bigcup_{l=1}^{k} \mathcal{S}[l]\right)}_{2}$$

In fact, since $\leq_{def}^{\mathcal{S}}$ is used to break ties not only within each $\mathcal{S}[k]$ but also among those worlds not appearing in any such set, the provided definition of a layered list actually just 'abbreviates' (but still it is equivalent to) a list that requires a full partition of W by not writing explicitly the set with the least priority.

Third, a layered list $\mathcal{S}$ has a semantic nature, as it is given in terms of subsets of the domain and binary relations over it. Of course, when it is intended to be applied to a given model, it can also be defined syntactically.

Definition 9. *A layered list $\mathcal{S}$ is defined syntactically within $\mathcal{L}^{PR}$ whenever each $\mathcal{S}[k]$ is given as a formula χ_k in $\mathcal{L}^{PR}$ and its default ordering $\leq_{def}^{\mathcal{S}}$ is given as a relational expression $\pi_{def}^{\mathcal{S}}$ in $\mathcal{L}^{PR}$. In such cases, Definition 8 is adjusted by writing $[\![\chi_k]\!]^M$ instead of $\mathcal{S}[k]$ and $\ulcorner \pi_{def}^{\mathcal{S}} \urcorner^M$ instead of $\leq_{def}^{\mathcal{S}}$, for M the model in which such layered list is applied. In such cases, $\leq_{\mathcal{S}}$ will be written as $\leq_{\mathcal{S}(M)}$.*

The next proposition makes possible the definition that follows it.

Proposition 1. *Let $\mathcal{S}$ be a layered list over W. If $\leq_{def}^{\mathcal{S}}$ is reflexive (transitive, total, respectively), then so is $\leq_{\mathcal{S}}$. For syntactically defined layered lists $\mathcal{S}$, if $\ulcorner \pi_{def}^{\mathcal{S}} \urcorner^M$ is reflexive (transitive, total, respectively), then so is $\leq_{\mathcal{S}(M)}$.*

Definition 10. *Let* $M = \langle W, \{\leq_i, \preccurlyeq_i\}_{i \in Ag}, V \rangle$ *be a PR model.*

- *Let* $\mathcal{S}$ *be a layered list whose default ordering is reflexive, transitive and total;[6] let* $j \in Ag$ *be an agent. The PR model* $\mathrm{gy}_{\mathcal{S}}^j(M) = \langle W, \{\leq'_i, \preccurlyeq_i\}_{i \in Ag}, V \rangle$ *is such that, for every agent* $i \in Ag$, $\leq'_i := \leq_{\mathcal{S}(M)}$ *if* $i = j$, *and* $\leq'_i := \leq_i$ *otherwise.*
- *Let* $\boldsymbol{\mathcal{S}}$ *be a list of* $|Ag|$ *layered lists whose default ordering are reflexive, transitive and total, with* $\boldsymbol{\mathcal{S}}_i$ *its* i*th element.[7] The PR model* $\mathrm{gy}_{\boldsymbol{\mathcal{S}}}(M) = \langle W, \{\leq'_i, \preccurlyeq_i\}_{i \in Ag}, V \rangle$ *is such that, for every agent* $i \in Ag$, $\leq'_i := \leq_{\boldsymbol{\mathcal{S}}_i(M)}$

On the Generality of the General Layered Upgrade. When layered lists are defined semantically, the general layered upgrade can build *any* conceivable *total, reflexive and transitive* preference ordering by simply using a layered list that spells out explicitly the desired output, using then the full Cartesian product as the default ordering.

When layered lists are restricted to syntactically definable ones, the power of the layered upgrade depends on the expressivity of the used language. Nevertheless, $\mathcal{L}^{PR}$ is expressive enough to define layered lists that replicate the behaviour of not only the general lexicographic (Definition 6) but also the conservative upgrade (Definition 7). This shows how the general layered upgrade is indeed a generalisation of the general lexicographic upgrade.

Proposition 2. *The general lexicographic upgrade is an instance of the general layered upgrade with* $\mathcal{S}$ *defined syntactically within* $\mathcal{L}^{PR}$.

Proof (Sketch). Let $M = \langle W, \{\leq_i, \preccurlyeq_i\}_{i \in Ag}, V \rangle$ *be a PR model. Take any lexicographic list* $\mathcal{R}$, *and let* $L_1, \ldots, L_m$ *be the layers it generates (with* L_1 *being the topmost) when applied over* M. *If the relational expression 1 is used for defining the default ordering, then in order to prove the proposition it is enough to provide* m *formulas* χ_k *such that* $L_k = [\![\chi_k]\!]^M$. *In order to do this, first observe how, if* $U = [\![\chi_U]\!]^M$, *then* $\mathrm{Max}_{\leq_i}(U) = [\![\chi_U \wedge [<_i]\neg\chi_U]\!]^M$. *Now, note how*

$$L_1 = \mathrm{Max}_{\leq_{\mathcal{R}[|\mathcal{R}|]}} \left(\mathrm{Max}_{\leq_{\mathcal{R}[|\mathcal{R}|-1]}} \left(\cdots \mathrm{Max}_{\leq_{\mathcal{R}[1]}} (W) \cdots \right) \right)$$

This and the previous observation suggest the following recursive definition:

$$\mu_1(\tau) := \tau \wedge [<_{\mathcal{R}[1]}]\neg\tau$$
$$\mu_2(\tau) := \mu_1(\tau) \wedge [<_{\mathcal{R}[2]}]\neg\mu_1(\tau)$$
$$\vdots$$
$$\mu_{|\mathcal{R}|}(\tau) := \mu_{|\mathcal{R}|-1}(\tau) \wedge [<_{|\mathcal{R}|}]\neg\mu_{|\mathcal{R}|-1}(\tau)$$

in which τ *is a parameter. Then, given* $W = [\![\top]\!]^M$, *it is not hard to see that*

$$\chi_1 := \mu_{|\mathcal{R}|}(\top) \qquad \textit{is such that} \qquad L_1 = [\![\chi_1]\!]^M$$

---

[6] If $\mathcal{S}$ is defined syntactically, then $\ulcorner \pi_{\mathrm{def}}^{\mathcal{S}} \urcorner^M$ should be reflexive, transitive and total.

[7] If $\boldsymbol{\mathcal{S}}_i$ is defined syntactically, then $\ulcorner \pi_{\mathrm{def}}^{\boldsymbol{\mathcal{S}}_i} \urcorner^M$ should be reflexive, transitive and total.

But then, since

$$L_2 = \text{Max}_{\leq_{\mathcal{R}[|\mathcal{R}|]}} \left(\text{Max}_{\leq_{\mathcal{R}[|\mathcal{R}|-1]}} \left(\cdots \text{Max}_{\leq_{\mathcal{R}[1]}} (W \setminus L_1) \cdots \right) \right)$$

it follows that

$$\chi_2 := \mu_{|\mathcal{R}|}(\top \wedge \neg\chi_1) \qquad \text{is such that} \qquad L_2 = [\![\chi_2]\!]^M$$

This process can be repeated. In its mth iteration, by observing

$$L_m = \text{Max}_{\leq_{\mathcal{R}[|\mathcal{R}|]}} \left(\text{Max}_{\leq_{\mathcal{R}[|\mathcal{R}|-1]}} \left(\cdots \text{Max}_{\leq_{\mathcal{R}[1]}} (W \setminus \bigcup_{k=1}^{m-1} L_k) \cdots \right) \right)$$

it follows that

$$\chi_m := \mu_{|\mathcal{R}|}(\top \wedge \bigwedge_{k=1}^{m-1} \neg\chi_k) \qquad \text{is such that} \qquad L_m = [\![\chi_m]\!]^M$$

The process stops here, as $W = L_1 \cup \cdots \cup L_m$, and thus further iterations will produce formulas χ such that $[\![\chi]\!]^M = \varnothing$.

Proposition 3. *The conservative upgrade is an instance of the general layered upgrade with S defined syntactically within $\mathcal{L}^{PR}$.*

Proof (Sketch). It is enough to provide the explicit definition of a syntactically defined layered list that does the job. Take any PR model M with domain W and observe how $[\![\bigwedge_{i' \in Ag}(i' \sqsubseteq_j i)]\!]^M = W$ iff $i = \text{mr}(j)$. Then,

$$\chi := \bigwedge_{i' \in Ag}(i' \sqsubseteq_j i) \rightarrow [<_i]\bot \qquad \text{implies} \qquad \text{Max}_{\leq_{\text{mr}(j)}}(W) = [\![\chi]\!]^M$$

Thus, a 'singleton list' with χ as its unique set and $\leq_j$ as its default relational expression induces the ordering generated by the conservative upgrade.

A more illuminating way to prove how the general layered upgrade indeed extends the general lexicographic one is by noticing that, while the general lexicographic upgrade cannot revert strict preferences when these are unanimous, the general layered can. More precisely, on the one hand,

Proposition 4. *For every PR model, if all agents put a given world strictly above another, then so does the ordering $\leq_{\mathcal{R}}$ induced by any lexicographic $\mathcal{R}$.*

Nevertheless, on the other hand,

Fact 1. *There are PR models in which all agents agree in the relative strict order between two worlds, and yet a general layered upgrade can reverse it.*

Proof. Take a frame with a single agent having a strict preference of w_1 over w_2. This order is switched by using a general layered upgrade with a singleton list given by $[>]\bot$ and with 1 being the relational expression for its default ordering.

The generality offered by the general layered upgrade might be welcomed from some perspectives, but it might not be completely desirable from others. For example, when the layered list is given syntactically by $\mathcal{L}^{PR}$, it allows the definition of 'unreasonable' preference upgrade policies. As an illustration, a singleton layered list for agent j with its set defined by $[>_{\mathrm{mr}(j)}]\bot$ (with $\mathrm{mr}(j)$ characterised as in the proof of Proposition 3) will move to the top of j's ordering those worlds that are, for her *most reliable* agent, the *least preferred* ones.

From this perspective, the general lexicographic upgrade has some advantages over the layered one: not only combines the current orderings in a 'natural' way, but also has some pleasant properties, as it respects unanimity not only over strict preferences, as shown before, but also over equal-preferability.

Proposition 5. *For every PR model, if all agents agree in that two worlds are equally preferred, then so does the ordering $\leq_{\mathcal{R}}$ induced by any lexicographic $\mathcal{R}$.*

A more detailed study of the expressivity of the general layered upgrade when the used list is syntactically defined within $\mathcal{L}^{PR}$ is left for future work.

Some Observations and a Comparison. In the literature one can find several operations describing changes in orderings among objects. In particular, there are several *dynamic epistemic logic* [10,11] proposals for orderings interpreted not only as preferences (so the operations represent preference change: e.g., [12,13]), but also as plausibility (so the operations are understood as forms of belief revision: e.g., [8,9,14,15]). It is worthwhile to discuss, albeit briefly, some key characteristics of the general layered upgrade and how it relates to existing frameworks.

A straightforward observation is that the general layered upgrade (*GLay*) only affects the ordering, keeping the domain intact (thus differing from, e.g., [14,15]). More interesting is the fact that, although it generalises the general lexicographic upgrade (*GLex*) of [3], *GLay* still has a lexicographic spirit: the sets in $\mathcal{S}$ actually define an ordering, and thus $\leq_{\mathcal{S}}$ is the result of a lexicographic upgrade with the order generated by the sets having the highest priority, and the default ordering being used only to 'break ties'.

A closer comparison between GLay and the plausibility action models (*PAM*) of [15] is also useful. They share the same spirit, as a *PAM* is a relational structure in which each ordered 'world' is associated to a formula, thus defining in this way an ordering among sets of worlds, just as the sets of a layered list in *GLay*. Moreover, a *PAM* acts over a plausibility model following the 'action priority' rule: the ordering in the resulting model is a combination of the one in initial model with that of the *PAM* in which the latter has the priority, exactly as *GLay* does when it prioritises the sets over the default ordering. In fact, the crucial difference between these frameworks might be simply the expressivity of the language used for both the initial ordering and the default one. With respect to the initial ordering, the language used in the *PAM* framework is equivalent to the fragment of $\mathcal{L}^{PR}$ in which the relational expressions are only $\leq_i$ and $\sim_i := \leq_i \cup \geq_i$; maybe more important, with respect to the default ordering, by construction *PAM* uses only the agent's preference relation, but *GLay* allows a full relational expression. Whether this difference in expressivity allows *GLay* to create orderings that cannot be defined by using *PAM* remains to be studied.

Note that the layers of a 'layered list' can be interpreted as levels of beliefs in such plausibility models, and also in the $KD45\text{-}O$ models of [16]. The basic difference here is the additional 'default ordering' which makes sense while describing preferences as it can be thought of as some preference ordering exogenously instilled in agents, which comes to the fore when absolutely necessary.

The Formal Language

Definition 11. *The language $\mathcal{L}^{PR}_{\{gy\}}$ extends $\mathcal{L}^{PR}$ with a modality $\langle gy^i_S \rangle$ for every agent $i \in Ag$ and every layered list S whose default ordering is reflexive, transitive and total. Given a PR model M, define*

$$[\![\langle gy^i_S \rangle \varphi]\!]^M := [\![\varphi]\!]^{gy^i_S(M)}$$

with $gy^i_S(M)$ as in Definition 10. Note how, by defining $[gy^i_S]\varphi := \neg\langle gy^i_S \rangle \neg\varphi$, then $[\![[gy^i_S]\varphi]\!]^M := [\![\varphi]\!]^{gy^i_S(M)}$ so $\langle gy^i_S \rangle \varphi \leftrightarrow [gy^i_S]\varphi$ is valid.

The modality $\langle gy^i_S \rangle$ allows to describe the effects of upgrading agent i's preferences via the general layered upgrade with S, keeping the preferences of the remaining agents as before. This definition can be extended to simultaneous upgrades by asking for a list $\boldsymbol{S}$ of layered lists and using a modality $\langle gy_{\boldsymbol{S}} \rangle$ whose semantic interpretation uses the operation $gy_{\boldsymbol{S}}(\cdot)$ of Definition 10.

For an axiom system, this paper provides valid formulas and validity-preserving rules indicating how to rewrite a formula using $\langle gy^i_S \rangle$ as a provably equivalent one in $\mathcal{L}^{PR}$. Then, while soundness follows from the validity and validity preserving properties of the rewriting tools, completeness follows from the completeness of the basic 'static' system (end of Subsection 2.1).[8]

Besides indicating how to translate atomic propositions, reliability formulas and their Boolean combinations, the rewriting formulas should indicate how to translate formulas involving modal operators of the form $\langle \pi \rangle$, where π can be any relational expression. Hence, given any relational expression in the model $gy^i_S(M)$, a 'matching' relational expression in the original model M should be provided. The layered relational transformer defined below, similar in spirit to the program transformers of [18] for providing rewriting axioms for regular *PDL*-expressions [7] (in their case, after the action-model updates of [19]), will capture this. However, in order to express within $\mathcal{L}^{PR}$ the effect of a general layered upgrade, the used layered list S must be syntactically defined in $\mathcal{L}^{PR}$: indeed, if either some $S[k]$ or else the default ordering $\leq^S_{\text{def}}$ is not $\mathcal{L}^{PR}$-definable, then the language cannot tell whether a world is in $S[k]$ or whether a pair satisfies $\leq^S_{\text{def}}$, and thus it cannot describe the upgrade's effects.

Definition 12 (Layered Relational Transformer). *Let M be a PR model with domain W; let i be an agent. Let S be a syntactically defined layered list over W for which each set $S[k]$ is characterised by a formula $\chi_{S[k]}$ in $\mathcal{L}^{PR}$ (i.e., $S[k] = [\![\chi_{S[k]}]\!]^M$) and whose default ordering $\leq^S_{\text{def}}$ is characterised by a relational*

---

[8] See Chapter 7 of [10] (cf. [17]) for an extensive explanation of this technique.

expression $\pi_{\text{def}}^{\mathcal{S}}$ *in* $\mathcal{L}^{PR}$ *(i.e.,* $\leq_{\text{def}}^{\mathcal{S}} = \ulcorner \pi_{\text{def}}^{\mathcal{S}} \urcorner^M$ *). Define* $\mathrm{N}_{\mathcal{S}}$ *as the formula satisfied by those worlds that do not appear in a set in* $\mathcal{S}$*, and* $\mathrm{N}_{\mathcal{S}}^{k}$ *as the formula satisfied by those worlds that do not appear in the sets* $\mathcal{S}[1], \ldots, \mathcal{S}[k]$ *for some index* k:[9]

$$\mathrm{N}_{\mathcal{S}} := \neg \bigvee_{k=1}^{|\mathcal{S}|} \chi_{\mathcal{S}[k]} \qquad\qquad \mathrm{N}_{\mathcal{S}}^{k} := \neg \bigvee_{l=1}^{k} \chi_{\mathcal{S}[l]}$$

A layered relational transformer $Ty_{\mathcal{S}}^{i}$ *is a function from relational expressions to relational expressions defined in the following way.*

$$Ty_{\mathcal{S}}^{i}(\leq_i) := \underbrace{\left(\pi_{\text{def}}^{\mathcal{S}} \cap \left(?(\mathrm{N}_{\mathcal{S}}, \mathrm{N}_{\mathcal{S}}) \cup \bigcup_{k=1}^{|\mathcal{S}|} ?(\chi_{\mathcal{S}[k]}, \chi_{\mathcal{S}[k]}) \right) \right)}_{1} \cup \underbrace{\bigcup_{k=1}^{|\mathcal{S}|} ?(\mathrm{N}_{\mathcal{S}}^{k}, \chi_{\mathcal{S}[k]})}_{2}$$

$$Ty_{\mathcal{S}}^{i}(\geq_i) := \underbrace{\left((\pi_{\text{def}}^{\mathcal{S}})^{-1} \cap \left(?(\mathrm{N}_{\mathcal{S}}, \mathrm{N}_{\mathcal{S}}) \cup \bigcup_{k=1}^{|\mathcal{S}|} ?(\chi_{\mathcal{S}[k]}, \chi_{\mathcal{S}[k]}) \right) \right)}_{1} \cup \underbrace{\bigcup_{k=1}^{|\mathcal{S}|} ?(\chi_{\mathcal{S}[k]}, \mathrm{N}_{\mathcal{S}}^{k})}_{2}$$

$$Ty_{\mathcal{S}}^{i}(1) := 1 \qquad\qquad\qquad Ty_{\mathcal{S}}^{i}(-\pi) := -Ty_{\mathcal{S}}^{i}(\pi)$$

$$Ty_{\mathcal{S}}^{i}(\leq_j) := \leq_j \quad for\ i \neq j \qquad Ty_{\mathcal{S}}^{i}(\pi \cup \sigma) := Ty_{\mathcal{S}}^{i}(\pi) \cup Ty_{\mathcal{S}}^{i}(\sigma)$$

$$Ty_{\mathcal{S}}^{i}(\geq_j) := \geq_j \quad for\ i \neq j \qquad Ty_{\mathcal{S}}^{i}(\pi \cap \sigma) := Ty_{\mathcal{S}}^{i}(\pi) \cap Ty_{\mathcal{S}}^{i}(\sigma)$$

$$Ty_{\mathcal{S}}^{i}(?(\psi_1, \psi_2)) := ?(\langle \text{gy}_{\mathcal{S}}^{i} \rangle \psi_1, \langle \text{gy}_{\mathcal{S}}^{i} \rangle \psi_2)$$

Intuitively, a layered relational transformer $Ty_{\mathcal{S}}^{i}$ takes a relational expression representing a relation in the model $\text{gy}_{\mathcal{S}}^{i}(M)$ and returns a matching relational expression representing a relation in the original model M. The cases for the basic relational expressions, $\leq_i$ and $\geq_i$, are the important ones. The first uses Definition 8 to establish that $\leq_i$ in $\text{gy}_{\mathcal{S}}^{i}(M)$ corresponds to $\leq_{\mathcal{S}(M)}$ in M; the second uses the same definition to indicate that $\geq_i$ in $\text{gy}_{\mathcal{S}}^{i}(M)$ is the converse of $\leq_{\mathcal{S}(M)}$ in M. The remaining cases take care of the constant 1, the basic relational expressions for agents other than i and of the relational test as well as the complement, union and intersection of relations. With $Ty_{\mathcal{S}}^{i}$ defined, it is possible now to provide the promised axiom system.

Theorem 2. *The axioms and rules on Table 1 together with those of the basic 'static' system (end of Subsection 2.1) provide a sound and complete axiom system (with i any agent) for* $\mathcal{L}_{\{\text{gy}\}}^{PR}$ *with respect to PR models.*

Proof (Sketch). The rule and the axioms for atomic propositions, reliability, negation and disjunction are standard for an operation without precondition that does not affect atomic propositions (and, in this case, neither reliability). The axiom for relational expressions, the key one, makes crucial use of the layered relational transformer, stating that there is a φ-world π-reachable from the evaluation point at $\text{gy}_{\mathcal{S}}^{i}(M)$ if and only if there is a $\langle \text{gy}_{\mathcal{S}}^{i} \rangle \varphi$-world $Ty_{\mathcal{S}}^{i}(\pi)$-reachable from the evaluation point at M. As an example, if π is $\leq_i$, then the axiom is

---

[9] The case with $|\mathcal{S}| = 0$ can be understood as a case in which each $\mathcal{S}[k]$ is the empty set, and thus $\chi_{\mathcal{S}[k]} = \perp$. In such case, $\mathrm{N}_{\mathcal{S}}$ becomes the always true $\top$.

Table 1. Axioms for $\mathcal{L}^{PR}_{\{gy\}}$ w.r.t. PR models.

$\vdash \langle gy^i_S \rangle \top$	$\vdash \langle gy^i_S \rangle (\varphi \vee \psi) \leftrightarrow (\langle gy^i_S \rangle \varphi \vee \langle gy^i_S \rangle \psi)$
$\vdash \langle gy^i_S \rangle p \leftrightarrow p$	$\vdash \langle gy^i_S \rangle (\varphi \to \psi) \leftrightarrow (\langle gy^i_S \rangle \varphi \to \langle gy^i_S \rangle \psi)$
$\vdash \langle gy^i_S \rangle j' \sqsubseteq_j j'' \leftrightarrow j' \sqsubseteq_j j''$	$\vdash \langle gy^i_S \rangle \langle \pi \rangle \varphi \leftrightarrow \langle Ty^i_S(\pi) \rangle \langle gy^i_S \rangle \varphi$
$\vdash \langle gy^i_S \rangle \neg \varphi \leftrightarrow \neg \langle gy^i_S \rangle \varphi$	From $\vdash \varphi$ infer $\vdash [gy^i_S] \varphi$

$$\langle gy^i_S \rangle \langle \leq_i \rangle \varphi \leftrightarrow \langle \underbrace{\left(\pi^S_{\text{def}} \cap \left(?(N_S, N_S) \cup \bigcup_{k=1}^{|S|} ?(\chi_{S[k]}, \chi_{S[k]}) \right) \right)}_{1} \cup \underbrace{\bigcup_{k=1}^{|S|} ?(N^k_S, \chi_{S[k]})}_{2} \rangle \langle gy^i_S \rangle \varphi$$

whose right-hand side, by using the axioms for $\cup$ and $?$ together with some commutation and distribution, is equivalent to

$$\underbrace{\langle \pi^S_{\text{def}} \cap ?(N_S, N_S) \rangle \langle gy^i_S \rangle \varphi \vee \bigvee_{k=1}^{|S|} \langle \pi^S_{\text{def}} \cap ?(\chi_{S[k]}, \chi_{S[k]}) \rangle \langle gy^i_S \rangle \varphi}_{1}$$

$$\vee \underbrace{\bigvee_{k=1}^{|S|} \left(N^k_S \wedge \langle 1 \rangle (\chi_{S[k]} \wedge \langle gy^i_S \rangle \varphi) \right)}_{2}$$

Thus, the axiom states that after a general layered upgrade for i with S there will be a $\leq_i$-reachable φ-world, $\langle gy^i_S \rangle \langle \leq_i \rangle \varphi$, if and only if before the operation the current world is not in S and can $\leq^S_{\text{def}}$-reach a world not in S that will satisfy φ after the operation (first disjunct on part 1^{10}), or else the current world is in some $S[k]$ and can $\leq^S_{\text{def}}$-reach a world also in $S[k]$ that will satisfy φ after the operation (second disjunct on part 1), or else there is a k such that the current world is not in the sets $S[1], \ldots, S[k]$ and there is a world in $S[k]$ that will satisfy φ after the operation (part 2). This is simply the unfolding of the definition of $\leq_S$ (Definition 8), and it emphasises the role played by the formulas characterising each $\chi_{S[k]}$ and the relational expression characterising $\leq^S_{\text{def}}$.

Observe how the simultaneous upgrade modality $\langle gy_S \rangle$, briefly sketched below Definition 11, is also axiomatised by the presented system as long as the relational transformer is changed by making the cases for each agent i relative to i's layered list S_i (thus removing the cases "for agents different from i").

Going back to the example discussed in Section 3, let p_i denote the fact that 'movie i is most preferred'. Consider the PR model M as given in Example 1, where agent a can be the most reliable agent for himself, and $V(p_i) = \{w_i\}$ for each i. Then one can easily show that $W_M = [\![\langle \leq_b \rangle p_2 \wedge \langle gy^b_S \rangle \langle \leq_b \rangle p_4]\!]^M$.

---

[10] Recall that $?(\varphi, \psi)$ describes the relation $[\![\varphi]\!]^M \times [\![\psi]\!]^M$. Hence, $\leq \cap ?(\psi, \psi)$ describes the restriction of $\ulcorner \leq \urcorner^M$ to the set of worlds satisfying ψ.

4   Conclusions and Further Work

The paper has introduced a general preference upgrade operation subsuming several reasonable upgrade policies, providing also a modality to describe its effects as well as its complete axiomatisation. As the motivation for this work comes from the modelling of the process of deliberation, the next step in this research project is the characterisation not only of those situations in which the repetitive application of (instances of) the defined operation leads to agents having the same preferences (preference unanimity), but also of those situations in which further applications of the operation do not make any difference (preference stability). Also interesting is an in-depth exploration of the power of such operation as well as an extensive and formal comparison with related frameworks. Finally, it would be meaningful to get a formal framework for decision making processes that combine the methods of deliberation and aggregation.

References

1. Dietrich, F., List, C.: Where do preferences come from? International Journal of Game Theory 42(3), 613–637 (2013)
2. Dryzek, J.S., List, C.: Social choice theory and deliberative democracy: A reconciliation. British Journal of Political Science 33, 1–28 (2003)
3. Ghosh, S., Velázquez-Quesada, F.R.: Agreeing to agree: Reaching unanimity via preference dynamics based on reliable agents. In: Bordini, R., Elkind, E., Weiss, G., Yolum, P. (eds.) AAMAS 2015, pp. 1491–1499 (2015)
4. Burgess, J.P.: Basic tense logic. In: Gabbay, D., Guenthner, F. (eds.) Handbook of Philosophical Logic, vol. II, pp. 89–133. Reidel (1984)
5. Goldblatt, R.: Logics of Time and Computation, 2nd edn. CSLI Lecture Notes, no. 7. Center for the Study of Language and Information, Stanford (1992)
6. Gargov, G., Passy, S.: A note on boolean modal logic. In: Petkov, P.P. (ed.) Mathematical Logic, pp. 299–309. Springer US (1990)
7. Harel, D., Kozen, D., Tiuryn, J.: Dynamic Logic. MIT Press, Cambridge (2000)
8. Rott, H.: Shifting priorities: Simple representations for 27 iterated theory change operators. In: Lagerlund, H., Lindström, S., Sliwinski, R. (eds.) Modality Matters: Twenty-Five Essays in Honour of Krister Segerberg. Uppsala Philosophical Studies, no. 53, pp. 359–384. University of Uppsala, Upsala (2006)
9. van Benthem, J.: Dynamic logic for belief revision. Journal of Applied Non-Classical Logics 17(2), 129–155 (2007)
10. van Ditmarsch, H., van der Hoek, W., Kooi, B.: Dynamic Epistemic Logic. Synthese Library, no. 337. Springer (2008)
11. van Benthem, J.: Logical Dynamics of Information and Interaction. Cambridge University Press (2011)
12. van Benthem, J., Liu, F.: Dynamic logic of preference upgrade. Journal of Applied Non-Classical Logics 17(2), 157–182 (2007)
13. Grüne-Yanoff, T., Hansson, S.O.: Preference Change. Theory and Decision Library, vol. 42. Springer (2009)
14. van Eijck, J., Wang, Y.: Propositional dynamic logic as a logic of belief revision. In: Hodges, W., de Queiroz, R. (eds.) WoLLIC 2008. LNCS (LNAI), vol. 5110, pp. 136–148. Springer, Heidelberg (2008), doi:10.1007/978-3-540-69937-8_13

15. Baltag, A., Smets, S.: A qualitative theory of dynamic interactive belief revision. In: Bonanno, G., van der Hoek, W., Wooldridge, M. (eds.) Logic and the Foundations of Game and Decision Theory (LOFT 7). Texts in Logic and Games, vol. 3, pp. 13–60. Amsterdam University Press, Amsterdam (2008)

16. Ghosh, S., de Jongh, D.: Comparing strengths of beliefs explicitly. Logic Journal of the IGPL 21(3), 488–514 (2013)

17. Wang, Y., Cao, Q.: On axiomatizations of public announcement logic. Synthese 190(18), 103–134 (2013)

18. van Benthem, J., van Eijck, J., Kooi, B.: Logics of communication and change. Information and Computation 204(11), 1620–1662 (2006)

19. Baltag, A., Moss, L.S., Solecki, S.: The logic of public announcements and common knowledge and private suspicions. In: Gilboa, I. (ed.) Proceedings of the 7th Conference on Theoretical Aspects of Rationality and Knowledge (TARK 1998), Evanston, IL, USA, July 22-24, pp. 43–56. Morgan Kaufmann (1998)

Informational Dynamics of 'Might' Assertions

Peter Hawke and Shane Steinert-Threlkeld

Department of Philosophy, Stanford University,
450 Serra Mall, Building 90, Standford, CA 94305, USA
{phawke,shanest}@stanford.edu

Abstract. We investigate, in a logical setting, the proposal that assertion primarily functions to express and coordinate doxastic states and
that 'might' fundamentally expresses *lack* of belief. We provide a formal
model of an agent's doxastic state and precise assertability conditions
for an associated formal language. We thereby prove that an arbitrary
assertion (including a complex of 'might' and 'factual' claims) always
succeeds in expressing a well-defined doxastic state. We then propose
a fully general and intuitive doxastic update operation as a model of
an agent coming to accept an arbitrary assertion. We provide reduction
axioms for some novel update operations related to this proposal.

1 Introduction

Consider the following conversation in ordinary language:

(1) Context: Mark hasn't been able to find his house keys in his pocket, his bag,
or on his nightstand. While searching, Mark looks out the window at his
partner Sue's car, but he sees no reason to think the keys could be there: it
is extremely rare that Sue uses his house keys without checking with him.
   a) M: I'm so annoyed. I must have accidentally left my keys on the bus.
   b) S: Actually, they might be in my car.
   c) M: Ah, OK. I'll go look.

Intuitively, Sue has raised for Mark the possibility that the keys are in her car,
which he had previously decided not to take 'seriously' even though he was aware
of it. He acknowledges that possibility and so goes to check her car.

Surprisingly, offering an explanation of this information flow has proven difficult. Most theorists have tried to provide a semantics for the word 'might' which,
when combined with a picture of assertion, will generate the right results.[1] The
simplest explanation would identify a particular piece of information that Sue
puts forward in (1b) and which Mark subsequently adopts. Note, however, that
the information that it's compatible with Sue's information that the keys are
in her car does not do the trick. If Mark accepts that information, he acquires
a belief about Sue, not about the keys. More sophisticated views posit *many*
pieces of information[2] or information whose truth depends on who assesses it.[3]

---

[1] The orthodox semantics belongs to Kratzer [1981, 2012]. See also Papafragou [2006].
[2] See von Fintel and Gillies [2011].
[3] See MacFarlane [2011a, 2014].

© Springer-Verlag Berlin Heidelberg 2015
W. van der Hoek et al. (Eds.): LORI 2015, LNCS 9394, pp. 143–155, 2015.
DOI: 10.1007/978-3-662-48561-3_12

Although we will not explicitly argue against these views here, their baroqueness merits initial hesitation.

Against this backdrop, expressivists argue that epistemic modals generally serve to express features of agents' doxastic states and assertion helps to co-ordinate on those features.[4] In this paper, we develop a formal model of the above information flow that captures these expressivist thoughts. In the next section, we motivate the view that 'might' expresses *lack* of belief (which we call abelief).[5] Following that, we provide a formal model of assertability and doxastic states that allows us to precisely identify the set of beliefs and abeliefs expressed by an arbitrary (and possibly complex) assertion. Then, we identify a *simultaneous update operation* for a given set of beliefs and abeliefs which is fully general, gives the right results in many cases, and reduces to natural updates in the particular case of bare indicative or bare 'might' assertions. After presenting this model, we demonstrate how our view handles epistemic contradictions, disagreement, and interactions with conjunction and disjunction.

2   Proposal: 'Might' as Abelief Coordinator

To make precise our view that assertion functions primarily to coordinate doxastic states, we need to say what doxastic states are. For us, such a state is a set of worlds W with a plausibility order $\succeq$. An agent believes that p if and only if p is true in all of the most plausible worlds. This natural model generalizes the standard modal semantics of belief in a way that allows conditional beliefs and various revision policies to be modeled.[6]

To warm up to our analysis of (1), consider a 'factual' version:

(2) Context: as in (1)
   a) M: I must have accidentally left my keys on the bus.
   b) S: They are in my car.
   c) M: Oh, OK. Thanks!

On our picture, Sue, in (2b), does two things: she *expresses* that she believes that the keys are on the table and *invites* Mark to modify his doxastic state so as to acquire that belief. When Mark accepts the assertion, he does so modify his state.

Let c, b, n, and p be the propositions that the keys are in Sue's car, in Mark's bag, on his nightstand, or in his pocket, respectively. We can model Mark's doxastic state with 5 worlds: $W = \{c, b, n, p, L\}$. In our abused notation, the worlds c, b, etc. are worlds in which only the corresponding proposition is true. L is a world in which the keys are 'lost' (i.e. left on the bus). Initially, Mark's doxastic state looks like this:[7]

---

[4] See Yalcin [2007, 2011] for expressivism about epistemic modals.

[5] An agent abelieves that p iff she does not believe that p. This is different from *disbeleiving* that p, which means believing that not-p.

[6] See, for instance, van Benthem [2011].

[7] $w \succ v$ means w is strictly more plausible than v. In our notation, b, n, p are all equally plausible.

(3) $L \succ c \succ b, n, p$

Because the unique most plausible world is an L world, Mark believes L, that the keys are lost. Upon accepting Sue's assertion, Mark's doxastic state becomes:

(4) $c \succ L \succ b, n, p$

The c world has been upgraded to be the unique most plausible world, and so Mark believes that the keys are in Sue's car.

What, then, about (1)? Sue's 'might' assertion in (1b) *expresses* that she abelieves that $\neg c$ and *invites* Mark to modify his doxastic state so as to acquire that abelief. After Mark accepts the assertion, his doxastic state looks like:

(5) $c, L \succ b, n, p$

Because there is a c world *among* the most plausible worlds, Mark no longer believes $\neg c$, i.e. he now abelieves $\neg c$.

In general, assertion functions to coordinate doxastic states by expressing a state and inviting one's interlocutors to adopt the same state. The two most fundamental such states are belief and abelief. Let us call assertions which primarily express beliefs B-assertions and those which primarily express abeliefs A-assertions. In general, then, we can say the following about the informational effect of accepting assertions of each type:

- B-assertion triggers *conservative revision*: $\uparrow p \,(\succeq)$ is just like $\succeq$ with the most plausible p-worlds made more plausible than all others
- A-assertion triggers *conservative contraction*: $\downarrow p \,(\succeq)$ is just like $\succeq$ with the most plausible p-worlds merged with the previous most plausible worlds

These notions of update – and our terminology – are not new: conservative revision is closely related to standard notions of *revision* in the AGM belief revision literature; while conservative contraction is closely related to standard notions of belief *contraction*.[8]

We note two points. First, given this picture of assertion, it would be very surprising if we had no means of expressing abelief. Secondly, viewing 'might' as expressing abelief provides a plausible model of its role in the dynamics of conversation. We substantiate this claim more below.

3   Two Problems for Mixed Assertions

While our previous story gave a precise and intuitive account of 'bare' B-assertions and A-assertions, it must be generalized to handle assertions of higher complexity, potentially mixing expressions of belief and abelief. A simple example: $p \wedge \Diamond q$.

---

[8] Conservative revision corresponds in a precise sense to *transitively relational partial meet revision* and conservative contraction corresponds to *transitively relational partial meet contraction*. See Hansson [2014], especially sect. 4, for an overview of these results. See Rott [2009] for a comprehensive list of belief update procedures, including those that appear in this paper.

Intuitively, an assertion thereof expresses belief in p and abelief in $\neg q$. But now consider an assertion of $p \vee \Diamond q$. What doxastic state is thereby expressed? Or, even worse: $(p \wedge \Diamond q) \vee \Diamond (s \wedge (\Diamond t \wedge \neg q))$? In the current section, we provide two logical frameworks (3.1-3.3) which together give precise answers (3.4-3.5) to the following two questions about an *arbitrary* assertion: (i) can it be understood to express a doxastic state (and, if so, what state is expressed)? (ii) what update operation is performed on acceptance?

3.1 Language

We work with a standard logical language containing: atomic proposition letters $(p, q, r, \dots)$; boolean operators $\neg, \vee, \wedge$; $\Diamond \varphi$ ("φ might be the case"); and $B\varphi$ ("the agent believes that φ").

3.2 Assertability Logic

Let $\mathbf{s}$ be an information set (a set of possible worlds). We will define what it means for a formula to be *assertable*[9] relative to an information set. In what follows, read $\mathbf{s} \Vdash \varphi$ as "φ is assertable relative to information $\mathbf{s}$". For the sake of simplicity, we save the case of $B\varphi$ for an extended version of this paper. Call the fragment of our language that ignores the B operator the *assertability language*.

Definition 1 (General Assertability Conditions). *Given a set of worlds W, an information state $\mathbf{s} \subseteq W$, and a valuation V:*

- $\mathbf{s} \Vdash p$ *iff:* $\forall w \in \mathbf{s}$: $w \in V(p)$
- $\mathbf{s} \Vdash \neg\varphi$ *iff:* $\forall w \in \mathbf{s}$: $\{w\} \nVdash \varphi$
- $\mathbf{s} \Vdash \varphi \wedge \psi$ *iff:* $\mathbf{s} \Vdash \varphi$ *and* $\mathbf{s} \Vdash \psi$
- $\mathbf{s} \Vdash \varphi \vee \psi$ *iff:* $\exists \mathbf{s}_1, \mathbf{s}_2$: $\mathbf{s} = \mathbf{s}_1 \cup \mathbf{s}_2$ *and* $\mathbf{s}_1 \Vdash \varphi$ *and* $\mathbf{s}_2 \Vdash \psi$
- $\mathbf{s} \Vdash \Diamond\varphi$ *iff:* $\mathbf{s} \nVdash \neg\varphi$

We intend these conditions to reflect compelling pre-theoretic intuitions. The final clause is, in particular, worth remarking on: this clause is inspired by the strongly felt illegitimacy of asserting both "it might be that φ" and "φ is not the case" in a single context (an intuition emphasized by Yalcin [2007, 2011]). Certain important consequences of these conditions are immediate:[10]

- s $\Vdash \Diamond\varphi$ iff $\exists w \in$ s : $\{w\} \Vdash \varphi$
- Relative to singletons $\{w\}$, this logic is classical
- Relative to singletons, $\Diamond\varphi$ and φ are equivalent

---

[9] It is not our goal to here offer an account that does full justice to our ordinary conception of assertion, nor the many facets of the theoretical role that assertion is intended to play in linguistic theorizing. For a more thoroughgoing discussion of assertion, see MacFarlane [2011b]. Our immediate goal is to offer a simple and natural account of when a sentence is assertable relative to a particular body of information, predominantly thought of as the belief worlds of a relevant agent.

[10] Our framework of assertability conditions is similar in technical spirit to the expressivist semantics of Lin [2013], a connection we do not detail here. At any rate, the formulation, conceptual underpinnings, dialectical role and technical consequences of the current framework diverge from that of Lin [2013] in significant ways.

3.3 Doxastic Logic

We now present a *truth-conditional semantics* for our language, with the intended purpose of making precise the manner of thinking about an agent's doxastic states that we have so far utilized in this paper. Our semantics is in the tradition of *dynamic doxastic logic*,[11] though we only add a dynamic component in the next section. Our goal is to have a precise language for describing the beliefs and abeliefs of an agent. In particular, it suits our purpose to exclude $\Diamond \varphi$ sentences from our semantics.

Definition 2 (Doxastic model). *A* doxastic model *is a tuple* $\mathcal{M} = \langle W, \{\succeq_w\}, V \rangle$ *where:*

- W *is a set of* worlds
- $\succeq_w$, *the* plausibility order *on W at w, is a* total pre-order *on W: a reflexive, transitive, total relation.*
- V *is a* valuation function *assigning a proposition (i.e. a set of worlds) to each atom p.*

Moreover, we require the orderings $\succeq_w$ to be reverse well-founded: *every non-empty $X \subseteq W$ has a maximal element.*[12]

For a given plausibility order $\succeq$, $\succ$ denotes its strict counterpart: $v \succ w$ iff $v \succeq w$ and $w \not\succeq v$. For $X \subseteq W$, we define $\text{BEST}_{\succeq}(X) := \{w \in X \mid \forall v \in X, w \succeq v\}$. This is the set of maximal, or 'best', worlds among X. We will denote by $\mathbf{b}_w$ the set of 'belief worlds' at w, that is the set of worlds maximal in $\succeq_w$, i.e. $\text{BEST}_{\succeq_w}(W)$. By the assumption of reverse well-foundedness, these sets are always non-empty.

Definition 3 (Static Semantics)

- $\mathcal{M}, w \vDash p$ *iff: $w \in V(p)$*
- $\mathcal{M}, w \vDash \neg \varphi$ *iff: $\mathcal{M}, w \nvDash \varphi$*
- $\mathcal{M}, w \vDash \varphi \wedge \psi$ *iff: $\mathcal{M}, w \vDash \varphi$ and $\mathcal{M}, w \vDash \psi$*
- $\mathcal{M}, w \vDash B\varphi$ *iff: for every $v \in \mathbf{b}_w$, $\mathcal{M}, v \vDash \varphi$*

Two more definitions will be useful in what follows. We will write $[\![\varphi]\!]^{\mathcal{M}} := \{w \in W \mid \mathcal{M}, w \vDash \varphi\}$ and omit the superscript when context allows. For brevity, we will also write $\text{BEST}_w(\varphi) := \text{BEST}_{\succeq_w}([\![\varphi]\!]^{\mathcal{M}})$.

3.4 From Assertion to Doxastic State Expression

In this section we work towards a theorem that addresses our first problem: what doxastic state is expressed by an arbitrary assertion? The theorem will state that for every assertable sentence there exists a well-defined doxastic state expressed by that sentence. What's more, the proof for this result supplies the ingredients for a method for constructing such a doxastic state, though we will not state such an algorithm explicitly.

---

[11] See van Benthem [2011] for an overview of this tradition.

[12] In terms of frame correspondence, we can impose this requirement via the Löb axiom $\Box (\Box p \to p) \to p$.

Definition 4 (Assertoric Equivalence). *We say that two sentences φ and ψ in the assertability language are* assertorically equivalent *just in case*

$$s \Vdash \varphi \text{ iff: } s \Vdash \psi$$

for every information state s (and every doxastic model $\mathcal{M}$).

Definition 5 ($\Diamond$-Free Formulae). *A sentence in our assertability language is $\Diamond$-free just in case it contains no occurrence of $\Diamond$.*

Lemma 1. *If φ is $\Diamond$-free, then $s \Vdash \varphi$ iff $s \subseteq \overline{V}(\varphi)$, where $\overline{V}$ is the unique extension of V interpreting $\neg$ as complement, $\wedge$ as intersection, and $\vee$ as union.*

Proof. By induction on $\Diamond$-free formulas (exercise).

Proposition 1 (Assertability Facts). *For any sentence φ in the assertability language, let φ^* be the sentence that results from φ by deleting every occurrence of $\Diamond$. Then:*

(1) $s \Vdash \Diamond\varphi$ iff: $s \Vdash \Diamond\varphi^*$. That is: $\Diamond$ is redundant in the scope of $\Diamond$.
(2) $s \Vdash \neg\varphi$ iff: $s \Vdash \neg\varphi^*$. That is: $\Diamond$ is redundant in the scope of $\neg$.
(3) If ψ^1 and ψ^2 are both $\Diamond$-free, then:

$$s \Vdash (\psi^1 \wedge \Diamond\varphi_1^1 \wedge \ldots \wedge \Diamond\varphi_m^1) \vee (\psi^2 \wedge \Diamond\varphi_1^2 \wedge \ldots \wedge \Diamond\varphi_n^2) \text{ iff:}$$
$$s \Vdash (\psi^1 \vee \psi^2) \wedge \Diamond(\psi^1 \wedge \varphi_1^1) \wedge \ldots \wedge \Diamond(\psi^1 \wedge \varphi_m^1) \wedge \Diamond(\psi^2 \wedge \varphi_1^2) \wedge \ldots \wedge \Diamond(\psi^2 \wedge \varphi_n^2)$$

Proof.
(1) $s \Vdash \Diamond\varphi$ is equivalent to $\exists w \in s : \{w\} \Vdash \varphi$. Further, for any ψ that does not contain a belief operator, $\{w\} \Vdash \Diamond\psi$ holds just in case $\{w\} \Vdash \psi$ holds. Hence: $\exists w \in s : \{w\} \Vdash \varphi$ is equivalent to $\exists w \in s : \{w\} \Vdash \varphi^*$.
(2) $s \Vdash \neg\varphi$ is equivalent to $\forall w \in s : \{w\} \nVdash \varphi$. Further, for any ψ that does not contain a belief operator, $\{w\} \Vdash \Diamond\psi$ holds just in case $\{w\} \Vdash \psi$ holds. Hence: $\forall w \in s : \{w\} \nVdash \varphi$ is equivalent to $\forall w \in s : \{w\} \nVdash \varphi^*$.
(3) We illustrate the proof with a particular instance. The general case uses Lemma 1. We show that

$$s \Vdash (p \wedge \Diamond q) \vee (r \wedge \Diamond s) \text{ iff: } s \Vdash (p \vee r) \wedge \Diamond(p \wedge q) \wedge \Diamond(r \wedge s)$$

$s \Vdash (p \wedge \Diamond q) \vee (r \wedge \Diamond s)$
iff: $\exists s_1, s_2 : s_1 \cup s_2 = s$, and $s_1 \Vdash p$, and $s_1 \Vdash \Diamond q$, and $s_2 \Vdash r$, and $s_2 \Vdash \Diamond s$
iff: $s \Vdash p \vee r$ and $\exists v_1 \in s : \{v_1\} \Vdash p \wedge q$ and $\exists v_2 \in s : \{v_2\} \Vdash r \wedge s$
iff: $s \Vdash (p \vee r) \wedge \Diamond(p \wedge q) \wedge \Diamond(r \wedge s)$ □

Lemma 2. *Let φ be a sentence in the assertability language. Then there exist sentences $\beta, \alpha_1, \ldots, \alpha_n$ (for some $n \geq 0$) such that:*

- *$\beta, \alpha_1, \ldots, \alpha_n$ contain no occurrences of $\Diamond$,*
- *$s \Vdash \varphi$ iff $s \Vdash \beta \wedge \Diamond\alpha_1 \wedge \ldots \wedge \Diamond\alpha_n$*

Proof. By induction on the complexity of formulae. The non-trivial cases (taking the assumption that φ is assertorically equivalent to $\beta \wedge \Diamond \alpha_1 \wedge \ldots \wedge \Diamond \alpha_n$ as the induction hypothesis):

- $\neg \varphi$: using fact 2 of proposition 1, we conclude that $\neg \varphi$ is assertorically equivalent to $\neg (\beta \wedge \alpha_1 \wedge \ldots \wedge \alpha_n)$.
- $\varphi_1 \vee \varphi_2$: assume that φ_1 is assertorically equivalent to $\beta^1 \wedge \Diamond \alpha_1^1 \wedge \ldots \wedge \Diamond \alpha_m^1$, and that φ_2 is assertorically equivalent to $\beta^2 \wedge \Diamond \alpha_1^2 \wedge \ldots \wedge \Diamond \alpha_n^2$. Now use fact 3 of proposition 1.
- $\Diamond \varphi$: fact 1 of proposition 1 shows that this is assertorically equivalent to $\Diamond (\beta \wedge \alpha_1 \wedge \ldots \wedge \alpha_n)$ □

Lemma 3. *Let φ be a $\Diamond$-free sentence in the assertability language and $\mathcal{M}$ and w be an arbitrary doxastic model and world. Then:*

$$\mathbf{b}_w \Vdash \varphi \text{ iff: } \mathcal{M}, w \models B\varphi$$

Proof. By induction on the complexity of formulae (exercise).   □

Lemma 4. *Let $\mathcal{M}$ and w be an arbitrary doxastic model and world. Then, for any φ in the assertability language:*

$$\mathbf{b}_w \Vdash \Diamond \varphi \text{ iff: } \mathcal{M}, w \models \neg B(\neg \varphi)$$

Proof. By induction on the complexity of formulae (exercise).   □

Definition 6 (Doxastic State Description). *A doxastic state description is a sentence of the form*

$$B\varphi \wedge \neg B(\neg \psi_1) \wedge \ldots \wedge \neg B(\neg \psi_n)$$

where φ and ψ_i, for $i \leq n$, are all $\Diamond$-free.

Definition 7 (Doxastic State Expression). *φ (in the assertability language) expresses doxastic state description δ just in case: for every doxastic model $\mathcal{M}$ and world w,*

$$\mathbf{b}_w \Vdash \varphi \text{ iff: } \mathcal{M}, w \models \delta$$

Theorem 1 (From Assertion to Doxastic State Expression). *For every sentence φ in the assertability language, there exists a doxastic state description δ_φ that is expressed by φ.*

Proof. $\mathbf{b}_w \Vdash \varphi$
  iff: $\mathbf{b}_w \Vdash \beta \wedge \Diamond \alpha_1 \wedge \ldots \wedge \Diamond \alpha_n$ [Lemma 2]
  iff: $\mathbf{b}_w \Vdash \beta$ and $\mathbf{b}_w \Vdash \Diamond \alpha_1$ and ... and $\mathbf{b}_w \Vdash \Diamond \alpha_n$
  iff: $\mathcal{M}, w \models B\beta$ and $\mathcal{M}, w \models \neg B(\neg \alpha_1)$ and ... and $\mathcal{M}, w \models \neg B(\neg \alpha_n)$ [Lemmas 3 and 4]
  iff: $\mathcal{M}, w \models B(\beta) \wedge \neg B(\neg \alpha_1) \wedge \ldots \wedge \neg B(\neg \alpha_n)$ □

3.5 Dynamics

Recall our second problem: for an arbitrary assertion, what update does an agent's doxastic state undergo upon acceptance of that assertion? For example, consider an assertion $p \wedge \Diamond q$. This expresses the doxastic state $Bp \wedge \neg B \neg q$ in the sense of Definition 7. How can an agent update her doxastic state so that this is an adequate description thereof? First, note that *sequentially* applying conservative revision and contraction will not work. Consider $W = \{w_1, w_2, w_3\}$ where w_1 satisfies p but not q, w_2 satisfies q but not p, and w_3 satisfies both p and q. Suppose the agent's doxastic state has the form

(6) $w_1 \succ w_2 \succ w_3$

Then $\uparrow p \, (\downarrow \neg q \, (\succeq))$ will be the same order (6). But at a world with that order, $B \neg q$ is true. On the other hand, $\downarrow \neg q \, (\uparrow p \, (\succeq))$ is

(7) $w_1, w_2 \succ w_3$

But at a world with this order, Bp is not true. Thus, $Bp \wedge \neg B \neg q$ is not an accurate description of either doxastic state. This example shows that the update to perform upon accepting a mixed assertion cannot simply be an iteration of our earlier updates. To address this issue, we enrich the language with expressions of the form $[\Uparrow \varphi] \psi$ with intended reading: "after conservative expansion by φ, ψ holds" and use this to define a simultaneous update operation.

Definition 8 (Conservative Expansion). *Given an order $\succeq$ and $X \subseteq W$, we denote by $\Uparrow X \, (\succeq)$ the conservative expansion of $\succeq$ by X, where: $\Uparrow X \, (\succeq)$ is the order that is just like $\succeq$ except with all of X made most plausible and all worlds in X made equally plausible to each other.*

We extend this to a model-changing operation as follows: $\mathcal{M} \Uparrow X$ is just like $\mathcal{M}$, except with each $\succeq_w$ replaced with $\Uparrow X \, (\succeq_w)$. We focus on the case where $X = \llbracket \varphi \rrbracket^{\mathcal{M}}$ for some formula in our language, in which case we will write $\mathcal{M} \Uparrow \varphi$, calling this the conservative expansion of $\mathcal{M}$ by φ.

Definition 9 (Dynamic Semantics). *The static semantics can be extended:*

– $\mathcal{M}, w \vDash [\Uparrow \varphi] \psi$ *iff*: $\mathcal{M} \Uparrow \varphi, w \vDash \psi$

Using this framework, we have the resources to define *conservative revision* and *conservative contraction* operations, respectively as follows:

i. $\uparrow \varphi \, (\succeq_w) := \Uparrow \mathrm{BEST}_w(\varphi) \, (\succeq_w)$
ii. $\downarrow \varphi \, (\succeq_w) := \Uparrow (\mathrm{BEST}_w(\varphi) \cup \mathrm{BEST}_w(\top)) \, (\succeq_w)$

Now, we can define an operation that tells us how to update on an arbitrary doxastic state description. Intuitively, it is the operation of *simultaneously performing* the conservative revisions and conversative contractions suggested by the set of beliefs and abeliefs expressed by that description.

Definition 10 (Simultaneous Update). *The simultaneous update to believe* β *and abelieve* $\alpha_1, \ldots, \alpha_n$ *is the following operation:*

$$[\Uparrow \beta, \alpha_1, \ldots, \alpha_n]\,(\succeq_w) := \left[\Uparrow \text{BEST}_w(\beta) \cup \bigcup_{1 \leq i \leq n} \text{BEST}_w\,(\neg\alpha_i \wedge \beta)\right](\succeq_w)$$

For all of conservative revision, conservative contraction, and simultaneous update, we can define the appropriate model-changing operations and extend the syntax with dynamic operators in exactly the same way as was done for conservative expansion above. Our definiton of simultaneous update has many attractive consequences. First, note that this update handles our earlier counterexample. $[\Uparrow p, \neg q]$ applied to the order in (6) yields

(8) $w_1, w_3 \succ w_2$

In a world with this order, however, $Bp \wedge \neg B \neg q$ is true. Moreover, this definition handles our motivating cases (1) and (2) with aplomb. If φ^* is a doxastic state description, we will abbreviate the above by $[\Uparrow \varphi^*]$. In the case when φ^* has no conjunct $B(\beta)$, replace β with $\top$. In the case when φ^* has no conjunct $\neg B\psi_i$, set $n = 1$ and $\alpha_1 = \bot$.

Proposition 2. *Let* φ *be a sentence in the assertability language. Then, for every model and order:*

  i. *If* φ *expresses no abeliefs, then* $[\Uparrow \varphi^*]\,(\succeq) = [\uparrow \beta]\,(\succeq)$
 ii. *If* φ *expresses a single abelief, then* $[\Uparrow \varphi^*]\,(\succeq) = [\text{1}\, \alpha_1]\,(\succeq)$

When working with dynamic operators like this, a natural question to ask is: is every sentence in the language with dynamic operators equivalent to some sentence in the static 'base' language? One usually provides a 'yes' answer to this question by giving *reduction axioms* which show how to push the dynamic operators to simpler subformulas. We can provide such axioms for many of our operators. Conservative revision is already well understood,[13] so we focus on conservative expansion and conservative contraction. We start with conservative expansion and the doxastic language. We must augment the language with an existential modality E and its dual universal modality U.

Proposition 3. *The following reduction axioms are valid for the class of doxastic models:*

$$
\begin{aligned}
[\Uparrow \varphi]\, p &\leftrightarrow p \\
[\Uparrow \varphi]\, \neg\psi &\leftrightarrow \neg\,[\Uparrow \varphi]\, \psi \\
[\Uparrow \varphi]\, \psi \wedge \chi &\leftrightarrow [\Uparrow \varphi]\, \psi \wedge [\Uparrow \varphi]\, \chi \\
[\Uparrow \varphi]\, B\psi &\leftrightarrow (E\varphi \wedge U\,(\varphi \rightarrow [\Uparrow \varphi]\, \psi)) \vee (\neg E\varphi \wedge B\,[\Uparrow \varphi]\, \psi) \\
[\Uparrow \varphi]\, E\psi &\leftrightarrow E\,[\Uparrow \varphi]\, \psi
\end{aligned}
$$

---

[13] See chapter 8 of van Benthem [2011], where it goes by 'conservative upgrade'.

For the case of conservative contraction, we extend the language with a *conditional belief* operator $B^\varphi \psi$ with the following semantics:

Definition 11 (Conditional Belief). *For a doxastic model $\mathcal{M}$ and world w:*

$$\mathcal{M}, w \vDash B^\varphi \psi \text{ iff: for every } v \in \text{BEST}_w(\varphi), \mathcal{M}, v \vDash \psi$$

Note that $B\varphi$ is the special case $B^\top \varphi$.

Proposition 4. *The following reduction axioms are valid for the class of doxastic models: those above for atoms, $\neg$, $\wedge$, and E but with $\uparrow \varphi$ and*

$$[\uparrow \varphi] B\psi \qquad\qquad \leftrightarrow \qquad\qquad B[\uparrow \varphi]\psi \wedge B^\varphi [\uparrow \varphi]\psi$$

This axiom makes good intuitive sense. After the update, the agent believes ψ iff the new best worlds are all ψ. The new best worlds are: the old best worlds merged with the old best φ worlds. The first conjunct handles the former and the second conjunct the latter. Of course, if we have conditional belief in our language, one would like a reduction axiom for that operator.

Theorem 2. *The following reduction axioms are valid for doxastic models:*

$$[\Uparrow \varphi] B^\chi \psi \quad \leftrightarrow \quad \left(\neg E(\varphi \wedge [\Uparrow \varphi]\chi) \wedge B^{[\Uparrow \varphi]\chi}[\Uparrow \varphi]\psi \right) \vee$$
$$(E(\varphi \wedge [\Uparrow \varphi]\chi) \wedge U(\varphi \wedge [\Uparrow \varphi]\chi \to [\Uparrow \varphi]\psi))$$

$$[\uparrow \varphi] B^\chi \psi \quad \leftrightarrow \quad \left(B^\varphi \neg [\uparrow \varphi]\chi \wedge B^{[\uparrow \varphi]\chi}[\uparrow \varphi]\psi \right) \vee$$
$$\left(\neg B^\varphi \neg [\uparrow \varphi]\chi \wedge B^{\varphi \wedge [\uparrow \varphi]\chi}[\uparrow \varphi]\psi \wedge \left(\neg B\neg [\uparrow \varphi]\chi \to B^{[\uparrow \varphi]\chi}[\uparrow \varphi]\psi \right) \right)$$

Proof. First, consider conservative expansion. $[\Uparrow \varphi] B^\chi \psi$ says: after making all of the φ worlds most plausible (and equally plausible), the best χ worlds are ψ worlds. We make a case distinction: (i) no φ worlds become χ-worlds or (ii) some φ worlds become χ-worlds. If (i), then the best χ-worlds after the update are the best worlds pre-update that *become* χ. We then need to check that those worlds become ψ worlds. That is what the first disjunct in the reduction axiom states. If (ii), the best χ-worlds post-update are exactly the current φ worlds that *become* χ worlds since all of the current φ worlds become best overall. We thus need to check that every φ world which becomes χ also becomes a ψ world. That's what the second disjunct in the recursiom axiom states.

Now, consider conservative contraction. $[\uparrow \varphi] B^\chi \psi$ says: after merging the best φ worlds with the best-overall worlds, the best χ worlds are ψ worlds. We again make a case distinction: (i) no best φ worlds become χ worlds or (ii) some best φ worlds become χ worlds. In case (i), the best χ worlds post-update are simply the best worlds that *become* χ. We need those to become ψ, which is just what the first disjunct states. In case (ii), the best χ worlds post-update come from two sources: (a) previous best φ worlds that become χ and (b) previous best-overall worlds that become χ. The conjunct

$$\neg B^\varphi \neg [\uparrow \varphi]\chi \wedge B^{\varphi \wedge [\uparrow \varphi]\chi}[\uparrow \varphi]\psi$$

handles case (a) by requiring that the best φ worlds that become χ also become ψ. The conjunct

$$\neg B\neg [\mathord{\uparrow} \varphi]\, \chi \to B^{[\mathord{\uparrow} \varphi]\chi}\, [\mathord{\uparrow} \varphi]\, \psi$$

handles case (b).   □

We can derive reduction axioms for full belief in Propositions 3 and 4 as special cases of the above.

Corollary 1. *The following reduction axioms are valid for doxastic models:*

$$[\mathord{\uparrow\uparrow} \varphi]\, B\psi \quad\leftrightarrow\quad (E\varphi \wedge U\,(\varphi \to [\mathord{\uparrow\uparrow} \varphi]\, \psi)) \vee (\neg E\varphi \wedge B\,[\mathord{\uparrow\uparrow} \varphi]\, \psi)$$
$$[\mathord{\uparrow} \varphi]\, B\psi \quad\leftrightarrow\quad B\,[\mathord{\uparrow} \varphi]\, \psi \wedge B^\varphi\,[\mathord{\uparrow} \varphi]\, \psi$$

Proof. We do the conservative contraction case and leave conservative expansion as an exercise. Substituting $\top$ for χ in the above reduction axiom yields:

$$[\mathord{\uparrow} \varphi]\, B\psi \;\leftrightarrow\; \left(B^\varphi\neg[\mathord{\uparrow} \varphi]\,\top \wedge B^{[\mathord{\uparrow} \varphi]\top}\,[\mathord{\uparrow} \varphi]\, \psi\right) \vee$$
$$\left(\neg B^\varphi\neg[\mathord{\uparrow} \varphi]\,\top \wedge B^{\varphi \wedge [\mathord{\uparrow} \varphi]\top}\,[\mathord{\uparrow} \varphi]\, \psi \wedge \left(\neg B\neg[\mathord{\uparrow} \varphi]\,\top \to B^{[\mathord{\uparrow} \varphi]\top}\,[\mathord{\uparrow} \varphi]\, \psi\right)\right)$$

Notice that the first disjunction is a contradiction: $B^\varphi\neg[\mathord{\uparrow} \varphi]\,\top$ is always false since every world satisfies $\top$. The conjunct $\neg B^\varphi\neg[\mathord{\uparrow} \varphi]\,\top$ is always true since it merely states the existence of a best φ world, which the assumption of reverse well-foundedness ensures. The conjunct $B^{\varphi \wedge [\mathord{\uparrow} \varphi]\top}\,[\mathord{\uparrow} \varphi]\, \psi$ simplifies to $B^\varphi\,[\mathord{\uparrow} \varphi]\, \psi$. Now, the antecedent of the conditional is trivially true since it merely asserts that there are best worlds, which again holds by reverse well-foundedness. The consequent simplifies to $B\,[\mathord{\uparrow} \varphi]\, \psi$. Thus, we are left with the desired equivalence

$$[\mathord{\uparrow} \varphi]\, B\psi \qquad\qquad\leftrightarrow\qquad\qquad B\,[\mathord{\uparrow} \varphi]\, \psi \wedge B^\varphi\,[\mathord{\uparrow} \varphi]\, \psi$$

as desired.   □

4   Some Welcome Consequences

Epistemic Contradictions. As emphasized by Yalcin [2007, 2011], statements of the following form (so-called 'epistemic contradictions') seem defective: "John is in his office. But it might be that John is not in his office". Fortunately, then, our assertability conditions immediately yield (for any information state $\mathbf{s}$):

$$\mathbf{s} \not\Vdash p \wedge \Diamond\neg p$$

This does not yet entirely deal with the observation that sentences that *embed* epistemic contradictions are notably defective, for a key case for Yalcin is sentences of the form "supposing that $p \wedge \Diamond\neg p$, then ...", and our current setup does not have resources capturing suppositional actions/operators. In the extended version of this paper, we add such operators to our language, and treat their assertability conditions in a fashion inspired by Yalcin's domain semantics, yielding pleasing results.

Disagreement. Consider the following variation of (1), where Mark does not accept Sue's assertion.

(9) Context: as before, except that Mark actually went out and checked Sue's car, where he did not find the keys.
   a) M: I'm so annoyed. I must have accidentally left my keys on the bus.
   b) S: They might be in my car.
   c) M: No, I already checked your car.

In such a case two things must be explained: (i) How is it possible to disagree with an assertion of $\Diamond c$? (ii) Why, when one does so disagree, are the reasons provided about the prejacent c itself? Our story provides natural answers to both questions. To (i): Mark's disagreement consists in *rejecting* Sue's invitation to update his doxastic state to incorporate a c-world. To (ii): Mark rejects this invitation because he thinks he has good reason to have already ruled out the c worlds (for example, having already checked the car). Thus, when explaining his disagreement, he will argue about the prejacent c itself.

Interactions with Conjunction and Disjunction. It has been observed that conjunction and disjunction display unusual behavior when connecting 'might' claims (Zimmermann [2000], Ciardelli et al. [2009]). Namely, 'or' and 'and' seem *equivalent* in this linguistic context: to say "John might be in his office or he might be at home" seems equivalent to saying "John might be in his office and he might be at home". It may be seen as a virtue then that our assertability conditions yield (for any information state $\mathbf{s}$):

$$\mathbf{s} \Vdash \Diamond p \wedge \Diamond q \text{ iff } \mathbf{s} \Vdash \Diamond p \vee \Diamond q$$

This is a consequence of proposition 1, fact 3. To see this, set $\psi^1 := \top$, $\psi^2 := \top$, $m = 1$ and $n = 1$ in the statement of the fact.

5   Conclusion and Further Work

We have developed the idea that 'might' fundamentally functions to express abelief, providing a formal model which explains the doxastic state expressed by and the update operation performed upon accepting an arbitrary assertion. Our theory can also handle various problematic phenomena involving 'might' that have proven tricky to accommodate in the context of other approaches.

Further work remains to be done. First, more needs to be said to relate our current results to the elaborate debate on the semantics and pragmatics of 'might' and other epistemic modals in the philosophy and linguistics literature. Second, there are various intriguing avenues for further technical results. It would be of interest to identify a complete axiomatization for our assertability logic. We also note that the motivation for our study of conservative contraction – that it captures the idea of "coming to take a possibility seriously" – resembles that for the *suggestion operation* introduced in van Benthem and Liu [2007], Liu [2011]. However, these operations have very different technical consequences. For example, conservative contraction preserves totality, while suggestion does not. It is of interest therefore to thoroughly contrast conservative contraction and suggestion as alternative proposals for 'might' updates.

References

Ciardelli, I., Groenendijk, J., Roelofsen, F.: Attention! Might in Inquisitive Semantics. In: Proceedings of Semantics and Linguistic Theory (SALT) 19(1), pp. 91–108 (2009)

Hansson, S.O.: Logic of Belief Revision. Stanford Encyclopedia of Philosophy (Winter) (2014), http://plato.stanford.edu/archives/win2014/entries/logic-belief-revision/

Kratzer, A.: The Notional Category of Modality. In: Eikmeyer, H.-J., Rieser, H. (eds.) Words, Worlds, and Context, pp. 38–74. Walter de Gruyter (1981)

Kratzer, A.: Modals and Conditionals. Oxford University Press (2012)

Lin, H.: Acceptance-Conditional Semantics & Modality-Disjunction Interaction. In: USC Deontic Modality Workshop (2013)

Liu, F.: Reasoning About Preference Dynamics. Springer, Dordrecht (2011)

MacFarlane, J.: Epistemic Modals are Assessment-Sensitive. In: Egan, A., Weatherson, B. (eds.) Epistemic Modality, pp. 144–179. Oxford University Press (2011a)

MacFarlane, J.: What Is Assertion?. In: Brown, J., Cappelen, H. (eds.) Assertion: New Philosophical Essays, pp. 79–97. Oxford University Press (2011b)

MacFarlane, J.: Assessment Sensitivity. Oxford University Press (2014), doi:10.1093/acprof:oso/9780199682751.001.0001

Papafragou, A.: Epistemic modality and truth conditions. Lingua 116(10), 1688–1702 (2006), doi:10.1016/j.lingua.2005.05.009

Rott, H.: Shifting Priorities: Simple Representations for Twenty-seven Iterated Theory Change Operators. In: Makinson, D., Malinkowski, J., Wansing, H. (eds.) Towards Mathematical Philosophy. Trends in Logic, vol. 28, pp. 269–296. Springer, Dordrecht (2009)

van Benthem, J.: Logical Dynamics of Information and Interaction. Cambridge University Press, Cambridge (2011)

van Benthem, J., Liu, F.: Dynamic logic of preference upgrade. Journal of Applied Non-Classical Logics 17(2), 157–182 (2007), doi:10.3166/jancl.17.157-182

von Fintel, K., Gillies, A.S.: Might Made Right. In: Egan, A., Weatherson, B. (eds.) Epistemic Modality, pp. 108–130. Oxford University Press, Oxford (2011)

Yalcin, S.: Epistemic Modals. Mind 116(464), 983–1026 (2007), doi:10.1093/mind/fzm983

Yalcin, S.: Nonfactualism About Epistemic Modality. In: Egan, A., Weatherson, B. (eds.) Epistemic Modality, pp. 295–332. Oxford University Press, Oxford (2011)

Zimmermann, T.E.: Free Choice Disjunction and Epistemic Possibility. Natural Language Semantics 8, 255–290 (2000)

A Poor Man's Epistemic Logic Based on Propositional Assignment and Higher-Order Observation

Andreas Herzig, Emiliano Lorini, and Faustine Maffre

University of Toulouse, IRIT, 118, Route de Narbonne, F-31062 Toulouse, France

Abstract. We introduce a dynamic epistemic logic that is based on what an agent can observe, including joint observation and observation of what other agents observe. This generalizes van der Hoek, Wooldridge and colleague's logics ECL-PC(PO) and LRC where it is common knowledge which propositional variables each agent observes. In our logic, facts of the world and their observability can both be modified by assignment programs. We show how epistemic operators can be interpreted in this framework and identify the conditions under which the principles of positive and negative introspection are valid. We also provide a sound and complete axiomatization and prove that the satisfiability problem is PSPACE-complete. Finally, we show how public and private announcements can be expressed and illustrate the latter by the gossip spreading problem.

1 Introduction

In recent years, several authors investigated how an epistemic logic could be grounded on the notion of visibility (or observability) of propositional variables, most prominently Epistemic Coalition Logic of Propositional Control with Partial Observability ECL-PC(PO) [12] and Logic of Revelation and Concealment LRC [11]. The idea is that each agent has a set of propositional variables she can observe: no different truth value is possible for her. The other way round, any combination of truth values of the non-observable variables is possible for her.

A disadvantage of these logics is that what each agent can see is common knowledge. This is a strong hypothesis that we are going to relax in the present paper. While in ECL-PC(PO) and LRC, visibility information is in terms of propositional variables associated to agents, we here consider propositional variables associated to *sequences* of agents. Syntactically, we represent this by means of atomic formulas that we call *visibility atoms*. They take the form $S_{i_1} S_{i_2} ... S_{i_n} p$, where p is a propositional variable and $i_1, i_2, ..., i_n$ are agents. When $n=0$ then we have nothing but a propositional variable. For $n=1$, the atom $S_{i_1} p$ reads "agent i_1 sees the value of the variable p", and for $n=2$, the second-order observation $S_{i_1} S_{i_2} p$ reads "agent i_1 sees whether i_2 sees the value of p"; and so on.

Our models are simply sets of visibility atoms. In order to guarantee positive and negative introspection we have to ensure that agents are always aware of what they see: for every agent i and propositional variable p, we require $S_i S_i p$

© Springer-Verlag Berlin Heidelberg 2015
W. van der Hoek et al. (Eds.): LORI 2015, LNCS 9394, pp. 156–168, 2015.
DOI: 10.1007/978-3-662-48561-3_13

to be in every valuation. We say that a valuation V is *introspective* when it contains every visibility atom having two consecutive S_i, such as $S_j S_i S_i S_k p$.

Visibility information allows to interpret epistemic operators: for propositional variables p, the formula $K_i p$ is true in a valuation V if V contains both p and $S_i p$. More generally, the truth condition for $K_i \varphi$ is based on a relation between valuations that can be defined from our visibility atoms: $V \sim_i V'$ if every atom that i sees in V has the same truth value in V and in V'. While the relations $\sim_i$ are reflexive everywhere, they are symmetric and transitive—and therefore equivalence relations—on the set of introspective valuations only. The truth condition for the epistemic operator then takes the standard form: $K_i \varphi$ is true in V if φ is true in every valuation related to V by $\sim_i$. The positive and negative introspection axioms $K_i \varphi \to K_i K_i \varphi$ and $\neg K_i \varphi \to K_i \neg K_i \varphi$ are valid in the set of introspective valuations. A further novelty of our approach as compared to existing visibility-based epistemic logics is that we also account for common knowledge: our language includes a special atomic formula for joint attention of the form JSp that reads "all agents jointly see the value of p". Metaphorically, joint attention about a propositional variable p can be understood as eye contact between the agents when observing p. Just as individual visibility, we generalize our account to higher-order visibility, adding a constraint on valuations that guarantees introspection of common knowledge. We moreover require that joint visibility implies individual visibility by imposing that $S_i p \in V$ whenever $JS p \in V$. We can then interpret a modal operator of common knowledge CK in the same way as the modal operator of individual knowledge.

Just as several existing proposals, we take inspiration from dynamic epistemic logics DEL [4] and add dynamics to our observation-based epistemic logic. Specifically, we adapt van der Hoek et al.'s logic LRC which has two update operations modifying visibility: revealing and concealing the value of a variable to some agent. These two primitives can however not be taken over as they stand because the naive update of a valuation may no longer be introspective. We exclude this by an appropriate definition of update. We relate our assignment programs to Dynamic Logic of Propositional Assignments DL-PA [10,3], which is a dialect of Propositional Dynamic Logic PDL [7] where PDL's abstract atomic programs are instantiated by assignments of truth values to atomic formulas. The benefit of that link is a PSPACE upper bound of the complexity of both satisfiability and model checking. Moreover, visibility updates can capture public and private announcements of visibility atoms and negations thereof.

We call our logic DEL-PAO: Dynamic Epistemic Logic of Propositional Assignment and Observation. The paper is organized as follows: sections 2 and 3 introduce language and semantics of DEL-PAO. Sections 4 and 5 contain an axiomatization and the complexity result. Section 6 illustrates our logic by two applications: the embedding of announcements and a modeling of the gossip spreading problem. Section 7 discusses related work and Section 8 concludes.[1]

---

[1] A long version of this paper including proofs and a further case study (the coordinated attack problem) is available at `http://www.irit.fr/~Andreas.Herzig/P/Lori15.html`.

2 Language

Let *Prop* be a countable non-empty set of propositional variables and let *Agt* be a finite non-empty set of agents. Atomic formulas of our language are sequences of visibility operators followed by propositional variables. The formal definition is as follows.

The set of *observability operators* is

$$OBS = \{S_i : i \in Agt\} \cup \{JS\},$$

where S_i stands for individual visibility of agent i and JS stands for joint visibility of all agents. The set of all sequences of visibility operators is noted OBS^* and the set of all non-empty sequences is noted OBS^+. We use $\sigma, \sigma', \ldots$ for elements of OBS^*. Finally, the set of atomic formulas is

$$ATM = \{\sigma\, p : \sigma \in OBS^*, p \in Prop\}.$$

The elements of that set are also called *visibility atoms*, or atoms for short. For example, $JS\, S_2\, q$ reads "all agents jointly see whether agent 2 sees the value of q"; in other words, there is joint attention in the group of all agents concerning 2's observation of q. We use $\alpha, \alpha', \ldots, \beta, \beta', \ldots$ for elements of ATM.

The language of DEL-PAO is then defined by the following grammar:

$$\pi ::= +\alpha \mid -\alpha \mid \pi; \pi \mid \pi \sqcup \pi \mid \varphi?$$
$$\varphi ::= \alpha \mid \neg\varphi \mid \varphi \wedge \varphi \mid K_i\varphi \mid CK\varphi \mid [\pi]\varphi$$

where α ranges over ATM and i over Agt.

Our atomic programs are assignments of truth values to atoms from ATM: $+\alpha$ makes α true and $-\alpha$ makes α false. Complex programs are constructed with dynamic logic operators: $\pi; \pi'$ is sequential composition, $\pi \sqcup \pi'$ is nondeterministic choice, and $\varphi?$ is test. Just as in dynamic logic, the formula $[\pi]\varphi$ reads "after every execution of π, φ is true". The formula $K_i\varphi$ reads "i knows that φ is true on the basis of what she observes", and $CK\varphi$ reads "all agents jointly know that φ is true on the basis of what they jointly observe". Our epistemic operators account for forms of individual and common knowledge that are respectively obtained via individual observation and joint observation of facts. This differs therefore conceptually from the classical operators of individual and common knowledge as studied in the area of epistemic logic [5]. We will come back to this in Section 3.4.

The other boolean operators $\top, \bot, \vee, \to$ and $\leftrightarrow$ are defined as usual, and $\widehat{K}_i\varphi$ abbreviates $\neg K_i \neg \varphi$. The program *skip* abbreviates $\top?$ and *fail* abbreviates $\bot?$. We also use the abbreviation π^k, for $k \geq 0$, inductively defined by $\pi^0 = skip$ and $\pi^{k+1} = \pi^k; \pi$.

The set of atomic formulas of ATM occurring in the formula φ is noted $ATM(\varphi)$; the set $ATM(\pi)$ is defined similarly. For example, $ATM(q?; +S_2\, p) = \{q, S_2\, p\}$ and $ATM([\pi]S_1\, JS\, p \to q) = \{q, S_2\, p, S_1\, JS\, p\}$. (So $JS\, p$ is not an atom of the latter.) The length of formulas φ and programs π, noted $length(\varphi)$ and $length(\pi)$, is the number of symbols used to write them down, where we do not count [,] and parentheses and consider that the length of JS, CK, agent names and propositional variables is 1. For example, $length(S_2\, S_2\, p) = 5$ and $length([+S_2\, p]JS\, p \wedge q) = 8$.

3 Semantics

We define valuations and stipulate constraints that are motivated by the requirement that visibility information should be introspective and that joint visibility
should imply individual visibility. We then define indistinguishability relations
between valuations and interpret formulas and programs.

3.1 Introspective Valuations

A *valuation* is a subset of the set of atoms ATM. A valuation $V \in 2^{ATM}$ is
introspective if and only if the following hold, for every $\alpha \in ATM$ and $i \in Agt$:

$$S_i\, S_i\, \alpha \in V \tag{C1}$$

$$JS\, JS\, \alpha \in V \tag{C2}$$

$$JS\, S_i\, S_i\, \alpha \in V \tag{C3}$$

$$\text{if } JS\, \alpha \in V, \text{ then } S_i\, \alpha \in V \tag{C4}$$

$$\text{if } JS\, \alpha \in V, \text{ then } JS\, S_i\, \alpha \in V \tag{C5}$$

The set of all introspective valuations is noted $INTR$.

(C1) is about introspection of individual sight: an agent always sees whether
she sees the value of an atom. (C2) requires the same for joint sight; indeed, if
$JS\,\alpha$ is true then $JS\,JS\,\alpha$ should be true by introspection, and if $JS\,\alpha$ is false
then all agents jointly see that at least one of them has broken eye contact. (C3)
forces the first to be common knowledge. (C4) guarantees that joint visibility
implies individual visibility. Together with (C2), (C5) guarantees that $JS\,\alpha \in V$
implies $JS\,\sigma\,\alpha \in V$ for $\sigma \in OBS^*$.[2] The constraints (C4) and (C5) ensure that
$JS\,\alpha \in V$ implies $\sigma\,\alpha \in V$ for $\sigma \in OBS^+$. This motivates the following relation
of *introspective consequence* between atoms: $\alpha \rightsquigarrow \beta$ iff either $\alpha = \beta$, or $\alpha = JS\,\alpha'$ and $\beta = \sigma\,\alpha'$ for some $\sigma \in OBS^+$.

Closure under introspective consequence characterizes introspective valuations.

Proposition 1. *A valuation $V \subseteq ATM$ is introspective if and only if, for every $\alpha, \beta \in ATM$ and $i \in Agt$:*

$$\sigma\, S_i\, S_i\, \alpha \in V \text{ for every } \sigma \in OBS^* \tag{1}$$

$$\sigma\, JS\, \alpha \in V \text{ for every } \sigma \in OBS^+ \tag{2}$$

$$\text{if } \alpha \in V \text{ and } \alpha \rightsquigarrow \beta \text{ then } \beta \in V \tag{3}$$

Call an atom $\alpha \in ATM$ is *valid in INTR* if and only if α belongs to every
valuation in $INTR$. By Proposition 1, α is valid in $INTR$ if and only if α is of
the form either $\sigma\, S_i\, S_i\, \alpha$ with $\sigma \in OBS^*$, or $\sigma\, JS\, \alpha$ with $\sigma \in OBS^+$.

Observe that we do not impose the constraint "if $\sigma\,\alpha \in V$ for every $\sigma \in OBS^*$
then $JS\,\alpha \in V$", which corresponds to the greatest fixed point definition of the
operator of common knowledge from shared knowledge. We will comment on
this in Section 3.4.

---

[2] We need (C2) when σ contains JS: in order to prove that $JS\,\alpha \in V$ implies
$JS\, S_i\, JS\, \alpha \in V$ we use that $JS\, JS\, \alpha \in V$ by (C2) and that $JS\, JS\, \alpha \in V$ implies
$JS\, S_i\, JS\, \alpha \in V$ by (C5).

3.2 Indistinguishability Relations

Two valuations are related by the indistinguishability relation for agent i, noted $\sim_i$, if every α that i sees has the same value. Similarly, we have a relation $\sim_{Agt}$ for joint indistinguishability. They are defined as follows:

$$V \sim_i V' \quad \text{iff } S_i\, \alpha \in V \text{ implies } V(\alpha) = V'(\alpha)$$
$$V \sim_{Agt} V' \quad \text{iff } JS\, \alpha \in V \text{ implies } V(\alpha) = V'(\alpha)$$

with $V(\alpha) = V'(\alpha)$ when either $\alpha \in V$ and $\alpha \in V'$, or $\alpha \notin V$ and $\alpha \notin V'$.

The binary relations $\sim_i$ and $\sim_{Agt}$ are reflexive. They are neither transitive nor symmetric: for example, $\emptyset \sim_i V$ for every $V \subseteq ATM$, while $V \not\sim_i \emptyset$ as soon as there is a p such that p and S_i are in V. However, both properties hold on valuations satisfying the introspection constraints (C1) and (C2).

Proposition 2. *The relation* $\sim_{Agt}$ *and every* $\sim_i$ *are equivalence relations on INTR.*

Lemma 1. *Let* $V \in INTR$, $V' \in 2^{ATM}$. *If* $V \sim_i V'$ *or* $V \sim_{Agt} V'$ *then* $V' \in INTR$.

3.3 Truth Conditions and Validity

Given an introspective valuation V, our update operations add or remove atoms from V. This requires some care: we want the resulting valuation to be introspective. For example, removing $S_i\, S_i\, p$ should be impossible. Another example is when V does not contain $S_i\, p$: then $V \cup \{JS\, p\}$ would violate (C4). So when adding an atom to V we also have to add all its *positive consequences*. Symmetrically, when removing an atom we also have to remove its *negative consequences*. Let us define the following:

$$Eff^+(\alpha) = \{\beta \in ATM : \alpha \rightsquigarrow \beta\}$$
$$Eff^-(\alpha) = \{\beta \in ATM : \beta \rightsquigarrow \alpha\}$$

Clearly, when V is introspective then both $V \cup Eff^+(\alpha)$ and $V \setminus Eff^-(\alpha)$ are so, too (unless α is valid). Now the truth conditions are as follows:

$$
\begin{aligned}
V &\models \alpha &&\text{iff } \alpha \in V \\
V &\models \neg\varphi &&\text{iff } V \not\models \varphi \\
V &\models \varphi \wedge \psi &&\text{iff } V \models \varphi \text{ and } V \models \psi \\
V &\models K_i\varphi &&\text{iff } V' \models \varphi \text{ for all } V' \text{ such that } V \sim_i V' \\
V &\models CK\varphi &&\text{iff } V' \models \varphi \text{ for all } V' \text{ such that } V \sim_{Agt} V' \\
V &\models [\pi]\varphi &&\text{iff } V' \models \varphi \text{ for all } V' \text{ such that } V R_\pi V'
\end{aligned}
$$

where R_π is a binary relation on valuations that is defined (by mutual recursion with the definition of $\models$) by:

$$
\begin{aligned}
V R_{+\alpha} V' &\quad \text{iff } V' = V \cup Eff^+(\alpha) \\
V R_{-\alpha} V' &\quad \text{iff } V' = V \setminus Eff^-(\alpha) \text{ and } \alpha \text{ is not valid in } INTR \\
V R_{\pi_1;\pi_2} V' &\quad \text{iff there is } U \text{ such that } V R_{\pi_1} U \text{ and } U R_{\pi_2} V' \\
V R_{\pi_1 \sqcup \pi_2} V' &\quad \text{iff } V R_{\pi_1} V' \text{ or } V R_{\pi_2} V' \\
V R_{\varphi?} V' &\quad \text{iff } V = V' \text{ and } V \models \varphi
\end{aligned}
$$

The relation R_π is defined just as in PDL for the program operators ;, $\sqcup$ and ?. The interpretation of assignments is designed in a way such that we stay in $INTR$: the program $+\alpha$ adds all the positive consequences of α; the program $-\alpha$ fails if α is valid in $INTR$ and otherwise removes all the negative consequences of α. For example, we never have $V R_{-S_1 S_1 p} V'$, i.e., the program $-S_1 S_1 p$ always fails. In contrast, the program $-S_1 S_2 p$ always succeeds, and we have $V R_{-S_1 S_2 p} (V \setminus \{S_1 S_2 p, JS S_2 p, JS p\})$ because the only atoms—beyond $S_1 S_2 p$ itself—whose consequence is $S_1 S_2 p$ are $JS S_2 p$ and $JS p$. Therefore $V \not\models [-S_1 S_2 p] JS p$ for every V.

Lemma 2. *Let $V \in INTR$ and $V R_\pi V'$. Then $V' \in INTR$.*

Proposition 3. *For every $V \in INTR$, $i \in Agt$ and program π, V is only related to valuations in $INTR$ by $\sim_i$, $\sim_{Agt}$ and R_π.*

When $V \models \varphi$ we say that V is a *model* of φ. The set of (not necessarily introspective) models of φ is noted $\|\varphi\|$. A formula φ is *satisfiable in $INTR$* if φ has an introspective model, i.e., if $\|\varphi\| \cap INTR \neq \emptyset$. For example, $JS p \wedge \neg S_i p$ has a model, but does not have an introspective model and is therefore unsatisfiable in $INTR$. A formula φ is *valid in $INTR$* if $INTR \subseteq \|\varphi\|$. We also say that φ is a *validity* of DEL-PAO . For example, $\neg[-S_1 S_2 p] JS p$ is valid in $INTR$. Note that $\neg\beta \rightarrow [+\alpha]\neg\beta$ is valid in $INTR$ if and only if $\alpha \not\rightsquigarrow \beta$.

Formulas without epistemic operators only depend on atoms occurring in it.

Proposition 4. *Let φ be without epistemic operators. Let $V, V' \in 2^{ATM}$ such that $V(\alpha) = V'(\alpha)$ for every $\alpha \in ATM(\varphi)$. Then $V \models \varphi$ if and only if $V' \models \varphi$.*

This proposition will be instrumental in the rest of the paper. Observe that it does not hold when φ contains epistemic operators. For example, the truth value of $K_i p$ depends on that of $S_i p$, which however does not occur in $ATM(K_i p)$.

3.4   Discussion

Both the operators of individual knowledge and the operator of common knowledge of DEL-PAO satisfy all the principles of the standard epistemic logic S5. There are also some further validities of DEL-PAO, for example the S5-invalid formula $K_i(p \vee q) \rightarrow (K_i p \vee K_i q)$; cf. the axiom $Red_{K,\vee}$ below. This is a strong principle: to give an example, if one knows that the butler or the gardener was the murderer then one knows which of them it was. It is however shared by all visibility-based epistemic logics.

Our common knowledge operator obeys the fixed point axiom: $CKp \rightarrow p \wedge \left(\bigwedge_{i \in Agt} K_i CKp\right)$. This is ensured by the fact that by constraints (C2) and (C4), the formula $\bigwedge_{i \in Agt} S_i JS p$ is valid in $INTR$. Our notion of common knowledge is however weaker than standard common knowledge because the induction axiom $\left(\varphi \wedge CK\left(\varphi \rightarrow \bigwedge_{i \in Agt} K_i \varphi\right)\right) \rightarrow CK\varphi$ is invalid in $INTR$. Beyond the technical reason for that choice (such an infinitary constraint cannot be captured by formula built from visibility atoms) we follow [13,9] and assume that such a principle is too strong for a logic of common knowledge.

4 Axiomatization

The axiomatization of DEL-PAO is given by:

- the axioms of CPL (Classical Propositional Logic);
- the reduction axioms for epistemic operators:

$$K_i \alpha \leftrightarrow S_i \alpha \wedge \alpha \qquad (Red_{K,\alpha})$$

$$CK \alpha \leftrightarrow JS \alpha \wedge \alpha \qquad (Red_{CK,\alpha})$$

$$K_i \neg \alpha \leftrightarrow S_i \alpha \wedge \neg \alpha \qquad (Red_{K,\neg})$$

$$CK \neg \alpha \leftrightarrow JS \alpha \wedge \neg \alpha \qquad (Red_{CK,\neg})$$

$$K_i(\varphi \wedge \varphi') \leftrightarrow K_i\varphi \wedge K_i\varphi' \qquad (Red_{K,\wedge})$$

$$CK(\varphi \wedge \varphi') \leftrightarrow CK\varphi \wedge CK\varphi' \qquad (Red_{CK,\wedge})$$

$$K_i\left(\bigvee_{\alpha \in A^+} \alpha \vee \bigvee_{\alpha \in A^-} \neg\alpha \right) \leftrightarrow \left(\bigvee_{\alpha \in A^+} K_i\alpha \right) \vee \left(\bigvee_{\alpha \in A^-} K_i\neg\alpha \right) \qquad (Red_{K,\vee})$$

$$CK\left(\bigvee_{\alpha \in A^+} \alpha \vee \bigvee_{\alpha \in A^-} \neg\alpha \right) \leftrightarrow \left(\bigvee_{\alpha \in A^+} CK\alpha \right) \vee \left(\bigvee_{\alpha \in A^-} CK\neg\alpha \right) \qquad (Red_{CK,\vee})$$

- the reduction axioms for dynamic operators:

$$[\pi; \pi']\varphi \leftrightarrow [\pi][\pi']\varphi \qquad (Red_;)$$

$$[\pi \sqcup \pi']\varphi \leftrightarrow [\pi]\varphi \wedge [\pi']\varphi \qquad (Red_\sqcup)$$

$$[\varphi?]\varphi' \leftrightarrow \varphi \rightarrow \varphi' \qquad (Red_?)$$

$$[+\alpha]\neg\varphi \leftrightarrow \neg[+\alpha]\varphi \qquad (Red_{+\alpha,\neg})$$

$$[-\alpha]\neg\varphi \leftrightarrow \begin{cases} \top & \text{if } \alpha \text{ is valid in } INTR \\ \neg[-\alpha]\varphi & \text{otherwise} \end{cases} \qquad (Red_{-\alpha,\neg})$$

$$[+\alpha](\varphi \wedge \varphi') \leftrightarrow [+\alpha]\varphi \wedge [+\alpha]\varphi' \qquad (Red_{+\alpha,\wedge})$$

$$[-\alpha](\varphi \wedge \varphi') \leftrightarrow [-\alpha]\varphi \wedge [-\alpha]\varphi' \qquad (Red_{-\alpha,\wedge})$$

$$[+\alpha]\beta \leftrightarrow \begin{cases} \top & \text{if } \alpha \rightsquigarrow \beta \\ \beta & \text{otherwise} \end{cases} \qquad (Red_{+\alpha})$$

$$[-\alpha]\beta \leftrightarrow \begin{cases} \top & \text{if } \alpha \text{ is valid in } INTR \\ \bot & \text{if } \alpha \text{ is not valid in } INTR \text{ and } \beta \rightsquigarrow \alpha \\ \beta & \text{otherwise} \end{cases} \qquad (Red_{-\alpha})$$

- the introspection axioms:

$$S_i S_i \alpha \qquad (Vis_{C1})$$

$$JS\, JS\, \alpha \qquad (Vis_{C2})$$

$$JS\, S_i S_i \alpha \qquad (Vis_{C3})$$

$$JS\, \alpha \rightarrow S_i \alpha \qquad (Vis_{C4})$$

$$JS\, \alpha \rightarrow JS\, S_i \alpha \qquad (Vis_{C5})$$

- the rule of Modus Ponens and the rules of inference for K_i, CK, and $[\pi]$:

$$\frac{\varphi \leftrightarrow \varphi'}{K_i\varphi \leftrightarrow K_i\varphi'} \qquad \frac{\varphi \leftrightarrow \varphi'}{CK\varphi \leftrightarrow CK\varphi'} \qquad \frac{\varphi \leftrightarrow \varphi'}{[\pi]\varphi \leftrightarrow [\pi]\varphi'}$$

Theorem 1. *The axiomatization of* DEL-PAO *is sound and complete.*

5 Complexity

Theorem 2. *The* DEL-PAO *satisfiability and* DEL-PAO *model checking problems are both* PSPACE-*complete.*

We devote the rest of the section to the proof of this result. We start by proving that all epistemic operators can be eliminated in polynomial time. We then show interreducibility of model and satisfiability checking. We finally establish lower and upper bounds by embedding QBF into DEL-PAO and DEL-PAO into DL-PA.

5.1 Elimination of Epistemic Operators

Let us define the following programs:

$$\pi_{i,\alpha} = S_i\,\alpha? \sqcup (\neg S_i\,\alpha?; (+\alpha \sqcup -\alpha))$$
$$\pi_{Agt,\alpha} = JS\,\alpha? \sqcup (\neg JS\,\alpha?; (+\alpha \sqcup -\alpha))$$

The first checks whether i sees α, and if not, varies the truth value of α; the second does the same but for joint visibility. Then for a set of atoms $A = \{\alpha_1, ..., \alpha_n\}$, we define:

$$\pi_{i,A} = \pi_{i,\alpha_1}; ...; \pi_{i,\alpha_n}$$
$$\pi_{Agt,A} = \pi_{Agt,\alpha_1}; ...; \pi_{Agt,\alpha_n}$$

We suppose that the program is *skip* if the set A is empty.

We did not impose any ordering on atoms in A; this will not influence the program execution. More details can be found in the long version of the paper.

Proposition 5. *Let* φ *be a* DEL-PAO *formula without epistemic operators. Then*

$$K_i\varphi \leftrightarrow [\pi_{i,ATM(\varphi)}]\varphi$$
$$CK\varphi \leftrightarrow [\pi_{Agt,ATM(\varphi)}]\varphi$$

are valid in INTR.

Proposition 5 can be turned into a procedure eliminating epistemic operators: it suffices to iterate the application of the equivalences, starting with the innermost operators.

Procedure 1. *While there is an epistemic operator in* φ:

1. *if there exists a subformula* $K_i\varphi'$ *such that* φ' *does not contain epistemic operators, replace* φ *by* $[\pi_{i,ATM(\varphi)}]\varphi'$;
2. *if there exists a subformula* $CK\varphi'$ *such that* φ' *does not contain epistemic operators, replace* φ *by* $[\pi_{Agt,ATM(\varphi)}]\varphi'$.

Proposition 6. *For every* DEL-PAO *formula* φ, *there exists a* DEL-PAO *formula* φ' *without epistemic operators such that* $\varphi \leftrightarrow \varphi'$ *is valid in INTR. The length of* φ' *is polynomial in length(φ).*

5.2   Model Checking and SAT Interreducible

For formulas without epistemic operators, satisfiability and model checking have the same complexity.

Proposition 7. *Let φ be a* DEL-PAO *formula without epistemic operators such that $ATM(\varphi) = \{\alpha_1, \ldots, \alpha_n\}$. Let $\pi = (+\alpha_1 \sqcup -\alpha_1); \ldots; (+\alpha_n \sqcup -\alpha_n)$. Then:*

- *if φ is satisfiable in INTR, then for every $V \in INTR$, $V \models \langle \pi \rangle \varphi$;*
- *if φ is unsatisfiable in INTR, then for every $V \in INTR$, $V \not\models \langle \pi \rangle \varphi$.*

The length of the program $(+\alpha_1 \sqcup -\alpha_1); \ldots; (+\alpha_n \sqcup -\alpha_n)$ is linear in $length(\varphi)$. It follows from Proposition 7 that the satisfiability problem can be reduced in polynomial time to model checking in a randomly chosen valuation.

Proposition 8. *Let φ be a* DEL-PAO *formula without epistemic operators. For $V \in INTR$, $V \models \varphi$ if and only if the formula*
$$\langle +\alpha_1; \ldots; +\alpha_n; -\beta_1; \ldots; -\beta_m \rangle \varphi$$
is satisfiable in INTR, where $ATM(\varphi) \cap V = \{\alpha_1, \ldots, \alpha_n\}$ and $ATM(\varphi) \setminus V = \{\beta_1, \ldots, \beta_n\}$.

The length of $+\alpha_1; \ldots; +\alpha_n; -\beta_1; \ldots; -\beta_m$ is again linear in $length(\varphi)$. It follows from Proposition 8 that the model checking problem can be polynomially reduced to the satisfiability problem.

We observe that from a practical point of view, model checking requires a finite valuation. For formulas without epistemic operators such valuations can always be obtained: due to Proposition 4 we have $V \models \varphi$ iff $V \cap ATM(\varphi) \models \varphi$.

5.3   Lower Bound

In DEL-PAO we can express Quantified Boolean Formulas (QBF), whose satisfiability problem is PSPACE-complete. Details can be found in the full version of the paper.

5.4   Dynamic Logic of Propositional Assignments

In order to establish the upper bound we will embed our logic into the star-free fragment of Dynamic Logic of Propositional Assignments DL-PA [10,3], whose satisfiability problem is PSPACE-complete. We briefly recall this logic.

Just as the language of DEL-PAO, the language of DL-PA has formulas and programs. They are defined by the following grammar:
$$\pi ::= +\alpha \mid -\alpha \mid \pi; \pi \mid \pi \sqcup \pi \mid \varphi?$$
$$\varphi ::= \alpha \mid \neg\varphi \mid \varphi \wedge \varphi \mid [\pi]\varphi$$

where α ranges over ATM and i over Agt. So the language has the same atoms as DEL-PAO, but no epistemic operators. Formulas are interpreted in valuations

$V \in 2^{ATM}$ in exactly the same way as in DEL-PAO, except that atomic programs do not take introspective consequences into account. We have:

$$VR_{+\alpha}V' \text{ iff } V' = V \cup \{\alpha\}$$
$$VR_{-\alpha}V' \text{ iff } V' = V \setminus \{\alpha\}$$

A counterpart of Proposition 4 holds for DL-PA.

Proposition 9 ([3], Proposition 1). *Let $V, V' \in 2^{ATM}$ such that $V(\alpha) = V'(\alpha)$ for every $\alpha \in ATM(\varphi)$. Then $V \models_{\text{DL-PA}} \varphi$ if and only if $V' \models_{\text{DL-PA}} \varphi$.*

5.5   Upper Bound

The final step is to polynomially translate non-epistemic DEL-PAO formulas and programs into DL-PA formulas and programs. The introspection constraints will be taken into account by translating DEL-PAO assignments into appropriate DL-PA programs.

Given an atom α and a set of relevant atoms $A \subseteq ATM$, let $\mathit{Eff}^+(\alpha) \cap A = \{\beta_1, ..., \beta_n\}$ and $\mathit{Eff}^-(\alpha) \cap A = \{\beta'_1, ..., \beta'_m\}$. Translate assignments of α as follows:

$$tr(+\alpha, A) = +\beta_1; ...; +\beta_n$$

$$tr(-\alpha, A) = \begin{cases} \mathit{fail} & \text{if } \alpha \text{ valid in } \mathit{INTR} \\ -\beta'_1; ...; -\beta'_m & \text{otherwise} \end{cases}$$

Again we suppose that the program is *skip* if the set $\{\beta_1, ..., \beta_n\}$ is empty.

We extend tr to complex programs and formulas by stipulating $tr(\alpha) = \alpha$ and $tr([\pi]\varphi) = [tr(\pi, ATM(\varphi))]tr(\varphi)$, and homomorphic otherwise. Note that $ATM(tr(\pi, A)) \subseteq A$ and $ATM(tr(\varphi)) \subseteq ATM(\varphi)$.

Proposition 10. *Let φ be a DEL-PAO formula without epistemic operators. Then we have $V \models_{\text{DEL-PAO}} \varphi$ if and only if $V \models_{\text{DL-PA}} tr(\varphi)$.*

The grande finale follows from propositions 6, 7, 8 and 10 and because $tr(\varphi)$ can be computed in time polynomial in $length(\varphi)$.

Theorem 3. *In DEL-PAO, both satisfiability and model checking are PSPACE-complete.*

6   Private Announcements and Spreading Gossip

Public Announcement Logic PAL [14] is a logic of the DEL family extending standard epistemic logic with an operator $[\psi!]$, such that $[\psi!]\varphi$ reads "after ψ is publicly and truthfully announced, φ is true". Its validities are axiomatized by means of the reduction axioms $[\psi!]p \leftrightarrow \psi \rightarrow p$, $[\psi!]\neg\varphi \leftrightarrow \psi \rightarrow \neg[\psi!]\varphi$, $[\psi!](\varphi \wedge \varphi') \leftrightarrow [\psi!]\varphi \wedge [\psi!]\varphi'$, and $[\psi!]K_i\varphi \leftrightarrow \psi \rightarrow K_i[\psi!]\varphi$.

We claim that we can express public announcements of literals as $p! = p?; +JS\,p$ and $\neg p! = \neg p?; +JS\,p$. We furthermore claim that we can express the public announcement of knowledge of atoms as $K_i p! = K_i p?; +JS\,p$. It can indeed be checked that with these definitions all the reduction axioms for PAL are valid in our logic (see the full version of the paper). Beyond that we can also easily model *private* announcements of the same kind of formulas. Read $j : \psi!$ as "ψ is privately announced to agent j". Then: $j : p! = p?; +S_j\, p$, $j : \neg p! = \neg p?; +S_j\, p$ and $j : K_i p! = K_i p?; +S_j\, p; +S_j\, S_i\, p$.

Let us illustrate this by the Spreading Gossip problem, of which a detailed study can be found in [4]. Six friends each know a secret. When they call each other, they exchange every secret that they know. The problem is to find how many calls are necessary to spread all secrets among all friends. It was proven ([1], among others) that the minimal number of calls is 8; for example, if we write ij the fact that i calls j (or that j calls i), the following sequence spreads all secrets: 12, 34, 56, 13, 45, 16, 24, 35 [4]. Let us model this with private announcements. With $Agt = \{i : 1 \le i \le 6\}$ and s_i meaning that i has the secret s_i, we define the program $Call_{ij}$, for $i, j \in Agt$, as:

$$Call_{ij} = ((S_i\, s_1?; j : s_1!) \sqcup \neg S_i\, s_1?); ...; ((S_i\, s_6?; j : s_6!) \sqcup \neg S_i\, s_6?);$$
$$((S_j\, s_1?; i : s_1!) \sqcup \neg S_j\, s_1?); ...; ((S_j\, s_6?; i : s_6!) \sqcup \neg S_j\, s_6?)$$

Our program expresses that i tells all she knows to j, and conversely; each call makes each atom known by one agent known to both. Then the formula

$$[Call_{12}; Call_{34}; Call_{56}; Call_{13}; Call_{45}; Call_{16}; Call_{24}; Call_{35}] \bigwedge_{i \in Agt} K_i \left(\bigwedge_{j \in Agt} s_j \right)$$

is true at the initial state V_0 defined as:

$$V_0 = \{\alpha : \alpha \text{ is valid in } INTR\} \cup \{s_i : i \in Agt\} \cup \{S_i\, s_i : i \in Agt\}.$$

This establishes that the above sequence is correct. Furthermore, the formula

$$\left\langle \left(\bigsqcup_{i,j \in Agt, i \neq j} \neg S_i\, s_j?; Call_{ij} \right)^8 \right\rangle \bigwedge_{i \in Agt} K_i \left(\bigwedge_{j \in Agt} s_j \right)$$

expresses that a more general protocol is correct. Finally, the formula

$$\left[\left(\bigsqcup_{i,j \in Agt, i \neq j} Call_{ij} \right)^7 \right] \neg \bigwedge_{i \in Agt} K_i \left(\bigwedge_{j \in Agt} s_j \right)$$

expresses that only 7 calls are not enough. Both are true at V_0.

Note that our modelling does not account for second-order knowledge. In order to do so we should modify the program $Call_{ij}$ in a way such that when $S_i\, s_1$ is true then not only $j : s_1!$ is performed, but also $i : S_j\, s_1!$. With that modelling we could check not only that everybody knows each secret, but also that everybody knows that everybody knows each secret. In the same vein, third-order knowledge can be attained by adding $j : S_i\, S_j\, s_1!$, and so on.

Beyond that, we may also want to model that $Call_{ij}$ leads to common knowledge of i and j. This requires the extension of DEL-PAO by visibility atoms with non-empty sets of agents as arguments. However, secrets can never become common knowledge of *all* agents. This can also be highlighted by the Two Generals' problem where common knowledge cannot be reached. Details are in the full version of the paper.

7 Related Work

As said in the introduction, our logic is in the tradition of several other logics developed in the past few years. In the logic ECL-PC(PO) [12], visibility is represented by a set of atoms for each agent, containing the variables the agent observes. This does not allow for higher-order observations such as "i observes whether j observes p". Instead and as already mentioned, the observational capabilities of each agent become common knowledge among all agents. The logic LRC [11] allows to express, as programs, that a variable is revealed to an agent or concealed from her. Semantically, formulas are interpreted over pointed models with a visibility set for each agent; revealing a variable p to an agent i will add p to i's visibility set, while concealing p will remove p from i's set. Just as in ECL-PC(PO), who sees what is common knowledge among all agents.

The logic of knowing whether [6] adds an operator standing for "i knows whether φ" to the language of standard epistemic logic, interpreted as "φ has the same value in all indistinguishable worlds for i". This can be compared to our visibility atoms S_i which express the same notion on atoms.

In Flatland Logic [2], visibility is further grounded on geometry in order to give semantics to epistemic operators: an agent can (or cannot) observe the positions of other agents and can reason about what they observe. Visibility can be higher-order and is also fully determined by geometric constraints. The main difference with our logic is that in Flatland Logic, agents see other agents instead of propositional variables.

8 Conclusion

We have introduced a dynamic epistemic logic of propositional assignment and observation DEL-PAO which accounts for higher-order and joint observation as well as updates thereof. It avoids the strong hypothesis of common knowledge of visibility that other observation-based epistemic logics make. It is remarkable that the addition of higher-order observability and in particular of joint observability comes without supplementary cost: both satisfiability and model checking remain PSPACE-complete. This contrasts with standard logics of common knowledge: there, satisfiability checking is ExpTime-hard [8].

A simple extension of our logic is to generalize the operator of common knowledge of all agents CK to operators taking any subset of Agt as arguments. It suffices to introduce visibility atoms $JS_J\alpha$, one per group of agents J. Another interesting generalization is to consider belief instead of knowledge. A way to achieve this is to replace S_i by two operators O_i and C_i, respectively meaning that i has an opinion on something and that i is correct on something. This requires other constraints on valuations that should match the properties of belief. Further possible extensions concern the dynamic part: following [10], one may add atoms representing that i controls some propositional variable p, in the sense that i can change the truth value of p at will. One may then associate to each assignment an author, which is the agent performing the assignment. As shown in [10], this allows to embed Coalition Logic of Propositional Control [12]. It remains to be worked out how this combines with higher-order observations.

Acknowledgments. We would like to thank François Schwarzentruber for useful comments and the anonymous reviewers for their thoughtful reading and comments.

References

1. Baker, B., Shostak, R.: Gossips and telephones. Discrete Mathematics 2(3), 191–193 (1972)
2. Balbiani, P., Gasquet, O., Schwarzentruber, F.: Agents that look at one another. Logic Journal of the IGPL 21(3), 438–467 (2013)
3. Balbiani, P., Herzig, A., Troquard, N.: Dynamic logic of propositional assignments: a well-behaved variant of PDL. In: Kupferman, O. (ed.) Proceedings of the 28th Annual IEEE/ACM Symposium on Logic in Computer Science, pp. 143–152 (2013)
4. van Ditmarsch, H., van der Hoek, W., Kooi, B.: Dynamic Epistemic Logic, 1st edn. Springer Publishing Company, Incorporated (2007)
5. Fagin, R., Halpern, J.Y., Moses, Y., Vardi, M.Y.: Reasoning about Knowledge. MIT Press (1995)
6. Fan, J., Wang, Y., van Ditmarsch, H.: Knowing whether. CoRR abs/1312.0 (2013)
7. Fischer, M.J., Ladner, R.E.: Propositional dynamic logic of regular programs. Journal of Computer and System Sciences 18(2), 194–211 (1979)
8. Halpern, J.Y., Moses, Y.: A guide to completeness and complexity for modal logics of knowledge and belief. Artificial Intelligence 54(3), 319–379 (1992)
9. Herzig, A.: Logics of knowledge and action: critical analysis and challenges. Journal of Autonomous Agents and Multi-Agent Systems, 1–35 (2014)
10. Herzig, A., Lorini, E., Troquard, N., Moisan, F.: A dynamic logic of normative systems. In: Proceedings of the 22nd International Joint Conference on Artificial Intelligence, pp. 228–233 (2011)
11. van der Hoek, W., Iliev, P., Wooldridge, M.: A logic of revelation and concealment. In: van der Hoek, W., Padgham, L., Conitzer, V., Winikoff, M. (eds.) Proceedings of the 11th International Conference on Autonomous Agents and Multiagent Systems, IFAAMAS, pp. 1115–1122 (2012)
12. van der Hoek, W., Troquard, N., Wooldridge, M.: Knowledge and control. In: Sonenberg, L., Stone, P., Tumer, K., Yolum, P. (eds.) Proceedings of the 10th International Conference on Autonomous Agents and Multiagent Systems, IFAAMAS, pp. 719–726 (2011)
13. Lorini, E., Herzig, A.: Direct and indirect common belief. In: Institutions, Emotions, and Group Agents, pp. 355–372. Springer, Netherlands (2014)
14. Plaza, J.: Logics of public communications. In: Emrich, M.L., Pfeifer, M.S., Hadzikadic, M., Ras, Z. (eds.) Proceedings of the 4th International Symposium on Methodologies for Intelligent Systems, pp. 201–216. Oak Ridge National Laboratory, ORNL/DSRD- 24 (1989)

Trace Semantics for IPDL

Fengkui Ju, Nana Cui, and Shujiao Li

School of Philosophy, Beijing Normal University, Beijing, China
fengkui.ju@bnu.edu.cn, {cvn.zzx52,lishujiao1121}@163.com

Abstract. Other than relation semantics, IPDL, the extension of PDL with intersection of actions, has a natural trace semantics where the interpretation of an action is a set of sequences of states. IPDL in trace semantics can describe paced concurrent games very well. Surprisingly, IPDL can be reduced to a sublanguage of it in which intersection connects only atomic actions.

Keywords: propositional dynamic logic, intersection, trace semantics, concurrency.

1 Introduction

Propositional Dynamic Logic (PDL), introduced in [4], is a formal system for reasoning about the behavior of programs of computers, e.g. *if ϕ holds now, then ψ will hold after the program α is executed in whatever way*. PDL is a modal logic. The language of it is an extension of the language of propositional logic with a set of modalities which can be intuitively viewed as programs. This set has a structure: composite programs are generated from atomic ones by the program constructors *composition*, *union* and *iteration*. The semantics of PDL is a *relation semantics*: programs are binary relations and program constructors are operations of binary relations. Modalities in PDL can also be viewed as actions of agents, therefore, PDL has applications in other areas as well, besides in computer science.

PDL has various variants introduced for different purposes. One of them is IPDL, the extension of PDL with *intersection* of programs, which is used to formalize the notion of *concurrency* in a way.

PDL deals with the input/output behavior of programs very well, but it can not deal with their progressive behavior, e.g., *ϕ is true at some point during the execution of the program α*. To solve this problem, *process logic* is developed as a mix of PDL and the temporal logic introduced in [11]. The core idea of process logic is that the intermediate states of computations should be taken into consideration. There are a variety of process logics including [9], [5] and [12]. The typical work among these is [5], which introduces a logic called Process Logic (PL). The language of PL is an extension of PDL with some temporal operators such as *until*[1]. In the semantics of PL, programs are not binary relations, but sets

---

[1] Strictly speaking, it is not that the language of PL is an extension of PDL, but that the language of PDL is definable in PL.

© Springer-Verlag Berlin Heidelberg 2015
W. van der Hoek et al. (Eds.): LORI 2015, LNCS 9394, pp. 169–181, 2015.
DOI: 10.1007/978-3-662-48561-3_14

of state sequences intuitively viewed as computation traces; truth of formulas is defined at state sequences, not at states like in relation semantics. By observing the semantics closely, it can be seen that the formulas of PDL are essentially evaluated at states, and state sequences make sense for only temporal formulas. Therefore, from [5], we can get such a semantics for PDL: programs are sets of state sequences and formulas are evaluated at states. It is different from relation semantics and may be called *trace semantics*.

Accordingly, IPDL has a trace semantics. This is the focus of this paper. In the sequel, firstly, we define trace semantics for IPDL in detail and point out where it makes a difference from relation semantics. Then we indicate by some examples that trace semantics has applications in *paced concurrent games*: IPDL in trace semantics captures the notion of concurrency better there than it does in relation semantics. We then show that trace semantics does not make any essential difference for PDL. At last, we show a fact which is not easy to be seen: IPDL in trace semantics can not say more than a sublanguage of it where intersection connects only atomic actions.

2   Trace Semantics for IPDL

2.1   Language and Relation Semantics

Let Π_0 be a countable set of atomic actions and Φ_0 a countable set of atomic propositions. Let a range over Π_0 and p over Φ_0. The sets Π_{IPDL} of actions and Φ_{IPDL} of propositions are defined simultaneously as follows:

$$\alpha ::= a \mid (\alpha; \alpha) \mid (\alpha \cap \alpha) \mid (\alpha \cup \alpha) \mid \alpha^* \mid \phi?$$
$$\phi ::= p \mid \top \mid \neg\phi \mid (\phi \wedge \phi) \mid \langle\alpha\rangle\phi$$

This language is the extension of PDL by adding $\alpha \cap \alpha$, the *intersection* of actions. To perform $\alpha; \beta$ is to perform α and then β. To perform $\alpha \cap \beta$ is to perform α and β at the same time. To perform $\alpha \cup \beta$ is to perform α or β. To perform α^* is to perform α a finite number (possibly zero) of times. To perform $\phi?$ is to test whether ϕ is the case, and if so, continue, or else halt. $\langle\alpha\rangle\phi$ means that there is a way to perform α s.t. ϕ is the case after α is done. Other routine propositional connectives, the falsity $\bot$ and the dual $[\alpha]\phi$ of $\langle\alpha\rangle\phi$ are defined in the usual way.

A *model* $\mathfrak{M}$ is a triple $(W, \{R_a \mid a \in \Pi_0\}, V)$ where

1. W is a nonempty set of states;
2. $R_a \subseteq W \times W$ for any atomic action a;
3. V is a function from Φ_0 to the power set of W.

Given a model $\mathfrak{M} = (W, \{R_a \mid a \in \Pi_0\}, V)$. R_α, the interpretation of α in $\mathfrak{M}$, and $\mathfrak{M}, w \Vdash_r \phi$, ϕ being true at w in $\mathfrak{M}$, are defined as follows:

1. (a) $R_{\beta;\gamma} = R_\beta; R_\gamma$;
   (b) $R_{\beta \cap \gamma} = R_\beta \cap R_\gamma$;
   (c) $R_{\beta \cup \gamma} = R_\beta \cup R_\gamma$;
   (d) $R_{\alpha^*} = \{(w,w) \mid w \in W\} \cup R_\alpha \cup R_{\alpha;\alpha} \cup \ldots$;
   (e) $R_{\phi?} = \{(w,w) \mid \mathfrak{M}, w \Vdash_r \phi\}$.
2. (a) $\mathfrak{M}, w \Vdash_r p \Leftrightarrow w \in V(p)$;
   (b) $\mathfrak{M}, w \Vdash_r \top$ always holds;
   (c) $\mathfrak{M}, w \Vdash_r \neg\phi \Leftrightarrow$ not $\mathfrak{M}, w \Vdash_r \phi$;
   (d) $\mathfrak{M}, w \Vdash_r (\phi \wedge \psi) \Leftrightarrow \mathfrak{M}, w \Vdash_r \phi$ and $\mathfrak{M}, w \Vdash_r \psi$;
   (e) $\mathfrak{M}, w \Vdash_r \langle\alpha\rangle\phi \Leftrightarrow$ there is a u s.t. $(w,u) \in R_\alpha$ and $\mathfrak{M}, u \Vdash_r \phi$.

Here the interpretations of actions are binary relations and the action constructors *composition, intersection, union* and *iteration* are operations of relations. This semantics is called *relation semantics*. A formula ϕ is *valid* iff for any model $\mathfrak{M}$ and any state w of $\mathfrak{M}$, $\mathfrak{M}, w \Vdash_r \phi$.

2.2 Trace Semantics

Let $\mathfrak{M} = (W, \{R_a \mid a \in \Pi_0\}, V)$ be a model defined as above. Let Δ_W be the set of the nonempty finite sequences of states in W. Define a partial binary function *ext* on Δ_W as this: $ext((u_0, \ldots, u_n), (v_0, \ldots, v_m)) = (u_0, \ldots, u_n, v_1, \ldots, v_m)$ if $u_n = v_0$, or else $ext((u_0, \ldots, u_n), (v_0, \ldots, v_m))$ is undefined. Define a function $\otimes$, called *fusion*, on the power set of Δ_W as this: $S \otimes S' = \{ext(C, C') \mid C \in S \,\&\, C' \in S'\}$ where S and S' are two sets of finite sequences. Here is an example of the fusion function: $S = \{(u_1, u_2), (u_3)\}$ and $S' = \{(u_3, u_5), (u_2, u_4, u_6)\}$; $S \otimes S' = \{(u_1, u_2, u_4, u_6), (u_3, u_5)\}$. S_α, the interpretation of α in $\mathfrak{M}$, and $\mathfrak{M}, w \Vdash_t \phi$, ϕ being true at w in $\mathfrak{M}$, are defined as what follows:

1. (a) $S_a = R_a$;
   (b) $S_{\beta;\gamma} = S_\beta \otimes S_\gamma$;
   (c) $S_{\beta \cap \gamma} = S_\beta \cap S_\gamma$;
   (d) $S_{\beta \cup \gamma} = S_\beta \cup S_\gamma$;
   (e) $S_{\alpha^*} = W \cup S_\alpha \cup S_{\alpha;\alpha} \cup \ldots$;
   (f) $S_{\phi?} = \{w \mid \mathfrak{M}, w \Vdash_t \phi\}$.
2. (a) $\mathfrak{M}, w \Vdash_t p \Leftrightarrow w \in V(p)$;
   (b) $\mathfrak{M}, w \Vdash_t \top$ always holds;
   (c) $\mathfrak{M}, w \Vdash_t \neg\phi \Leftrightarrow$ not $\mathfrak{M}, w \Vdash_r \phi$;
   (d) $\mathfrak{M}, w \Vdash_t (\phi \wedge \psi) \Leftrightarrow \mathfrak{M}, w \Vdash_r \phi$ and $\mathfrak{M}, w \Vdash_r \psi$;
   (e) $\mathfrak{M}, w \Vdash_t \langle\alpha\rangle\phi \Leftrightarrow$ there is a sequence $(x_0, \ldots, x_n) \in S_\alpha$ s.t. $x_0 = w$ and $\mathfrak{M}, x_n \Vdash_t \phi$.

It can be verified that $\mathfrak{M}, w \Vdash_t [\alpha]\phi \Leftrightarrow$ for any sequence $(x_0, \ldots, x_n) \in S_\alpha$, if $x_0 = w$, then $\mathfrak{M}, x_n \Vdash_t \phi$. Note that the interpretation of the test $\phi?$ is a set of states, not a partial identity relation like in relation semantics. However, it can be verified that $\langle\phi?\rangle\psi$ is equivalent to $\phi \wedge \psi$, like in relation semantics. Here the interpretations of actions are sets of sequences and the action constructors are operations of sets of sequences. This semantics is called *trace semantics*. A formula ϕ is *valid* iff for any model $\mathfrak{M}$ and any state w of $\mathfrak{M}$, $\mathfrak{M}, w \Vdash_t \phi$.

Strictly speaking, a structure $(W, \{R_a \mid a \in \Pi_0\}, V)$ does not completely specify a model of relation semantics and the complete specification is $(W, \{R_a \mid a \in \Pi_0\}, ;, \cap, \cup, {}^*, V)$ where $;, \cap, \cup$ and * are operations of composition, intersection, union and iteration. Similarly, $(W, \{R_a \mid a \in \Pi_0\}, V)$ does not completely specify a model of trace semantics either and the complete specification is $(W, \{R_a \mid a \in \Pi_0\}, \otimes, \cap, \cup, {}^*, V)$.

Does trace semantics make a difference from relation semantics for IPDL? *Yes*. This can be illustrated by the model in Figure 1 where we do not specify a valuation. It can be verified that $R_{(a;b) \cap (c;d)} = \{(w_1, w_4)\}$ but $S_{(a;b) \cap (c;d)} = \emptyset$. Then $w_1 \Vdash_r \langle (a;b) \cap (c;d) \rangle \top$ but $w_1 \not\Vdash_t \langle (a;b) \cap (c;d) \rangle \top$. The two semantics are not equivalent.

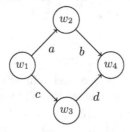

Fig. 1. A model showing the difference of the two semantics

Technically, the constructor ";" is interpreted as the operation of fusion in trace semantics, while in relation semantics, it is interpreted as the operation of composition. Intuitively, trace semantics considers the intermediate states of performing an action while relation semantics does not. This is the essential difference between the two semantics.

3   Applications

Here are the presuppositions of trace semantics. Transiting from a state to another costs one unit of time. An atomic action is a set of binary tuples; therefore, performing it costs one unit of time. A state sequence represents a succession of transitions. An action is a set of state sequences each of which represents a possible process of performing this action. A test is a set of states; hence, performing it does not cost any time. Intersection in trace semantics formalizes such a type of concurrency: *cobegining* and *coending*. The action $\alpha \cap \beta$ is more accurately read as this: to perform it is to start doing α and β at the same time and finish doing them at the same time. The properties of trace semantics are coincident with this reading. For example, doing $b; c$ costs twice time as doing a, so there is no way to start and finish a and $b; c$ at the same time. It can be seen that $S_{a \cap (b;c)}$ is empty in any models.

One application area of trace semantics is what we called *paced concurrent games* where players make their moves in a paced way. These games are very common in daily life and a typical example of them is the *rock-paper-scissors* (RPS) game. This game can be played in different ways and here is one way to play it: to start the game, the two players speak the words "scissors, paper, rock" aloud together; at "rock", they throw their hands; during the game, they speak "rock" aloud each time they throw their hands. Here are a few things which we want to point out. Firstly, when the game starts, the two players throw their hands at the same time. Speaking those words together helps them do this. Secondly, during the game, the two players throw their hands in a fixed rhythm. How they speak the words "scissors, paper, rock" together before the game helps them decide the rhythm and speaking the word "rock" during the game serves for the same purpose. Finally, the two players finish their actions at the same time. All these are used to guarantee that the game is played fairly.

IPDL in trace semantics can abstractly describe this sort of games very well. Suppose that 1 and 2 are the two players in the RPS game. We use a_i, b_i and c_i to denote the play i's actions of showing the hand gestures of rock, paper and scissors. We use p_i to express that the player i wins. Then $[(a_1; b_1) \cap (a_2; a_2)]p_1$ says that the player 1 wins whenever the player 1 shows rock and then paper and the play 2 shows rock twice. $[(a_1 \cap a_2); (b_1 \cap a_2)]p_1$ says that the player 1 wins whenever both of them show rock and then 1 shows paper and 2 shows rock. Intuitively, these two formulas express the same meaning; it can be verified that $[(a_1; b_1) \cap (a_2; a_2)]p_1 \leftrightarrow [(a_1 \cap a_2); (b_1 \cap a_2)]p_1$ is valid in trace semantics. Note that this formula is invalid in relation semantics. Let $\alpha = (a_1 \cap a_2) \cup (b_1 \cap b_2) \cup (c_1 \cap c_2)$. Then $[\alpha^*](\neg p_1 \wedge \neg p_2)$ says that nobody would win if the two players always show the same hand gestures. Let $\beta = (a_1 \cap b_2) \cup (a_1 \cap c_2) \cup (b_1 \cap a_2) \cup (b_1 \cap c_2) \cup (c_1 \cap a_2) \cup (c_1 \cap b_2)$. Then $[\alpha^*; \beta](p_1 \vee p_2)$ says that somebody would win once they show different hand gestures.

The language of IPDL does not contain any specific ingredients related with games; therefore, it just describes paced games in an abstract way. However, when augmented with some core notions of games, we think that it can say meaningful things about them. This is a future work for us and [1] would be a good reference object there.

4 Making No Difference for PDL

Let Π_{PDL} be the set of actions of PDL which does not have intersection of actions. Let Φ_{PDL} be the set of propositions of PDL. Actually, trace semantics is equivalent to relation semantics w.r.t. PDL.

As Π_{PDL} and Φ_{PDL} are defined in a mutually recursive way, we can not prove that the formulas in Φ_{PDL} have some property just by induction on the structure of them, and we need to consider the actions in Π_{PDL} as well. The function defined as follows helps us do induction on the actions and propositions at the same time:

Definition 1 (Complexity of Actions and Formulas). *Define a function* $c : \Pi_{PDL} \cup \Phi_{PDL} \to \mathbf{N}$ *as follows:*

1. $a^c = 0$
2. $(\alpha; \beta)^c = \alpha^c + \beta^c + 1$
3. $(\alpha \cup \beta)^c = \alpha^c + \beta^c + 1$
4. $(\alpha^*)^c = \alpha^c + 1$
5. $(\phi?)^c = \phi^c + 1$
6. $p^c = 0$
7. $\top^c = 0$
8. $(\neg\psi)^c = \psi^c + 1$
9. $(\psi \wedge \chi)^c = \psi^c + \chi^c + 1$
10. $(\langle\alpha\rangle\psi)^c = \alpha^c + \psi^c + 1$

The following proposition gives a sufficient condition for trace semantics and relation semantics to be equivalent w.r.t. PDL:

Proposition 1. $\mathfrak{M} = (W, \{R_a \mid a \in \Pi_0\}, V)$ *is a model.* $\mathfrak{M}, w \Vdash_t \phi \Leftrightarrow \mathfrak{M}, w \Vdash_r \phi$ *for any* ϕ *if the action constructors* $;, \cup$ *and* * *preserve the following properties*[2]:

1. $(x_0, \ldots, x_n) \in S_\alpha$ *implies* $(x_0, x_n) \in R_\alpha$;
2. $(u, v) \in R_\alpha$ *implies that there are* $x_0, \ldots, x_n$ *s.t.* $x_0 = u$, $x_n = v$ *and* $(x_0, \ldots, x_n) \in S_\alpha$.

Proof. Suppose that the three action constructors preserve the two properties. It suffices to show by induction on k that for any $k \in \mathbf{N}$, the following two statements hold:

(a) for any action α of complexity k, (i) $(x_0, \ldots, x_n) \in S_\alpha$ implies $(x_0, x_n) \in R_\alpha$ and (ii) $(u, v) \in R_\alpha$ implies that there are $x_0, \ldots, x_n$ s.t. $x_0 = u$, $x_n = v$ and $(x_0, \ldots, x_n) \in S_\alpha$;
(b) for any proposition ϕ of complexity k, $\mathfrak{M}, w \Vdash_t \phi \Leftrightarrow \mathfrak{M}, w \Vdash_r \phi$.

Case $\alpha^c = 0$. Then $\alpha = a$ for some atomic action a. Since $S_a = R_a$, the statement (a) is the case.

Case $\phi^c = 0$. Then $\phi = p$ for some atomic proposition p or $\phi = \top$. Clearly the statement (b) is the case.

Case $\alpha^c = k$ $(0 < k)$. Since the action constructors $;, \cup$ and * preserve the two properties, the statement (a) holds in the **subcases** $\alpha = \beta; \gamma$, $\alpha = \beta \cup \gamma$

---

[2] More formally, the first property should be stated as follows: for any n and $x_0, \ldots, x_n$, if $(x_0, \ldots, x_n) \in S_\alpha$, then there are y_0 and y_1 s.t. $y_0 = x_0, y_1 = x_n$ and $(y_0, y_1) \in R_\alpha$; the second property should be stated as follows: for any u, v, if $(u, v) \in R_\alpha$, then there are a n and $x_0, \ldots, x_n$ s.t. $x_0 = u$, $x_n = v$ and $(x_0, \ldots, x_n) \in S_\alpha$. By saying that an action constructor, say $\cup$, preserves a property, say the first one, we mean this: if for any $x_0, \ldots, x_n$, $(x_0, \ldots, x_n) \in S_\beta$ implies $(x_0, x_n) \in R_\beta$, and for any $x_0, \ldots, x_n$, $(x_0, \ldots, x_n) \in S_\gamma$ implies $(x_0, x_n) \in R_\gamma$, then for any $x_0, \ldots, x_n$, $(x_0, \ldots, x_n) \in S_{\beta \cup \gamma}$ implies $(x_0, x_n) \in R_{\beta \cup \gamma}$.

and $\alpha = \beta^*$. It remains to handle the **subcase** $\alpha = \psi?$. Suppose $x \in S_{\psi?}$. Then $\mathfrak{M}, x \Vdash_t \psi$. Since $\psi^c < k$, we get $\mathfrak{M}, x \Vdash_r \psi$ by the inductive hypothesis. Then $(x, x) \in R_{\psi?}$. Suppose $(u, v) \in R_{\psi?}$. Then $u = v$ and $\mathfrak{M}, u \Vdash_r \psi$. By the inductive hypothesis, we have $\mathfrak{M}, u \Vdash_t \psi$. Then $u \in S_{\psi?}$.

  Case $\phi^c = k$ $(0 < k)$. The **subcases** $\phi = \neg\psi$ and $\phi = \psi \wedge \chi$ are easy to go through and we only give a proof for the **subcase** $\phi = \langle \alpha \rangle \psi$. Suppose $\mathfrak{M}, w \Vdash_t \langle \alpha \rangle \psi$. Then there is a sequence $(x_0, \ldots, x_n) \in S_\alpha$ s.t. $x_0 = w$ and $\mathfrak{M}, x_n \Vdash_t \psi$. Since $\alpha^c < k$ and $\psi^c < k$, we have $(x_0, x_n) \in R_\alpha$ and $\mathfrak{M}, x_n \Vdash_r \psi$ by the inductive hypothesis. Then $\mathfrak{M}, w \Vdash_r \langle \alpha \rangle \psi$. Suppose $\mathfrak{M}, w \Vdash_r \langle \alpha \rangle \psi$. Then there is a u s.t. $(w, u) \in R_\alpha$ and $\mathfrak{M}, u \Vdash_r \psi$. By the inductive hypothesis, there are $x_0, \ldots, x_n$ s.t. $x_0 = w$, $x_n = u$, $(x_0, \ldots, x_n) \in S_\alpha$ and $\mathfrak{M}, u \Vdash_t \psi$. Then $\mathfrak{M}, w \Vdash_t \langle \alpha \rangle \psi$.

If we add to this proposition an extra requirement that the action constructor $\cap$ preserves the two properties as well, we would get a sufficient condition for the two semantics to be equivalent w.r.t. IPDL. This result will be used later at Section 5.3.

Lemma 1. *The action constructors* $;, \cup$ *and* * *preserve the following two properties:*

1. $(x_0, \ldots, x_n) \in S_\alpha$ *implies* $(x_0, x_n) \in R_\alpha$;
2. $(u, v) \in R_\alpha$ *implies that there are* $x_0, \ldots, x_n$ *s.t.* $x_0 = u$, $x_n = v$ *and* $(x_0, \ldots, x_n) \in S_\alpha$.

By this lemma and Proposition 1, we know that for PDL, trace semantics does not make a difference from relation semantics:

Proposition 2. $\mathfrak{M} = (W, \{R_a \mid a \in \Pi_0\}, V)$ *is a model. Then for any* ϕ *in* Φ_{PDL}, $\mathfrak{M}, w \Vdash_t \phi \Leftrightarrow \mathfrak{M}, w \Vdash_r \phi$.

The action constructor $\cap$ does not generally preserve the second property mentioned in Proposition 1, although it does preserve the first one. This is the essential reason that trace semantics is not equivalent to relation semantics w.r.t. IPDL.

5   Reduction of IPDL to iPDL

Suppose that there is no test. Compound actions are built up from atomic actions; performing atomic actions costs the same time; therefore, the performance of compound actions can be decomposed into a sequence of atomic steps each of which is a performance of some atomic actions at the same time. This give us a feeling that IPDL in trace semantics might be reduced to a sublanguage of it in which $\cap$ connects only atomic actions. However, two things make us worry whether this is the case: test behaves differently from atomic actions and the generation power of iteration might not be strong enough. In fact, the reduction can be done and we now are going to prove it. As we will see in the proof, our worry makes sense: test and iteration do cause difficulties there.

5.1   Computation Sequences

For any atomic actions $a_1, \ldots, a_n$, we call $a_1 \cap \ldots \cap a_n$ an I-action. For any tests $\phi_1?, \ldots, \phi_n?$, we call $\phi_1? \cap \ldots \cap \phi_n?$ a T-action. $\gamma_1 \cap \ldots \cap \gamma_n$ is called an IT-action if each γ_i is an I-action or a T-action. Note that the intersection of two IT-actions is still an IT-action.

Definition 2 (R-seqs). $\beta_1; \ldots; \beta_n$ *is called a rough computation sequence (a r-seq) if each β_i is an IT-action.*

In computer science, a computation sequence represents a sequence of atomic steps of computation. R-seqs can be understood in a similar way; what is special here is that performing an IT-action should be treated as an atomic step. Note that some r-seqs can not be performed anyway, as they might contain empty IT-actions like $a \cap p?$.

Definition 3 (Merge of R-seqs). *Let $\alpha = \gamma_1; \ldots; \gamma_n$ and $\beta = \delta_1; \ldots; \delta_m$ be two r-seqs. $\alpha \sqcap \beta$, the merge of α and β, is defined as this: $\alpha \sqcap \beta = (\gamma_1 \cap \delta_1); \ldots; (\gamma_n \cap \delta_n)$ if $m = n$, or else $\alpha \sqcap \beta$ is undefined.*

$\alpha \sqcap \beta$ might not be equivalent to $\alpha \cap \beta$. For instance, $(a; \top?) \cap (\top?; b)$ is not always empty, but $(a; \top?) \sqcap (\top?; b)$, which equals to $(a \cap \top?); (\top? \cap b)$, is. Note that if both α and β are IT-actions, then $\alpha \sqcap \beta = \alpha \cap \beta$.

Definition 4 (R-seqs of Actions). $CS(\alpha)$, *the set of r-seqs of α, is defined as follows:*

1. $CS(a) = \{a\}$
2. $CS(\phi?) = \{\phi?\}$
3. $CS(\alpha; \beta) = \{\gamma; \delta \mid \gamma \in CS(\alpha) \ \& \ \delta \in CS(\beta)\}$
4. $CS(\alpha \cap \beta) = \{\gamma \sqcap \delta \mid \gamma \in CS(\alpha) \ \& \ \delta \in CS(\beta)\}$
5. $CS(\alpha \cup \beta) = CS(\alpha) \cup CS(\beta)$
6. $CS(\alpha^*) = \{\top?\} \cup CS(\alpha) \cup CS(\alpha; \alpha) \cup \ldots$

The function CS defined here is an extension of the function CS defined in the literature of PDL, for instance, in [6]. For any α of PDL, $CS(\alpha)$ consists of all the ways of performing α: $S_\alpha = \bigcup\{S_\delta \mid \delta \in CS(\alpha)\}$ in any models. However, this is not the case here. For example, let $\alpha = (a; \top?) \cap (\top?; b)$. It can be verified that $CS(\alpha) = \{(a \cap \top?); (\top? \cap b)\}$ and $\bigcup\{S_\delta \mid \delta \in CS(\alpha)\}$ is always empty, but S_α is not. The reason comes from the way we define the merge of r-seqs. We define it this way for technical reasons. When merge is not essentially involved, the result holds.

Lemma 2. $CS(\alpha) = \{\alpha\}$ *for any IT-action α.*

Lemma 3. α *is an action in which the occurrences of $\cap$ outside of any tests connect only IT-actions. Then for any model, $S_\alpha = \bigcup\{S_\delta \mid \delta \in CS(\alpha)\}$.*

Proof. We put an induction on α. Other cases can be handled easily; the special case is $\alpha = \beta \cap \gamma$ where β and γ are IT-actions. As $\beta \cap \gamma$ is still an IT-action, by Lemma 2, $CS(\beta \cap \gamma) = \{\beta \cap \gamma\}$. Then $S_{\beta \cap \gamma} = \bigcup \{S_\delta \mid \delta \in CS(\beta \cap \gamma)\}$.

The function CS is *blind* to the internal structure of tests; $\beta \sqcap \gamma = \beta \cap \gamma$ for any IT-actions β and γ; hence, if the occurrences of $\cap$ in α outside any tests connect only IT-actions, then $CS(\alpha)$ does not really involve the operation of merge.

There is something unnatural with r-seqs. Computation sequences defined as follows are more intuitive:

Definition 5 (F-seqs). $\beta_1; \ldots; \beta_n$ *is called a fine computation sequence (f-seqs) if each β_i is either an I-action or a T-action.*

In principle, f-seqs can be performed, as they have no empty actions like $a \cap p?$. Note that for any r-seq $\beta_1; \ldots; \beta_n$, if $S_{\beta_1;\ldots;\beta_n}$ is not empty in some model, then $\beta_1; \ldots; \beta_n$ is a f-seq. Define $dep(\beta_1; \ldots; \beta_n)$, the *depth* of the f-seq $\beta_1; \ldots; \beta_n$, as the number of the I-actions in $\beta_1, \ldots, \beta_n$. For example, $dep((a \cap b); p?; (c \cap d)) = 2$. It can be seen that for any f-seq $\beta_1; \ldots; \beta_n$ and model, if $S_{\beta_1;\ldots;\beta_n}$ is not empty, then all the sequences in $S_{\beta_1;\ldots;\beta_n}$ contain $dep(\beta_1; \ldots; \beta_n) + 1$ elements.

Definition 6 (Matching F-seqs). *F-seqs α and β match if $\alpha \sqcap \beta$ is a f-seq as well.*

That two f-seqs match means that the merge of them can also be performed. In the sequel, we use $\alpha \equiv \beta$ to denote that α and β have the same interpretation in any models.

Lemma 4. *F-seqs $\gamma_1; \ldots; \gamma_n$ and $\delta_1; \ldots; \delta_n$ match. Then $(\gamma_1; \ldots; \gamma_n) \cap (\delta_1; \ldots; \delta_n) \equiv (\gamma_1; \ldots; \gamma_n) \sqcap (\delta_1; \ldots; \delta_n)$.*

This is a very important property which trace semantics does not share with relation semantics.

Definition 7 (Cushioned Replicas of Actions). α^σ, *the cushioned replica of α, is the result of replacing each atomic action a in α by $(\top?)^*; a; (\top?)^*$.*

Since $a \equiv (\top?)^*; a; (\top?)^*$, $\alpha \equiv \alpha^\sigma$. Compared to $CS(\alpha)$, $CS(\alpha^\sigma)$ might contain a lot of redundant r-seqs. For example, $CS(a)$ is the single set $\{a\}$, but $CS(a^\sigma)$ is the infinite set $\{a, \top?; a, a; \top?, \top?; a; \top?, \ldots\}$. There are so many redundant r-seqs in $CS(a^\sigma)$ that we have the following result:

Lemma 5. α *and β are two actions. For any f-seqs δ in $CS(\alpha^\sigma)$ and τ in $CS(\beta^\sigma)$, if $dep(\delta) = dep(\tau)$, then there are f-seqs δ' in $CS(\alpha^\sigma)$ and τ' in $CS(\beta^\sigma)$ s.t. $\delta \equiv \delta'$, $\tau \equiv \tau'$ and δ' and τ' match.*

We give an example to illustrate why this is the case. Suppose that $a; b; p? \in CS(\alpha^\sigma)$ and $c; q?; d \in CS(\beta^\sigma)$. $a; b; p?$ and $c; q?; d$ do not match. $a; \top?; b; p?$ is in $CS(\alpha^\sigma)$ and $c; q?; d; \top?$ is in $CS(\beta^\sigma)$. $a; \top?; b; p?$ and $c; q?; d; \top?$ match.

Previously, we mentioned that $CS(\alpha)$ might not consist of all the ways of performing α. Actually, $CS(\alpha^\sigma)$ does.

Proposition 3. $S_\alpha = \bigcup\{S_\delta \mid \delta \in CS(\alpha^\sigma)\}$.

Proof. We put an induction on α. We only handle the **case** $\alpha = \beta \cap \gamma$ and skip others. Then $\alpha^\sigma = \beta^\sigma \cap \gamma^\sigma$. By the inductive hypothesis, $S_\beta = \bigcup\{S_\delta \mid \delta \in CS(\beta^\sigma)\}$ and $S_\gamma = \bigcup\{S_\delta \mid \delta \in CS(\gamma^\sigma)\}$.

Let $(w_0, \ldots, w_n) \in S_{\beta \cap \gamma}$. Then $(w_0, \ldots, w_n) \in S_\beta \cap S_\gamma$. Then $(w_0, \ldots, w_n) \in \bigcup\{S_\delta \mid \delta \in CS(\beta^\sigma)\}$ and $(w_0, \ldots, w_n) \in \bigcup\{S_\delta \mid \delta \in CS(\gamma^\sigma)\}$. There is a f-seq $\delta \in CS(\beta^\sigma)$ s.t. $(w_0, \ldots, w_n) \in S_\delta$ and there is a f-seq $\tau \in CS(\gamma^\sigma)$ s.t. $(w_0, \ldots, w_n) \in S_\tau$. Then $dep(\delta) = dep(\tau) = n$. By Lemma 5, there are f-seqs δ' in $CS(\beta^\sigma)$ and τ' in $CS(\gamma^\sigma)$ s.t. $\delta \equiv \delta'$, $\tau \equiv \tau'$ and δ' and τ' match. Then $(w_0, \ldots, w_n) \in S_{\delta'}$ and $(w_0, \ldots, w_n) \in S_{\tau'}$. Then $(w_0, \ldots, w_n) \in S_{\delta' \cap \tau'}$. As δ' and τ' match, $\delta' \sqcap \tau'$ is defined. Then by Lemma 4, $(w_0, \ldots, w_n) \in S_{\delta' \sqcap \tau'}$. Then $\delta' \sqcap \tau' \in CS(\beta^\sigma \cap \gamma^\sigma)$. Then $(w_0, \ldots, w_n) \in \bigcup\{S_\delta \mid \delta \in CS(\beta^\sigma \cap \gamma^\sigma)\}$.

Let $(w_0, \ldots, w_n) \in \bigcup\{S_\delta \mid \delta \in CS(\beta^\sigma \cap \gamma^\sigma)\}$. Then $(w_0, \ldots, w_n) \in S_{\delta \sqcap \tau}$ for some $\delta \in CS(\beta^\sigma)$ and $\tau \in CS(\gamma^\sigma)$. Then δ and τ match. By Lemma 4, $(w_0, \ldots, w_n) \in S_{\delta \cap \tau}$. Then $(w_0, \ldots, w_n) \in S_\delta$ and $(w_0, \ldots, w_n) \in S_\tau$. Then $(w_0, \ldots, w_n) \in \bigcup\{S_\delta \mid \delta \in CS(\beta^\sigma)\}$ and $(w_0, \ldots, w_n) \in \bigcup\{S_\delta \mid \delta \in CS(\gamma^\sigma)\}$. Then $(w_0, \ldots, w_n) \in S_\beta$ and $(w_0, \ldots, w_n) \in S_\gamma$. Then $(w_0, \ldots, w_n) \in S_{\beta \cap \gamma}$.

It is implied that $\alpha \equiv \alpha'$ if $CS(\alpha^\sigma) = CS(\alpha'^\sigma)$.

5.2   Reduction of IPDL to itPDL

Given Π_0 and Φ_0 as in Section 2.1. Simultaneously define the set Π_{itPDL} of actions and the set Φ_{itPDL} of propositions as follows:

$$\gamma ::= a \mid \phi? \mid (\gamma \cap \gamma)$$
$$\alpha ::= \gamma \mid (\alpha; \alpha) \mid (\alpha \cup \alpha) \mid \alpha^*$$
$$\phi ::= p \mid \top \mid \neg\phi \mid (\phi \wedge \phi) \mid \langle\alpha\rangle\phi$$

The difference between Π_{itPDL} and Π_{IPDL} is that $\cap$ has a restricted connecting ability in Π_{itPDL}: it connects only IT-actions. Actually, Φ_{itPDL} has the same expressivity with Φ_{IPDL} in trace semantics. We now are going to prove it.

We are going to use the following notions in the usual sense: *regular expressions*, *regular languages* and *correspondence between regular expressions and regular languages*. We are also going to use such a result: regular languages are closed under intersection. For the definitions of these notions and the proof of this result, we refer to [7]. Usually, alphabets consist of simple symbols. We in the sequel use alphabets in a generalized sense; they may take IT-actions like $(a \cap p?)$ as letters. Given an alphabet Γ. For any regular expression α over Γ, $CS(\alpha)$, defined as above, is the regular language corresponding to α.

Let Γ and Σ be two sets of actions. Define $\Gamma \barwedge \Sigma = \{\alpha \cap \beta \mid \alpha \in \Gamma \ \& \ \beta \in \Sigma\}$. What follows is a crucial lemma:

Lemma 6. *Γ and Σ are two finite alphabets consisting of IT-actions. For any regular expressions α over Γ and β over Σ, there is a regular expression λ over $\Gamma \barwedge \Sigma$ s.t. $CS(\lambda) = CS(\alpha \cap \beta)$.*

Proof. Let $\Gamma = \{\gamma_1, \ldots, \gamma_n\}$ and $\Sigma = \{\delta_1, \ldots, \delta_m\}$. For any $i \leq n$, let $\gamma_i' = (\gamma_i \cap \delta_1) \cup \ldots \cup (\gamma_i \cap \delta_m)$. For any $i \leq m$, let $\delta_i' = (\gamma_1 \cap \delta_i) \cup \ldots \cup (\gamma_n \cap \delta_i)$. Let α' be the result of replacing each γ_i in α by γ_i'. Let β' be the result of replacing each δ_i in β by δ_i'. Then both α' and β' are regular expressions over the alphabet $\Gamma \wedge \Sigma$.

It can be seen that $CS(\alpha') = \bigcup \{CS(\gamma_{j_1}'; \ldots; \gamma_{j_k}') \mid \gamma_{j_1}; \ldots; \gamma_{j_k} \in CS(\alpha)\}$ and $CS(\beta') = \bigcup \{CS(\delta_{h_1}'; \ldots; \delta_{h_l}') \mid \delta_{h_1}; \ldots; \delta_{h_l} \in CS(\beta)\}$. By Lemma 2, we know that for any $\gamma_{j_1}, \ldots, \gamma_{j_k}$ in Γ and $\delta_{h_1}, \ldots, \delta_{h_k}$ in Σ: (i) if $(\gamma_{j_1} \cap \delta_{h_1}); \ldots; (\gamma_{j_k} \cap \delta_{h_k}) \in CS(\alpha')$, then $\gamma_{j_1}; \ldots; \gamma_{j_k} \in CS(\alpha)$; (ii) if $(\gamma_{j_1} \cap \delta_{h_1}); \ldots; (\gamma_{j_k} \cap \delta_{h_k}) \in CS(\beta')$, then $\delta_{h_1}; \ldots; \delta_{h_k} \in CS(\beta)$. By Lemma 2 again, we know that for any $\gamma_{j_1}, \ldots, \gamma_{j_k}$ in Γ and $\delta_{h_1}, \ldots, \delta_{h_k}$ in Σ: (i) if $\gamma_{j_1}; \ldots; \gamma_{j_k} \in CS(\alpha)$, then $(\gamma_{j_1} \cap \delta_{h_1}); \ldots; (\gamma_{j_k} \cap \delta_{h_k}) \in CS(\alpha')$; (ii) if $\delta_{h_1}; \ldots; \delta_{h_k} \in CS(\beta)$, then $(\gamma_{j_1} \cap \delta_{h_1}); \ldots; (\gamma_{j_k} \cap \delta_{h_k}) \in CS(\beta')$.

We claim $CS(\alpha \cap \beta) = CS(\alpha') \cap CS(\beta')$. Let $\zeta_1; \ldots; \zeta_k \in CS(\alpha \cap \beta)$. Then there are $\gamma_{j_1}, \ldots, \gamma_{j_k}$ in Γ and $\delta_{h_1}, \ldots, \delta_{h_k}$ in Σ s.t. $\gamma_{j_1}; \ldots; \gamma_{j_k} \in CS(\alpha)$, $\delta_{h_1}; \ldots; \delta_{h_k} \in CS(\beta)$ and $\zeta_1; \ldots; \zeta_k = (\gamma_{j_1}; \ldots; \gamma_{j_k}) \sqcap (\delta_{h_1}; \ldots; \delta_{h_k}) = (\gamma_{j_1} \cap \delta_{h_1}); \ldots; (\gamma_{j_k} \cap \delta_{h_k})$. Then $\zeta_1; \ldots; \zeta_k \in CS(\alpha')$ and $\zeta_1; \ldots; \zeta_k \in CS(\beta')$. Then $\zeta_1; \ldots; \zeta_k \in CS(\alpha') \cap CS(\beta')$. Now suppose $\zeta_1; \ldots; \zeta_k \in CS(\alpha') \cap CS(\beta')$. Then $\zeta_1; \ldots; \zeta_k \in CS(\alpha')$ and $\zeta_1; \ldots; \zeta_k \in CS(\beta')$. Then there are $\gamma_{j_1}, \ldots, \gamma_{j_k}$ in Γ and $\delta_{h_1}, \ldots, \delta_{h_k}$ in Σ s.t. $\zeta_1; \ldots; \zeta_k = (\gamma_{j_1} \cap \delta_{h_1}); \ldots; (\gamma_{j_k} \cap \delta_{h_k})$. As $\zeta_1; \ldots; \zeta_k \in CS(\alpha')$, $\gamma_{j_1}; \ldots; \gamma_{j_k} \in CS(\alpha)$. As $\zeta_1; \ldots; \zeta_k \in CS(\beta')$, $\delta_{h_1}; \ldots; \delta_{h_k} \in CS(\beta)$. Since $(\gamma_{j_1} \cap \delta_{h_1}); \ldots; (\gamma_{j_k} \cap \delta_{h_k}) = (\gamma_{j_1}; \ldots; \gamma_{j_k}) \sqcap (\delta_{h_1}; \ldots; \delta_{h_k})$, we know $\zeta_1; \ldots; \zeta_k \in CS(\alpha \cap \beta)$.

By Lemma 2, we know that both $CS(\alpha')$ and $CS(\beta')$ are regular languages over $\Gamma \wedge \Sigma$. Since regular languages are closed under intersection, $CS(\alpha') \cap CS(\beta')$ is a regular language over $\Gamma \wedge \Sigma$. Then there is a regular expression λ over $\Gamma \wedge \Sigma$ s.t. $CS(\lambda) = CS(\alpha') \cap CS(\beta')$. Then $CS(\lambda) = CS(\alpha \cap \beta)$.

For any action α, a set Σ is called the *immediate ingredient set* of α if Σ consists of (i) the atomic actions of α not inside any tests, (ii) the outermost tests in α and (iii) $\top?$. Such a Σ is a set of IT-actions. For any set Σ of actions, define $\Sigma_\cap^n = \{\gamma_1 \cap \ldots \cap \gamma_i \mid 1 \leq i \leq n \ \& \ \gamma_1, \ldots, \gamma_i \in \Sigma\}$. We see $\Sigma \subseteq \Sigma_\cap^n$. Note that if Σ is a set of IT-actions, then $\Sigma_\cap^n$ is a set of IT-action as well.

Lemma 7. *α is an action in Π_{IPDL}. n is the number of the occurrences of $\cap$ in α. Σ is the immediate ingredient set of α. Then there is a regular expression α' over the alphabet $\Sigma_\cap^n$ s.t. $\alpha \equiv \alpha'$.*

Proof. We put an induction on α. We present a proof for only the **case** $\alpha = \beta \cap \gamma$. Let m and n be the numbers of the occurrences of $\cap$ in β and γ respectively. Then $m + n + 1$ is the number of the occurrences of $\cap$ in $\beta \cap \gamma$. Let Δ and Γ be the immediate ingredient sets of β and γ respectively. Then $\Sigma = \Delta \cup \Gamma$.

By the inductive hypothesis, there are regular expressions β' over $\Delta_\cap^m$ and γ' over $\Gamma_\cap^n$ s.t. $\beta \equiv \beta'$ and $\gamma \equiv \gamma'$. Then β'^σ and γ'^σ are regular expressions over $\Delta_\cap^m$ and $\Gamma_\cap^n$. By Lemma 6, there is a regular expression λ over $\Delta_\cap^m \wedge \Gamma_\cap^n$ s.t. $CS(\lambda) = CS(\beta'^\sigma \cap \gamma'^\sigma)$. It can be seen that $\Delta_\cap^m \wedge \Gamma_\cap^n \subseteq (\Delta \cup \Gamma)_\cap^{m+n+1}$. Then λ is a regular expression over $(\Delta \cup \Gamma)_\cap^{m+n+1}$. Then λ is an action in which the occurrences of $\cap$ not inside any tests connect only IT-actions. By Lemma 3, we know that for any model, $S_\lambda = \bigcup \{S_\delta \mid \delta \in CS(\lambda)\}$. Then for any model, with the help of Proposition 3, we have such an equation: $S_\lambda = \bigcup \{S_\delta \mid \delta \in CS(\lambda)\} = \bigcup \{S_\delta \mid \delta \in CS(\beta'^\sigma \cap \gamma'^\sigma)\} = \bigcup \{S_\delta \mid \delta \in CS(\beta' \cap \gamma')^\sigma\} = S_{\beta' \cap \gamma'} = S_{\beta \cap \gamma}$. Then we know that $\lambda \equiv \beta \cap \gamma$.

Proposition 4. *For any $\alpha \in \Pi_{IPDL}$, there is a $\alpha' \in \Pi_{itPDL}$ s.t. $\alpha \equiv \alpha'$.*

Proof. Let $\alpha \in \Pi_{IPDL}$. Let n be the number of the occurrences of $\cap$ in α. Let Σ be the immediate ingredient set of α. By Lemma 7, there is a regular expression β over $\Sigma^n_\cap$ s.t. $\alpha \equiv \beta$. β has such a feature: the occurrences of $\cap$ not inside any tests connect only IT-actions. Then we go inside an outermost test of β and do the same thing. And so on. Finally, we get a α' s.t. $\alpha \equiv \alpha'$ and all the occurrences of $\cap$ in α' connect only IT-actions. Then α' is in Π_{itPDL}.

The notions of *rough computation sequences*, defined in Definition 2, and *cushioned replicas*, defined in Definition 7, are used to handle test. The proof of Proposition 4 would be shorter if there is no test.

5.3   Reduction of itPDL to iPDL

There are three types of IT-actions in Π_{itPDL}: I-actions, T-actions and mixed actions containing I-actions and T-actions. Here are two facts about Π_{itPDL}: (i) mixed actions are always empty; (ii) the intersection of finite tests is equivalent to a test: $\phi_1? \cap \ldots \cap \phi_n? \equiv (\phi_1 \wedge \ldots \wedge \phi_n)?$. This implies that Φ_{itPDL} can be reduced to Φ_{iPDL}, which is defined as what follows:

$$\gamma ::= a \mid (\gamma \cap \gamma)$$
$$\alpha ::= \gamma \mid (\alpha; \alpha) \mid (\alpha \cup \alpha) \mid \alpha^* \mid \phi?$$
$$\phi ::= p \mid \top \mid \neg\phi \mid (\phi \wedge \phi) \mid \langle \alpha \rangle \phi$$

In this language, $\cap$ connects only atomic actions. We call the logic of this language in trace semantics as iPDL.

As mentioned at the end of Section 4, the reason that trace semantics makes a difference from relation semantics w.r.t. IPDL lies in that $\cap$ does not preserve the following property: $(u, v) \in R_\alpha$ implies that there are $x_0, \ldots, x_n$ s.t. $x_0 = u$, $x_n = v$ and $(x_0, \ldots, x_n) \in S_\alpha$. However, when connecting only I-actions, $\cap$ preserves it. By an extended version of Proposition 1, we have the following result:

Proposition 5. *For any ϕ of Φ_{iPDL}, $\mathfrak{M}, w \Vdash_t \phi \Leftrightarrow \mathfrak{M}, w \Vdash_r \phi$.*

This means that the logic IPDL in trace semantics can be reduced to the logic iPDL in relation semantics.

6   Future Work

There are a few other open questions for us besides the one mentioned at Section 3. Doing α and β at the same time may mean a few different things: (i) starting

α and β at the same time; (ii) finishing α and β at the same time; (iii) starting and finishing α and β at the same time; (iv) doing α during doing β. We in this paper formalize the third reading. We think that others can also be caught in trace semantics. This is an issue we want to pursue in the future. In addition, we want to look at this work against the backgrounds of Concurrent PDL due to [10] and Process Algebra initiated by [8], [3] and [2], which represent two main directions in the research of concurrency.

Acknowledgment. This research was supported by the National Social Science Foundation of China (No. 12CZX053) and the Fundamental Research Funds for the Central Universities (No. SKZZY201304). The second author was also supported by the Scientific Research Foundation for the Returned Overseas Chinese Scholars, State Education Ministry (No. KJZXCJ2015185) and the Fundamental Research Funds for the Central Universities (No. 105585GK). We would like to thank Johan van Benthem, Valentin Goranko, Yanjing Wang and the anonymous referees for their useful comments and suggestions.

References

1. van Benthem, J., Ghosh, S., Liu, F.: Modelling simultaneous games in dynamic logic. Synthese 165(2), 247–268 (2008)
2. Bergstra, J.A., Klop, J.W.: Process algebra for synchronous communication. Information and control 60(1), 109–137 (1984)
3. Brookes, S.D., Hoare, C.A.R., Roscoe, A.W.: A theory of communicating sequential processes. J. ACM 31(3), 560–599 (1984)
4. Fischer, M.J., Ladner, R.E.: Propositional dynamic logic of regular programs. Journal of Computer and System Sciences 18(2), 194–211 (1979)
5. Harel, D., Kozen, D., Parikh, R.: Process logic: Expressiveness, decidability, completeness. Journal of Computer and System Sciences 25(2), 144–170 (1982)
6. Harel, D., Kozen, D., Tiuryn, J.: Dynamic Logic. MIT Press (2000)
7. Hopcroft, J., Motwani, R., Ullman, J.: Introduction to Automata Theory, Languages, and Computation. Prentice Hall (2006)
8. Milner, R.: A calculus of communicating systems. Springer (1980)
9. Nishimura, H.: Descriptively complete process logic. Acta Informatica 14(4), 359–369 (1980)
10. Peleg, D.: Concurrent dynamic logic. In: Proceedings of the Seventeenth Annual ACM Symposium on Theory of Computing, pp. 232–239. ACM (1985)
11. Pnueli, A.: The temporal logic of programs. In: Proceedings of 18th Symposium on Foundations of Computer Science, pp. 46–57 (1977)
12. Vardi, M., Wolper, P.: Yet another process logic. In: Clarke, E., Kozen, D. (eds.) Logic of Programs 1983. LNCS, vol. 164, pp. 501–512. Springer, Heidelberg (1984)

A Decidable Temporal Relevant Logic for Time-Dependent Relevant Human Reasoning

Norihiro Kamide

Teikyo University, Faculty of Science and Engineering,
Department of Information and Electronic Engineering,
Toyosatodai 1-1, Utsunomiya, Tochigi 320-8551, Japan
drnkamide08@kpd.biglobe.ne.jp

Abstract. In this paper, a Gentzen-type sequent calculus STRW is introduced for a new temporal relevant logic TRW which is obtained from the positive contraction-less relevant logic by adding some temporal operators. The cut-elimination and completeness theorems for STRW are proved. STRW is shown to be decidable and to have relevance principle.

1 Introduction

Formalizing and implementing time-dependent relevant human reasoning in computer systems is gaining increasing importance in the fields of computer science because such reasoning is required for modeling and verifying sophisticated software agents in computer systems. Such formalization and implementation require a decidable temporal relevant logic (or deductive system) that can suitably represent time-dependency and relevancy in human reasoning. The representation of time-dependency and relevancy in human reasoning requires a construction of a logic that combines a *temporal logic* [12] and a *relevant logic (or relevance logic)* [1].

In this paper, a Gentzen-type sequent calculus STRW for a new temporal relevant logic TRW is introduced by extending a sequent calculus SRW^+ for the positive fragment RW^+ of the *contraction-less relevant logic* RW [1,7,5,6]. Cut-elimination theorem and completeness theorem (with respect to an extended *Routley-Meyer semantics*) are proved for STRW, and STRW is also shown to be decidable and to have *relevance principle* (or *variable sharing property*). These theorems and properties are proved using some theorems for embedding STRW (or TRW) into SRW^+ (or RW^+). The embedding-based proof method for proving these theorems is analogous to the proof method for proving some theorems in a *paraconsistent (or inconsistency-tolerant) relevant logic* RWP (and its Gentzen-type sequent calculus SRWP) proposed in [10].

The *positive contraction-less relevant logic* RW^+, which is a base logic for both RWP and TRW, is known to be a typical "decidable" relevant logic [1,7,5,6]. The logic RW^+ can appropriately represent "relevant" human reasoning in the sense that it has the relevance principle. The statement of the relevance principle in the propositional case is presented as follows: If a formula $\alpha \rightarrow \beta$ is provable,

W. van der Hoek et al. (Eds.): LORI 2015, LNCS 9394, pp. 182–194, 2015.
DOI: 10.1007/978-3-662-48561-3_15

then there exists a propositional variable p such that $p \in V(\alpha) \cap V(\beta)$ where $V(\alpha)$ denotes the set of all propositional variables in α. It is known that this principle does not hold for classical logic and intuitionistic logic. For more detailed philosophical discussions on the relevance principle and relevant logics, see e.g., [1].

As mentioned, the logic RW^+ can appropriately represent "relevant" human reasoning, but it cannot represent "time-dependent" human reasoning. Thus, we need to extend RW^+ by adding some temporal operators used in some temporal logics. One of the standard temporal logics is the *linear-time temporal logic* (LTL) [12], which is known to be useful for verifying, specifying and modeling concurrent systems. The logic TRW (or its Gentzen-type sequent calculus STRW) proposed in this paper adopts some "bounded time" versions of the standard temporal operators X (next-time), G (globally in the future) and F (eventually in the future) used in LTL. Such bounded temporal operators used in TRW have the bounded time domain which is useful for obtaining an efficient or decidable temporal logic.

Some motivations on the bounded time domain of TRW are explained as follows. Although the standard LTL has an infinite (unbounded) time domain, i.e., the set ω of all natural numbers, TRW has a bounded time domain which is restricted by a fixed positive integer l, i.e., the set $\omega_l := \{x \in \omega \mid x \leq l\}$. Despite the restriction on the time domain, TRW can prove almost all the typical temporal axioms of LTL, such as a temporal induction axiom. It is also known that to restrict the time domain is a technique that may be applied to obtain a decidable or efficient fragment of LTL [8]. Restricting the time domain implies not only some purely theoretical merits as mentioned above, but also some practical merits for describing temporal databases [4] and for implementing an efficient model checking algorithm, called *bounded model checking* [3]. These practical merits are important due to the fact that there are problems in computer science and artificial intelligence where only a finite fragment of the time sequence is of interest [4].

The contents of this paper are then summarized as follows.

In Section 2, a Gentzen-type sequent calculus SRW^+ for RW^+ and a Gentzen-type sequent calculus STRW for the temporal relevant logic TRW are introduced. A theorem for syntactically embedding STRW into SRW^+ is proved. The cut-elimination theorem, decidability, and relevance principle for STRW are proved using this syntactical embedding theorem.

In Section 3, completeness theorem with respect to an extended Routley-Meyer semantics is proved for STRW (or TRW). A Routley-Meyer semantics for RW^+ is reviewed, and an extended Routley-Meyer semantics is introduced for TRW. A theorem for semantically embedding TRW into RW^+ is proved. The completeness theorem (with respect to the extended semantics) for STRW is proved using both the semantical and syntactical embedding theorems.

In Section 4, this paper is concluded, and some remarks on the idea of bounded-time domain are given.

2   Cut-Elimination and Decidability

The language used in this paper is introduced below. *Formulas* are constructed from propositional variables, $\rightarrow$ (implication), $\wedge$ (conjunction), $\vee$ (disjunction) and the following bounded linear-time temporal operators G (globally), F (eventually) and X (next). Lower-case letters $p, q,...$ are used to represent propositional variables, Greek lower-case letters $\alpha, \beta, ...$ are used to represent formulas, and Greek capital letters $\Gamma, \Delta, ...$ are used to represent finite (possibly empty) sequences of formulas or bunches. We write $A \equiv B$ to indicate the syntactical identity between A and B.

For any $\sharp \in \{G, F, X\}$, an expression $\sharp\Gamma$ is used to denote the sequence $\langle \sharp\gamma \mid \gamma \in \Gamma \rangle$. The symbol ω is used to represent the set of natural numbers. The symbol ω_l with a fixed positive integer l is used to represent the set $\{i \in \omega \mid i \leq l\}$. The symbols $\leq$ (or $\geq$) and $<$ (or $>$) are respectively used to represent the linear and strict linear orders on ω (or a finite subset of ω). An expression $X^i\alpha$ for any $i \in \omega$ is defined inductively by $X^0\alpha \equiv \alpha$ and $X^{n+1}\alpha \equiv X^nX\alpha$. Lower-case letters i, j and k are used to denote any natural numbers. Expressions $\bigwedge\{\alpha_i \mid i \in \omega_l\}$ and $\bigvee\{\alpha_i \mid i \in \omega_l\}$ are used to represent $\alpha_0 \wedge \alpha_1 \wedge \cdots \wedge \alpha_l$ and $\alpha_0 \vee \alpha_1 \vee \cdots \vee \alpha_l$, respectively.

Following [5,7], we give some definitions below. *Bunches* are inductively defined by (1) any formula is a bunch, and (2) for $n \geq 2$, if X_i is a bunch for $i = 1, ..., n$, then both sequences $(X_1, ..., X_n)$ and $(X_1; ...; X_n)$ are bunches. Bunches of the forms $(X_1, ..., X_n)$ and $(X_1; ...; X_n)$ are respectively called *intensional* and *extensional*. Each bunch X_i is called an *immediate constituent* of $(X_1, ..., X_n)$ and $(X_1; ...; X_n)$. For the sake of simplicity, we assume that immediate constituents of an intensional (and an extensional) bunch are not intensional (and extensional, respectively). Thus, a bunch of the form $(X; (Y; Z); W)$ is identified with the bunch $(X; Y; Z; W)$. In other words, intensional bunches and extensional bunches must appear alternatingly in a given bunch. We will omit parentheses when no confusion will occur.

In the following, capital letters X, Y and Z etc. with or without subscripts denote bunches. *Subbunches* of a given bunch Z can be defined in the usual way. We will sometimes pay special attention to a particular occurrence of a subbunch X of Z. In such a case, the occurrence X is called a *bunch occurrence* of X (in Z) which is indicated. An expression $\Gamma(X)$ is used to denote a bunch with an indicated bunch occurrence of X in it. *Sequents* are expressions of the form $X \Rightarrow \gamma$ where X is a (possibly empty) bunch and γ is a formula.

The expression of the form $L \vdash S$ means that the sequent S is provable in a sequent calculus L. We will sometimes omit L in this expression. A rule R of inference is said to be *admissible* in a sequent calculus L if the following condition

$$\frac{S_1 \cdots S_n}{S}$$

is satisfied: for any instance of R, if $L \vdash S_i$ for all i, then $L \vdash S$.

A sequent calculus STRW for TRW is then defined as follows.

Definition 1 (STRW). *Let l be a fixed positive integer (called a time bound). The initial sequents of* STRW *are of the form: for any propositional variable p,*

$$X^i p \Rightarrow X^i p.$$

The cut rule of STRW *is of the form:*

$$\frac{X \Rightarrow \alpha \quad \Gamma(\alpha) \Rightarrow \gamma}{\Gamma(X) \Rightarrow \gamma} \text{ (cut).}$$

The intensional and extensional structural rules of STRW *are of the form:*

$$\frac{\Gamma(Y, X) \Rightarrow \gamma}{\Gamma(X, Y) \Rightarrow \gamma} \text{ (I-ex)} \qquad \frac{\Gamma(Y; X) \Rightarrow \gamma}{\Gamma(X; Y) \Rightarrow \gamma} \text{ (E-ex)}$$

$$\frac{\Gamma(X; X) \Rightarrow \gamma}{\Gamma(X) \Rightarrow \gamma} \text{ (E-co)} \qquad \frac{\Gamma(X) \Rightarrow \gamma}{\Gamma(X; Y) \Rightarrow \gamma} \text{ (E-wk)}$$

where (E-wk) *has the proviso that $\Gamma(X)$ is non-empty.*
The logical inference rules of STRW *are of the form:*

$$\frac{Y \Rightarrow X^i \alpha \quad \Gamma(X^i \beta) \Rightarrow \gamma}{\Gamma(X^i(\alpha \to \beta), Y) \Rightarrow \gamma} \text{ (}\to\text{l)} \qquad \frac{Y, X^i \alpha \Rightarrow X^i \beta}{Y \Rightarrow X^i(\alpha \to \beta)} \text{ (}\to\text{r)}$$

$$\frac{\Gamma(X^i \alpha; X^i \beta) \Rightarrow \gamma}{\Gamma(X^i(\alpha \wedge \beta)) \Rightarrow \gamma} \text{ (}\wedge\text{l)} \qquad \frac{Y \Rightarrow X^i \alpha \quad Z \Rightarrow X^i \beta}{Y; Z \Rightarrow X^i(\alpha \wedge \beta)} \text{ (}\wedge\text{r)}$$

$$\frac{\Gamma(X^i \alpha) \Rightarrow \gamma \quad \Gamma(X^i \beta) \Rightarrow \gamma}{\Gamma(X^i(\alpha \vee \beta)) \Rightarrow \gamma} \text{ (}\vee\text{l)}$$

$$\frac{Y \Rightarrow X^i \alpha}{Y \Rightarrow X^i(\alpha \vee \beta)} \text{ (}\vee\text{r1)} \qquad \frac{Y \Rightarrow X^i \beta}{Y \Rightarrow X^i(\alpha \vee \beta)} \text{ (}\vee\text{r2).}$$

The specific temporal inference rules of STRW *are of the form: for any $k \in \omega_l$,*

$$\frac{\Gamma(X^l \alpha) \Rightarrow \gamma}{\Gamma(X^{i+l} \alpha) \Rightarrow \gamma} \text{ (Xl)} \qquad \frac{Y \Rightarrow X^l \alpha}{Y \Rightarrow X^{i+l} \alpha} \text{ (Xr)}$$

$$\frac{\Gamma(X^{i+k} \alpha) \Rightarrow \gamma}{\Gamma(X^i G\alpha) \Rightarrow \gamma} \text{ (Gl)} \qquad \frac{\{ Y \Rightarrow X^{i+j} \alpha \}_{j \in \omega_l}}{Y \Rightarrow X^i G\alpha} \text{ (Gr)}$$

$$\frac{\{ \Gamma(X^{i+j} \alpha) \Rightarrow \gamma \}_{j \in \omega_l}}{\Gamma(X^i F\alpha) \Rightarrow \gamma} \text{ (Fl)} \qquad \frac{Y \Rightarrow X^{i+k} \alpha}{Y \Rightarrow X^i F\alpha} \text{ (Fr).}$$

Some remarks concerning the definition of STRW are given as follows.

1. The sequents of the form $X^i \alpha \Rightarrow X^i \alpha$ for any formula α are provable in cut-free STRW. This can be shown by induction on α.

2. (Gr) and (Fl) in STRW have $l + 1$ premises. By (Xl) and (Xr), the nesting of the outermost occurrence of X in a formula can be bounded by l. In (Gl) and (Fr), the number k is bounded by l.

3. Strictly speaking, STRW is just the sequent calculus parameterized by a fixed positive integer l, and hence such a calculus should precisely be denoted as STRW[l]. But, when we don't need to specify such an integer l, we will use the name "STRW" instead of the concrete name "STRW[l]". Indeed, for example, STRW[2] is different from STRW[1]: $p \wedge Xp \Rightarrow Gp$ is provable in STRW[1], but it is not provable in STRW[2].

4. Let STRW[ω] be the system obtained from STRW[l] by replacing l with ω and deleting the inference rules (Xl) and (Xr). Then, STRW[ω] may be regarded as a relevant logic version of Kawai's sequent calculus LT_ω [11] for (unbounded) *linear-time temporal logic* (LTL). It was shown in [11] that the completeness (w.r.t. Kripke semantics) and cut-elimination theorems hold for the first-order version of LT_ω.

5. Let SXRW be the system obtained from STRW[l] by deleting the specific temporal inference rules (Xl), (Xr), (Gl), (Gr), (Fl) and (Fr). Then, SXRW may be regarded as a relevant logic version of a sequent calculus for *Prior's tomorrow tense logic* [13,14].

Proposition 2. *The following rule is admissible in cut-free STRW:*

$$\frac{Y \Rightarrow \gamma}{XY \Rightarrow X\gamma} \text{ (Xregu)}.$$

An expression $\alpha \Leftrightarrow \beta$ means the sequents $\alpha \Rightarrow \beta$ and $\beta \Rightarrow \alpha$.

Proposition 3. *The following sequents are provable in cut-free* STRW: *for any formulas α, β and any $i \in \omega$,*

1. $X^i(\alpha \circ \beta) \Leftrightarrow X^i\alpha \circ X^i\beta$ *where* $\circ \in \{\rightarrow, \wedge, \vee\}$,
2. $G\alpha \Rightarrow \alpha$,
3. $G\alpha \Rightarrow X\alpha$,
4. $G\alpha \Rightarrow XG\alpha$,
5. $G\alpha \Rightarrow GG\alpha$,
6. $\alpha; G(\alpha \rightarrow X\alpha) \Rightarrow G\alpha$ *(temporal induction)*,
7. $X^{i+l}\alpha \Leftrightarrow X^l\alpha$ *(bounded next-time)*,
8. $G\alpha \Leftrightarrow \alpha \wedge X\alpha \wedge \cdots \wedge X^l\alpha$ *(bounded globally)*,
9. $F\alpha \Leftrightarrow \alpha \vee X\alpha \vee \cdots \vee X^l\alpha$ *(bounded eventually)*.

In order to show a syntactical embedding theorem for STRW, a sequent calculus SRW$^+$ [5,6,7] for RW$^+$ is defined below.

Definition 4 (SRW$^+$). SRW$^+$ *is obtained from* STRW *by deleting the specific temporal inference rules* (Xl), (Xr), (Gl), (Gr), (Fl), (Fr) *and deleting all the occurrences* X^i *in the initial sequents and logical inference rules.*

We will sometimes use the same names for the inference rules of both STRW and SRW$^+$ if we have no confusion. If we need to distinguish the inference rules of STRW and of SRW$^+$, we impose a superscript notation to the objective inference rule, e.g., $(\to\text{r})^{\text{STRW}}$ for the STRW inference rule and $(\to\text{r})^{\text{SRW}^+}$ for the SRW$^+$ inference rule.

Some remarks concerning the definition of SRW$^+$ are given as follows.

1. The sequents of the form $\alpha \Rightarrow \alpha$ for any formula α are provable in cut-free SRW$^+$. This can be proved by induction on α.
2. A sequent calculus for the positive fragment R$^+$ [6] of the relevant logic R is obtained from SRW$^+$ by adding the following intensional structural rule (I-co):

$$\frac{\Gamma(X, X) \Rightarrow \gamma}{\Gamma(X) \Rightarrow \gamma} \text{ (I-co).}$$

3. The relevant logics R$^+$ and R are undecidable [16]. The contraction-less relevant logics RW$^+$ and RW are decidable [7,2]. These decidability results for RW$^+$ and RW were proved by Giambrone [7] and Brady [2] using some Gentzen-type sequent calculi.

An expression $V(\alpha)$ denotes the set of all propositional variables in a formula α. We then have the following theorems (see e.g., [5,6,7]).

Proposition 5. *We have:*

1. *The rule* (cut) *is admissible in cut-free* SRW$^+$.
2. SRW$^+$ *is decidable.*
3. *If* SRW$^+$ $\vdash \alpha \Rightarrow \beta$, *then there exists a propositional variable p such that $p \in V(\alpha) \cap V(\beta)$.*

We then define a translation of STRW into SRW$^+$ below.

Definition 6. *We fix a countable non-empty set Φ of propositional variables, and define the sets $\Phi_i := \{p_i \mid p \in \Phi\}$ ($i \in \omega$) of propositional variables where $p_0 := p$, i.e., $\Phi_0 = \Phi$. The language $\mathcal{L}^t$ of STRW is defined using Φ, $\to$, $\wedge$, $\vee$, X, G and F. The language $\mathcal{L}$ of SRW$^+$ is defined using $\bigcup_{i \in \omega} \Phi_i$, $\to$, $\wedge$ and $\vee$.*

A mapping f from $\mathcal{L}^t$ to $\mathcal{L}$ is defined by: for any $i \in \omega$,

1. $f(\mathrm{X}^i p) := p_i \in \Phi_i$ *for any $p \in \Phi$ (especially, $f(p) := p \in \Phi_0$),*
2. $f(\mathrm{X}^i(\alpha \circ \beta)) := f(\mathrm{X}^i \alpha) \circ f(\mathrm{X}^i \beta)$ *where $\circ \in \{\to, \wedge, \vee\}$,*
3. $f(\mathrm{X}^{i+l} \alpha) := f(\mathrm{X}^l \alpha)$,
4. $f(\mathrm{X}^i \mathrm{G} \alpha) := \bigwedge\{f(\mathrm{X}^{i+j} \alpha) \mid j \in \omega_l\}$,
5. $f(\mathrm{X}^i \mathrm{F} \alpha) := \bigvee\{f(\mathrm{X}^{i+j} \alpha) \mid j \in \omega_l\}$.

Strictly speaking, the mapping f is strongly dependent on l, i.e., f should precisely be denoted as f_l. In fact, $f_3(\mathrm{G}p)$ and $f_5(\mathrm{G}p)$ are different. But, for the sake of brevity, f is used instead of f_l in the following.

An expression $f(X)$ (or $f(\Gamma)$) denotes the result of replacing every occurrence of a formula α in X (or Γ, respectively) by an occurrence of $f(\alpha)$.

We then obtain a weak theorem for syntactically embedding STRW into SRW$^+$.

Theorem 7 (Weak syntactical embedding). *Let f be the mapping defined in Definition 6. Then:*

1. *If* STRW $\vdash X \Rightarrow \gamma$, *then* SRW$^+$ $\vdash f(X) \Rightarrow f(\gamma)$.
2. *If* SRW$^+$ $-$ (cut) $\vdash f(X) \Rightarrow f(\gamma)$, *then* STRW $-$ (cut) $\vdash X \Rightarrow \gamma$.

Proof. • (1): By induction on the proofs P of $X \Rightarrow \gamma$ in STRW. We distinguish the cases according to the last inference of P, and show some cases.

1. Case ($X^i p \Rightarrow X^i p$ for any propositional variable p): The last inference of P is of the form: $X^i p \Rightarrow X^i p$. In this case, we obtain $f(X^i p) \Rightarrow f(X^i p)$, i.e., $p_i \Rightarrow p_i$ ($p_i \in \Phi_i$) by the definition of f. This is an initial sequent of SRW$^+$.
2. Case (Xl): The last inference of P is of the form:

$$\frac{\Gamma(X^l \alpha) \Rightarrow \gamma}{\Gamma(X^{i+l} \alpha) \Rightarrow \gamma} \text{ (Xl).}$$

By induction hypothesis, we have SRW$^+$ $\vdash f(\Gamma)(f(X^l \alpha)) \Rightarrow f(\gamma)$, and $f(X^l \alpha)$ coincides with $f(X^{i+l} \alpha)$ by the definition of f. Thus, we obtain the required fact: SRW$^+$ $\vdash f(\Gamma)(f(X^{i+l} \alpha)) \Rightarrow f(\gamma)$.

3. Case (Fr): The last inference of P is of the form:

$$\frac{Y \Rightarrow X^{i+k} \alpha}{Y \Rightarrow X^i F \alpha} \text{ (Fr).}$$

By induction hypothesis, we have SRW$^+$ $\vdash f(Y) \Rightarrow f(X^{i+k} \alpha)$, and hence obtain the required fact:

$$\vdots$$
$$f(Y) \Rightarrow f(X^{i+k} \alpha)$$
$$\vdots \ (\vee r1)^{\text{SRW}^+}, (\vee r2)^{\text{SRW}^+}$$
$$f(Y) \Rightarrow \bigvee \{f(X^{i+j} \alpha) \mid j \in \omega_l\}$$

where $\bigvee \{f(X^{i+j} \alpha) \mid j \in \omega_l\}$ coincides with $f(X^i F \alpha)$ by the definition of f, and $f(X^{i+k} \alpha)$ is in the multiset $\{f(X^{i+j} \alpha) \mid j \in \omega_l\}$. The case $i > l$ is also included in this proof. In such a case, $f(X^{i+k} \alpha)$ and $\bigvee \{f(X^{i+j} \alpha) \mid j \in \omega_l\}$

mean $f(X^l \alpha)$ and $\overbrace{f(X^l \alpha) \vee f(X^l \alpha) \vee \cdots \vee f(X^l \alpha)}^{l}$, respectively.

4. Case (Gr): The last inference of P is of the form:

$$\frac{\{ Y \Rightarrow X^{i+j} \alpha \}_{j \in \omega_l}}{Y \Rightarrow X^i G \alpha} \text{ (Gr).}$$

By induction hypothesis, we have SRW$^+$ $\vdash f(Y) \Rightarrow f(X^{i+j} \alpha)$ for all $j \in \omega_l$. Let Φ be the multiset $\{f(X^{i+j} \alpha) \mid j \in \omega_l\}$. We then obtain the required fact:

$$\vdots$$
$$\{ f(Y) \Rightarrow f(X^{i+j} \alpha) \}_{f(X^{i+j} \alpha) \in \Phi}$$
$$\vdots \ (\wedge r)^{\text{SRW}^+}$$
$$f(Y) \Rightarrow \bigwedge \Phi$$

where $\bigwedge \Phi$ coincides with $f(\mathrm{X}^i \mathrm{G}\alpha)$ by the definition of f.

• (2): By induction on the proofs Q of $f(X) \Rightarrow f(\gamma)$ in SRW$^+$. We distinguish the cases according to the last inference of Q, and show some cases.

1. Case ($\rightarrow$r): The last inference of Q is of the form:

$$\frac{f(Y) \Rightarrow f(\mathrm{X}^i\alpha) \quad f(\Gamma)(f(\mathrm{X}^i\beta)) \Rightarrow f(\gamma)}{f(\Gamma)(f(\mathrm{X}^i(\alpha\rightarrow\beta)), f(Y)) \Rightarrow f(\gamma)} \ (\rightarrow\mathrm{l})^{\mathrm{SRW}^+}$$

where $f(\mathrm{X}^i(\alpha\rightarrow\beta))$ coincides with $f(\mathrm{X}^i\alpha)\rightarrow f(\mathrm{X}^i\beta)$ by the definition of f. By induction hypothesis, we have STRW $\vdash Y \Rightarrow \mathrm{X}^i\alpha$ and STRW $\vdash \Gamma(\mathrm{X}^i\beta) \Rightarrow \gamma$. Then, we obtain the required fact:

$$\frac{\vdots \qquad \vdots}{\begin{array}{cc} Y \Rightarrow \mathrm{X}^i\alpha & \Gamma(\mathrm{X}^i\beta) \Rightarrow \gamma \end{array}}$$
$$\frac{}{\Gamma(\mathrm{X}^i(\alpha\rightarrow\beta)), Y) \Rightarrow \gamma} \ (\rightarrow\mathrm{l})^{\mathrm{STRW}}.$$

2. Case ($\wedge$r): The last inference of Q is of the form:

$$\frac{f(Y) \Rightarrow f(\mathrm{X}^i\alpha) \quad f(Z) \Rightarrow f(\mathrm{X}^i\beta)}{f(Y); f(Z) \Rightarrow f(\mathrm{X}^i(\alpha \wedge \beta))} \ (\wedge\mathrm{r})^{\mathrm{SRW}^+}$$

where $f(\mathrm{X}^i(\alpha\wedge\beta))$ coincides with $f(\mathrm{X}^i\alpha)\wedge f(\mathrm{X}^i\beta)$ by the definition of f. By induction hypothesis, we have STRW $\vdash Y \Rightarrow \mathrm{X}^i\alpha$ and STRW $\vdash Z \Rightarrow \mathrm{X}^i\beta$. Then, we obtain the required fact:

$$\frac{\vdots \qquad \vdots}{\begin{array}{cc} Y \Rightarrow \mathrm{X}^i\alpha & Z \Rightarrow \mathrm{X}^i\beta \end{array}}$$
$$\frac{}{Y; Z \Rightarrow \mathrm{X}^i(\alpha \wedge \beta)} \ (\wedge\mathrm{r})^{\mathrm{STRW}}.$$

∎

Theorem 8 (Cut-elimination). *The rule* (cut) *is admissible in cut-free* STRW.

Proof. Suppose STRW $\vdash X \Rightarrow \gamma$. Then, we have SRW$^+ \vdash f(X) \Rightarrow f(\gamma)$ by Theorem 7 (1), and hence SRW$^+ -$ (cut) $\vdash f(X) \Rightarrow f(\gamma)$ by Proposition 5 (1). By Theorem 7 (2), we obtain STRW $-$ (cut) $\vdash X \Rightarrow \gamma$. ∎

Theorem 9 (Syntactical embedding). *Let f be the mapping defined in Definition 6. Then:*

1. STRW $\vdash X \Rightarrow \gamma$ *iff* SRW$^+ \vdash f(X) \Rightarrow f(\gamma)$.
2. STRW $-$ (cut) $\vdash X \Rightarrow \gamma$ *iff* SRW$^+ -$ (cut) $\vdash f(X) \Rightarrow f(\gamma)$.

Proof. • (1). ($\Longrightarrow$): By Theorem 7 (1). ($\Longleftarrow$): Suppose SRW$^+$ $\vdash f(X) \Rightarrow f(\gamma)$. We then have SRW$^+$ $-$ (cut) $\vdash f(X) \Rightarrow f(\gamma)$ by Proposition 5 (1). Thus, we obtain STRW $-$ (cut) $\vdash X \Rightarrow \gamma$ by Theorem 9 (2). Therefore we have STRW $\vdash$ $X \Rightarrow \gamma$.

• (2). ($\Longrightarrow$): Suppose STRW $-$ (cut) $\vdash X \Rightarrow \gamma$. Then we have STRW $\vdash$ $X \Rightarrow \gamma$. We then obtain SRW$^+$ $\vdash f(X) \Rightarrow f(\gamma)$ by Theorem 9 (1). Therefore we obtain SRW$^+$ $-$ (cut) $\vdash f(X) \Rightarrow f(\gamma)$ by Proposition 5 (1). ($\Longleftarrow$): By Theorem 9 (2). ∎

Theorem 10 (Decidability). STRW *is decidable.*

Proof. By Proposition 5 (2), for each α, it is possible to decide if $f(\alpha)$ is provable in SRW$^+$. Then, by Theorem 9, STRW is decidable. ∎

Theorem 11 (Relevance principle). *If* STRW $\vdash \alpha \Rightarrow \beta$, *then there exists a propositional variable p such that* $p \in V(\alpha) \cap V(\beta)$.

Proof. Suppose that STRW $\vdash \alpha \Rightarrow \beta$. Then, we have SRW$^+$ $\vdash f(\alpha) \Rightarrow f(\beta)$ by Theorem 9. By Proposition 5 (3), we have the fact that there exists a propositional variable q such that $q \in V(f(\alpha)) \cap V(f(\beta))$. In this fact, it can be seen that q is of the form $q = f(X^i p) = p_i$. Therefore we have the required fact that there exists a propositional variable p such that $p \in V(\alpha) \cap V(\beta)$. ∎

3   Completeness

The Routley-Meyer semantics for RW$^+$ is presented below.

Definition 12. *An* RW$^+$*-frame is a structure* $\langle M, R, 0 \rangle$ *such that*

1. *M is a nonempty set,*
2. *R is a ternary relation on M,*
3. *$0 \in M$,*
4. *the following conditions hold:*
   (a) *$a \leq a$,*
   (b) *$(a \leq b$ and $b \leq c)$ imply $a \leq c$,*
   (c) *$(Rabc$ and $a' \leq a)$ imply $Ra'bc$,*
   (d) *$Rabc$ implies $Rbac$,*
   (e) *R^2abcd implies R^2acbd,*
   where the binary relation $\leq$ on M is defined by $a \leq b$ iff $R0ab$, and the 4-ary relation R^2 on M is defined by R^2abcd iff $\exists x \in M$ $[Rabx$ and $Rxcd]$.

Definition 13. *An* RW$^+$*-model is a structure* $\langle M, R, 0, \models \rangle$ *such that*

1. *$\langle M, R, 0 \rangle$ is an RW$^+$-frame,*
2. *$\models$ is a relation from M to propositional variables satisfying the following hereditary condition: for any propositional variable p, if $a \models p$ and $a \leq b$, then $b \models p$.*

The relation $\models$ on the RW^+-model $\langle M, R, 0, \models\rangle$ is inductively extended to formulas by:

1. $a \models \alpha{\rightarrow}\beta$ *iff* $\forall b, c \in M$ *[if $Rabc$ and $b \models \alpha$, then $c \models \beta$]*,
2. $a \models \alpha \wedge \beta$ *iff* $a \models \alpha$ *and* $a \models \beta$,
3. $a \models \alpha \vee \beta$ *iff* $a \models \alpha$ *or* $a \models \beta$.

A formula α is said to be true *in an RW^+-model $\langle M, R, 0, \models\rangle$ iff $0 \models \alpha$. A formula α is said to be* RW^+*-valid iff α is true in all RW^+-models.*

The hereditary condition can be extended to formulas: For any formula α, if $a \models \alpha$ and $a \leq b$, then $b \models \alpha$.

We then have the following completeness theorem (see e.g., [1,5,15]).

Proposition 14 (Completeness). *For any formula α,*

$$\text{SRW}^+ \vdash \Rightarrow \alpha \text{ iff } \alpha \text{ is } RW^+\text{-valid.}$$

Next, we introduce an extended Routley-Meyer semantics with the timed satisfaction relations $\models^i$ $(i \in \omega)$.

Definition 15. *A TRW-model is a structure $\langle M, R, 0, \{\models^i\}_{i\in\omega}\rangle$ such that*

1. *$\langle M, R, 0\rangle$ is an RW^+-frame,*
2. *$\models^i$ $(i \in \omega)$ are relations from M to propositional variables satisfying the following hereditary condition: for any propositional variable p and any $i \in \omega$, if $a \models^i p$ and $a \leq b$, then $b \models^i p$.*

The relations $\models^i$ $(i \in \omega)$ on the TRW-model $\langle M, R, 0, \{\models^i\}_{i\in\omega}\rangle$ are inductively extended to formulas by:

1. *$a \models^i \alpha{\rightarrow}\beta$ iff $\forall b, c \in M$ [if $Rabc$ and $b \models^i \alpha$, then $c \models^i \beta$],*
2. *$a \models^i \alpha \wedge \beta$ iff $a \models^i \alpha$ and $a \models^i \beta$,*
3. *$a \models^i \alpha \vee \beta$ iff $a \models^i \alpha$ or $a \models^i \beta$,*
4. *$a \models^i \text{X}\alpha$ iff $a \models^{i+1} \alpha$,*
5. *$a \models^i \text{G}\alpha$ iff $\forall j \geq i$ with $j \in \omega_l$ [$a \models^j \alpha$],*
6. *$a \models^i \text{F}\alpha$ iff $\exists j \geq i$ with $j \in \omega_l$ [$a \models^j \alpha$],*
7. *for any $k \in \omega$, $a \models^{l+k} \alpha$ iff $a \models^l \alpha$.*

A formula α is said to be true *in a TRW-model $\langle M, R, 0, \{\models^i\}_{i\in\omega}\rangle$ iff $0 \models^0 \alpha$. A formula α is said to be* TRW-valid *iff α is true in all TRW-models.*

The hereditary condition can be extended to formulas: for any formula α and any $i \in \omega$, if $a \models^i \alpha$ and $a \leq b$, then $b \models^i \alpha$.

Lemma 16. *Let f be the mapping defined in Definition 6. For any TRW-model $\langle M, R, 0, \{\models^i\}_{i\omega}\rangle$, we can construct an RW^+-model $\langle M, R, 0, \models\rangle$ such that for any formula α in $\mathcal{L}^t$ and any $i \in \omega$,*

$$a \models^i \alpha \text{ iff } a \models f(\text{X}^i\alpha).$$

Proof. Let Φ be a non-empty set of propositional variables and Φ_i be the set $\{p_i \mid p \in \Phi\}$ of propositional variables. Suppose that $\langle M, R, 0, \{\models^i\}_{i \in \omega}\rangle$ is a TRW-model such that $\models^i$ $(i \in \omega)$ are relations from M to Φ. Suppose that $\langle M, R, 0, \models\rangle$ is a structure such that $\models$ is a relation from M to $\bigcup_{i \in \omega} \Phi_i$. Suppose moreover that $\models^i$ $(i \in \omega)$ and $\models$ satisfy the following conditions: $\forall i \in \omega. \forall p \in \Phi$ $[a \models^i p$ iff $a \models p_i]$. Then, the lemma is proved by induction on the complexity of α.

- Base step:

    Case ($\alpha \equiv p$: propositional variable): $\alpha \equiv p \in \Phi$. $a \models^i p$ iff $a \models p_i$ iff $a \models f(X^i p)$ (by the definition of f).

- Induction step:

1. Case ($\alpha \equiv \beta \to \gamma$): $a \models^i \beta \to \gamma$ iff $\forall b, c \in M$ [if $Rabc$ and $b \models^i \beta$, then $c \models^i \gamma$] iff $\forall b, c \in M$ [if $Rabc$ and $b \models f(X^i \beta)$, then $c \models f(X^i \gamma)$] (by induction hypothesis) iff $a \models f(X^i \beta) \to f(X^i \gamma)$ iff $a \models f(X^i(\beta \to \gamma))$ (by the definition of f).

2. Case ($\alpha \equiv \beta \wedge \gamma$): $a \models^i \beta \wedge \gamma$ iff $a \models^i \beta$ and $a \models^i \gamma$ iff $a \models f(X^i \beta)$ and $a \models f(X^i \gamma)$ (by induction hypothesis) iff $a \models f(X^i \beta) \wedge f(X^i \gamma)$ iff $a \models f(X^i(\beta \wedge \gamma))$ (by the definition of f).

3. Case ($\alpha \equiv \beta \vee \gamma$): $m \models^i \beta \vee \gamma$ iff $a \models^i \beta$ or $a \models^i \gamma$ iff $a \models f(X^i \beta)$ or $a \models f(X^i \gamma)$ (by induction hypothesis) iff $a \models f(X^i \beta) \vee f(X^i \gamma)$ iff $a \models f(X^i(\beta \vee \gamma))$ (by the definition of f).

4. Case ($\alpha \equiv X\beta$):
    Subcase ($i \leq l - 1$): $a \models^i X\beta$ iff $a \models^{i+1} \beta$ iff $a \models f(X^{i+1}\beta)$ (by induction hypothesis) iff $a \models f(X^i X\beta)$.
    Subcase ($i \geq l$): $a \models^i X\beta$ iff $a \models^l \beta$ iff $a \models f(X^l \beta)$ (by induction hypothesis) iff $a \models f(X^i X\beta)$ (by the definition of f).

5. Case ($\alpha \equiv G\beta$):
    Subcase ($i \leq l$): $a \models^i G\beta$ iff $\forall j \geq i$ with $j \in \omega_l$ $[a \models^j \beta]$ iff $\forall j \geq i$ with $j \in \omega_l$ $[a \models f(X^j \beta)]$ (by induction hypothesis) iff $\forall k \in \omega_l$ $[a \models f(X^{i+k}\beta)]$ iff $a \models \gamma$ for all $\gamma \in \{f(X^{i+k}\beta) \mid k \in \omega_l\}$ iff $a \models \bigwedge\{f(X^{i+k}\beta) \mid k \in \omega_l\}$ iff $a \models f(X^i G\beta)$ (by the definition of f).
    Subcase ($i > l$): $a \models^i G\beta$ iff $a \models^l G\beta$ iff $\forall j \geq l$ with $j \in \omega_l$ $[a \models^j \beta]$ iff $\forall j \geq l$ with $j \in \omega_l$ $[a \models f(X^j \beta)]$ (by induction hypothesis) iff $\forall k \in \omega_l$ $[a \models f(X^{l+k}\beta)]$ iff $a \models \gamma$ for all $\gamma \in \{f(X^{l+k}\beta) \mid k \in \omega_l\}$ iff $a \models \bigwedge\{f(X^{l+k}\beta) \mid k \in \omega_l\}$ iff $a \models f(X^l G\beta)$ (by the definition of f) iff $a \models f(X^i G\beta)$ (by the definition of f with $i > l$). $\blacksquare$

Lemma 17. *Let f be the mapping defined in Definition 6. For any RW^+-model $\langle M, R, 0, \models\rangle$, we can construct a TRW-model $\langle M, R, 0, \{\models^i\}_{i \in \omega}\rangle$ such that for any formula α in $\mathcal{L}^t$ and any $i \in \omega$,*

$$a \models^i \alpha \text{ iff } a \models f(X^i \alpha).$$

Proof. Similar to the proof of Lemma 16. ∎

Theorem 18 (Semantical embedding). *Let f be the mapping defined in Definition 6. For any formula α,*

α *is TRW-valid iff* $f(\alpha)$ *is RW^+-valid.*

Proof. By Lemmas 16 and 17. ∎

Theorem 19 (Completeness). *For any formula α,*

STRW $\vdash \Rightarrow \alpha$ *iff* α *is TRW-valid.*

Proof. STRW $\vdash \Rightarrow \alpha$ iff SRW$^+$ $\vdash \Rightarrow f(\alpha)$ (by Theorem 9) iff $f(\alpha)$ is RW$^+$-valid (by Proposition 14) iff α is TRW-valid (by Theorem 18). ∎

4 Conclusions and Remarks

In this paper, the sequent calculus STRW of the new temporal relevant logic TRW was introduced, and the cut-elimination and decidability theorems for STRW were proved using a theorem for syntactically embedding STRW into a sequent calculus SRW$^+$ of the well-known positive contraction-less relevant logic RW$^+$. The relevance principle for STRW was also proved using the syntactical embedding theorem. The extended Routley-Meyer semantics with timed satisfaction relations was introduced for STRW, and a theorem for semantically embedding TRW into RW$^+$ was proved. The completeness theorem with respect to this extended semantics was proved using both the syntactical and semantical embedding theorems. It was thus shown in this paper that STRW (and TRW) is a plausible temporal relevant logic for time-dependent relevant human reasoning.

Finally in this paper, some remarks on some bounded time domain approaches to temporal logics are given. To restrict the time domain in temporal logics is not a new idea introduced in this paper. Such an idea was discussed in [3,4,8,9]. In [9], Gentzen-type sequent calculi BLTL and FBLTL (for propositional and first-order bounded linear-time temporal logics) and the corresponding Robinson-type resolution calculi RC and FRC, respectively, were introduced based on ω_l. Some theorems for embedding BLTL and FBLTL into (propositional and first-order, respectively) classical logic were proved in [9]. In [4], by using and introducing a bounded time domain and the notion of bounded validity, bounded tableaux calculi (with temporal constraints) for propositional and first-order LTLs were studied by Cerrito, Mayer and Prand.

Acknowledgments. We would like to thank the anonymous referees for their valuable comments. This work was supported by JSPS KAKENHI Grant (C) 26330263 and by Grant-in-Aid for Okawa Foundation for Information and Telecommunications.

References

1. Anderson, A.R., Belnap, N.D., et al.: Entailment: the logic of relevance and necessity, vol. 1. Princeton University Press (1975)
2. Brady, R.T.: The Gentzenization and decidability of RW. Journal of Philosophical Logic 19, 35–73 (1990)
3. Biere, A., Cimatti, A., Clarke, E.M., Strichman, O., Zhu, Y.: Bounded model checking. Advances in Computers 58, 118–149 (2003)
4. Cerrito, S., Mayer, M.C., Prand, S.: First order linear temporal logic over finite time structures. In: Ganzinger, H., McAllester, D., Voronkov, A. (eds.) LPAR 1999. LNCS, vol. 1705, pp. 62–76. Springer, Heidelberg (1999)
5. Dunn, J.M.: Relevance logic and entailment. In: Gabbay, D., Guenthner, F. (eds.) Handbook of Philosophical Logic, vol. 3, pp. 117–224. Reidel Publishing Company (1986)
6. Dunn, J.M.: Consecution formulation of positive R with co-tenability and t. In: Anderson, A.R., Belnap, N.D. (eds.) Entailment: the Logic of Relevance and Necessity, vol. 1, pp. 381–391. Princeton University Press (1975)
7. Gambrone, S.: TW_+ and RW_+ are decidable. Journal of Philosophical Logic 14, 235–254 (1985)
8. Hodkinson, I., Wolter, F., Zakharyaschev, M.: Decidable fragments of first-order temporal logics. Annals of Pure and Applied Logic 106, 85–134 (2000)
9. Kamide, N.: Bounded linear-time temporal logic: A proof-theoretic investigation. Annals of Pure and Applied Logic 163(4), 439–466 (2012)
10. Kamide, N.: Formalizing inconsistency-tolerant relevant human reasoning: A decidable paraconsistent relevant logic with constructible falsity. In: Proce. of the 2013 IEEE International Conference on Systems, Man, and Cybernetics, pp. 1865–1870 (2013)
11. Kawai, H.: Sequential calculus for a first order infinitary temporal logic. Zeitschrift für Mathematische Logik und Grundlagen der Mathematik 33, 423–432 (1987)
12. Pnueli, A.: The temporal logic of programs. In: Proceedings of the 18th IEEE Symposium on Foundations of Computer Science, pp. 46–57 (1977)
13. Prior, A.N.: Time and modality. Clarendon Press, Oxford (1957)
14. Prior, A.N.: Past. present and future. Clarendon Press, Oxford (1967)
15. Routley, R., Meyer, R.K.: Semantics of entailment 1. In: Leblanc, H. (ed.) Truth, Syntax and Modality. Studies in Logic and the Foundations of Mathematics, vol. 68, pp. 199–243. North Holland, Amsterdam (1973)
16. Urquhart, A.: The undecidability of entailment and relevant implication. Journal of Symbolic Logic 49(4), 1059–1073 (1984)

Introspection, Normality and Agglomeration

Dominik Klein[1,3], Norbert Gratzl[2], and Olivier Roy[1]

[1] University of Bayreuth
[2] LMU Munich
[3] University of Bamberg

Abstract. This paper explores a non-normal logic of beliefs for bound-edly rational agents. The logic we study stems from the epistemic-doxastic system developed by Stalnaker [1]. In that system, if knowledge is not positively introspective then beliefs are not closed under conjunction. They are, however, required to be pairwise consistent, a requirement that has been called agglomerativity elsewhere. While bounded agglom-erativity requirements, i.e., joint consistency for every n-tuple of beliefs up to a fixed n, are expressible in that logic, unbounded agglomerativ-ity is not. We study an extension of this logic of beliefs with such an unbounded agglomerativity operator, provide a sound and complete ax-iomatization for it, show that it has a sequent calculus that enjoys the admissibility of cut, that it has the finite model property, and that it is decidable.

Robert Stalnaker [1] has proposed a logic of knowledge and belief where the latter turns out to be definable as the epistemic possibility of knowledge.

$$B\varphi \leftrightarrow \langle K \rangle K\varphi \qquad\qquad \text{(BePK)}$$

He takes knowledge to be an S4 modality, with the usual interpretation in re-flexive and transitive Kripke frames. Beliefs, on the other hand, are essentially defined through their relation with knowledge, except for the assumption that beliefs are consistent, the D-axiom.[1] See Table 1 for the axioms capturing that relation. One of the crucial axioms is (KB), stating that belief implies the belief that one knows. Thus, beliefs in this model are interpreted as absolute subjec-tive certainty [2, 3]. We should emphasize one further property of this model: Notably, while knowledge in this model is not introspective, belief is, both posi-tively and negatively. The reason for this is simple: knowledge presupposes truth, a property we don't have direct access to. Belief, on the other hand, is a mental state that we do have privileged and immediate access to. In particular, we know whether we have such mental states or not.

Notably, the above formula (BePK) is a theorem of this logic. The derivation from left to right goes as follows. Assuming that $B\varphi$ holds, we start by invoking (SB), $B\varphi \to BK\varphi$. From there, we arrive at $\langle B \rangle K\varphi$ and finally at $\langle K \rangle K\varphi$

---

[1] Throughout this paper, we will refer to axioms by putting their names in parentheses. Thus (D) stands for the D-axiom, (K) for the K-axiom and so on.

© Springer-Verlag Berlin Heidelberg 2015
W. van der Hoek et al. (Eds.): LORI 2015, LNCS 9394, pp. 195–206, 2015.
DOI: 10.1007/978-3-662-48561-3_16

Table 1. Stalnaker's axioms. K is an $S4$ modality

PI	$\vdash B\varphi \to KB\varphi$
NI	$\vdash \neg B\varphi \to K\neg B\varphi$
KB	$\vdash K\varphi \to B\varphi$
D	$\vdash B\varphi \to \langle B\rangle\varphi$
SB	$\vdash B\varphi \to BK\varphi$

using (D) and (KB), respectively. In the other direction, we start by assuming $\langle K\rangle K\varphi$. We can derive $\langle K\rangle K\varphi \to \langle K\rangle B\varphi$ using (KB) and the fact that K is a normal modality. One application of (NI) and our assumption then gives us $B\varphi$.

Finally, we finish with two observations. First, we remark that the resulting belief in Stalnaker's system is a KD45 operator if the underlying knowledge modality is S4. Crucially, the 4 axiom has not been used for either direction of the above derivation.

Our second observation is that, given that knowledge is an S4 modality, the axiom system consisting of (BePK) and (D) is *equivalent* to Stalnaker's axiom system given in Table 1. Furthermore, under (BePK), the D-axiom for belief translates to the .2 axiom for knowledge: $\vdash \langle K\rangle K\varphi \to K\langle K\rangle\varphi$.

Stalnaker explicitly rejects negative introspection for knowledge. This axiom, he argues, precludes being mistaken about what one knows. But this seems an implausible assumption: One of the necessary conditions for knowledge is truth. And as we don't always have direct access to truth, it may not be irrational to be in error regarding what one knows. On the other hand, in that very paper Stalnaker "provisionally" accepts positive introspection for knowledge [1, p.173], i.e. the 4 axiom, also known as the KK-principle. Yet, also this principle has been criticized, notably by Williamson [4, chap.5] in recent years, who argues that the axiom fails even as a requirement of rationality in contexts involving vagueness or margins of errors.

This paper starts with observing that leaving out positive introspection for knowledge while maintaining (BePK) results in an interesting, non-normal logic for beliefs (Section 1). Beliefs in this logic are not closed under conjunction. They are otherwise normal, with the additional requirement that they should be pairwise consistent, which we show in Section 2. In Section 3 we consider possible philosophical interpretations for this belief operator. We show that the resulting belief notion cannot be interpreted as belief-as-high-enough credence, even though failure of closure under conjunction has often been motivated on that ground. We rather argue that the operator captures a plausible, baseline rationality requirement, which we call agglomerativity. The agglomerativity requirement can be strengthened incrementally. In Section 4 we show, however, that the limit of this strengthening, which we call unbounded agglomerativity, is not definable using only belief operators. It is, however, completely axiomatizable and the resulting, extended logic of belief and unbounded agglomerativity has the finite model property. We show in Section 6 that this logic has a cut-free sequent calculus, and that it is decidable.

1 Non-normality from Failure of Introspection

What happens to beliefs when knowledge is not positively introspective? That is, what logic of belief do we get when knowledge is a KT modality? Recall, first, that (BePK) and (D) are together equivalent to the axiom system in Table 1, given that knowledge is an S4 modality. This equivalence fails if knowledge is only a KT.2 modality. The two systems are different. For the remaining of this paper we will look at the second one, i.e. a KT logic for knowledge, together with the belief operator interpreted using (BePK) and (D). In that logic (D) for beliefs still boils down to endorsing the .2 axiom for knowledge. For reasons that become clearer later in this paper, we will call this logic MUD logic. Further, note that by (BePK) we can treat B as a derived operator, leaving K as the only primitive operator. Thus, model theoretically, we are working in reflexive frames having the Church-Rosser property.

If knowledge is not positively introspective,[2] however, belief is not a normal modality anymore. We understand normality here in the technical sense, as an operator satisfying necessitation and distributing over conjunctions as follows:

$$B\varphi \wedge B\psi \leftrightarrow B(\varphi \wedge \psi) \qquad \text{(}\wedge\text{-Dist)}$$

The left-to-right direction, i.e. closure under conjunction for beliefs, fails together with positive introspection for knowledge. Figure 1 illustrates this with a simple counter-example. This model displays the knowledge relation for a KT.2 knowledge operator. At w_1 we have both $\langle K \rangle Kp$ and $\langle K \rangle Kq$ but not $\langle K \rangle K(p \wedge q)$, thus the derived belief operator (via BePK) is not normal.

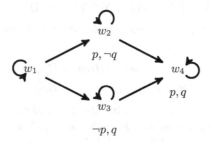

Fig. 1. A KT.2 model in which the corresponding belief is not closed under conjunction

This belief operator is otherwise normal. It validates necessitation and closure under arbitrary union, that is the right-to-left direction of ∧-Dist and necessitation. The latter can be encapsulated using a standard regularity rule [5,6].

---

[2] Note that on the level of frames, non-introspective knowledge translates to the fact that frames need not be transitive.

$$\frac{\vdash \varphi}{\vdash B\varphi} \ (NEC) \qquad \frac{\vdash \varphi \to \psi}{\vdash B\varphi \to B\psi} \ (REG)$$

Observation 1. *(NEC), (REG) and (D) are sound with respect to KT.2 models where B is interpreted as $\langle K \rangle K$.*

2   KT.2, $\langle K \rangle K$, and MUD Logic

From now on we will focus on the belief part of MUD logic. Following standard usage of notation, we will refer to the set of formulas that only contain B as modal operator as the "B fragment" of our logic. The first question we ask is a technical one. Do necessitation, closure under logical implication and D give a complete axiomatization for the belief operator defined above? The answer is no. (NEC), (REG) and (D), together with the usual propositional logic apparatus, are *not* complete for the belief part of MUD logic.

Recall the question we are asking. We are not looking at Stalnaker's logic of beliefs, i.e. one that starts with two primitive operators, B and K, with their relation constrained by the axioms in Table 1. Rather, we look at an epistemic logic where K is a KT modality, and define B as $\langle K \rangle K$, adding the (D) axiom for B. The question we ask is then: what is the sound and complete logic of the $\langle K \rangle K$ fragment?

(NEC), (REG) and (D) are sound and complete with respect to the class of what we call MUD *neighborhood* frames, or MUD-frames for short[3]. These are frames $\langle W, n \rangle$, where W is a set of worlds and $n : W \to \mathcal{P}(\mathcal{P}(W))$ is the neighborhood function, satisfying the following conditions:

– For all w, if $X \in n(w)$ and $X \subseteq Y$ then $Y \in n(w)$. (Monotonicity)
– For all w, $W \in n(w)$. (contains the Unit)
– If $X \in n(w)$ then for all $Y \in n(w)$, $X \cap Y \neq \emptyset$. (D)

That MUD-logic is sound but incomplete for the $\langle K \rangle K$ fragment can be shown by observing that:

$$\vdash_{KT.2} \langle B \rangle (p \to Bp)$$

$$\nvdash_{NEC,REG,D} \langle B \rangle (p \to Bp)$$

For the proof the first claim, start with the following theorem of KT: $Kp \to \langle K \rangle \langle K \rangle Kp$. This is equivalent to $\langle K \rangle \neg p \vee \langle K \rangle \langle K \rangle Kp$. K being normal, the latter is in turn equivalent to $\langle K \rangle (p \to \langle K \rangle Kp)$. One application of necessitation give us the required formula, using (BePK): $K \langle K \rangle (p \to \langle K \rangle Kp)$. It is easy to construct a counter-model to the validity of that formula in MUD-frames.

Observe that this counter-example to completeness is a second-order formula. The $\langle B \rangle$ scopes over a B. We conjecture that MUD logic is complete for the first-order, i.e. non-embedded belief fragment. What is the complete logic of that full fragment is still open. From now on we focus on the belief operator B in MUD-logic. This system has a cut-free sequent calculus. We will show the latter in an extended language (Section 6), motivated by the sort of bounded rationality requirements on beliefs this operator suggests.

---

[3] See [6] for some background on neighborhood frames.

3   What Kind of Belief Is This?

3.1   Not Belief with High Enough Credence

One tempting interpretation of the belief operator at hand, motivated by the Lockean thesis [7], is in terms of high enough credence or subjective probability. That is, a formula should be believed if its credence is above a given threshold $t > 1/2$, i.e.,

$$B\varphi \text{ iff } p(\varphi) \geq t > 1/2$$

for some given probability measure p. It is well-known that such an operator would not be closed under intersection. And such an operator would validate all axioms and rules for the current logic of beliefs:

Observation 2. *(NEC), (REG) and (D) are sound for B interpreted as "probability strictly above 1/2".*

To see this, take a probability measure over a σ-algebra and let be the set $\mathcal{X}$ of measurable sets that have probability $> 1/2$. It is immediately clear that $\mathcal{X}$ satisfies (NEC) and (REG). To see that $\mathcal{X}$ satisfies (D), let X and Y in $\mathcal{X}$. Since $p(X)$ and $p(Y)$ are stricly larger than 0.5, we have $X \cap Y \neq \emptyset$, showing that (D) holds.

(NEC), (REG) and (D) are however not complete with respect to that interpretation. There are models of that logic which cannot be equipped with a probability measure in such a way that the belief operator respects the equivalence above. We show a stronger result. Rather than focusing on a threshold of .5, we show that for every treshold ϵ there is some MUD model that cannot be equipped with a probability measure in such a way that every believed proposition has at least probability ϵ. Furthermore, this example will also be such that there is some proposition of probability *at least* $(1-\epsilon)$ that the agent disbelieves. So let us give our example. We fix a size n of the counterexample, where this n will depend on the ϵ chosen. Then, we construct the following model:

- Possible worlds: All w_{ij} with $1 \leq i < j \leq n$
- For each $k \leq n$ define the set T_k as

$$T_k = \{w_{ij} | i = k \text{ or } j = k\}$$

- Each world w_{ij} has the same neighborhoods N. These are given by $N = n(w_{ij}) = \uparrow \{T_1, \ldots, T_n\}$, the upward closed set generated by $T_1 \ldots T_n$.

To illustrate this with an example, for $n = 4$ we have

$$W = \begin{matrix} w_{12} & w_{13} & w_{14} \\ & w_{23} & w_{24} \\ & & w_{34} \end{matrix}$$

$$T_1 = \{w_{12}, w_{13}, w_{14}\}$$
$$T_2 = \{w_{12}, w_{23}, w_{24}\}$$
$$T_3 = \{w_{13}, w_{23}, w_{34}\}$$
$$T_4 = \{w_{14}, w_{24}, w_{34}\}$$

Finally, we set $N = \uparrow \{T_1, T_2, T_3, T_4\}$.

So let us offer some interpretation for this example. We assume a lottery with n tickets. However, this lottery has a slight peculiarity. In this lottery, there are always exactly *two* winning tickets. Thus, at the world w_{13}, the first and the third ticket wins. For any ticket i, let T_i be the proposition that ticket i wins. Then any two T_i and T_j are jointly compatible. In particular, the model above validates (REG), (NEC) and (D), thus it is a model of the belief fragment of MUD-logic.

Yet, as there are only two winners, not every T_i can be assigned a high probability. More particularly, if n grows, we are guaranteed to find some T_i that receives a low probability, no matter how we choose to assign probabilities. To be a bit more explicit about this argument, assume we want to find some way of equipping W with a probability function that makes the T_i the agent believes as probable as possible. In particular, we are interested in the probability distribution that maximizes $min_i prob(T_i)$, that is, we want to make the *most improbable* proposition that the agent still believes as probable as possible. It is not difficult to see that the probability distribution maximizing $min_i prob(T_i)$ assigns equal weight to all worlds. Since there are $(n-1)(n-2)$ many worlds, this probability distribution will assign a weight of $\frac{1}{(n-1)(n-2)}$ to every world w_{ij}. Since the T_i all have cardinality $n-1$, they each receive a weight of $\frac{1}{n-2}$. In particular, if n becomes large, the agent will believe some proposition that is extremely implausible.

The example shows that the notion of belief defined above is not strong enough to enforce the "belief with high enough credence" interpretation. Now, a natural question to ask is: What additional constraints on beliefs would be required? We leave that question open for future work. Instead, we argue now that B provides a plausible approximation of consistency requirements for resource-bounded agents.

3.2   Belief with Pairwise Agglomeration

The present logic constitutes a plausible *weakening* of the rationality requirement of consistency for beliefs. Resource-bounded agents may believe φ and believe ψ without being thereby required to put these two together and form the belief that $\varphi \wedge \psi$. The classical cluttering argument against closure can be used to support that idea [8]. Inference might be costly in time and energy, and storing the conclusions of these inferences might take valuable space. Resource-bounded agents often have better to do than closing their beliefs under conjunction, and they are certainly not required to clutter their minds with the trivial consequences of what they believe. So they are not necessarily irrational if they believe φ and believe ψ without believing $\varphi \wedge \psi$.

For such agents, the constraint imposed by the axiom D is more plausible. It merely requires a minimal level of consistency in beliefs. They should be *pairwise agglomerable*.[4] On this account, an agent cannot rationally believe φ and believe ψ while not considering it possible for them to be jointly satisfied. To see that

---

[4] We take this terminology from the literature on intentions, e.g. [9,10].

this is enforced by D in the neighborhood model, just recall that $\langle B \rangle \varphi$ is true at a state just in case φ is consistent with every other belief of the agent. So D requires of that agent to be *able* to consistently agglomerate any pair of her beliefs without running the risk of believing absurdities.

Pairwise agglomerativity is a plausible baseline requirement of rationality, even for resource-bounded agents. Note that the cluttering argument itself has no grip on pairwise agglomerativity. Agglomerativity or minimal consistency is no closure operation. It does not require the agents to add anything to their belief set in order to escape the charge of irrationality. So cluttering, and related pragmatic considerations (time, energy, storage) are unlikely to undercut that requirement. On the other hand, an agent who violates that requirement has beliefs that are mutually contradictory. But beliefs arguably aim at truth. And mutually contradictory believes can never be true together.

It should be emphasized that, even though the present logic provides a plausible alternative to full closure under conjunction for beliefs, it does not give the full picture of rational requirements for resource-bounded agents. Agents are by and large logically omniscient in this system. They believe all tautologies and all the logical consequences of each of their individual beliefs. The logic studied here is merely a testing ground for different consistency requirements.

4   From Bounded to Unbounded Agglomerativity

Pairwise agglomeration is permissive. An agent's belief can be pairwise agglomerable while only some or even no single combination of three (different) beliefs of hers are jointly satisfiable. This holds true, for instance, in the model presented above, built to exclude the "high threshold" interpretation. There, the intersection of any three belief sets T_i is empty.

Pairwise agglomerativity can be generalized to *bounded agglomerativity*, that is agglomerativity for every n-tuple of beliefs up to a fixed n. Such a generalization is definable in the present logic of beliefs:

$$(B\varphi_1 \wedge ... \wedge B\varphi_{n-1}) \rightarrow \langle B \rangle(\varphi_1 \wedge ... \wedge \varphi_{n-1}) \qquad (n\text{-AGG})$$

In neighborhood frames bounded agglomerativity means that for any state, any n-tuple of neighborhoods of that state is jointly consistent. (n-AGG) corresponds to that frame property.

Pairwise agglomerativity is a bottom line requirement, and the demands of bounded agglomerativity increase as n gets larger. Obviously the agent must be able to put together larger collections of beliefs. The problem of contracting (or revising) one's beliefs in order to re-establish bounded agglomerativity becomes also more intricate as n grows. In the general case the agent will be faced with more possible options to solve failures of agglomerativity.

But the arguments in favor of seeing pairwise agglomerativity as a plausible rational requirement for resource-bounded agents carries over to the bounded case, even as n grows larger. Again, as this is not a closure requirement, practical considerations such as non-cluttering have much less weight. And, since

n-tuples of non-agglomerable beliefs are not satisfiable, agglomerability seems a reasonable rationality requirement.

5   The Logic of BAGG Frames

By the same reasoning, bounded agglomerativity for all n, or *unbounded agglomerativity*, can thus be seen as a plausible candidate requirement too, even for resource-bounded agents. Just as with finite agglomerativity, also the unbounded version is not a closure operation, thus it does not fall prey to a cluttering argument. On the other hand, unbounded agglomerativity is a rational criterion to demand, since the failure of unbounded agglomerativity gets in the way of truth, the aim of belief. In this section we study an extension of MUD logic that can express this requirement.

First, we note that unbounded agglomerativity is not definable in MUD logic. To see this, we define a new modal operator $\Box$ and let $\Box\varphi$ be true at a state w in a neighborhood model if and only if

$$\bigcap_{X \in n(w)} X \subseteq ||\varphi||$$

where $||\varphi||$ is the extension of φ in that model. Thus, in a MUD-model, $\Box\varphi$ means that φ holds true in the intersection of all belief sets. In particular, φ is compatible with every single belief the agent has.

Observation 3. $\Box$ *is not definable in the present logic of belief.*

We relegate the proof of this observation to the full paper. So to express unbounded agglomerativity we need to enrich our logic of beliefs with this new operator. As usual, we write $\Diamond$ for the dual of $\Box$, i.e. $\Diamond = \neg\Box\neg$. The new logic, BAGG logic (for *B*elief and unbounded *Agg*lomerativity) is axiomatized by KT.2 for the knowledge modality, (D) for belief and the two axioms displayed in table 2 for the new modality. In particular, $\Box$ is a normal modality. As an intersection modality it is completely axiomatized in a similar fashion as distributed knowledge. See Table 2.

Table 2. The additional axioms for the BAGG logic

$B\varphi \rightarrow \Box\varphi$	Belief Consistency
$\Box\varphi \rightarrow \Diamond\varphi$	Unbounded Agglomerativity

Theorem 1. *The axioms in Table 2, together with K for $\Box$, Necessitation for B and $\Box$ and REG for B are sound and complete with respect to the class of BAGG neighborhood frames.*

Again, due to space restrictions, we omit the proof of this theorem. As desired, this logic is strong enough to derive all instances of (n-AGG). Indeed, from $B\varphi_1 \wedge \ldots \wedge B\varphi_n$ one gets $\Box(\varphi_1 \wedge \ldots \wedge \varphi_n)$ using Belief consistency and the normality of $\Box$, from which one application of Unbounded Agglomerativity gives $\Diamond(\varphi_1 \wedge \ldots \wedge \varphi_n)$, and one more application of Belief Consistency gives $\langle B \rangle(\varphi_1 \wedge \ldots \wedge \varphi_n)$. So, in particular, D for B is derivable in that system.

Finally, this logic has the finite model property. The proof is omitted for reasons of space. We do however show in the next section that this logic is decidable and has a cut-free sequent calculus.

6   Proof Theory of BAGG

We now present a sequent calculus formulation of BAGG.[5] We base it on Gentzen's propositional (multiplicative) fragment LK. It has the familiar initial sequent, i.e. $\varphi \Longrightarrow \varphi$, with φ is atomic, and the usual structural and logical rules. The present sequent calculus follows [11]; and uses the context-free version of Cut, i.e.

$$\frac{\Gamma \Longrightarrow \Delta, \varphi \quad \varphi, \Phi \Longrightarrow \Psi}{\Gamma, \Phi \Longrightarrow \Delta, \Psi} \; Cut$$

For B we use the following rule. Notice the crucial constraint that the antecedent Γ contains at most one formula. With Γ empty this rule captures (NEC), and (REG) otherwise.

$$|\Gamma| \leq 1 \; \frac{\Gamma \Longrightarrow \varphi}{B\Gamma \Longrightarrow B\varphi} \; \text{(1-Reg)}$$

The main rule of the system is the following.

$$|\Phi| \leq 1 \; \frac{\Gamma, \Delta \Longrightarrow \Phi}{B\Gamma, \Box\Delta \Longrightarrow \Box\Phi} \; \text{(B}\Box\text{)}$$

With Γ empty this rule gives both (NEC) and (K) for $\Box$. With Δ empty it derives Belief Consistency, and with both Γ and Φ empty we obtain (D) for $\Box$. Both (1-REG) and ($B\Box$) are sound, so the system is deductively equivalent to the logic of BAGG frames. With the structural rule of contraction, we prove so-called mix-elimination. Mix is slight modification of Cut:

$$\text{(Multicut, Mix)} \; \frac{\Gamma \Longrightarrow \Delta, F^n \quad F^m, \Phi \Longrightarrow \Psi}{\Gamma, \Phi \Longrightarrow \Delta, \Psi} \; (n, m > 0; n, m \in \mathbb{N})$$

Call LK-BAGG the sequent calculus LK augmented with the rules above.

Theorem 2. LK-*BAGG enjoys the cut-elimination theorem, and so has the sub-formula property and BAGG is consistent.*

---

[5] We assume familiarity with LK, and both the standard logical and the structural rules.

The logic of the B alone is a fragment of LK- BAGG . With a slight modification of the argument we get cut-elimination for this system as well. The proof for both follows the usual "road to cut" sketched in [12]. Details are omitted for reasons of space. We outline one case illustrating the interaction between B and $\Box$. Here both mix-formulas are principal.

$$\frac{\dfrac{\Gamma \Longrightarrow \varphi}{B\Gamma \Longrightarrow \Box\varphi} \quad \dfrac{\varphi \Longrightarrow \psi}{\Box\varphi \Longrightarrow \neg\Box\neg\psi}}{B\Gamma \Longrightarrow \neg\Box\neg\psi} \qquad \text{transforms to...} \qquad \frac{\dfrac{\dfrac{\dfrac{\Gamma \Longrightarrow \varphi \quad \varphi \Longrightarrow \psi}{\Gamma \Longrightarrow \psi}}{\Gamma, \neg\psi \Longrightarrow}}{B\Gamma, \Box\neg\psi \Longrightarrow}}{B\Gamma \Longrightarrow \neg\Box\neg\psi}$$

LK- BAGG is decidable. To show that we follow [13] and [14]. Suppose that a sequent $\Gamma \Longrightarrow \Delta$ is derivable in BAGG; then by the cut elimination theorem, there is a cut-free derivation of it that also satisfies the subformula property. Furthermore, there are no *redundancies* in the derivation of $\Gamma \Longrightarrow \Delta$, i.e. no sequent of the form $\Phi \Longrightarrow \Psi$ occurs twice in the derivation. Call a sequent is 1-*reduced* iff every formula in the antecedent (succedent) occurs exactly once in the antecedent (succedent). A sequent $\Gamma^* \Longrightarrow \Delta^*$ is called a *contraction* iff it is derivable from a sequent $\Gamma \Longrightarrow \Delta$ by application(s) of contractions and permutations.

Lemma 1. *For every sequent* $\Gamma \Longrightarrow \Delta$, *there is a (1-)reduced sequent of* $\Gamma \Longrightarrow \Delta$ *such that: if* LK- *BAGG* $\vdash \Gamma^* \Longrightarrow \Delta^*$ *iff* LK- *BAGG* $\vdash \Gamma \Longrightarrow \Delta$.

The procedure for transforming such sequents in reduced sequents is effective. Now, take any reduced sequent $\Gamma \Longrightarrow \Delta$. In view of the above lemmata, if there is a derivation of the sequent, then the derivation consists only of reduced sequents. The number of reduced sequents is finite; since the derivations have no redundancies, the number of derivations is finite. Hence, the search procedure produces all possible derivations; if there is a derivation that begins with initial sequents, then there is a derivation of $\Gamma \Longrightarrow \Delta$, otherwise there is no derivation of it. This gives rise to the following theorem:

Theorem 3. LK- *BAGG is decidable.*

7   A Probabilistic Interpretation for BAGG?

Before concluding, it is worth observing that the logic of beliefs with unbounded agglomerativity also bars a probabilistic interpretation, but for different reasons than the logic of beliefs alone.

First, we observe that BAGG neighborhood models are easily produced from probabilistic measures on countable sets. In particular, comparing this with the results in section 2.5 of Leitgeb's [15], we see that agglomerative belief is a strictly weaker concept than stable belief.

Lemma 2. *Let* W *be a finite or countable set and let* $\mu : \mathcal{P}(W) \to [0 : 1]$ *be a probability function. Then* W *with neighborhood function* $n(w) = \{Y \subseteq W | \mu(Y) > t\}$ *is a BAGG frame iff there is some* $w \in W$ *with* $\mu(w) \geq 1 - t$.

Clearly, the set $B = \{Y \subseteq W | \mu(Y) > t\}$ is upward closed. Now, it remains to see whether we have $\bigcap_{X \in B} X \neq \emptyset$. First, assume that there is some w with $p(w) \geq 1 - t$. Then we have for every $Y \subseteq W$ that $Prob(Y) > t$ implies $w \in Y$. In particular $w \in \bigcap_{X \in B} X$. For the other direction assume that there is no such w. Then for every $y \in W$ we have $Prob(y) < 1 - t$ and therefore $Prob(W - \{y\}) > t$. Thus, all sets of the form $W - \{y\}$ are in B - but $\bigcap_{y \in W} W - \{y\} = \emptyset$.

In the other direction, we get the following partial correspondence results:

Lemma 3. *Let $\mathcal{M}, w$ be a finite or countable BAGG frame. Then for every $t \in (0; 1)$ there is a probability function $Prob : W \to [0; 1]$ such that $B \in n(w)$ implies $Prob(B) > t$*

Notice, however, that this lemma does not generalize to an if and only if: There are BAGG models M, w such that no possible probability function $Prob : \mathcal{P}(W) \to [0 : 1]$ and no t satisfies:

$$Prob(X) > t \Leftrightarrow X \in n(w).$$

To see this, recall the example from section 4. There, we had a lottery with n tickets and two winners. We modeled this with a set of worlds $W = \{w_{ij} | i < j \leq n\}$ where w_{ij} is the world in which i and j are the winners. Further, we had belief sets $T_k = \{w_{ij} | i = k \text{ or } j = k\}$ for each k. And we showed that no matter which probability distribution we pick, there needs to be some k such that T_k is less probable than $W - T_k$. Now, we expand this model to a model W' by adding a further world w_c in which the lottery is called off. We also expand the T_k to T'_K by adding the world w_c to each belief set - i.e. T'_k expresses that k wins or the lottery is called off. That is, in M' we have $n(w) = \uparrow \{T'_1, \ldots, T'_n\}$: Now, clearly $\bigcap T'_i = \{w_c\}$, thus M' is a BAGG -model. On the other hand, arguing as above, we can see that for every possible probability distribution on W' there is some k such that $Prob(T'_k) < Prob((W' - T'_k) \cup \{w_c s\})$ But we have $T'_k \in n(w)$ and $(W' - T'_k) \cup \{w_c\} \notin n(W)$, thus there can be no t such that $Prob(X) > t \Leftrightarrow X \in n(w)$.

So, taken together, these lemmas tell us that probability measures can generate BAGG frames under rather weak conditions, and that BAGG frames are easily related to measurable algebras. In fact too easily so to sustain a plausible probabilistic interpretation. Lemma 2 and 3 hold for *arbitrary* threshold t. The starting point of this logic, however, was a notion of beliefs as absolute subjective certainty. This was embodied by Stalnaker's (SB) axiom: $B\varphi \to BK\varphi$. We do not see how this constraint can be made plausible except for a very high threshold, certainly higher than 0.5. The present logic is not strong enough to exclude this interpretation. Adding unbounded agglomerativity to the logic of beliefs makes it more amendable to a probabilistic interpretation. But it is not strong enough to rule out implausible ones.

8   Conclusion

We have studied a logic for beliefs stemming from from leaving out positive introspection for knowledge in Stalnaker's [1] system. We have argued for two philosophical points. First, this logic allows to formulate plausible rationality requirements on beliefs for resource-bounded agents: pairwise, bounded and unbounded agglomerativity. Second, this logic is not strong enough to support a plausible probabilistic interpretation. On the technical side, the main contributions are: sound and complete axiomatization of both the original logic of beliefs and its extension to unbounded agglomerativity; cut-free sequent calculus for both logics and decidability results.

Obvious next steps are applications of this logic to classical problems and questions in game theory, for instance epistemic characterization results or agreement theorems. For this we need two things. First we need to extend this logic to multi-agent cases, and study common belief for such a non-normal modality. Second we need to introduce dynamics. For both the work in [16] will be of relevance.

References

1. Stalnaker, R.: On logics of knowledge and belief. Philosophical Studies 128, 169–199 (2006)
2. Baltag, A., Bezhanishvili, N., Özgün, A., Smets, S.: The topology of belief, belief revision and defeasible knowledge. In: Grossi, D., Roy, O., Huang, H. (eds.) LORI. LNCS, vol. 8196, pp. 27–40. Springer, Heidelberg (2013)
3. Özgün, A.: Topological models for belief and belief revision. Master's thesis, Universiteit van Amsterdam (2013)
4. Williamson, T.: Knowledge and its Limits. Oxford UP (2000)
5. Chellas, B.F.: Modal Logic: An Introduction. Cambridge University Press (1980)
6. Pacuit, E.: Neighborhood semantics for modal logic. Course Notes (2007)
7. Leitgeb, H.: The stability theory of belief. Philosophical Review 123, 131–171 (2014)
8. Harman, G.: Change in view: Principles of reasoning. MIT Press (1986)
9. Shoham, Y.: Logical theories of intention and the database perspective. Journal of Philosophical Logic 38, 633–647 (2009)
10. Roy, O.: Interpersonal coordination and epistemic support for intentions with we-content. Economics and Philosophy 26, 345–367 (2010)
11. Buss, S.: An Introduction to Proof Theory. In: Buss, S. (ed.) Handbook of Proof Theory, pp. 1–78. Elsevier (1998)
12. Negri, S., von Plato, J.: Structural Proof Theory. Cambridge Universtiy Press (2001)
13. Gentzen, G.: Untersuchungen über das logische Schließen. Mathematische Zeitschrift 39, 176–210, 405–431 (1934-1935)
14. Ono, H., et al.: Proof-theoretic methods in nonclassical logic; an introduction. Theories of Types and Proofs 2, 207–254 (1998)
15. Leitgeb, H.: Reducing belief simpliciter to degrees of belief. Annals of Pure and Applied Logic 164, 1338–1389 (2013)
16. Ma, M., Sano, K.: How to update neighborhood models. In: Grossi, D., Roy, O., Huang, H. (eds.) LORI. LNCS, vol. 8196, pp. 204–217. Springer, Heidelberg (2013)

On the Expressivity of First-Order Modal Logic with "Actually"*

Alexander W. Kocurek

Group in Logic and the Methodology of Science,
University of California, Berkeley,
Berkeley, CA, 94720, United States
akocurek@berkeley.edu

Abstract. Many authors have noted that a number of English modal sentences cannot be formalized into standard first-order modal logic. Some widely discussed examples include "There could have been things other than there actually are" and "Everyone who's actually rich could have been poor." In response, many authors have introduced an "actually" operator @ into the language of first-order modal logic. It is occasionally noted that some of the example sentences still cannot be formalized with @ if one allows only actualist quantifiers, and embedded versions of these example sentences cannot be formalized even with possibilist quantifiers and @. The typical justification for these claims is to observe that none of the most plausible candidate formalizations succeed. In this paper, we prove these inexpressibility results by using a modular notion of bisimulation for first-order modal logic with "actually" and other operators. In doing so, we will explain in what ways these results do or do not generalize to more expressive modal languages.

Keywords: first-order modal logic, actually, two-dimensional semantics, actualist and possibilist quantification, expressivity, bisimulation.

1 Introduction

Despite all of its strengths, first-order modal logic faces fundamental limitations in expressive power. Some classic examples demonstrating this include:

(E) There could have been things other than there actually are.[1]

(R) Everyone who's actually rich could have been poor.[2]

The first says that there is a possible world where something exists that doesn't actually exist. The second, on one reading, says that there's a possible world where everyone that is rich in the actual world is poor in that world. It has

---

* Special thanks to Wes Holliday and anonymous reviewers for their helpful comments and suggstions for improving this paper.
[1] Originally from [11, p. 31].
[2] Originally from [5, p. 34].

W. van der Hoek et al. (Eds.): LORI 2015, LNCS 9394, pp. 207–219, 2015.
DOI: 10.1007/978-3-662-48561-3_17

been shown using (rather complicated) Henkin-style constructions that even very simple sentences like (E) and (R) cannot be expressed in first-order modal logic with actualist quantifiers (i.e., quantifiers ranging over existents) [13]. Using possibilist quantifiers (i.e., quantifiers ranging over all possible objects) and an existence predicate, (E) can be expressed, but (R) is still inexpressible [19].

In response to these expressive limitations, a number of authors have considered introducing an "actually" operator @ into the language [6,7,11,12,14]. They then point out that in the presence of @ and possibilist quantifiers (where Π is the universal possibilist quantifier) we can formalize (R) as:

$$\Diamond\Pi x\,(@\mathsf{Rich}(x) \to \mathsf{Poor}(x)). \tag{1}$$

However, if we replace the Π above with an actualist quantifier $\forall$, (1) would yield the wrong result [4,8]. For then (1) would only require that there is a world w where everyone in w who is actually rich is poor in w, whereas (R) requires that everyone in the actual world who's actually rich is poor in w.

It has also been noted that even with possibilist quantifiers, sentences like:

(NE) Necessarily, there could have been other things than those that existed.

(NR) Necessarily, the rich could have all been poor.

remain inexpressible [4,5,11,17]. For instance, on one reading, (NR) says that in all possible worlds w, there's a possible world v where everyone rich in w is poor in v. But, for instance, formalizing (NR) as

$$\Box\Diamond\Pi x\,(@\mathsf{Rich}(x) \to \mathsf{Poor}(x)) \tag{2}$$

will yield the wrong result. This says that for all worlds w, there's a world v such that everyone that's actually rich (not rich in w) is poor in v. One could try to add more operators to the language, but problems keep cropping up [1,5].

These inexpressibility claims are often justified in the literature by example: all of the most straightforward attempts at formalizing these English sentences fail. While this style of argument may be convincing, it does not constitute a proof of these expressive limitations. Furthermore, the only proofs known in the literature involve quite complicated and indirect Henkin constructions that are limited to specific languages. In this paper, we will provide a single proof method for generating these inexpressibility proofs for a wide variety of quantified modal languages using a suitable modular notion of bisimulation for first-order modal logic. For concreteness, we'll focus on the proofs for the inexpressibility of (R) and (NR), which have proven more difficult than (E) and (NE). In passing, we will see how these inexpressibility results do, and do not, generalize to more powerful modal languages.

2 First-Order Modal Logic

First, we'll need to get clear about what exactly we're taking first-order modal logic to be. The details below are fairly standard, with the exception that our semantics is two-dimensional (to account for the actuality operator @). While we've picked a particularly simple formulation of first-order modal logic, these inexpressibility results apply to a wide range of formulations.[3]

The signature for our first-order modal language $\mathcal{L}^{1M}$ contains:

- VAR $= \{x_1, x_2, x_3, \ldots\}$ (the set of *(object) variables*);
- PRED$^n = \{P_1^n, P_2^n, P_3^n, \ldots\}$ for each $n \geqslant 1$ (the set of *n-place predicates*);

The set of *formulas in $\mathcal{L}^{1M}$* or $\mathcal{L}^{1M}$*-formulas* is defined recursively:

$$\varphi ::= P^n(y_1, \ldots, y_n) \mid \neg \varphi \mid (\varphi \wedge \varphi) \mid \Box \varphi \mid \forall x \varphi$$

where $P^n \in$ PRED^n for any $n \geqslant 1$, and $x, y_1, \ldots, y_n \in$ VAR. The usual abbreviations for $\vee$, $\rightarrow$, $\exists$, and $\Diamond$ apply. We may drop parentheses for readability. If the free variables of φ are among $y_1, \ldots, y_n$, we may write "$\varphi(y_1, \ldots, y_n)$" to indicate this.

Let $S_1, \ldots, S_n$ be some new symbols with well-defined syntax. We'll indicate the language obtained from $\mathcal{L}^{1M}$ by adding $S_1, \ldots, S_n$ as $\mathcal{L}^{1M}(S_1, \ldots, S_n)$. Some symbols that might be added include:

$$\varphi ::= \cdots \mid y_1 \approx y_2 \mid @\varphi \mid {\downarrow}\varphi \mid \mathcal{F}\varphi \mid \forall_@ x \varphi \mid \Pi x \varphi$$

where $\approx$ is the identity relation, @ is an "actually" operator, $\downarrow$ is a diagonalization operator [16] that does the opposite of @, $\mathcal{F}$ is a "fixedly" operator [7], and $\forall_@$ is a quantifier over all actual objects. In what follows, $\mathcal{L}$ will just be any arbitrary $\mathcal{L}^{1M}(S_1, \ldots, S_n)$ where $S_1, \ldots, S_n$ are among the symbols above.

Definition 1 (First-Order Modal Models). *An $\mathcal{L}^{1M}$-model or modal model is an ordered tuple $\mathcal{M} = \langle W, R, D, \delta, I \rangle$ where:*

- *W is a nonempty set (the **state space**);*
- *$R \subseteq W \times W$ (the **accessibility relation**);*
- *D is a nonempty set (the **(global) domain**);*
- *$\delta \colon W \rightarrow \wp(D)$ is a function (the **local domain assignment**), where for each $w \in W$, $\delta(w)$ is the **local domain of w**;*
- *I is a function (the **interpretation function**) such that for each $P^n \in$ PRED^n, $I(P^n, w) \subseteq D^n$.*

By convention, where $\mathcal{M}$ is a modal model, we'll say that $\mathcal{M}$'s state space is $W^\mathcal{M}$, $\mathcal{M}$'s accessibility relation is $R^\mathcal{M}$, etc. We'll let $R[w] := \{v \in W \mid wRv\}$.

Let $\mathcal{M}$ be an $\mathcal{L}^{1M}$-model. A *variable assignment for $\mathcal{M}$* is a function assigning members of its global domain to variables. Let the set of variable assignments on $\mathcal{M}$ be VA($\mathcal{M}$). If a variable assignment g for $\mathcal{M}$ agrees with a variable assignment g' for $\mathcal{M}$ on every variable except possibly x, then g and g' are *x-variants*, $g \sim_x g'$. The variable assignment $g[x \mapsto a]$, or g_a^x, is the x-variant of g that sends x to a.

Some notation: if $\alpha_1, \ldots, \alpha_n$ is a sequence (of terms, objects, etc.), we may write "$\overline{\alpha}$" in place of "$\alpha_1, \ldots, \alpha_n$". $\overline{\alpha}$ is assumed to be of the appropriate length,

---

[3] See [10] for a tree of such formulations.

whatever that is in a given context. When f is some unary function, we may write "$f(\overline{\alpha})$" in place of "$f(\alpha_1), \ldots, f(\alpha_n)$". We'll let $|\overline{\alpha}|$ be the length of $\overline{\alpha}$.

Since we want to consider operators like @, our semantics will be two-dimensional (as suggested in e.g., [7, pp. 4-5]). That is, indices will have to contain two worlds. The first world is to be interpreted as the world "considered as actual", and the second as the world of evaluation.

Definition 2 (Satisfaction). *The **satisfaction relation**, $\Vdash$, is defined recursively, for all $\mathcal{L}^{1M}$-models $\mathcal{M} = \langle W, R, D, \delta, I \rangle$, all $w, v \in W$ and all $g \in VA(\mathcal{M})$:*

$$\mathcal{M}, w, v, g \Vdash P^n(\overline{x}) \quad \Leftrightarrow \quad \langle g(\overline{x}) \rangle \in I(P^n, v)$$

$$\mathcal{M}, w, v, g \Vdash x \approx y \quad \Leftrightarrow \quad g(x) = g(y)$$

$$\mathcal{M}, w, v, g \Vdash \neg\varphi \quad \Leftrightarrow \quad \mathcal{M}, w, v, g \nVdash \varphi$$

$$\mathcal{M}, w, v, g \Vdash \varphi \wedge \psi \quad \Leftrightarrow \quad \mathcal{M}, w, v, g \Vdash \varphi \text{ and } \mathcal{M}, w, v, g \Vdash \psi$$

$$\mathcal{M}, w, v, g \Vdash \Box\varphi \quad \Leftrightarrow \quad \forall v' \in R[v] \colon \mathcal{M}, w, v', g \Vdash \varphi$$

$$\mathcal{M}, w, v, g \Vdash @\varphi \quad \Leftrightarrow \quad \mathcal{M}, w, w, g \Vdash \varphi$$

$$\mathcal{M}, w, v, g \Vdash \downarrow\varphi \quad \Leftrightarrow \quad \mathcal{M}, v, v, g \Vdash \varphi$$

$$\mathcal{M}, w, v, g \Vdash \mathcal{F}\varphi \quad \Leftrightarrow \quad \forall w' \in R[w] \colon \mathcal{M}, w', v, g \Vdash \varphi$$

$$\mathcal{M}, w, v, g \Vdash \forall x\varphi \quad \Leftrightarrow \quad \forall a \in \delta(v) \colon \mathcal{M}, w, v, g_a^x \Vdash \varphi$$

$$\mathcal{M}, w, v, g \Vdash \forall_@ x\varphi \quad \Leftrightarrow \quad \forall a \in \delta(w) \colon \mathcal{M}, w, v, g_a^x \Vdash \varphi$$

$$\mathcal{M}, w, v, g \Vdash \Pi x\varphi \quad \Leftrightarrow \quad \forall a \in D \colon \mathcal{M}, w, v, g_a^x \Vdash \varphi.$$

If $|\overline{x}| \leq |\overline{a}|$, then $\mathcal{M}, w, v \Vdash \varphi[\overline{a}]$ if for all $g \in VA(\mathcal{M})$, $\mathcal{M}, w, v, g_{\overline{a}}^{\overline{x}} \Vdash \varphi(\overline{x})$.

3 The Two-Sorted Language

In order to prove our inexpressibility results, we need to translate ordinary English sentences like (R) into a correspondence language. This language is just a two-sorted first order language: one sort for objects, and one sort for worlds.

The signature for our two-sorted first-order language $\mathcal{L}^{2S}$ contains VAR plus:

- SVAR $= \{s_1, s_2, s_3, \ldots\}$ (the set of **state variables**).
- PRED$^{n/m} = \left\{ P_1^{n/m}, P_2^{n/m}, P_3^{n/m}, \ldots \right\}$ for each $n, m \geq 1$ (the set of n/m-**place predicates**).

For a predicate $P^{n/m}$, n is the object-arity, while m is the state-arity. Thus, $P^{n/m}$ takes exactly n object variables and m state variables as arguments.[4]

The set of **formulas in** $\mathcal{L}^{2S}$ or $\mathcal{L}^{2S}$-**formulas** is defined recursively:

$$\varphi ::= P^{n/m}(y_1, \ldots, y_n; s_1, \ldots, s_m) \mid E(x; s_1) \mid R(s_1, s_2) \mid \neg\varphi \mid (\varphi \wedge \varphi) \mid \forall x\varphi \mid \forall s\varphi$$

where $P^{n/m} \in$ PRED$^{n/m}$, $x, y_1, \ldots, y_n \in$ VAR, and $s, s_1, \ldots, s_m \in$ SVAR.

---

[4] We'll use ";" to separate object variables and state variables.

For instance, here are the intended formalizations of (R) and (NR), where s^* is meant to be interpreted as the actual world:

$$\exists t \ (R(s^*, t) \land \forall x \ (\text{Rich}(x; s^*) \land \text{Poor}(x; t))) \tag{3}$$

$$\forall s \ (R(s^*, s) \to \exists t \ (R(s, t) \to \forall x \ (\text{Rich}(x; s) \land \text{Poor}(x; t)))). \tag{4}$$

Definition 3 (Two-Sorted Models). *An $\mathcal{L}^{2S}$-model or two-sorted model is an ordered tuple $\mathfrak{M} = \langle W, D, V \rangle$ where W and D are nonempty sets, and V is a function (the valuation function) such that:*

- *for each $P^{n/m} \in PRED^{n/m}$, $V(P^{n/m}) \subseteq D^n \times W^m$;*
- *$V(E) \subseteq D \times W$;*
- *$V(R) \subseteq W \times W$.*

We are usually interested in the correspondence between $\mathcal{L}^{2S}$ and $\mathcal{L}^{1M}$-models.

Definition 4 (Model Correspondents). *Let $M = \langle W, R, D, \delta, I \rangle$ be a $\mathcal{L}^{1M}$-model. A two-sorted correspondent of M is a $\mathcal{L}^{2S}$-model $\mathfrak{M} = \langle W, D, V \rangle$ such that:*

- *for all $P \in PRED^{n/1}$, $V(P) = \{\langle \overline{a}; w \rangle \mid \langle \overline{a} \rangle \in I(P, w)\}$;*
- *$V(E) = \{\langle a; w \rangle \in D \times W \mid a \in \delta(w)\}$;*
- *$V(R) = R$.*

The satisfaction and consequence relations $\models$ for $\mathcal{L}^{2S}$ are just the standard ones for first-order logic with two sorts. We can now translate in the standard way every $\mathcal{L}^{1M}$-formula into $\mathcal{L}^{2S}$.

Definition 5 (Standard Translation). *Let φ be a $\mathcal{L}$-formula, and let $s, t \in SVAR$. The standard translation of φ wrt $\langle s, t \rangle$, $ST_{s,t}(\varphi)$, is defined recursively:*

$$ST_{s,t}(P^n(\overline{x})) = P^n(\overline{x}; t) \qquad\qquad ST_{s,t}(@\varphi) = ST_{s,s}(\varphi)$$

$$ST_{s,t}(x \approx y) = x \approx y \qquad\qquad ST_{s,t}(\downarrow\varphi) = ST_{t,t}(\varphi)$$

$$ST_{s,t}(\neg\varphi) = \neg ST_{s,t}(\varphi) \qquad\qquad ST_{s,t}(\forall x \varphi) = \forall x \ (E(x; t) \to ST_{s,t}(\varphi))$$

$$ST_{s,t}(\varphi \land \psi) = ST_{s,t}(\varphi) \land ST_{s,t}(\psi) \qquad ST_{s,t}(\forall_@ x \varphi) = \forall x \ (E(x; s) \to ST_{s,t}(\varphi))$$

$$ST_{s,t}(\Box\varphi) = \forall t' \ (R(t, t') \to ST_{s,t'}(\varphi)) \qquad ST_{s,t}(\Pi x \varphi) = \forall x \ ST_{s,t}(\varphi)$$

$$ST_{s,t}(\mathcal{F}\varphi) = \forall s' \ (R(s, s') \to ST_{s',t}(\varphi))$$

where t' is the next state variable not occurring anywhere in $ST_{s,t}(\varphi)$.

Lemma 6 (Translation). *Let M be an $\mathcal{L}^{1M}$-model, $\mathfrak{M}$ a two-sorted correspondent for M, $w, v \in W^M$, $g \in VA(M)$, $\mathfrak{g} \in VA(\mathfrak{M})$ (where $\mathfrak{g}(x) = g(x)$ for $x \in VAR$), $s, t \in SVAR$, and φ an $\mathcal{L}$-formula. Then $M, w, v, g \Vdash \varphi$ iff $\mathfrak{M}, \mathfrak{g}_{w,v}^{s,t} \models ST_{s,t}(\varphi)$.*

Proof. An easy induction on formulas.   □

With this result, we can define expressivity in the following manner:

Definition 7 (Expressivity). *A set of $\mathcal{L}$-formulas $\Gamma(\overline{x})$ **expresses** an $\mathcal{L}^{2S}$-formula $\alpha(\overline{x}; s, t)$ if α is equivalent (in the two-sorted language) to $ST_{s,t}(\Gamma)$ ($= \{ST_{s,t}(\varphi) \mid \varphi \in \Gamma\}$). A set of $\mathcal{L}$-formulas $\Gamma(\overline{x})$ **diagonally expresses** an $\mathcal{L}^{2S}$-formula $\alpha(\overline{x}; s)$ if α is equivalent to $ST_{s,s}(\Gamma)$.*

In what follows, we will focus on diagonal expressivity for simplicity, noting that the results below apply equally to the more general notion of expressibility.

4   Bisimulation

We now come to the notion of a bisimulation for ordinary first-order modal logic. This notion can be found in, e.g., [2,9,18,20]. However, we add clauses designed to ensure modal equivalence for formulas involving new symbols like @.

Definition 8 (Bisimulation). *Let $\mathcal{M}$ and $\mathcal{N}$ be $\mathcal{L}^{1M}$-models. An $\mathcal{L}^{1M}$-**bisimulation between $\mathcal{M}$ and $\mathcal{N}$** is a nonempty multigrade relation Z (so without a fixed arity) such that for all $w, v \in W^{\mathcal{M}}$, all $w', v' \in W^{\mathcal{N}}$, all fininite $\overline{a} \in D^{\mathcal{M}}$, and all finite $\overline{b} \in D^{\mathcal{N}}$, where $|\overline{a}| = |\overline{b}| = n$, we have that $Z(w, v, \overline{a}; w', v', \overline{b})$ implies:*

(Atomic) $\forall m \in \mathbb{N} \forall P^m \in PRED^m \, \forall \overline{\alpha}, \overline{\beta}$ *where* $|\overline{\alpha}| = |\overline{\beta}| = m$, *if for each i, there is a $j \leqslant n$ such that $\alpha_i = a_j$ and $\beta_i = b_j$, then:* $\langle \overline{\alpha} \rangle \in I^{\mathcal{M}}(P^m, v)$ *iff* $\langle \overline{\beta} \rangle \in I^{\mathcal{N}}(P^m, v')$
(Zig) $\forall u \in R^{\mathcal{M}}[v] \, \exists u' \in R^{\mathcal{N}}[v']: Z(w, u, \overline{a}; w', u', \overline{b})$
(Zag) $\forall u' \in R^{\mathcal{N}}[v'] \, \exists u \in R^{\mathcal{M}}[v]: Z(w, u, \overline{a}; w', u', \overline{b})$
(Forth) $\forall \alpha \in \delta^{\mathcal{M}}(v) \, \exists \beta \in \delta^{\mathcal{N}}(v'): Z(w, v, \overline{a}, \alpha; w', v', \overline{b}, \beta)$
(Back) $\forall \beta \in \delta^{\mathcal{N}}(v') \, \exists \alpha \in \delta^{\mathcal{M}}(v): Z(w, v, \overline{a}, \alpha; w', v', \overline{b}, \beta)$.

We may write "$\mathcal{M}, w, v, \overline{a} \leftrightarrows \mathcal{N}, w', v', \overline{b}$" *to indicate that there is a bisimulation Z between $\mathcal{M}$ and $\mathcal{N}$ such that $Z(w, v, \overline{a}; w', v', \overline{b})$ (where possibly $|\overline{a}| = |\overline{b}| = 0$). The notion of an $\mathcal{L}^{1M}(S_1, \ldots, S_n)$-**bisimulation between $\mathcal{M}$ and $\mathcal{N}$** is defined similarly, except one must add the condition(s) below corresponding to each S_i:*

(Eq) $\forall n, m \leqslant |\overline{a}| : a_n = a_m$ *iff* $b_n = b_m$
(Act) $Z(w, w, \overline{a}; w', w', \overline{b})$
(Diag) $Z(v, v, \overline{a}; v', v', \overline{b})$
(Fixedly-Zig) $\forall u \in R^{\mathcal{M}}[w] \, \exists u' \in R^{\mathcal{N}}[w']: Z(u, v, \overline{a}; u', v', \overline{b})$
(Fixedly-Zag) $\forall u' \in R^{\mathcal{N}}[w'] \, \exists u \in R^{\mathcal{M}}[w]: Z(u, v, \overline{a}; u', v', \overline{b})$
($\forall_@$-Forth) $\forall \alpha \in \delta^{\mathcal{M}}(w) \, \exists \beta \in \delta^{\mathcal{N}}(w'): Z(w, v, \overline{a}, \alpha; w', v', \overline{b}, \beta)$
($\forall_@$-Back) $\forall \beta \in \delta^{\mathcal{N}}(w) \, \exists \alpha \in \delta^{\mathcal{M}}(w'): Z(w, v, \overline{a}, \alpha; w', v', \overline{b}, \beta)$
(Π-Forth) $\forall \alpha \in D^{\mathcal{M}} \, \exists \beta \in D^{\mathcal{N}}: Z(w, v, \overline{a}, \alpha; w', v', \overline{b}, \beta)$
(Π-Back) $\forall \beta \in D^{\mathcal{N}} \, \exists \alpha \in D^{\mathcal{M}}: Z(w, v, \overline{a}, \alpha; w', v', \overline{b}, \beta)$.

The (Act), for instance, can be derived as follows. Suppose we introduced a relation $R_@ \subseteq W^2 \times W^2$ into models, and that we treated @ as a normal box operator. We could derive the truth conditions for @ by restricting to the class of models where $wvR_@w'v'$ iff $w = w' = v'$. Then the usual zig-zag clauses for @ just reduce to (Act). The same method applies to the other modal operators.

The standard results regarding bisimulations all carry over straightforwardly:

Definition 9 (Modal Equivalence). *Let M and N be $\mathcal{L}^{1M}$-models, where $w, v \in W^M$, $w', v' \in W^N$, $\overline{a} \in D^M$, and $\overline{b} \in D^N$ (where $|\overline{a}| = |\overline{b}|$). Then $\langle M, w, v, \overline{a}\rangle$ and $\langle N, w', v', \overline{b}\rangle$ are $\mathcal{L}$-equivalent or modally equivalent if for all $\mathcal{L}$-formulas $\varphi(\overline{x})$ (where $|\overline{x}| \leqslant |\overline{a}|$), $M, w, v \Vdash \varphi[\overline{a}]$ iff $N, w', v' \Vdash \varphi[\overline{b}]$. In such a case, we may write "$M, w, v, \overline{a} \equiv_{S_1, \dots, S_n} N, w', v', \overline{b}$", where $\mathcal{L} = \mathcal{L}^{1M}(S_1, \dots, S_n)$.*

Theorem 10 (Bisimulation Implies Modal Equivalence). *Suppose M and N are $\mathcal{L}^{1M}$-models, where $w, v \in W^M$, $w', v' \in W^N$, $\overline{a} \in D^M$, and $\overline{b} \in D^N$, such that $M, w, v, \overline{a} \leftrightarroweq_{S_1, \dots, S_n} N, w', v', \overline{b}$. Then $M, w, v, \overline{a} \equiv_{S_1, \dots, S_n} N, w', v', \overline{b}$.*

Corollary 11 (Translation Implies Invariance). *Let $\varphi(\overline{x}; s, t)$ be an $\mathcal{L}^{2S}$-formula. If φ is equivalent to the translation of some $\mathcal{L}^{1M}(S_1, \dots, S_n)$-formula, and if $M, w, v, \overline{a} \leftrightarroweq_{S_1, \dots, S_n} N, w', v', \overline{b}$, then for any two-sorted correspondents $\mathfrak{M}$ and $\mathfrak{N}$, $\mathfrak{M} \models \varphi[\overline{a}; w, v]$ iff $\mathfrak{N} \models \varphi[\overline{b}; w', v']$. Equivalently, if $\langle M, w, v, \overline{a}\rangle$ and $\langle N, w', v', \overline{b}\rangle$ have two-sorted correspondents that disagree on φ, then φ is not expressible as a $\mathcal{L}^{1M}(S_1, \dots, S_n)$-formula.*

5 Inexpressibility

We now turn to showing that (R) is not expressible in $\mathcal{L}^{1M}(@)$—in fact, not even in $\mathcal{L}^{1M}(\approx, @, \downarrow, \mathcal{F})$. We'll also show that (NR) is not expressible in $\mathcal{L}^{1M}(\approx, @, \Pi)$. In both cases, we construct two bisimilar models that disagree on the two-sorted formalization of the English sentence in question, and then invoke Corollary 11. We start by presenting a proof that $\mathcal{L}^{1M}(@)$ cannot express (3).

Let $\mathbb{N}^- := \mathbb{Z} - \mathbb{N}$. Our two models M_1 and M_2 are pictured in Figure 1. The global domain of each model is just $\mathbb{Z}$ and the accessibility relation is universal throughout. The world w is our actual world, where every positive integer is rich (top half of circle), and every negative integer is poor (bottom half of circle). For each nonempty finite subset S of $\mathbb{N}$, there is a world v_S where the members of S don't exist, and otherwise the rich and the poor are flipped with respect to w; so at v_S, the negative integers are rich, and the positive integers not in S are poor, and the positive integers in S don't exist. The extension of all other predicates is empty. The only difference betweem M_1 and M_2 is that M_2 includes an additional world $v_\varnothing$, where no integer fails to exist, and where the rich and poor are completely flipped with respect to w.

$\langle M_2, w, w\rangle$ satisfies (R), but not $\langle M_1, w, w\rangle$. But it turns out that $M_1, w, w \equiv_@ M_2, w, w$. In fact, $M_1, w, w \leftrightarroweq_@ M_2, w, w$. The reason is that each v-world looks isomorphic relative to first-order logic to every other v-world since $\mathcal{L}^{1M}(@)$ can only quantify over the existent objects. So at any given stage of construction of our bisimulation, we can treat each link between worlds and elements as if they're partial segments of an isomorphism between the two worlds considered as first-order models. Of course, we need to make sure that when we shift to new worlds, the elements linked still constitute a partial segment of an isomorphism between the new worlds. But as we'll see, this can be done.

Theorem 12 (Inexpressibility of (R)). *$M_1, w, w \leftrightarroweq_@ M_2, w, w$. But $M_2, w, w \Vdash (3)$ even though $M_1, w, w \not\Vdash (3)$. Hence, (3) is not expressible in $\mathcal{L}^{1M}(@)$.*

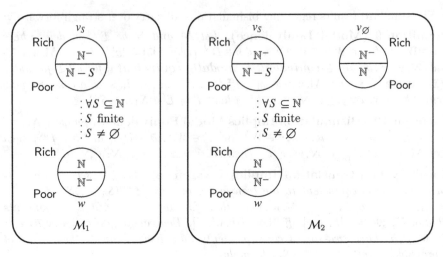

Fig. 1. $\mathcal{L}^{1M}(@)$-bisimilar models disagreeing on (R). The top half of each circle satisfies Rich, while the bottom half satisfies Poor; at each v_S, the members of S do not exist.

We show explicitly in the appendix how to construct a bisimulation between $\langle \mathcal{M}_1, w, w \rangle$ and $\langle \mathcal{M}_2, w, w \rangle$ in stages. Keeping track of the details is tedious, but the idea is simple. Basically, bisimulations are back-and-forth games that we might have to move to another accessible world to continue playing. So we just need to check that no matter where we move the game in one model, we can find a matching spot to move the game in the other model to keep playing.

Proof (Sketch). Our game starts at $\langle \mathcal{M}_1, w, w \rangle$ and $\langle \mathcal{M}_2, w, w \rangle$. Clearly, if we just play the back-and-forth game there, we'll eventually build an isomorphism. Let's suppose, after moving the game around a bit, we're now playing the back-and-forth game at $\langle \mathcal{M}_1, w, u_1 \rangle$ and $\langle \mathcal{M}_2, w, u_2 \rangle$, having linked $\bar{a} \in D_1$ to $\bar{b} \in D_2$, where a_i is positive iff b_i is. We'll show that no matter where we move the game in one model, we can move the game somewhere in the other model to keep playing. That is, we'll make sure that, wherever we move, if we want to extend the sequence of elements with a new a (that exists at our new location in $\mathcal{M}_1$), we can find a matching b (that exists at our new location in $\mathcal{M}_2$) such that a is positive iff b is (and similarly if we want to extend the sequence of elements with a new b that exists at our new location in $\mathcal{M}_2$).

Suppose first we move from u_1 to w in $\mathcal{M}_1$. It's easy to show that we can match that move in $\mathcal{M}_2$ by moving from u_2 to w.

Now suppose we move from u_1 to some v_S in $\mathcal{M}_1$. We need to match the move in $\mathcal{M}_2$ with some $v_{S'}$, but we need to do so in such a way so that $\langle \mathcal{M}_1, w, v_S \rangle$ and $\langle \mathcal{M}_2, w, v_{S'} \rangle$ don't disagree over existence between $\bar{a}$ and $\bar{b}$: we don't want a_i to exist at v_S but for b_i not to exist at $v_{S'}$. To get around this, let T be any finite set with the same cardinality as S such that $a_i \in S$ iff $b_i \in T$. Then it's straightforward to show that we can match the move to v_S in $\mathcal{M}_1$ with a move to v_T in $\mathcal{M}_2$. Similarly if we move from u_2 to either w or some v_S in $\mathcal{M}_2$.

Finally, suppose we move from u_2 to $v_\varnothing$ in $\mathcal{M}_2$. The only way to match that move in $\mathcal{M}_1$ is to move to some v_S. We can do this as long as we make sure that $\langle \mathcal{M}_1, w, v_S \rangle$ and $\langle \mathcal{M}_2, w, v_\varnothing \rangle$ don't disagree over existence between $\bar{a}$ and $\bar{b}$. But they won't disagree so long as we move to a v_S where no $a_i \in S$. So if $S \cap \{\bar{a}\} = \varnothing$, then we can match the move from u_2 to $v_\varnothing$ with a move from u_1 to v_S and continue playing. At each stage, it's easy to check that wherever we keep playing, we'll only match positives to positives, and negatives to negatives.   □

Again, the reason this strategy works is essentially because, modulo what exists, the v_S's and $v_\varnothing$ look like isomorphic first-order models, so linked elements can be treated as partial isomorphisms between the worlds. In particular, when we move to $v_\varnothing$, because only finitely many elements are linked at a time, we can always find a matching v_S where all of the linked elements exist, and just keep extending the partial isomorphism as usual. A similar strategy applies in showing that $\mathcal{M}_1, w, w \leftrightarrows_{\approx, \mathsf{E}, @, \downarrow, \mathcal{F}} \mathcal{M}_2, w, w$, though the details are messier.

However, this strategy fails when we try to show that $\mathcal{M}_1, w, w \leftrightarrows_{\approx, @, \Pi} \mathcal{M}_2, w, w$. This shouldn't be surprising, since (R) can be expressed as (1). But it's instructive to see why the proof above fails. Consider what happens when we try to guarantee the Forth clause. When we move from u_2 to $v_\varnothing$ in $\mathcal{M}_2$, we try to match that move in $\mathcal{M}_1$ by moving from u_1 to some v_S where $S \cap \{\bar{a}\} = \varnothing$. But the Π-Forth clause says that for any object $a \in D_1$ that we pick, there must be a matching $b \in D_2$. But if we pick a non-existent in v_S, we can be forced to end the game. Since every integer exists at $v_\varnothing$, we must pick a b that exists at $v_\varnothing$. But then by the Back clause, if we picked b again, we would need to match that pick with an a' that exists in v_S. But by the Eq clause, $a' = a$, and a doesn't exist in v_S. So we can't match that pick, and the game is over.

Now we'll show that even $\mathcal{L}^{1M}(\approx, @, \Pi)$ can't express (NR). Consider the two models $\mathcal{N}_1$ and $\mathcal{N}_2$ pictured in Figure 2. Again, the global domain of both models is $\mathbb{Z}$, and the accessibility relation is universal. This time, however, all of $\mathbb{Z}$ exists at every world. Our actual world this time is z, where no integer is either rich or poor. For every finite set $S \subseteq \mathbb{N}$, there's a world $v_{\mathbb{N}-S}$ where all the positive integers are rich except for S, and where all other integers are poor (so our old w is now just $v_\mathbb{N}$). And for every *nonempty* finite set $S \subseteq \mathbb{N}$, there's a world v_S like before, where the rich and poor are flipped with respect to $v_{\mathbb{N}-S}$. Again, the only difference between $\mathcal{N}_1$ and $\mathcal{N}_2$ is the presence of $v_\varnothing$ in $\mathcal{N}_2$, where every negative number is rich, and every positive number is poor.

$\langle \mathcal{N}_1, z, z \rangle$ and $\langle \mathcal{N}_2, z, z \rangle$ both agree that (3) is true. But they disagree on whether (4) is true; without the presence of $v_\varnothing$, there is no world for $v_\mathbb{N}$ (our old w) where everyone rich in $v_\mathbb{N}$ is poor. Furthermore, $\mathcal{N}_1, z, z \equiv_{\approx, @, \Pi} \mathcal{N}_2, z, z$. For even when we take existence into account, all of the v-worlds are isomorphic to one another. So as long as we're careful to move to the right worlds, we can always keep playing as if we're building an isomorphism between the worlds where the game is taking place. Thus:

Theorem 13 (Inexpressibility of (NR)). $\mathcal{N}_1, z, z \leftrightarrows_{\approx, @, \Pi} \mathcal{N}_2, z, z$. *But* $\mathcal{N}_2, z, z \Vdash$ (4) *while* $\mathcal{N}_1, z, z \nVdash$ (4). *Hence,* (4) *is not expressible in* $\mathcal{L}^{1M}(\approx, @, \Pi)$.

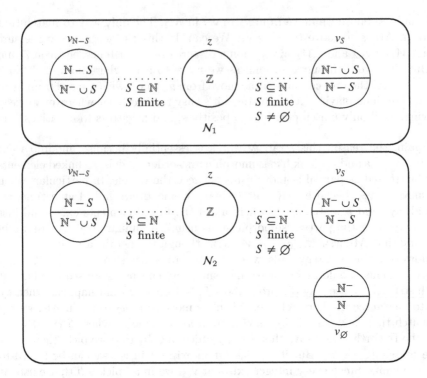

Fig. 2. $\mathcal{L}^{1M}(@, \Pi)$-bisimilar models disagreeing on (NR).

The proof is similar to the one before. For instance, suppose we're playing the back-and-forth game at $\langle \mathcal{N}_1, z, u_1 \rangle$ and $\langle \mathcal{N}_2, z, u_2 \rangle$, having linked $\bar{a}$ to $\bar{b}$, and suppose we make a move in $\mathcal{N}_2$ from u_2 to $v_\varnothing$. Then once again, the only matching move we can make in $\mathcal{N}_1$ is from u_1 to some v_S. But we can do this in general, so long as we pick an S that is disjoint from $\{\bar{a}\}$; since in that case, a_i will be in the extension of Rich at v_S iff b_i is in the extension of Rich in $v_\varnothing$. Similar reasoning as that above will show that matching moves can always be made no matter where we jump in the model.

Unlike in the case of (R), however, this inexpressibility proof doesn't extend to languages with $\downarrow$ or with $\mathcal{F}$. We can express (4) in either language with:

$$\Box \downarrow \Diamond \Pi x \, (@\mathrm{Rich}(x) \to \mathrm{Poor}(x)) \tag{5}$$

$$\mathcal{F}@\Diamond \Pi x \, (@\mathrm{Rich}(x) \to \mathrm{Poor}(x)). \tag{6}$$

But more complicated sentences can be constructed that reveal the expressive limitations of even languages with $\downarrow$ and $\mathcal{F}$.

6 Conclusion

It has often been noted, without proof, that (R) and (NR) are not expressible in $\mathcal{L}^{1M}$, even when one adds an actually operator [4,5,8,11,17]. Proofs of this claim can be found in [13,14], but involve rather complicated Henkin constructions that don't seem to illuminate the source of inexpressibility. In this paper, we've provided a simpler and more convenient method of proving inexpressibility results in $\mathcal{L}^{1M}$ using a modular notion of bisimulation. We've seen that inexpressibility proofs via bisimulation are illuminating as they reveal the ways in which $\mathcal{L}^{1M}$ can be insensitive to the location of certain back-and-forth games.

Some questions naturally arise from these results. First, is there a more general formal characterization of sentences like (E), (R), (NE), and (NR)? One syntactic characterization was proposed in [15], but it's open to debate whether this characterization is an appropriate one, or whether there is also a nice model-theoretic characterization of this class.

Second, is there a language weaker than $\mathcal{L}^{2S}$ that can express these kinds of sentences? It has been argued by [4] that adding second-order quantifiers suffices. In [5], Cresswell defined a language (that happens to be a notational variant of a quantified hybrid language without state variables as formulas) which he argued also suffices to express these kinds of sentences.[5] In both cases, heuristic arguments are given in support of the claim that these languages can express any sentence of the same kind as (E), (R), (NE), and (NR). But without an answer to the first question, no formal proof of these claims can provided.[6]

A   Proof of Theorem 12

To prove Theorem 12, we introduce a helpful definition:

Definition 14 (Partial Isomorphism). *Let M and N be modal models, let $w, v \in W^M$, and let $w', v' \in W^N$. A **partial $\mathcal{L}^{1M}$-isomorphism between** $\langle M, w, v \rangle$ **and** $\langle N, w', v' \rangle$ is a finite injective map $\rho \colon D \to D'$ such that:*

(Predicate) $\forall m \forall P^m \in PRED^m \, \forall a_1, \ldots, a_m \in \text{dom}\,(\rho) : \langle a_1, \ldots, a_m \rangle \in I^M(P^m, v)$
   iff $\langle \rho(a_1), \ldots, \rho(a_m) \rangle \in I^N(P^m, v')$
(Existence) $\forall a \in \text{dom}\,(\rho) : a \in \delta^M(v)$ *iff* $\rho(a) \in \delta^N(v')$.

The set of partial isomorphisms between $\langle M, w, v \rangle$ and $\langle N, w', v' \rangle$ will be $PAR^{M,w,v}_{N,w',v'}$. When the M and N are clear, we'll drop mention of them.

Now, at stage 0, set $Z_0 = \{\langle w, w; w, w \rangle\}$. Next, define the following:

$$Z_i^{\text{Act}} = \{\langle w, w, \overline{a}; w, w, \rho(\overline{a}) \rangle \mid \exists u, u' : \langle w, u, \overline{a}; w, u', \rho(\overline{a}) \rangle \in Z_i \text{ and } \rho \in \text{PAR}_{w,u'}^{w,u} \}$$

$$Z_i^{\text{Zig}} = \left\{ \langle w, v_S, \overline{a}; w, v_{\rho'[S]}, \rho'(\overline{a}) \rangle \;\middle|\; \begin{array}{l} \exists u, u' : \langle w, u, \overline{a}; w, u', \rho(\overline{a}) \rangle \in Z_i, \text{ where} \\ \rho \in \text{PAR}_{w,u'}^{w,u} \text{ and } \rho \subseteq \rho' \in \text{PAR}_{w,v_{\rho'[S]}}^{w,v_S} \\ \text{and dom}\,(\rho') \supseteq S \end{array} \right\}$$

$$Z_i^{\text{Zag}} = \left\{ \langle w, v_{\rho'^{-1}[S]}, \overline{a}; w, v_S, \rho'(\overline{a}) \rangle \;\middle|\; \begin{array}{l} \exists u, u' : \langle w, u, \overline{a}; w, u', \rho(\overline{a}) \rangle \in Z_i, \text{ where} \\ \rho \in \text{PAR}_{w,u'}^{w,u} \text{ and } \rho \subseteq \rho' \in \text{PAR}_{w,v_S}^{w,v_{\rho'^{-1}[S]}} \\ \text{and ran}\,(\rho') \supseteq S \end{array} \right\}$$

---

[5] However, [5] also shows that if $\approx$ is dropped and R is universal, then this language is as expressively powerful as $\mathcal{L}^{2S}$ without $\approx$.

[6] See [15] for one possible formal answer to this question.

$$\cup \left\{ \langle w, v_S, \bar{a}; w, v_\varnothing, \rho(\bar{a}) \rangle \;\middle|\; \begin{array}{l} \exists u, u': \; \langle w, u, \bar{a}; w, u', \rho(\bar{a}) \rangle \in Z_i \text{ and} \\ \rho \in \mathrm{PAR}^{w,u}_{w,u'}, \text{ where } S \cap \mathrm{dom}(\rho) = \varnothing \end{array} \right\}$$

$$Z_i^{\mathrm{Forth}} = \left\{ \langle w, u, \bar{a}, b; w, u', \rho'(\bar{a}), \rho'(b) \rangle \;\middle|\; \begin{array}{l} \langle w, u, \bar{a}; w, u', \rho(\bar{a}) \rangle \in Z_i \text{ where } \rho \subseteq \rho' \in \\ \mathrm{PAR}^{w,u}_{w,u'} \text{ and } b \in \delta_1(u) \cap \mathrm{dom}(\rho') \end{array} \right\}$$

$$Z_i^{\mathrm{Back}} = \left\{ \langle w, u, \bar{a}, {\rho'}^{-1}(b); w, u', \rho'(\bar{a}), b \rangle \;\middle|\; \begin{array}{l} \langle w, u, \bar{a}; w, u', \rho(\bar{a}) \rangle \in Z_i, \text{ where } \rho \subseteq \\ \rho' \in \mathrm{PAR}^{w,u}_{w,u'} \text{ and } b \in \delta_2(u') \cap \mathrm{ran}(\rho') \end{array} \right\}.$$

Then set: $Z_{i+1} = Z_i \cup Z_i^{\mathrm{Act}} \cup Z_i^{\mathrm{Zig}} \cup Z_i^{\mathrm{Zag}} \cup Z_i^{\mathrm{Forth}} \cup Z_i^{\mathrm{Back}}$. Finally, set $Z = \bigcup_{i \in \omega} Z_i$.

Lemma 15 (Trivial Observations). *If $\langle w, u, \bar{a}; w, u', \bar{b} \rangle \in Z_i$, then $u = w$ iff $u' = w$, and if $\rho \in \mathrm{PAR}^{w,u}_{w,u'}$, then $a_i \in \mathbb{N}$ iff $\rho(a_i) \in \mathbb{N}$.*

Lemma 16 (Partial Isomorphisms in Z). *For all $i \geq 0$, and all $\langle w, u, \bar{a}; w, u', \bar{b} \rangle \in Z_i$, there is a partial $\mathcal{L}^{1M}$-isomorphism ρ between $\langle \mathcal{M}_1, w, u \rangle$ and $\langle \mathcal{M}_2, w, u' \rangle$ such that $\rho(a_k) = b_k$ for $1 \leq k \leq |\bar{a}|$.*

Proof (Sketch). By induction on i. This clearly holds for the $i = 0$ case. Now suppose that every member of Z_i has the stated property. Show that for each condition C, every member of Z_i^{C} has the property, from which it will follow that every member of Z_{i+1} has the property. This is automatically guaranteed for Z_i^{Zig}, Z_i^{Back}, and Z_i^{Forth}. Z_i^{Zag} is almost immediate, but elements from the second listed set must be checked. Checking Z_i^{Act} is tedious, but straightforward. $\qquad \square$

We now turn to the proof of Theorem 12.

Proof (Theorem 12). Let $\langle w, u, \bar{a}; w, u', \bar{b} \rangle \in Z$. Then $\langle w, u, \bar{a}; w, u', \bar{b} \rangle \in Z_i$. By Lemma 16, there's a partial isomorphism ρ between $\langle \mathcal{M}_1, w, u \rangle$ and $\langle \mathcal{M}_2, w, u' \rangle$ such that $\rho(\bar{a}) = \bar{b}$. Hence, (Atomic) is met. As for the other conditions:

Act: By definition of Z_i^{Act}, $\langle w, w, \bar{a}; w, w, \bar{b} \rangle \in Z_{i+1}$. ✓

Zig: If u moves to w, then this case is covered by the Act-case. So suppose instead u moves to some v_S. It suffices to show that $\langle w, v_S, \bar{a}; w, v_{\rho'[S]}, \rho'(\bar{a}) \rangle \in Z_i^{\mathrm{Zig}}$ for some suitable $\rho' \supseteq \rho$. If $S \subseteq \mathrm{dom}(\rho)$, then let $\rho' = \rho$. Otherwise, let $b_1, \ldots, b_n \in S - \mathrm{dom}(\rho)$. Pick the least $b'_1, \ldots, b'_n \in \mathbb{N} - \mathrm{ran}(\rho)$ and set $\rho' = \rho \cup \{\langle b_i, b'_i \rangle \mid 1 \leq i \leq n\}$. It suffices to show that $\rho' \in \mathrm{PAR}^{w,v_S}_{w,v_{\rho'[S]}}$. Let $a \in \mathrm{dom}(\rho')$. By Lemma 15, $a \in I_1(\mathrm{Rich}, v_S)$ iff $\rho(a) \in I_2(\mathrm{Rich}, v_{\rho'[S]})$. As for Poor, $a \in I_1(\mathrm{Poor}, v_S)$ iff $a \in \mathbb{N} - S$ iff (by injectivity) $\rho'(a) \in \mathbb{N} - \rho'[S]$ iff $a \in I_2(\mathrm{Poor}, v_{\rho'[S]})$. ✓

Zag: We just need to check the case where u' moves to $v_\varnothing$. But by definition, for any S such that $S \cap \mathrm{dom}(\rho) = \varnothing$ (which will exist since $\mathrm{dom}(\rho)$ is finite), $\langle w, v_S, \bar{a}; w, v_\varnothing, \rho(\bar{a}) \rangle \in Z_i^{\mathrm{Zag}}$. ✓

Forth: Let $b \in \delta(u)$. WLOG, assume $b \notin \mathrm{dom}(\rho)$. If $b \in \mathbb{N}^-$, then just let b' be the least element in $\mathbb{N}^- - \mathrm{ran}(\rho)$. Otherwise, let b' be the least element in $\mathbb{N} - \mathrm{ran}(\rho)$. There are only three cases to consider:

(i) $u = u' = w$. By Lemma 15, $b \in I_1(\mathrm{Rich}, w)$ iff $b' \in I_2(\mathrm{Rich}, w)$. ✓

(ii) $u = v_S$ and $u' = v_{\rho[S]}$. Then $b \notin S$ and thus $b' \notin \rho'[S]$ (since, according to our construction, $S \subseteq \mathrm{dom}(\rho)$, and so $\rho[S] = \rho'[S]$). So $b \in I_1(\mathrm{Rich}, v_S)$ iff $b' \in I_2(\mathrm{Rich}, v_{\rho[S]})$. ✓

(iii) $u = v_S$ and $u' = v_\varnothing$. Since $b \in \delta_1(u)$, (Existence) is still upheld. And again, $b \in I_1(\text{Rich}, v_S)$ iff $b' \in I_2(\text{Rich}, v_\varnothing)$. ✓

Back: As above, except if $b' \in \mathbb{N}$, you pick the least $b \in \mathbb{N} - \text{dom}(\rho)$ in all cases except where $u = v_S$ and $u' = v_\varnothing$, in which case, you pick the least $b \in \mathbb{N} - (\text{dom}(\rho) \cup S)$. ✓ □

References

1. van Benthem, J.F.A.K.: Tense Logic and Standard Logic. Logique et Analyse 80, 395–437 (1997)
2. van Benthem, J.F.A.K.: Frame Correspondences in Modal Predicate Logic. In: Proofs, Categories and Computations: Essays in Honor of Grigori Mints, pp. 1–14. College Publications, London (2010)
3. Blackburn, P., de Rijke, M., Venema, Y.: Modal Logic. Cambridge University Press, Cambridge (2001)
4. Bricker, P.: Quantified Modal Logic and the Plural De Re. Midwest Studies in Philosophy, 372–394 (1989)
5. Cresswell, M.J.: Entities and Indicies. Kluwer Academic Publishers, Dordrecht (1990)
6. Crossley, J.N., Humberstone, I.L.: The Logic of "Actually". Reports on Mathematical Logic 8, 11–29 (1977)
7. Davies, M., Humberstone, L.: Two Notions of Necessity. Philosophical Studies 38, 1–30 (1980)
8. Fara, M., Williamson, T.: Counterparts and actuality. Mind 114, 1–30 (2005)
9. Fine, K.: Model theory for modal logic - Part III Existence and predication. Journal of Philosophical Logic 10, 293–307 (1981)
10. Garson, J.W.: Handbook of Philosophical Logic, vol. 3, pp. 267–323. Springer (2001)
11. Hazen, A.P.: Expressive Completeness in Modal Language. Journal of Philosophical Logic 5(1), 25–46 (1976)
12. Hazen, A.P.: Actuality and Quantification. Notre Dame Journal of Formal Logic 31(4), 498–508 (1990)
13. Hodes, H.T.: Some Theorems on the Expressive Limitations of Modal Languages. Journal of Philosophical Logic 13(1), 13–26 (1984)
14. Hodes, H.T.: Axioms of Actuality. Journal of Philosophical Logic 13(1), 27–34 (1984)
15. Kocurek, A.W.: The Problem of Cross-world Predication (2015) (Manuscript)
16. Lewis, D.K.: Counterfactuals and Comparative Possibility. Journal of Philosophical Logic 2, 418–446 (1973)
17. Sider, T.: Logic for Philosophy. Oxford University Press, Oxford (2010)
18. Sturm, H., Wolter, F.: First-Order Expressivity for S5-Models: Modal vs. Two-Sorted Languages. Journal of Philosophical Logic, 571–591 (July 2001)
19. Wehmeier, K.F.: World Travelling and Mood Swings. In: Löwe, B., Malzkorn, W., Räsch, T. (eds.) Foundations of the Formal Sciences II, pp. 257–260 (2001)
20. Yanovich, I.: Expressive Power of "Now" and "Then" Operators. Journal of Logic, Language, and Information 24, 65–93 (2015)

Causal Models and the Ambiguity of Counterfactuals

Kok Yong Lee

Department of Philosophy,
National Chung Cheng University,
Min-Hsiung, Taiwan
kokyonglee.mu@gmail.com

Abstract. Counterfactuals are inherently ambiguous in the sense that the same counterfactual may be true under one mode of counterfactualization but false under the other. Many have regarded the ambiguity of counterfactuals as consisting in the distinction between forward-tracking and backtracking counterfactuals. This is incorrect since the ambiguity persists even in cases not involving backtracking counterfactualization. In this paper, I argue that causal modeling semantics has the resources enough for accounting for the ambiguity of counterfactuals. Specifically, we need to distinguish two types of causal manipulation, which I call "intervention" and "extrapolation" respectively. To intervene in a causal model M is to change M's structural equations in some specific ways, while to extrapolate M is to change the value assignment of M's variables in some specific ways. I argue that intervention and extrapolation offer a natural explanation for the ambiguity of counterfactuals.

Keywords: Causal Model, Counterfactual Conditional, Intervention, Extrapolation, Backtracking.

1 Introduction

Counterfactual conditionals (hereafter 'counterfactuals') are inherently ambiguous in the sense that the same counterfactual may be true under one mode of counterfactualization but false under the other. Many have regarded the ambiguity of counterfactuals as consisting in the distinction between forward-tracking and backtracking counterfactuals. This is incorrect since the ambiguity persists even in cases not involving backtracking counterfactualization. In this paper, I argue that causal modeling semantics has the resources enough for accounting for the ambiguity of counterfactuals. Specifically, we need to distinguish two types of causal manipulation, which I call "intervention" and "extrapolation" respectively.

The following consists of four sections. Section 2 explains the ambiguity of counterfactuals. Section 3 introduces the crux of the causal modeling semantics of counterfactuals. Section 4 introduces two types of submodels. Section 5 explains the ambiguity of counterfactuals in terms of these two types of submodels.

© Springer-Verlag Berlin Heidelberg 2015
W. van der Hoek et al. (Eds.): LORI 2015, LNCS 9394, pp. 220–229, 2015.
DOI: 10.1007/978-3-662-48561-3_18

2   The Ambiguity of Counterfactuals

Consider the following case:

Ask. Jack had a quarrel with Jim yesterday, and Jack is still mad at Jim. When Jack is not mad, he is a generous person, who will help his friend if asked for a favor. Jim, on the other hand, is a prideful person, who will not ask someone for help after having a quarrel with this person. As a result, Jim does not ask Jack for help. (cf. Lewis 1979, 456; also Downing 1958)

Let '>' stand for the counterfactual-conditional connective, 'A > C' for the counterfactual *If A had obtained, C would have obtained*, 'Ask > Help' for *If Jim had asked Jack for help, Jack would have helped him.*[1] "Ask > Help" is ambiguous in the sense that it seems true under one mode of counterfactualization but false under the other. Under what we may call *forward-tracking* counterfactualization, "Ask > Help" seems false: if Jim were to ask Jack for help, he would have been rejected, since Jack is mad at him, and Jack is not generous when he is mad. Under what we may call *backtracking* counterfactualization, by contrast, "Ask > Help" seems true: Jim is a prideful person; he will not ask Jack for help after quarreling with him yesterday. Hence, if Jim were to ask Jack for help, it must be that they did not quarrel yesterday. If so, Jack would not be mad at Jim, and would have helped him.

Cases like *Ask* are theoretically significant, as they indicate that counterfactuals are *inherently ambiguous*, that is, the same counterfactual "A > C" can be true under one kind of counterfactual counterfactualization, while false under the other. The prominent treatment of the truth condition of counterfactuals has been relatively silent on its ambiguous nature.

The Lewis-Stalnaker possible-worlds semantics, at least in its orthodox form, does not admit that counterfactuals are ambiguous in such a way (cf. Stalnaker 1968; Lewis 1973). A standard reply to the problem is to dismiss backtracking counterfactuals as non-standard, by contending that we do not ordinarily perform backtracking counterfactualization (Lewis 1979, 458).

The possible-worlds semantics has suffered from some serious objections, which I will leave aside (cf. Schaffer 2004; Pruss 2003; also see Lee forthcoming). For the present purposes, it is important to point out that the problem of the ambiguity of counterfactuals cannot be evaded by arguing that we do not ordinarily reason in a backtracking manner. First, the contention that backtracking counterfactualization is non-ordinary is controversial (cf. Bennett 1984). Second, and more importantly, the ambiguity of counterfactuals runs deeper than the temporal aspect of counterfactuals, since it persists even in cases not involving backtracking counterfactuals.

---

[1] Throughout this paper, propositions (events) are denoted by italic sentences.

As Jonathan Bennett points out, *Ask* can be modified so that it does not contain a yesterday's quarrel:

Ask*. Jim knows that Jack is mad at him but he has no idea why. When Jack is not mad, he is a generous person, who will help his friend if asked for a favor. Jim, on the other hand, is a prideful person, who will not ask someone for help if this person is mad at him. As a result, Jim does not ask Jack for help. (Cf. Bennett 2003, 206)

Intuitively, "Ask > Help" is as ambiguous in *Ask** as in *Ask*. On the one hand, we may reason that since Jack is mad at Jim, and he is not generous when he is mad, Jack would not have helped Jim if Jim were to ask him for help. On the other hand, we may reason that if Jim were to ask Jack for help, it must be that Jack were not mad at him, since Jim is a prideful person, who will not ask someone for help knowing that this person is mad at him. If Jack were not mad at Jim, he would be his usual generous self and would have helped Jim.

Hence, as in *Ask*, "Ask > Help" is both true and false in *Ask**, depending on which mode of counterfactualization is in play. However, no backtracking counterfactualization is required in determining the truth-value of "Ask > Help" in *Ask**, since *Ask** does not presuppose a yesterday's quarrel.

I conclude that counterfactuals are inherently ambiguous. The distinction between forward-tracking and backtracking counterfactuals is in fact a special case of the distinction between two types of counterfactuals manifested by *Ask**. Elsewhere, I have argued that the causal modeling semantics of counterfactuals has the resources enough for accounting for the distinction between forward-tracking and backtracking counterfactuals (Lee forthcoming). In what follows, I will further argue that the causal modeling semantics can handle the inherent ambiguity of counterfactuals.

3   The Causal Modeling Semantics

A causal model is a mathematical entity aiming at representing the causal relations of the events in a scenario. To illustrate, let us construct a causal model for *Ask**.

Formally speaking, a causal model M is a triple $<V, S, A>$. V is a finite set of variables, $\{V_1, V_2, \ldots V_n\}$. These are *variables for events* in the scenario that M is supposed to represent. The causal model K* for *Ask** naturally contains the following set of variables:

MAD represents whether or not Jack is mad at Jim.[2]
PRIDE represents whether or not Jim is a prideful person.
ASK represents whether or not Jim asks Jack for help.
HELP represents whether or not Jack helps Jim.

Each V_i in V admits a range of values (finite or infinite). In the simplest cases such as K*, the variables admits only two possible values, i.e., "Yes" or "No". It is

---

[2]   I use uppercase letters to stand for variables for events.

customary to use '$V_i = v_i$' to stand for *The variable V_i has the value v_i*. For *binary* variables such as MAD, PRIDE, ASK, and HELP, we may use '1' and '0' to stand for YES and NO respectively. For instance, "MAD = 1" means that Jack is mad at Jim, and "PRIDE = 0" means that Jim is not a prideful person.

S, the second element of a causal model, is a set of structural equations specifying the causal relations among variables. For each variable V_i in *V*, *S* contains at most one structural equation of the following form:

$$V_i \Leftarrow f_i(PA_i).$$

The meaning of '$\Leftarrow$' is two-fold. For one thing, "$X \Leftarrow Y$" means that X is causally determined by Y, i.e., whether or not X obtains causally depends on whether or not Y obtains. For another, "$X \Leftarrow Y$" indicates that X takes on the value of Y. 'PA_i' stands for the set of V_i's *parents* (i.e., causes), which is a subset of *V*; V_i is a *child* (i.e., effect) of PA_i. F_i is a function that maps PA_i to $\{0,1\}$, for binary variables. We may further regard f_i as truth functions with truth and falsity being represented by 1 and 0 respectively. For the sake of convenience, we may even treat variables on the right-hand side of a structural equation as propositions such that "X" means X = 1, and that "~X" means X = 0.

Naturally, K*'s *S** consists of:

$$ASK \Leftarrow (\sim PRIDE \lor \sim MAD)$$

$$HELP \Leftarrow (ASK \land \sim MAD)$$

In words, "ASK $\Leftarrow$ (~PRIDE $\lor$ ~MAD)" indicates that whether or not Jim will ask Jack for help depends causally on whether or not Jim is a prideful person and on whether or not Jim is mad at Jack. Jim will ask Jack for help *iff* either Jim is not a prideful person or Jack is not mad at hm.[3] "HELP $\Leftarrow$ (ASK $\land$ ~MAD)" indicates that whether or not Jack will help Jim depends causally on whether or not Jim asks Jack for help and on whether or not Jim is mad at Jack. Jack will help Jim iff Jim asks Jack for help, and Jack is not mad at Jim.

There is no structural equation for MAD and PRIDE, meaning that their causes (parents) are not specified in K*. A causal model consists of two kinds of variables: *endogenous* variables, whose causes are specified by structural equations, and *exogenous* variables, whose causes are not so specified. The values of exogenous variables are *given* to the model. For instance, in K*, both MAD and PRIDE are *stipulated* to take on the value 1.

---

[3] The biconditional holds in *Ask**. That is, we assume that none of the conditions sabotaging the if-direction of the biconditional (such as Jim has temporally lost his capacity to communicate with others) holds. Nor does any of the condition sabotaging the only-if direction (such as Jim has been coerced into asking Jack for help). In Galles and Pearl's term, these are "inhibiting" and "triggering abnormalities" respectively (Galles and Pearl 1998). In other words, when constructing causal models, we assume that such inhibiting and triggering abnormalities do not hold.

A, the third element of a causal model, is a function of assigning a value to each variable in the model (cf. Hiddleston 2005; Briggs 2012). For each exogenous variable, *A* simply assigns a value to it. For each endogenous variable, *A* assigns a value to it based on the value of exogenous variables and the set of structural equations *S*. For instance, K^*'s A^* is as follows:

$$A^*(\text{ASK}) = A^*(\text{HELP}) = 0, \text{ and}$$
$$A^*(\text{MAD}) = A^*(\text{PRIDE}) = 1.^4$$

In words, in *Ask**, Jim is mad at Jack, Jim is a prideful person, Jim does not ask Jack for help, and Jack does not help Jim.

Causal models are instructive in characterizing the truth condition of counterfactuals. Suppose that we manipulate the causal structures of M in some way. More precisely, we may alter either some of M's structural equations or its value assignment, or both. We thus generate a certain *submodel* M' of M. If a causal model M represents a scenario s, a submodel M' of M naturally represents a "counterfactual" scenario s' of s, which is a scenario containing information about what would have happened if s had been different. Following an intuitive line, we may define the *c*ausal *m*odeling account of the truth condition of counterfactuals as:

(CM)   "A > C" is true in a causal model M iff "C" is true in certain submodels M'.

In what follows, I argue that the causal modeling semantics has the resources enough for accounting for the ambiguity of counterfactuals. The crux is that there are two types of submodels. "A > C" may have different truth-values provided that "C" is true in one type of submodel M' but false in the other type of submodel M".

4   Intervention vs. Extrapolation

A submodel M' is generated from manipulating the causal structures of M. That there are two types of submodels implies that there are two distinct types of manipulation. I will call them "intervention" and "extrapolation" respectively. The distinction was first brought to my attention by David Galles and Judea Pearl's distinction between *doing* and *seeing* (Galles and Pearl 1998, 159). But they do not develop it in the following way. As I will argue, the distinction between intervention and extrapolation offers a natural explanation for the ambiguity of counterfactuals.

Intervention has been featured in the prominent causal modeling semantics for counterfactuals (cf., e.g., Galles and Pearl 1998; Pearl 2000; Briggs 2012). Let M (= $<V, S, A>$) be a causal model, B be the sentential form '$C_1 = c_1 \wedge \ldots \wedge C_k = c_k$'[5], *VB*

---

[4] Calculation: MAD = 1 and PRIDE = 1 (by assumption); if PRIDE = 1 and MAD = 1, then ASK = 0 (by ASK $\Leftarrow$ (~PRIDE $\vee$ ~MAD)); if ASK = 0 then HELP = 0 (by HELP $\Leftarrow$ (ASK $\wedge$ ~MAD)).

[5] I follow Galles and Pearl (1998) here. For causal modeling semantics that deals with more complex antecedents, see Halpern (2000) and Briggs (2012).

be the set of variables that are in B. An *intervention in M with respect to B* generates a submodel M_B ($=<V_B, S_B, A_B>$) such that:

(i) $V_B = V$.

(ii) $S_B = S$ except that for each C_j in VB, S_B replaces the structural equation $C_j \Leftarrow f_j(PA_j)$ of S with $C_j = c_j$, if C_j is endogenous.

(iii) $A_B = A$ except that (a) for each C_k in VB, A_B assigns the value c_k to C_k if C is exogenous, and that (b) for each V_l in ($V_B \backslash VB$), A_B assigns the value c_l to C_l based on the value of C_k and S_B.

At its core, to intervene in a causal model M with respect to B ($C_1 = c_1 \wedge \dots \wedge C_k = c_k$) is to set each C_i in VB to take on the value c_i. Specifically, if C_i is endogenous, intervention disconnects the causal connections between C_i and its parents. The resulting submodel thus specifies the causal relations among variables differently. The values of B's children (causal effects) are calculated accordingly.

Suppose that we intervene in K* with respect to (ASK = 1). This generates the submodel $K*_{(ASK=1)}$, whose set of variables $V*_{(ASK=1)}$ is identical to K*'s. K*'s set of structural equations $S*_{(ASK=1)}$, by contrast, consists of the following:

ASK $\Leftarrow$ 1

HELP $\Leftarrow$ (ASK $\wedge$ ~MAD)

"ASK $\Leftarrow$ 1" means that ASK is to set to have the value 1 so that ASK no longer causally depends on the values of its parents (i.e., PRIDE and MAD). In other words, intervention disconnects the causal relations of ASK to its parents by stipulating the variable to take on the value 1.

As a result, $K*_{(ASK=1)}$'s value assignment $A*_{(ASK=1)}$ is that:

$A*_{(ASK=1)}(HELP) = 0$, and

$A*_{(ASK=1)}(ASK) = A*_{(ASK=1)}(PRIDE) = A*_{(ASK=1)}(MAD) = 1.$[6]

$A*$ and $A*_{(ASK=1)}$ are different, but notice that intervening in a causal model does not necessarily lead to different value assignments.

Let us introduce extrapolation, which, by contrast, has received little attention from philosophers. Let M ($=<V, S, A>$) be a causal model, B be the sentential form '$C_1 = c_1 \wedge \dots \wedge C_m = c_m$', and VB be the set of variables that are in B. An *extrapolation on M with respect to B* generates a submodel M^B ($=<V^B, S^B, A^B>$) of M such that:

(i) $V^B = V$.

(ii) $S^B = S$.

(iii) $A^B = A$ except that (a) for each C_i in VB, A^B assigns the value c_i to C_i, and that (b) for each V_i in ($V^B \backslash VB$), A^B assigns a value v_i to V_i based on the value of C_i and S^B.

---

[6] Calculation: MAD) = 1 and PRIDE = 1 (by assumption); ASK = 1 (by intervention); If ASK = 1 and MAD = 1, then HELP = 0 (by HELP $\Leftarrow$ (ASK $\wedge$ ~MAD)).

To extrapolate a causal model M with respect to B ($C_1 = c_1 \wedge \ldots \wedge C_m = c_m$) also sets each C_i in VB to take on the value c_i. But unlike intervention, extrapolation preserves the causal relations between C_i and its parents. The values of C_i's children *and* parents are calculated accordingly.

Suppose that we extrapolate K* with respect to (MAD = 0). The extrapolation generates the submodel $K^{*(MAD=0)}$. $K^{*(MAD=0)}$ and K* consist of the same sets of variables and structural equations, namely, $V^{*(MAD=0)} = V^*$, and $S^{*(MAD=0)} = S^*$. $A^{*(MAD=0)}$, by contrast, is as follows:

$$A^{*(MAD=0)}(MAD) = 0, \text{ and}$$
$$A^{*(MAD=0)}(PRIDE) = A^{*(MAD=0)}(ASK) = A^{*(MAD=0)}(HELP) = 1.^{7}$$

Extrapolation necessarily leads to different value assignments.

While intervention and extrapolation give rise to two types of submodels, they do not always determine a unique submodel.[8] Essentially, this indicates that the truth condition of counterfactuals is *context-sensitive*. That is, when more than one submodels are available, the context will determine which submodels are relevant to determining the truth-values of counterfactuals.

Let us call the submodels M' determined by the context *the relevant submodels*. Intervention and extrapolation give rise to different types of relevant submodels. Therefore, CM should be *disambiguated* into:

(CM$_{IN}$) "A > C" is true$_{IN}$ in M iff "C" is true in the relevant submodels M_A.
(CM$_{EX}$) "A > C" is true$_{EX}$ in M iff "C" is true in the relevant submodels M^A.

Let us introduce some terminology. On CM$_{IN}$ and CM$_{EX}$, the truth condition of counterfactuals is determined by two modes of counterfactualization, related to intervention and extrapolation, as indicated by the subscripts. Call them "intervention-counterfactualization" (or "IN-counterfactualization") and "extrapolation-counterfactualization" (or "EX-counterfactualization") respectively. "A > C" can be true under IN-counterfactualization, but false under EX-counterfactualization, and *vice versa*. Hence, we distinguish counterfactuals being true by IN-counterfactualization ('true$_{IN}$') from counterfactuals being true by EX-counterfactualization ('true$_{EX}$'), and, correspondingly, "intervention-counterfactuals" (or "IN-counterfactual") from "extrapolation-counterfactuals" (or "EX-counterfactuals"). The truth conditions of IN-counterfactuals and EX-counterfactuals are defined by CM$_{IN}$ and CM$_{EX}$ respectively.

---

[7] Calculation: MAD = 0 by extrapolation; PRIDE = 1 (by assumption); if MAD = 0, then ASK = 1 (by ASK $\Leftarrow$ (~PRIDE $\vee$ ~MAD)). If ASK = 1 and MAD = 0, then HELP = 1 (by HELP $\Leftarrow$ (ASK $\wedge$ ~MAD)).

[8] As noted in Footnote 9, extrapolating K* with respect to (ASK = 1) gives rise to two submodels. Moreover, intervening in a causal model M with respect to a disjunction arguably also gives rise to more than one submodels (cf. Briggs 2012). For a related discussion, see Hiddleston (2005, 650ff.).

5   Explaining the Ambiguity

CM_{IN} and CM_{EX} give the correct verdicts with respect to *Ask**. Intervening in K* with respect to (ASK = 1) gives rise to a unique submodel $K*_{(ASK=1)}$. On CM_{IN}, "ASK = 1 > HELP = 1" is true$_{IN}$ in K* iff "HELP = 1" is true in $K*_{(ASK=1)}$. As noted above, $A*_{(ASK=1)}(HELP) = 0$. It follows that "ASK = 1 > HELP = 1" is not true$_{IN}$ in K*, as desired.

By contrast, suppose that we extrapolate K* with respect to (ASK = 1). The extrapolation generates a submodel whose value assignments are as follows:

$$A*^{(ASK=1)}(MAD) = 1, \text{ and}$$
$$A*^{(ASK=1)}(PRIDE) = A*^{(ASK=1)}(ASK) = A*^{(ASK=1)}(HELP) = 1.^9$$

On CM_{AL}, "ASK = 1 > HELP = 1" is true$_{EX}$ in K* iff "HELP" is true in $K*^{(ASK=1)}$. Since $A*^{(ASK=1)}(HELP) = 1$, it follows that "ASK = 1 > HELP = 1" is true$_{EX}$ in K*, as desired.

In words, the causal modeling semantics correctly predicts that "Ask > Help" is ambiguous in *Ask**. Interpreted as an IN-counterfactual, the counterfactual is false. The same counterfactual will be true, however, if interpreted as an EX-counterfactual.

Not only does the distinction between intervention and extrapolation give the correct verdicts, it offers a natural explanation of the ambiguity of counterfactuals manifested by cases like *Ask**. When counterfactualizing that "Ask > Help" is not true in *Ask**, we focus solely on the causal effect of the event of Jim asking Jack for help (i.e., Ask) *per se*. We ignore the causal relations between Ask and its causes (parents). In particular, we do not attempt to rationalize how Ask could have happened in the first place. For instance, we ignored that the fact that Jim being a prideful person (i.e., Pride) prevents Ask from obtaining, and we simply stipulate that Ask holds without a specific story of how Ask could have happened in the first place (in many cases, such stories should not even be given). This mode of counterfactualization is nicely captured by intervention. Intervening in M with respect to ($C_i = c_i$) generates a submodel $M_{Ci=ci}$ that contains information necessary for understanding the causal effect of C_i (Galles and Pearl 1998). $M_{Ci=ci}$ *mutilates* all the causal influences of C_i's

---

9   Calculation: PRIDE = 1 (by assumption); ASK = 1 (by extrapolation); if ASK = 1 and PRIDE = 1, then MAD = 0 (by ASK $\Leftarrow$ (~PRIDE $\lor$ ~MAD)); if ASK = 1 and MAD = 0, then HELP = 1 (by HELP $\Leftarrow$ (ASK $\land$ ~MAD)).

Here, we touch on issues of the context-sensitivity of CM_{EX}. $A*^{(ASK=1)}$ as listed here has held (PRIDE = 1) fixed. It is by holding fixed (PRIDE = 1) that we can deduce (MAD = 0). If we hold fixed (MAD = 1) instead, we will have deduced (HELP = 0): MAD = 1 (by assumption); ASK = 1 (by extrapolation); if MAD = 1 and ASK = 1, then PRIDE = 0 (by ASK $\Leftarrow$ (~PRIDE $\lor$ ~MAD)); if ASK = 1 and MAD = 1, then HELP = 0 (by HELP $\Leftarrow$ (ASK $\land$ ~MAD)).

In other words, extrapolating K* with respect to (ASK = 1) is context-sensitive. Specifically, either PRIDE = 0 or MAD = 1 needs to be held fixed. Based on the natural reading of *Ask**, the relevant submodel in play here holds (PRIDE = 1) fixed (also see Footnote 10).

parents (i.e., PA_i) have on C_i, while *stipulating* that C_i takes on the value c_i. $M_{C_i=c_i}$ gives us a clear picture of $(C_i = c_i)$'s causal effect in M.

By contrast, when counterfactualizing that "Ask > Help" is true in *Ask**, we focus on rationalizing how Ask could have happened in the first place. In particular, we use the causal relations among events to help us determine under what condition Ask could have happened. For instance, we reason that Jim must not be mad at Jack (i.e., ~Mad) if Ask is to obtain, since Pride prevents Ask from obtaining if Mad has obtained. This mode of counterfactualization is represented nicely by EX-counterfactualization.[10] Extrapolating a causal model M with respect to $(C_i = c_i)$ generates a submodel $M^{C_i=c_i}$ that contains all information necessary for knowing under what condition $(C_i = c_i)$ could have happened in M. $M^{C_i=c_i}$ assigns the values of its variables in a way that preserves all the causal relations among events in M. $M^{C_i=c_i}$ thus gives us a story of what else needs to change if C_i is to have the value of c_i in M.

Acknowledgements. I am grateful to three anonymous reviewers for their comments. The present work has received funding from the Ministry of Science and Technology (MOST) of Taiwan (R.O.C.) (MOST 103-2410-H-194-125).

References

1. Bennett, J.: Counterfactuals and Temporal Direction. The Philosophical Review 93(1), 57–91 (1984)
2. Bennett, J.: A Philosophical Guide to Conditionals. Clarendon Press, Oxford (2003)
3. Briggs, R.: Interventionist Counterfactuals. Philosophical Studies 160(1), 139–166 (2012)
4. Downing, P.B.: Subjunctive Conditionals, Time Order, and Causation. Proceedings of the Aristotelian Society 59, 125–140 (1958)
5. Galles, D., Pearl, J.: An Axiomatic Characterization of Causal Counterfactuals. Foundations of Science 3(1), 151–182 (1998)
6. Halpern, J.Y.: Axiomatizing Causal Reasoning. Journal of Artificial Intelligence Research 12(1), 317–337 (2000)
7. Hiddleston, E.: A Causal Theory of Counterfactuals. Noûs 39(4), 632–657 (2005)

---

[10] EX-counterfactualization is context-sensitive in a way parallel with the context-sensitivity of extrapolation. Specifically, in *Ask**, there are two ways to EX-counterfactualize what would have happened if Jim were to ask Jack for help. On the one hand, we might EX-counterfactualize that if Jim were to ask Jack for help, it must be that Jim had somehow swallowed his pride, since Jack had been mad at him, and as a prideful fellow, Jim would not have asked Jack for help. On the other hand, we might EX-counterfactualize that if Jim were to ask Jack for help, it must be that Jim was not mad at him, since as a prideful person, Jim would not ask Jack for help given that Jack was mad at him. In a sense, both are legitimate Ex-counterfactualization. But the present context demands that we adopt the latter, since this is the most natural way to interpret *Ask**. This explains why in modeling *Ask**, we should hold (PRIDE = 1) fixed.

8. Lee, K.Y.: Motivating the Causal Modeling Semantics of Counterfactuals, or, Why We Should Favor the Causal Modeling Semantics over the Possible-Worlds Semantics. In: Yang, S.C.-M., Deng, D.-M., Lin, H. (eds.) Structural Analysis of Non-Classical Logics (forthcoming)
9. Lewis, D.: Counterfactuals. Blackwell, Malden (1973)
10. Lewis, D.: Counterfactual Dependence and Time's Arrow. Noûs 13(4), 455–476 (1979)
11. Pearl, J.: Causality: Models, Reasoning, and Inference. Cambridge University Press, Cambridge (2000)
12. Pruss, A.R.: David Lewis's Counterfactual Arrow of Time. Noûs 37(4), 606–637 (2003)
13. Schaffer, J.: Counterfactuals, Causal Independence and Conceptual Circularity. Analysis 64(4), 299–309 (2004)
14. Sloman, S.A.: Casual Models: How People Think about the World and Its Alternatives. Oxford University Press, Oxford (2009)
15. Stalnaker, R.: A Theory of Conditional. In: Harper, W.L., Stalnaker, R., Pearce, G. (eds.) Ifs: Conditionals, Belief, Decision, Chance, and Time, pp. 41–55. D. Reidel Publishing Company, Boston (1968)

Tableaux for Single-Agent Epistemic PDL with Perfect Recall and No Miracles

Yanjun Li

Department of Philosophy, Peking University, China
Faculty of Philosophy, University of Groningen, The Netherlands
Y.J.Li@rug.nl

Abstract. Epistemic propositional dynamic logic (EPDL) is a combination of epistemic logic and propositional dynamic logic. The properties, perfect recall and no miracles, capture the interactions between actions and knowledge. In this paper, we present a tableau-based decision procedure for deciding satisfiability of single-agent EPDL with perfect recall and no miracles. We prove the soundness and completeness of the tableau procedure with respect to models with perfect recall and no miracles.

1 Introduction

Temporal epistemic logic (TEL) [5], [10] is a logic for reasoning about information and its development over time by virtue of the combination of temporal and epistemic operators. When we consider reasoning about information and its developing over programs, the natural way is to combine epistemic logic (EL) [9] and propositional dynamic logic (PDL) [6], as in [12,13,14]. In this paper, we call the combination of EL and PDL as epistemic propositional dynamic logic (EPDL).

This paper focus on the single-agent EPDL with perfect recall (PR) and no miracles (NM) [8], [16], which are properties that capture the interactions between actions and knowledge. PR means that for each action a, if executing a at s results in t and the agent can tell s from all such s' that executing a at s' can result in t', the agent then can also tell t from t'. In other words, the agent can tell t from t' since he perfectly remembers that he can tell states resulting in t from states resulting in t'. NM means that if the agent cannot tell s and s' apart, the agent will not be able to distinguish states resulting from executing a at s or s'. Intuitively, the miracle here means that the agent cannot distinguish two states initially but nevertheless he can distinguish the states resulting from executing the same action on these two states.

To motivate EPDL with PR and NM, we observe that it is natural to think of certain knowledge changes over actions in terms of EPDL models with PR and NM. Let us consider the following example (it is a simplified version of the Monty Hall problem). There are three doors: behind one door is a car; behind the others are goats. Initially the agent does not know what is behind the doors. Now an action happens, that is, one door with a goat is opened. Subsequently, the agent

W. van der Hoek et al. (Eds.): LORI 2015, LNCS 9394, pp. 230–242, 2015.
DOI: 10.1007/978-3-662-48561-3_19

Fig. 1. Knowledge development

knows that the car is not behind the opened door, but he still does not know behind which door is the car. We use A to denote that the car is behind the first door, and Ab to denote that the car is behind the first door and the second door is opened. The action a means to open the first door. The others are similar. The model pictured in Figure 1 represents the knowledge development in this example, and it has the properties of PR and NM.

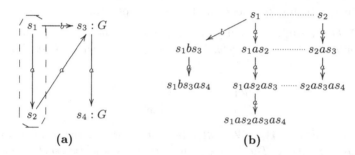

(a)   **(b)**

Fig. 2. Knowledge evolution in conformant planning

Furthermore, EPDL with PR and NM is a natural way to model the conformant planning problems (cf. [18]). Conformant planning is the problem of finding a linear plan (a sequence of actions) that is guaranteed to achieve a goal in presence of uncertainty about the initial state (cf. [15]). Considering the example depicted in Figure 2a, the initial uncertainty set is $\{s_1, s_2\}$ and the goal set is $\{s_3, s_4\}$. A *knowledge state* is a subset of the state space, which records the uncertainty during the execution of a plan, e.g., $\{s_1, s_2\}$ is an initial knowledge state. In order to make sure a goal is achieved eventually, it is crucial to track the transitions of knowledge states during the execution of the plan. Figure 2b, which sketches an EPDL model with PR and NM, displays the knowledge development over actions. From Figure 2b, we can see that the action sequence aa is a solution.[1] In [18], it is shown that the existence of a solution for a conformant planning problem can be expressed in the language of EPDL. Therefore, the satisfiability for EPDL with PR and NM is interesting.

This paper presents a tableau-based decision procedure for deciding satisfiability of single-agent EPDL in models with PR and NM. We build on methods that is developed in [4], [7], [11], [17] and is an adaptation of the method in [1,2] which

---

[1] All sequences b, ba and aaa are not solutions since they are not executable at s_2.

introduced a tableau-based procedure for single-agent linear TEL with perfect recall and no learning. What is new in this paper is 1) we simplify certain algorithms shown in [2]; 2) EPDL has branching actions and this combining with the property NM adds difficulty to the tableau procedure; 3) compared to TEL which has only one kind of temporal transition, *next*, and simpler eventualities, programs in EPDL are structured and can be extremely complicated.

The paper is organized as follows. Section 2 introduces the language, model and semantics of EPDL. Section 3 presents the tableau procedure for EPDL with PR and NM. Section 4 proves the soundness and completeness of the tableau procedure, and we conclude in Section 5 and point to future work.

2   Epistemic Propositional Dynamic Logic

This section will present the language, model and semantics of EPDL, and define the properties of perfect recall and no miracles. The language of EPDL is constructed by combining knowledge and program operators.

Definition 1 (Language). *Let Φ_0 and Π_0 be two countably infinite sets of propositions and actions, respectively. The language of EPDL is defined in BNF as follows:*

$$\phi ::= \top \mid p \mid \neg\phi \mid (\phi \wedge \phi) \mid [\pi]\phi \mid K\phi$$
$$\pi ::= a \mid ?\phi \mid \pi;\pi \mid \pi+\pi \mid \pi^*$$

where $p \in \Phi_0$ and $a \in \Pi_0$. We will often omit parentheses when doing so ought not cause confusion. The language has expressions of two sorts: formulas ϕ and programs π. The set of all programs is denoted Π, and the set of all formulas is denoted Φ. As usual, we use the following abbreviations: $\bot := \neg\top$, $\phi \vee \psi := \neg(\neg\phi \wedge \neg\psi), \phi \to \psi := \neg\phi \vee \psi, \langle a\rangle\phi := \neg[a]\neg\phi, \hat{K}\phi := \neg K\neg\phi$.

Definition 2 (Model). *A model $\mathcal{M}$ is a tuple $\langle S^{\mathcal{M}}, \{R_a^{\mathcal{M}} \mid a \in \Pi_0\}, R^{\mathcal{M}}, V^{\mathcal{M}}\rangle$, where $S^{\mathcal{M}}$ is a nonempty set of states, $R_a^{\mathcal{M}}$ is a binary relation on $S^{\mathcal{M}}$, $R^{\mathcal{M}}$ is an equivalence relation on $S^{\mathcal{M}}$ and $V^{\mathcal{M}} : \Phi_0 \to \mathcal{P}(S^{\mathcal{M}})$ is a function. A pointed model is a pair $(\mathcal{M}, s)$ consisting of a model $\mathcal{M}$ and a state $s \in S^{\mathcal{M}}$.*

Given a model $\mathcal{M}$, we also write $(s,t) \in R_a^{\mathcal{M}}$ as $s \xrightarrow{a}_{\mathcal{M}} t$ or $t \in R_a^{\mathcal{M}}(s)$, and write $(s,t) \in R^{\mathcal{M}}$ as $s \sim_{\mathcal{M}} t$ or $t \in R^{\mathcal{M}}(s)$. If the model $\mathcal{M}$ is obvious from the context, we omit it as an index.

Definition 3 (Semantics). *Given a pointed model $(\mathcal{M}, s)$ and a formula ϕ, we write $\mathcal{M}, s \vDash \phi$ to mean $(\mathcal{M}, s)$ satisfies ϕ. The satisfaction relation $\vDash$ is defined as usual by combining the semantics of EL and that of PDL:*

$$
\begin{array}{ll}
\mathcal{M}, s \vDash \top & \\
\mathcal{M}, s \vDash p & \Longleftrightarrow s \in V(p) \\
\mathcal{M}, s \vDash \neg\phi & \Longleftrightarrow \mathcal{M}, s \nvDash \phi \\
\mathcal{M}, s \vDash \phi \wedge \psi & \Longleftrightarrow \mathcal{M}, s \vDash \phi \text{ and } \mathcal{M}, s \vDash \phi \\
\mathcal{M}, s \vDash K\phi & \Longleftrightarrow s \sim t \text{ implies } \mathcal{M}, t \vDash \phi \\
\mathcal{M}, s \vDash [\pi]\phi & \Longleftrightarrow s \xrightarrow{\pi} t \text{ implies } \mathcal{M}, t \vDash \phi
\end{array}
\qquad
\begin{array}{ll}
\xrightarrow{a} & = R_a^{\mathcal{M}} \\
\xrightarrow{?\phi} & = \{(s,s) \mid \mathcal{M}, s \vDash \phi\} \\
\xrightarrow{\pi_1+\pi_2} & = \xrightarrow{\pi_1} \cup \xrightarrow{\pi_2} \\
\xrightarrow{\pi_1;\pi_2} & = \xrightarrow{\pi_1} \circ \xrightarrow{\pi_2} \\
\xrightarrow{\pi^*} & = (\xrightarrow{\pi})^*.
\end{array}
$$

A formula ϕ is satisfiable if $\mathcal{M}, s \vDash \phi$ for some model $\mathcal{M}$ and a state $s \in S^{\mathcal{M}}$.

Note that each program π can be viewed as a set of *computation sequences*, denoted $\mathcal{L}(\pi)$, which are sequences of actions in Π_0 and formulas in Φ. The sequence set $\mathcal{L}(\pi)$ is as follows.

$$\mathcal{L}(a) = \{a\}$$
$$\mathcal{L}(?\phi) = \{\phi\}$$
$$\mathcal{L}(\pi; \pi') = \{\sigma\eta \mid \sigma \in \mathcal{L}(\pi) \text{ and } \eta \in \mathcal{L}(\pi')\}$$
$$\mathcal{L}(\pi + \pi') = \mathcal{L}(\pi) \cup \mathcal{L}(\pi')$$
$$\mathcal{L}(\pi^*) = \{\epsilon\} \cup \bigcup_{n>0}(\mathcal{L}(\underbrace{\pi; \cdots ; \pi}_{n})) \text{ where } \epsilon \text{ is empty sequence}$$

Proposition 1. *For each model $\mathcal{M}$ and each program $\pi \in \Pi$, we have that* $\xrightarrow{\pi} = \bigcup_{\sigma \in \mathcal{L}(\pi)} \xrightarrow{\sigma}$.

Definition 4 (Properties of model). *A model $\mathcal{M}$ has the property of*

– *Perfect Recall (PR) if for all $a \in \Pi_0$ and all $s, t, t' \in S^{\mathcal{M}}$, $s \xrightarrow{a} t$ and $t \sim t'$ imply that there exists $s' \in S^{\mathcal{M}}$ such that $s \sim s'$ and $s' \xrightarrow{a} t'$.*
– *No Miracles (NM) if for all $a \in \Pi_0$ and all $s, s', t, t' \in S^{\mathcal{M}}$, $s \xrightarrow{a} t$, $s \sim s'$ and $s' \xrightarrow{a} t'$ imply $t \sim t'$.*

Intuitively, PR means that the agent can tell t from t' since he perfectly remembers that he can tell states resulting in t from states resulting in t'. NM means there are no such miracles that the agent cannot distinguish two states initially but nevertheless he can distinguish the states resulting from executing the same action on these two states.

In the rest of this paper, we will always assume that all the models have the properties of PR and NM. Moreover, since it follows by bisimulation invariance that EPDL has tree model property (cf. [3]), we also assume all the models have the property that for each $a \in \Pi_0$, $s \xrightarrow{a} t$ and $s' \xrightarrow{a} t$ imply that $s = s'$. Next we will focus on the problem whether a given formula is satisfiable in models with PR and NM. This problem is tackled by building a tableau from an input formula and deciding the tableau is open or not. Therefore, the key is how to build a proper tableau from an input formula.

3 Tableaux for EPDL with PR and NM

This section will present how to construct a tableau from an input formula. To deal with the complication arising with interacting actions and knowledge, the tableau procedure will act on *bubbles*. A bubble, defined below, is such a set of states that represents a possible epistemic cluster, and it is epistemically sufficient and knowledge-consistent. Eventually, it turns out to be that each relevant formula of the form $\neg K\phi$ will be realized within bubbles. The tableau procedure, then, will only need to focus on the realization of formulas in the shape of $\neg[\pi]\phi$ and the properties of PR and NM. Before the tableau procedure, we introduce some terminologies.

Let $\mathcal{FL}(\phi)$ be the Fisher-Ladner closure generated by ϕ. Note that $\mathcal{FL}(\phi)$ is finite for all $\phi \in \Phi$. Given a formula set Γ, we also use $\mathcal{FL}(\Gamma)$ to mean $\bigcup_{\phi \in \Gamma} \mathcal{FL}(\phi)$. Next we categorise formulas as α/β-formulas, each with two components, as shown in Table 1.

Table 1. α- and β-formulas

α	$\neg\neg\phi$	$\phi \wedge \psi$	$[\pi_1;\pi_2]\phi$	$\neg[\pi_1;\pi_2]\phi$	$\neg[?\psi]\phi$	$[\pi_1+\pi_2]\phi$	$[\pi^*]\phi$	$K\phi$
α_1	$\{\phi\}$	$\{\phi\}$	$\{[\pi_1][\pi_2]\phi\}$	$\{\neg[\pi_1][\pi_2]\phi\}$	$\{\psi\}$	$\{[\pi_1]\phi\}$	$\{\phi\}$	$\{\phi\}$
α_2	$\emptyset$	$\{\psi\}$	$\emptyset$	$\emptyset$	$\{\neg\phi\}$	$\{[\pi_2]\phi\}$	$\{[\pi][\pi^*]\phi\}$	$\emptyset$
β	$\neg(\phi \wedge \psi)$		$\neg[\pi_1+\pi_2]\phi$		$\neg[?\psi]\phi$		$\neg[\pi^*]\phi$	
β_1	$\{\neg\phi\}$		$\{\neg[\pi_1]\phi\}$		$\{\neg\psi\}$		$\{\neg\phi\}$	
β_2	$\{\neg\psi\}$		$\{\neg[\pi_2]\phi\}$		$\{\phi\}$		$\{\phi, \neg[\pi][\pi^*]\phi\}$	

Definition 5 (Fully expanded set). *A finite set Δ of formulas is fully expanded if it satisfies the following conditions.*

- *Δ is not patently inconsistent, i.e. it does not contain both ϕ and $\neg\phi$;*
- *$\alpha \in \Delta$ implies $\alpha_1 \subseteq \Delta$ and $\alpha_2 \subseteq \Delta$;*
- *$\beta \in \Delta$ implies $\beta_1 \subseteq \Delta$ or $\beta_2 \subseteq \Delta$;*
- *For each $K\phi \in \mathcal{FL}(\psi)$ where $\psi \in \Delta$, either $K\phi \in \Delta$ or $\neg K\phi \in \Delta$.*

A fully expanded set will be called a *state*. Given a finite set of formulas Γ, let $\mathsf{S}(\Gamma)$ be the set of states generated by Γ (the procedure are omitted due to the lack of space). When $\Gamma = \{\phi\}$, we also write it as $\mathsf{S}(\phi)$. For each $\Delta \in \mathsf{S}(\Gamma)$, we have that $\Delta \subseteq \mathcal{FL}(\Gamma)$. Given a formula set Δ, let $K(\Delta) = \{K\phi \mid K\phi \in \Delta\}$, $Epi(\Delta) = K(\Delta) \cup \{\neg K\phi \mid \neg K\phi \in \Delta\}$ and $[a]^-(\Delta) = \{\phi \mid [a]\phi \in \Delta\}$.

Definition 6 (Bubble). *Any finite and non-empty set of states is called a pre-bubble. A bubble is a pre-bubble B such that:*

- *B is epistemically sufficient, i.e. for each $\Delta \in B$ and each $\neg K\phi \in \Delta$, there exists a set $\Delta' \in B$ such that $\neg\phi \in \Delta'$;*
- *B is knowledge-consistent, i.e. $K(\Delta) = K(\Delta')$ for all $\Delta, \Delta' \in B$.*

Definition 7 (Satisfiability of a (pre-)bubble). *Given a (pre-)bubble B and a pointed model $(\mathcal{M}, s)$, let $[s]$ be the equivalence class containing s, i.e. $\{s' \in S^{\mathcal{M}} \mid s \sim s'\}$. If there is a surjective function $f : [s] \to B$ such that $\mathcal{M}, s \models f(s)$, we say B is satisfied by $(\mathcal{M}, s)$ and f, written as $\mathcal{M}, s \models^f B$. If there are such a pointed model and such a function, we say B is satisfiable (or sat).*

A tableau $\mathcal{T}$ is a graph with bubbles and labelled edges $\xrightarrow{a}$ with labels from $\mathbf{\Pi_0}$. As mentioned above, a bubble is an epistemic cluster. Let B and B' be two bubbles. Intuitively, $B \overset{a}{\rightarrow} B'$ means that there are $\Delta \in B$ and $\Delta' \in B'$ such that $\Delta \overset{a}{\rightarrow} \Delta'$. Thus, we annotate $\Delta \in B$ with $\Delta' \in B'$, indicating Δ is the a-predecessor of Δ'. The states in the first layer will be annotated with $\emptyset$ since they have no predecessors.

Let B and B' be two bubbles from a tableau such that $B \overset{a}{\rightarrow} B'$. The property of PR requires that each state in B' has an a-predecessor in B. In other words, there must be 'enough' states in B. From the construction of the tableau below, we will see that each state $\Delta' \in B'$ is annotated with a state $\Delta \in B$ and $[a]^{-}(\Delta) \subseteq \Delta'$. The property of NM requires that all formulas of the form $\neg[a]\phi$ from B will be realized in B', which is also guaranteed from the construction of the tableau.

The tableau procedure for EPDL with PR and NM consists two phases: a construction phase and an elimination phase. In the construction phase, a pre-tableau with pre-bubbles and bubbles is established. In the elimination phase, the pre-tableau is pruned to an initial tableau and then a final tableau.

3.1   Construction of the Pre-tableau

Each state $\Delta \in \mathsf{S}(\phi)$ is a potential expansion of ϕ. For any state belonging to the same epistemic cluster as Δ, $Epi(\Delta)$ is the minimal set of formulas that it contains, so there exists $P \subseteq \mathsf{S}(Epi(\Delta))$ such that $\{\Delta\} \cup P$ has enough states. The intuition of the construction from an input ϕ is that initially we make $\{\Delta\} \cup P$ a pre-bubble for each $\Delta \in \mathsf{S}(\phi)$ and each $P \subseteq \mathsf{S}(Epi(\Delta))$. To continue the construction, we need to realize formulas of the forms $\neg K\psi$ and $\neg[a]\psi$ and keep the properties of PR and NM. The procedure EXTENDTOBUBBLES extends a pre-bubble to all potential bubbles, and this procedure realizes all $\neg K\psi$ in a bubble. The procedure SUCCESSORPREBUBBLES realizes $\neg[a]\psi$ in a bubble. What is more, PR and NM are guaranteed in these procedures.

The construction of the pre-tableau for ϕ works as follows.

1. For each $\Delta \in \mathsf{S}(\phi)$ and each $P \subseteq \mathsf{S}(Epi(\Delta))$, make the pre-bubble $\{\Delta\} \cup P$ as a node and annotate each state in it with $\emptyset$.
2. Expand each pre-bubble A into bubbles by calling the procedure EXTEND-TOBUBBLES(A), which is presented in Algorithm 3. For each bubble $B' \in$ EXTENDTOBUBBLES(A), add B' as a node if B' is not already there. We then produce an arrow $A \dashrightarrow B'$.
3. For each bubble B, if there exists a formula $\neg[a]\psi \in \Delta$ for some $\Delta \in B$ and some $a \in \mathbf{\Pi_0}$, produce successor pre-bubbles by calling the procedure SUCCESSORPREBUBBLES(B, a), which is presented in Algorithm 4. For each pre-bubble $A \in$ SUCCESSORPREBUBBLES(B, a), add A as a node if A is not already there. We then produce an arrow $B \overset{a}{\rightarrow} A$.
4. Repeat steps 2 and 3 until no new bubbles or pre-bubbles are created.

Since each bubble or pre-bubble in the construction is a subset of $\mathcal{P}(\mathcal{FL}(\phi))$, the pre-tableau for ϕ is a finite graph, and the construction of the pre-tableau will terminate. The following are all the procedures, which are adaptations from [2], and we will also show that each procedure functions well.

KNOWLEDGECONSISTENT. The procedure KNOWLEDGECONSISTENT takes a pre-bubble A as input and returns a set of knowledge-consistent pre-bubbles. To make the pre-bubble A knowledge-consistent, it seems that each $\Delta \in A$ should be extended to $\Delta \cup \Sigma$ for some $\Sigma \in \mathsf{S}(Epi(A) \setminus Epi(\Delta))$. However, this is not sufficient to preserve the satisfiability. For example, let $\mathcal{M}, s \models^f A$ with $[s] = \{s, s', t\}$, $A = \{\Delta, \Delta'\}$ and $f = \{s, s' \mapsto \Delta, t \mapsto \Delta'\}$. Let $Epi(A) = Epi(\Delta')$. If $\mathcal{M}, s \models \Sigma$ and $\mathcal{M}, s' \models \Sigma'$ for some different states $\Sigma, \Sigma' \in \mathsf{S}(Epi(A) \setminus Epi(\Delta))$, then to make A knowledge-consistent and preserve the satisfiability of A, we need to extend A to $A' = \{\Delta \cup \Sigma, \Delta \cup \Sigma', \Delta'\}$. For Algorithm 1, it follows that A' is in the returned set. We then can have $\mathcal{M}, s \models^{f'} A'$ with $f' = \{s \mapsto \Delta \cup \Sigma, s' \mapsto \Delta \cup \Sigma', t \mapsto \Delta'\}$.

The algorithm is presented in Algorithm 1. It can be shown that each pre-bubble A' returned by KNOWLEDGECONSISTENT(A) is knowledge-consistent and that A is sat implies some returned pre-bubble is also sat.

Algorithm 1. Making a pre-bubble knowledge-consistent

  input : Pre-bubble A
  output: A set of pre-bubbles

1 **Procedure** KNOWLEDGECONSISTENT(A)
2 | Let $\Delta_1, \cdots, \Delta_n$ be all the sets in A where $n = |A|$;
3 | $epi \leftarrow Epi(A)$;
4 | $\Gamma \leftarrow \{K\phi \in K(\Delta_i) \mid 1 \le i \le n\}$;
5 | **if** $K(\Delta_i) = \Gamma$ *for each* $1 \le i \le n$ **then return** A;
6 | **else**
7 | | **for** $i \leftarrow 1$ *to* n **do**
8 | | | $epi' \leftarrow epi \setminus Epi(\Delta_i)$;
9 | | | $Alt_i \leftarrow \{\Delta_i \cup \Sigma \mid \Sigma \in \mathsf{S}(epi'), \Delta_i \cup \Sigma$ is not patently inconsistent.$\}$;
   | | | /* We annotate $\Delta_i \cup \Sigma$ with the annotation of Δ_i. */
10 | | | **if** $Alt_i = \emptyset$ **then return** $\emptyset$;
11 | $Q \leftarrow \emptyset$;
12 | Let $P(Alt_i)$ be the set of all non-empty subsets of Alt_i for each $1 \le i \le n$;
13 | **foreach** $(D_1, \cdots, D_n) \in P(Alt_1) \times \cdots \times P(Alt_n)$ **do**
14 | | Add $\bigcup\{D_1, \cdots, D_n\}$ to Q;
15 | **return** Q;

EPISTEMICALLYSUFFICIENT. The procedure of EPISTEMICALLYSUFFICIENT takes a pre-bubble A as input and returns a set of epistemically sufficient pre-bubbles. To make A epistemically sufficient, it seems that for each unrealized $\neg K\phi$ from A, we need to extend A by adding some $\Sigma \in \mathsf{S}(\neg\phi)$. However, there are several problems.

Firstly, if A is knowledge-consistent, then to preserve the knowledge consistence, it seems that we should add some $\Sigma \in \mathsf{S}(Epi(A) \cup \{\neg\phi\})$ to A. Secondly, if $B \xrightarrow{a} A$ it follows by PR that Σ needs to have an a-predecessor $\Delta \in B$. This means we should add $\Delta' \cup \Sigma$ to A for certain $\Delta' \in \mathsf{S}([a]^-(\Delta))$. We assume that

A has enough states, which means that all such Δ' are already in A. Therefore, if we want to extend A with some $\Sigma \in \mathsf{S}(Epi(A) \cup \{\neg\phi\})$, it seems that we only need to add $\Delta' \cup \Sigma$ to A for certain $\Delta' \in A$.

Finally, if there are more than one unrealized formulas $\neg K\phi_1 \cdots \neg K\phi_n$ from A such that they are realized by the same state, then to make A epistemically sufficient and preserve the satisfiability of A, we should add $\Delta' \cup \Sigma$ for some $\Delta' \in A$ and some $\Sigma \in \mathsf{S}(Epi(A) \cup \{\neg\phi_1 \cdots \neg\phi_2\})$. For example, let $\mathcal{M}, s \models^f A$ with $[s] = \{s, s'\}$, $A = \{\Delta, \Delta'\}$ and $f = \{s \mapsto \Delta, s' \mapsto \Delta'\}$. Let A be knowledge-consistent, and there are formulas $\neg K\phi_1 \in \Delta$, $\neg K\phi_2 \in \Delta'$ which are unrealized in A. If $\mathcal{M}, s \models \neg\phi_1 \wedge \neg\phi_2$ and $\mathcal{M}, s' \models \phi_1 \wedge \phi_2$, we have $\mathcal{M}, s \models \Sigma$ for some $\Sigma \in \mathsf{S}(Epi(A) \cup \{\neg\phi_1, \neg\phi_2\})$. To make the pre-bubble A epistemically sufficient and preserve the satisfiability of A, we need to extend A to $A' = \{\Delta \cup \Sigma, \Delta'\}$. From Algorithm 2, it follows that A' is in the returned set. We then have $\mathcal{M}, s \models^{f'} A'$ with $f' = \{s \mapsto \Delta \cup \Sigma, s' \mapsto \Delta'\}$.

The algorithm is presented in Algorithm 2. It can be shown that A is knowledge-consistent implies each pre-bubble A' returned by EPISTEMICALLY-SUFFICIENT(A) is a bubble and that A is sat implies some returned bubble is also sat.

EXTENDTOBUBBLES. The procedure EXTENDTOBUBBLES(A), which is presented in Algorithm 3, extends a pre-bubble A to bubbles. We firstly make the pre-bubble A knowledge-consistent (KNOWLEDGECONSISTENT(A)) and then make each knowledge-consistent pre-bubble $A' \in$ KNOWLEDGECONSISTENT(A) epistemically sufficient (EPISTEMICALLYSUFFICIENT(A')). The properties of the procedures KNOWLEDGECONSISTENT and EPISTEMICALLYSUFFICIENT guarantee that each $B \in$ EXTENDTOBUBBLESA is a bubble extension of A and that A is sat implies some returned bubble is also sat.

SUCCESSORPREBUBBLES. The procedure SUCCESSORPREBUBBLES takes a bubble B and an action $a \in \mathbf{\Pi_0}$ as input and returns a set of a-successor prebubbles for B. If a pre-bubble A is returned, it follows by NM that all $\neg[a]\phi$ from B is realized in A. Similar to the case in EPISTEMICALLYSUFFICIENT, formulas $\neg[a]\phi_1 \cdots \neg[a]\phi_n \in \Delta \in B$ might be realize by the same state, so there might be state $\Sigma \in \mathsf{S}([a]^-(\Delta) \cup \{\neg\phi_1 \cdots \neg\phi_n\})$ in some returned successor pre-bubble.

Furthermore, the property PR requires that each epistemic cluster has enough states. Step 1 of the construction of the pre-tableau guarantees that certain bubble in the first layer has enough states. To preserve PR, our strategy is that if the bubble B has enough states and $\{\Omega_1, \cdots, \Omega_k\}$ is the state set realizing all $\neg[a]\phi$ from B, let the pre-bubble $\{\Omega_1, \cdots, \Omega_k\} \cup P$ is in the returned set for each $P \subseteq H$ where $H = \bigcup_{\Delta \in B} \mathsf{S}([a]^-(\Delta))$. This guarantees certain returned pre-bubble has enough states.

The algorithm is presented in Algorithm 4. It can be shown that, for each $A \in$ SUCCESSORPREBUBBLESB, all $\neg[a]\phi$-formulas from states in B is realized in A and the annotation of each state in A is some state in B. What is more, if B is sat then some returned A is also sat.

Algorithm 2. Making a pre-bubble epistemically sufficient

---

input : Pre-bubble A
output: A set of bubbles

1 **Procedure** EPISTEMICALLYSUFFICIENT(A)
2    $epi \leftarrow Epi(A)$;
3    $\Gamma \leftarrow \{\neg K\phi \in epi \mid \neg\phi \notin \Delta \text{ for each } \Delta \in A\}$;
4   **if** $\Gamma = \emptyset$ **then return** A;
5   **else**
6   Let $\neg K\phi_1, \cdots, \neg K\phi_n$ be the formulas in Γ where $n = |\Gamma|$;
7   **for** $i \leftarrow 1$ *to* n **do**
8        $Alt_i \leftarrow \{\Delta \cup \Sigma \mid \Sigma \in \mathsf{S}(epi \cup \{\neg\phi_i\}), \Delta \in A \text{ and } \Delta \cup \Sigma \text{ is not}$
          patently inconsistent.$\}$;   /* We annotate $\Delta \cup \Sigma$ with the
          annotation of Δ. */
9   **if** $Alt_i = \emptyset$ **then return** $\emptyset$;
10    $EAlt_i \leftarrow Alt_i$ for each $1 \le i \le n$;
11   **foreach** $(\Delta_1 \cup \Sigma_1, \cdots, \Delta_n \cup \Sigma_n) \in Alt_1 \times \cdots \times Alt_n$ **do**
12   **foreach** $\{i_1, \cdots, i_k\} \subseteq \{1, \cdots, n\}$ **do**
13   **if** *all* Δ_{i_j} *are the same where* $1 \le j \le k$ *and* $\Delta_{i_1} \cup \Sigma_{i_1} \cdots \Sigma_{i_k}$ *is not*
         patently inconsistent **then**
14          $\Omega \leftarrow \Delta_{i_1} \cup \Sigma_{i_1} \cdots \Sigma_{i_k}$;   /* Annotate Ω with the annotation
          of Δ_{i_1}. */
15   Add Ω to $EAlt_{i_1} \cdots EAlt_{i_k}$
16    $Q \leftarrow \emptyset$;
17   **foreach** $(\Omega_1, \cdots, \Omega_n) \in EAlt_1 \times \cdots \times EAlt_n$ **do**
18   **foreach** $P \subseteq A$ **do**
19        $A' \leftarrow P \cup \{\Omega_i \mid 1 \le i \le n\}$;
20   Add A' to Q;
21   **return** Q;

Algorithm 3. Expanding a pre-bubble to bubbles

---

input : A pre-bubble A
output: A set of bubbles

1 **Procedure** EXTENDTOBUBBLES(A)
2    $Q \leftarrow \emptyset$;
3   **foreach** $A' \in$ KNOWLEDGECONSISTENT(A) **do**
4   **foreach** $A'' \in$ EPISTEMICALLYSUFFICIENT(A') **do**
5   Add A'' to Q;
6   **return** Q;

Algorithm 4. Making successor pre-bubbles for a bubble

---

input : A bubble B and an action a

output: A set of pre-bubbles

1   **Procedure** SUCCESSORPREBUBBLES(B, a)

2   Let $\Delta_1, \cdots, \Delta_n$ be all the sets in B such that there exists a $\neg[a]$-formula in Δ_i for each $1 \leq i \leq n$;

3   Let $\neg[a]\phi_{i_1}, \cdots, \neg[a]\phi_{i_m}$ be all the $\neg[a]$-formulas in Δ_i for each $1 \leq i \leq n$;

4   **for** $i \leftarrow 1$ *to* n **do**

5   **for** $j \leftarrow 1$ *to* m **do**

6            $Alt_{i,j} \leftarrow \{\Delta' \cup \Gamma \mid \Gamma \in \mathsf{S}(\neg\phi_{i_j}), \Delta' \in \mathsf{S}([a]^-(\Delta_i)), \Delta' \cup \Gamma$ is not patently inconsistent.$\}$;   /* Annotate $\Delta' \cup \Gamma$ with Δ_i. */

7   **if** $Alt_{i,j} = \emptyset$ **then return** $\emptyset$;

8      $h \leftarrow 1_m + \cdots + n_m$; /* Then we can arrange all Alt as $Alt_1 \cdots Alt_h$. */

9      $EAlt_i \leftarrow Alt_i$ for each $1 \leq i \leq h$;

10   **foreach** $(\Delta'_1 \cup \Gamma_1, \cdots, \Delta'_h \Gamma_h) \in Alt_1 \times \cdots \times Alt_h$ **do**

11   **foreach** $\{i_1, \cdots, i_k\} \subseteq \{1, \cdots, h\}$ **do**

12   **if** *all* Δ'_{i_j} *are the same where* $1 \leq j \leq k$ *and* $\Delta_{i_1} \cup \Gamma_{i_1} \cdots \Gamma_{i_k}$ *is not patently inconsistent* **then**

13              $\Omega \leftarrow \Delta'_{i_1} \cup \Gamma_{i_1} \cdots \Gamma_{i_k}$;   /* Annotate Ω with the annotation of Δ_{i_1}. */

14   Add Ω to $EAlt_{i_1} \cdots EAlt_{i_k}$

15   Let $B = \{\Delta_1 \cdots, \Delta_k\}$ where $k = |B|$;

16   **for** $l \leftarrow 1$ *to* k **do**

17   **if** $[a]^-(\Delta_l) = \emptyset$ **then**

18            $AAlt_l = \{\{\top\}\}$;   /* Annotate $\{\top\}$ with Δ_l. */

19   **else**

20            $AAlt_l \leftarrow \{\mathsf{S}([a]^-(\Delta_l))\}$;   /* Annotate each state in $AAlt_l$ with Δ_l. */

21      $H \leftarrow AAlt_1 \cup \cdots \cup AAlt_k$;

22      $Q \leftarrow \emptyset$;

23   **foreach** $(\Omega_1, \cdots, \Omega_h) \in EAlt_1 \times \cdots \times EAlt_h$ **do**

24   **foreach** $P \subseteq H$ **do**

25   Add $\{\Omega_1, \cdots, \Omega_h\} \cup P$ to Q;

26   **return** Q;

3.2   Construction of the Initial and Final Tableau

From the construction of the pre-tableau, we know that if the pre-bubble A is an a-successor of a bubble B for some $a \in \Pi_0$ and the bubble B' is a bubble-extension of A, namely $B \xrightarrow{a} A \dashrightarrow B'$, then the annotation of each state $\Delta' \in B'$ is some state $\Delta \in B$. The initial tableau is produced from the pre-tableau by removing the pre-bubbles and redirecting the arrows. For example, if $B \xrightarrow{a} A \dashrightarrow B'$, we then delete the pre-bubble A and add an a-arrow from B to B', namely $B \xrightarrow{a} B'$.

To check the realization of the formula of the form $\neg[\pi]\phi$, we add action arrows between states in two successive bubbles, and add reflective ψ-arrows at states.

For example, if $B \xrightarrow{a} B'$ and the annotation of the state $\Delta' \in B$ is some state $\Delta \in B$ then we add an a-arrow $\Delta \dashrightarrow^{a} \Delta'$. For each bubble B and each state $\Delta \in B$, if $\phi \in \Delta$ then we add an arrow $\Delta \dashrightarrow^{\phi} \Delta$

Definition 8 (State path). *Given a tableau* $\mathcal{T}$*, a state sequence* $(\Delta_i)_{0 \leq i \leq n}$ *and a computation sequence* $(\alpha_i)_{1 \leq i \leq n}$*, we say* $(\Delta_{i-1} \alpha_i \Delta_i)_{1 \leq i \leq n}$ *is a* state path *from* $\mathcal{T}$ *if there is a bubble sequence* $(B_i)_{0 \leq i \leq n}$ *such that 1)* $\Delta_i \in B_i$*; 2)* $B_{i-1} \xrightarrow{\alpha_i} B_i$ *if* $\alpha_i \in \Pi_0$*; 3)* $\Delta_{i-1} \dashrightarrow^{\alpha_i} \Delta_i$*. Especially,* (Δ) *is also a state path for each* $\Delta \in B \in \mathcal{T}$*.*

Definition 9 (Realization). *Given a tableau* $\mathcal{T}$*, a state path* $(\Delta_{i-1} \alpha_i \Delta_i)_{1 \leq i \leq n}$ *from* $\mathcal{T}$ *and* $\neg[\pi]\phi \in \Delta \in B$*, we say* $(\Delta_{i-1} \alpha_i \Delta_i)_{1 \leq i \leq n}$ realizes $\neg[\pi]\phi \in \Delta \in B$ *if* $\Delta_0 = \Delta$*,* $\neg\phi \in \Delta_n$ *and* $(\alpha_i)_{1 \leq i \leq n} \in \mathcal{L}(\pi)$*.*

The final tableau $\mathcal{T}$ is obtained by the following procedure.

1. Let $\mathcal{T}_0$ be the initial tableau.
2. If there is $\neg[\pi]\phi \in \Delta \in B \in \mathcal{T}_n$ such that it cannot be realized by any state path from $\mathcal{T}_n$, then let $\mathcal{T}_{n+1} = \mathcal{T}_n \setminus \{B\}$.
3. Repeat steps 2 until no bubbles are deleted, i.e. $\mathcal{T}_{n+1} = \mathcal{T}_n$.

Definition 10. *The final tableau* $\mathcal{T}$ *for* ϕ *is* open *if there is a bubble B in* $\mathcal{T}$ *and* $\Delta \in B$ *such that* $\phi \in \Delta$*.*

4   Soundness and Completeness

In this section, we will show that the tableau constructed in the previous section is proper, i.e. it is sound and complete. Due to the lack of space, we omit the proofs. We leave them to a complete version of the paper.

Theorem 1 (Soundness). *If* ϕ *is sat, the final tableau* $\mathcal{T}$ *for* ϕ *is open.*

The sketch of the proof is that if the input formula ϕ is sat, then there is also a satisfiable state $\Delta \in \mathsf{S}(\phi)$. Consequently there is a satisfiable bubble B in the initial tableau such that $\Delta \in B$. The nutshell is to show the bubble B will survive in the final tableau. The idea is to prove that all satisfiable bubbles will survive in the final tableau.

Theorem 2 (Completeness). *If the final tableau* $\mathcal{T}$ *for* ϕ *is open,* ϕ *is sat.*

To show the completeness, we need to construct a model on which ϕ is satisfied if the final tableau for ϕ is open. The sketch is that we derive a deterministic tree of bubbles from the open final tableau such that each formula of the form $\neg[\pi]\phi$ is realized on the tree. We then can show that this deterministic bubble tree is a model with PR and NM and that ϕ is satisfied on it.

5 Conclusion

This paper extended and adapted the incremental tableau procedure sketched in [1,2] to work for EPDL with the properties of PR and NM which capture the interactions between knowledge and actions. Therefore, this paper developed a practically implementable method for deciding the satisfiability of EPDL formula in models with PR and NM. For future directions, we could extend and adapt this method for the other combination of the interaction properties, such as perfect recall, no miracles and no learning. The other direction is to investigate the complexity of the satisfiability in EPDL with these properties.

Acknowledgements. The author acknowledges Barteld Kooi, Jan Albert van Laar and Yanjing Wang for their valuable comments which helped to improve the paper, and also thanks the anonymous reviewers for their comments. The work was supported by China Scholarship Council.

References

1. Ajspur, M., Goranko, V.: Tableaux-based decision method for single-agent linear time synchronous temporal epistemic logics with interacting time and knowledge. In: Lodaya, K. (ed.) Logic and Its Applications. LNCS, vol. 7750, pp. 80–96. Springer, Heidelberg (2013)
2. Ajspur, M.L.: Tableau-based decision procedures for epistemic and temporal epistemic logics. PhD thesis, Doctoral School of Communication, Business and Information Technologies, Roskilde University, Denmark (October 2013)
3. Blackburn, P., de Rijke, M., de Venema, Y.: Modal Logic. Cambridge University Press, New York (2001)
4. Cerrito, S., David, A., Goranko, V.: Optimal tableaux-based decision procedure for testing satisfiability in the alternating-time temporal logic ATL+. In: Demri, S., Kapur, D., Weidenbach, C. (eds.) IJCAR 2014. LNCS, vol. 8562, pp. 277–291. Springer, Heidelberg (2014), doi:10.1007/978-3-319-08587-6-21
5. Fagin, R., Halpern, J., Moses, Y., Vardi, M.: Reasoning about knowledge. MIT Press, Cambridge (1995)
6. Fischer, M.J., Ladner, R.E.: Propositional dynamic logic of regular programs. Journal of Computer and System Sciences 18(2), 194–211 (1979)
7. Goranko, V., Shkatov, D.: Tableau-based decision procedures for logics of strategic ability in multiagent systems. ACM Transactions on Computational Logic 11(1) (2009)
8. Halpern, J.Y., van der Meyden, R., Vardi, M.Y.: Complete axiomatizations for reasoning about knowledge and time. SIAM Journal on Computing 33(3), 674–703 (2004)
9. Hintikka, J.: Knowledge and belief: an introduction to the logic of the two notions. Cornell University Press (1962)
10. Parikh, R., Ramanujam, R.: A knowledge based semantics of messages. Journal of Logic, Language and Information 12(4), 453–467 (2003)
11. Pratt, V.R.: A practical decision method for propositional dynamic logic: Preliminary report. In: Lipton, R.J., Burkhard, W.A., Savitch, W.J., Friedman, E.P., Aho, A.V. (eds.) Proceedings of the 10th Annual ACM Symposium on Theory of Computing, San Diego, California, USA, May 1-3, pp. 326–337. ACM (1978)

12. Schmidt, R.A., Tishkovsky, D.: Combining dynamic logic with doxastic modal logics. In: Balbiani, P., Suzuki, N.-Y., Wolter, F., Zakharyaschev, M. (eds.) Advances in Modal Logic 4, Papers from the Fourth Conference on "Advances in Modal logic," held in Toulouse (France), pp. 371–392. King's College Publications (October 2002)

13. Schmidt, R.A., Tishkovsky, D.: On combinations of propositional dynamic logic and doxastic modal logics. Journal of Logic, Language and Information 17(1), 109–129 (2008)

14. Schmidt, R.A., Tishkovsky, D., Hustadt, U.: Interactions between Knowledge, Action and Commitment within Agent Dynamic Logic. Studia Logica 78(3), 381–415 (2004)

15. Smith, D.E., Weld, D.S.: Conformant graphplan. In: Mostow, J., Rich, C. (eds.) Proceedings of the Fifteenth National Conference on Artificial Intelligence and Tenth Innovative Applications of Artificial Intelligence Conference, AAAI 1998, IAAI 1998, Madison, Wisconsin, USA, July 26-30, pp. 889–896. AAAI Press/The MIT Press (1998)

16. van Benthem, J., Gerbrandy, J., Hoshi, T., Pacuit, E.: Merging frameworks for interaction. Journal of Philosophical Logic 38(5), 491–526 (2009)

17. Wolper, P.: The tableau method for temporal logic: an overview. Logique et Analyse 28(110-111), 119–136 (1985)

18. Yu, Q., Li, Y., Wang, Y.: A dynamic epistemic framework for conformant planning. In: Ramanujam, R. (ed.) Proceedings of the 15th Conference on Theoretical Aspects of Rationality and Knowledge, TARK 2015, Pittsburgh, USA, June 4-6, pp. 249–259 (2015)

Formulating Semantics of Probabilistic Argumentation by Characterizing Subgraphs

Beishui Liao[1,2] and Huaxin Huang[1]

[1] Center for the Study of Language and Cognition,
Zhejiang University, Hangzhou 310028, P.R. China
[2] University of Luxembourg, Luxembourg

Abstract. The existing approaches to formulate the semantics of probabilistic argumentation are based on the notion of possible world. Given a probabilistic argument graph with n nodes, 2^n subgraphs are constructed and their extensions under a given semantics are computed. Then, the probability of a set of arguments E being an extension is equal to the sum of the probabilities of all subgraphs each of which has the extension E. Since in many cases, computing the extensions of a subgraph is computationally expensive, these approaches are fundamentally inefficient or infeasible. In order to cope with this problem, the present paper proposes a novel approach to formulate the semantics of probabilistic argumentation by charactering subgraphs w.r.t. an extension. The results show that under some semantics (admissible, complete, stable), the probability of a set of arguments E being an extension can be obtained without computing the extensions of subgraphs, while under some other semantics (preferred, grounded), only partial computation of extensions is needed.

Keywords: Probabilistic Argumentation, Semantics, Computational Complexity, Computational Efficiency, Characterized Subgraphs.

1 Introduction

In the past two decades, argumentation has been a very active research area in the field of knowledge representation and reasoning, as a nonmonotonic formalism to handle inconsistent and incomplete information by means of constructing, comparing and evaluating arguments. In 1995, Dung proposed a notion of abstract argumentation framework [1], which can be viewed as a directed graph (called *argument graph*, or defeat graph) $G = (A, R)$, in which A is a set of arguments and $R \subseteq A \times A$ is a set of attacks. Given an argument graph, a fundamental problem is to determine which arguments can be regarded as justified. According to [1], extension-based semantics is a formal way to answer this question. Here, an extension represents a set of arguments that are considered to be acceptable (i.e. able to survive the conflict) together, under a certain semantics which is defined according to a set of evaluation criteria [2]. Dung's abstract argumentation theory lays a concrete foundation for the development of various argument systems.

© Springer-Verlag Berlin Heidelberg 2015
W. van der Hoek et al. (Eds.): LORI 2015, LNCS 9394, pp. 243–254, 2015.
DOI: 10.1007/978-3-662-48561-3_20

However, in classical argumentation theory, the uncertainty of arguments and/or attacks is not considered. So, it could be regarded as a purely qualitative formalism. But, in the real world, arguments and/or attacks are often uncertain. So, in recent years, the importance of combining argumentation and uncertainty has been well recognized, and probability-based argumentation is gaining momentum [3–7]. In a *probabilistic argument graph* (or PrAG in brief), each argument is assigned with a probability, denoting the likelihood of the argument appearing in the graph[1]. Similar to classical argumentation theory, given a PrAG, a basic problem is to define the status of arguments. The existing approaches are based on the notion of possible world [3–5, 8]. Given a PrAG with n nodes, 2^n subgraphs are constructed (each subgraph corresponds to a possible world of arguments appearing in the graph). Then, the extensions of each subgraph is computed according to classical argumentation semantics. Since in many cases, computing the extensions of a subgraph is computationally expensive, these approaches are fundamentally inefficient or infeasible [9, 10]. This gives rise to the following research problem:

Research Problem. How to formulate the semantics of probabilistic argumentation (i.e., the probability of a set of arguments being an extension), such that the computation of extensions of subgraphs can be avoided or decreased?

In order to cope with this problem, the present paper proposes a novel approach to formulate the semantics of probabilistic argumentation by *characterizing subgraphs w.r.t. an extension*. In this approach, the probability of a set of arguments E being an extension is obtained by identifying a set of subgraphs that have the extension E according to some conditions, rather than by directly computing the extensions of all subgraphs.

The rest of this paper is organized as follows. In Section 2, we review the notions of abstract argumentation and probabilistic abstract argumentation. In Section 3, we introduce a novel notion: characterized subgraphs w.r.t. an extension, with specific definitions and properties. In Section 4, semantics of probabilistic argumentation is introduced on the basis of the notion of characterized subgraphs. Then, in Section 5, we conclude the paper and point out some future work.

2   Preliminaries

2.1   Classical Abstract Argumentation

The notions of (classical) abstract argumentation are originally introduced in [1], including abstract argumentation framework (called *argument graph*, or *classical argument graph*, in this paper) and extension-based semantics.

---

[1] A probabilistic argument graph can be defined by assigning probabilities to arguments [3, 4, 8], or attacks [7], or both arguments and attacks [5]. For simplicity, in this paper, we only consider the probabilistic argument graph in which only arguments are associated with probabilities.

An argument graph is a directed graph $G = (A, R)$, in which A is a set of nodes representing arguments and R is a set of edges representing attacks between the arguments.

Definition 1. *An argument graph is a tuple* $G = (A, R)$, *where A is a set of arguments, and $R \subseteq A \times A$ is a set of attacks. For convenience, sometimes we use $args(G)$ to denote A.*

As usual, we say that argument $\alpha \in A$ attacks argument $\beta \in A$ iff $(\alpha, \beta) \in R$. If $E \subseteq A$ and $\alpha \in A$ then we say that α attacks E iff there exists $\beta \in E$ such that α attacks β, that E attacks α iff there exists $\beta \in E$ such that β attacks α, and that E attacks E' iff there exist $\beta \in E$ and $\alpha \in E'$ such that β attacks α. Given $G = (A, R)$, for $\alpha \in A$ we write α_G^- for $\{\beta \mid (\beta, \alpha) \in R\}$; for $E \subseteq A$ we write E_G^- for $\{\beta \mid \exists \alpha \in E : (\beta, \alpha) \in R\}$ and E_G^+ for $\{\beta \mid \exists \alpha \in E : (\alpha, \beta) \in R\}$. Formally, we have the following formulas.

$$\alpha_G^- = \{\beta \mid (\beta, \alpha) \in R\} \tag{1}$$
$$E_G^- = \{\beta \mid \exists \alpha \in E : (\beta, \alpha) \in R\} \tag{2}$$
$$E_G^+ = \{\beta \mid \exists \alpha \in E : (\alpha, \beta) \in R\} \tag{3}$$

If without confusion, we write α^-, E^- and E^+ for α_G^-, E_G^- and E_G^+ respectively.

Given an argument graph, according to certain evaluation criteria, sets of arguments (called *extensions*) are identified as acceptable together. Two important notions for the definitions of various kinds of extensions are *conflict-freeness* and *acceptability* of arguments.

Definition 2. *Let $G = (A, R)$ be an argument graph, and $E \subseteq A$ be a set of arguments.*

- *E is* conflict-free *iff $\nexists \alpha, \beta \in E$, such that $(\alpha, \beta) \in R$.*
- *An argument $\alpha \in A$ is* acceptable *w.r.t. (defended by) E, iff $\forall (\beta, \alpha) \in R$, $\exists \gamma \in E$, such that $(\gamma, \beta) \in R$.*

Based on the above two notions, several classes of (classical) extensions can be defined as follows.

Definition 3. *Let $G = (A, R)$ be an argument graph, and $E \subseteq A$ a set of arguments.*

- *E is* admissible *iff E is conflict-free, and each argument in E is acceptable w.r.t. E.*
- *E is* preferred *iff E is a maximal (w.r.t. set-inclusion) admissible set.*
- *E is* complete *iff E is admissible, and each argument that is acceptable w.r.t. E is in E.*
- *E is* grounded *iff E is the minimal (w.r.t. set-inclusion) complete extension.*
- *E is* stable *iff E is conflict-free, and each argument in $A \backslash E$ is attacked by E.*

In this paper, for convenience, we use $\sigma \in \{ad, co, pr, gr, st\}$ to represent a semantics (admissible, complete, preferred, grounded or stable). An extension under semantics σ is called a σ-extension. The set of σ-extensions of G is denoted as $\mathcal{E}_\sigma(G)$. In $G = (A, R)$, if $A = R = \varnothing$, then $\mathcal{E}_\sigma(G) = \{\varnothing\}$.

Example 1. Let $G_1 = (A_1, R_1)$ be an argument graph illustrated as follows.

$$a \longleftrightarrow b \longrightarrow c \longleftrightarrow d \circlearrowleft$$

According to Definition 3, G_1 has four admissible sets: $\varnothing$, $\{a\}$, $\{b\}$ and $\{a, c\}$, in which $\varnothing$, $\{b\}$ and $\{a, c\}$ are complete extensions, $\{b\}$ and $\{a, c\}$ are preferred extensions, $\{a, c\}$ is the only stable extension, $\varnothing$ is the unique grounded extension.

2.2 Probabilistic Abstract Argumentation

The notions of probabilistic abstract argumentation are defined by combining the notions of classical abstract argumentation and that of probabilistic theory, including probabilistic argument graph and its semantics.

According to [8], we have the following definition.

Definition 4. *A probabilistic argument graph (or PrAG for short) is a triple $G^p = (A, R, p)$ where $G = (A, R)$ is an argument graph and $p : A \to [0, 1]$ is a probability function assigning to every argument $\alpha \in A$ a probability $p(\alpha)$ that α appears (and hence a probability $1 - p(\alpha)$ that α does not).*

In existing literature, the semantics of a PrAG is defined according to the notion of possible world. Given a PrAG, a possible world represents a scenario consisting of some subset of the arguments and attacks in the graph. So, given a PrAG with n nodes, there are 2^n subgraphs. A subgraph induced by a set $A' \subseteq A$ is represented as $G' = (A', R')$, in which $R' = R \cap (A' \times A')$. Under a semantics $\sigma \in \{ad, co, pr, gr, st\}$, the extensions of each subgraph are computed according to the definition of classical argumentation semantics. Then, the probability that a set of arguments $E \subseteq A$ is a σ-extension, denoted as $p(E^\sigma)$, is the sum of the probability of each subgraph for which E is a σ-extension.

In order to calculate the probability of each subgraph, it is desirable to assume independence of arguments. In [8], the reason why independence can be assumed is provided. For an argument α in a graph G^p, $p(\alpha)$ is treated as the probability that α is a justified point (i.e. each is a self-contained, internally valid, contribution) and therefore should appear in the graph, and $1 - p(\alpha)$ is the probability that α is not a justified point and so should not appear in the graph. So, one may assume that the probability of one argument appearing in a graph is independent of the probability of some other arguments appearing.

Throughout this paper, we assume the independence of arguments appearing in a graph. In [11], the authors proposed an approach to relax independence assumptions in probabilistic argumentation. However, this aspect of research is out of the scope of the present paper.

For simplicity, let us abuse the notation, using $p(\bar{\alpha})$ to denote $1 - p(\alpha)$. Then, the probability of subgraph G', denoted $p(G')$, can be defined as follows.

$$p(G') = (\Pi_{\alpha \in A'} p(\alpha)) \times (\Pi_{\alpha \in A \setminus A'} p(\bar{\alpha})) \tag{4}$$

Given a PrAG $G^p = (A, R, p)$, let $Q_\sigma(E)$ denote the set of subgraphs of G^p, each of which has an extension E under a given semantics $\sigma \in \{ad, co, pr, gr, st\}$. Based on formula (4), $p(E^\sigma)$ is defined as follows [8].

$$p(E^\sigma) = \Sigma_{G' \in Q_\sigma(E)} p(G') \tag{5}$$

Example 2. Let $G_1^p = (A_1, R_1, p)$ be a PrAG (illustrated as follows), where $p(a) = 0.5$, $p(b) = 0.8$, $p(c) = 0.4$ and $p(d) = 0.5$.

$$a \longleftrightarrow b \longrightarrow c \longleftrightarrow d \circlearrowright$$
$$0.5 \qquad 0.8 \qquad 0.4 \qquad 0.5$$

The subgraphs of G_1^p are presented in Table 1. According to formula (5), there are 5 preferred extensions with non-zero probability:

$$p(\varnothing^{pr}) = p(G_1^{15}) + p(G_1^{16}) = 0.06$$
$$p(\{a\}^{pr}) = p(G_1^3) + p(G_1^4) + p(G_1^7) + p(G_1^8) = 0.3$$
$$p(\{b\}^{pr}) = p(G_1^1) + p(G_1^2) + p(G_1^3) + p(G_1^4) + p(G_1^9) + p(G_1^{10})$$
$$\qquad\qquad + p(G_1^{11}) + p(G_1^{12}) = 0.8$$
$$p(\{c\}^{pr}) = p(G_1^{13}) + p(G_1^{14}) = 0.04$$
$$p(\{a, c\}^{pr}) = p(G_1^1) + p(G_1^2) + p(G_1^5) + p(G_1^6) = 0.2$$

This example shows that according to the existing possible world-based approach, in order to compute the probability of a set of arguments being an extension, we have to compute the extensions of each subgraph, which in many cases is computationally expensive.

3   Characterized Subgraphs w.r.t. an Extension

In this section, we introduce an approach to formulate the semantics of probabilistic argumentation by characterizing subgraphs w.r.t. an extension, such that the probability of a set of arguments being an extension can be evaluated without computing (or with less computation of) the extensions of subgraphs.

First, let us introduce a novel notion σ-*subgraph w.r.t. an extension*: If a subgraph has a σ-extension E, then it is called a σ-subgraph w.r.t. E. Formally, we have the following definition.

Definition 5. *Let $G^p = (A, R, p)$ be a PrAG, $G' = (A', R')$ be a subgraph of G^p where $A' \subseteq A$ and $R' = R \cap (A' \times A')$, and $E \subseteq A$ be a set of arguments. We say that G' is a σ-subgraph of G^p w.r.t. E, iff G' has a σ-extension E, where $\sigma \in \{ad, co, pr, gr, st\}$.*

Table 1. Subgraphs of G_1^p

Subgraphs		Probability of subgraph	Preferred extensions
G_1^1	$a \leftrightarrow b \rightarrow c \leftrightarrow d \circlearrowleft$	0.08	$\{b\}, \{a,c\}$
G_1^2	$a \leftrightarrow b \rightarrow c$	0.08	$\{b\}, \{a,c\}$
G_1^3	$a \leftrightarrow b \quad d \circlearrowleft$	0.12	$\{a\}, \{b\}$
G_1^4	$a \leftrightarrow b$	0.12	$\{a\}, \{b\}$
G_1^5	$a \quad c \leftrightarrow d \circlearrowleft$	0.02	$\{a,c\}$
G_1^6	$a \quad c$	0.02	$\{a,c\}$
G_1^7	$a \quad d \circlearrowleft$	0.03	$\{a\}$
G_1^8	a	0.03	$\{a\}$
G_1^9	$b \rightarrow c \leftrightarrow d \circlearrowleft$	0.08	$\{b\}$
G_1^{10}	$b \rightarrow c$	0.08	$\{b\}$
G_1^{11}	$b \quad d \circlearrowleft$	0.12	$\{b\}$
G_1^{12}	b	0.12	$\{b\}$
G_1^{13}	$c \leftrightarrow d \circlearrowleft$	0.02	$\{c\}$
G_1^{14}	c	0.02	$\{c\}$
G_1^{15}	$d \circlearrowleft$	0.03	$\{\}$
G_1^{16}		0.03	$\{\}$

Example 3. Consider G_1^p in Example 2. Given $E_1 = \{a\}$, G_1^3, G_1^4, G_1^7 and G_1^8 are preferred subgraphs of G_1^p.

Then, given a PrAG $G^p = (A, R, p)$, a set of arguments $E \subseteq A$ and a semantics $\sigma \in \{ad, co, pr, gr, st\}$, a function (called *subgraph characterization function*) is used to map E to a set of σ-subgraphs of G^p w.r.t. E.

Definition 6. *Let $G^p = (A, R, p)$ be a PrAG. Let $\mathbb{G} = \{(B, R \cap (B \times B)) \mid B \in 2^A\}$ be the set of all subgraphs of G^p. A subgraph characterization function under a semantics $\sigma \in \{ad, co, pr, gr, st\}$ (denoted as ρ^σ) is defined as a mapping:*

$$\rho^\sigma : 2^A \rightarrow 2^{\mathbb{G}} \tag{6}$$

such that given $E \in 2^A$, for all $G' \in \rho^\sigma(E)$, G' is a σ-subgraph of G^p w.r.t. E.

In Definition 6, the function ρ^σ can be further specified in terms of some specific conditions.

First, under admissible semantics, we have the following theorem.

Theorem 1. *Let $G^p = (A, R, p)$ be a PrAG, $G' = (A', R')$ be a subgraph of G^p where $A' \subseteq A$ and $R' = R \cap (A' \times A')$, and $E \subseteq A$ be a conflict-free set of arguments. G' is an admissible subgraph of G^p w.r.t. E, iff the following conditions hold:*

- *$E \subseteq A'$, which means that all arguments in E appear in G'; and*
- *$(E^-\backslash E^+) \cap A' = \varnothing$, which means that every argument in $E^-\backslash E^+$ does not appear in G'.*

Proof. According to Definition 5, we only need to verify that E is an admissible extension of G'. Since $(E^-\backslash E^+) \cap A' = \varnothing$, every argument in E is defended by E. Since E is conflict-free, it follows that E is admissible.

According to theorem 1, the function ρ^{ad} is specified as follows:

$$\rho^{ad}(E) = \{(A', R \cap (A' \times A')) \mid A' \in 2^A : (E \subseteq A') \wedge ((E^-\backslash E^+) \cap A' = \varnothing)\} \quad (7)$$

Example 4. Consider G_1^p in Example 2 again. According to formula (7), there are eight admissible subgraphs w.r.t. $\{a\}$: $G_1^1, G_1^2, \ldots, G_1^8$ (as shown in the third column of Table 2), i.e., $\rho^{ad}(\{a\}) = \{G_1^1, G_1^2, \ldots, G_1^8\}$.

Second, according to the relationship between complete extension and admissible extension, it holds that w.r.t. a given conflict-free set of arguments, every complete subgraph is an admissible subgraph, but not vice versa. However, in an admissible subgraph w.r.t. a set E, if all arguments acceptable w.r.t. E is in E, then the admissible subgraph is a complete subgraph, in which E is a complete extension of the subgraph. Formally, we have the following theorem.

Theorem 2. *Let $G^p = (A, R, p)$ be a PrAG, $E \subseteq A$ be a conflict-free set of arguments, and $G' = (A', R')$ be an admissible subgraph of G^p w.r.t. E. Let $G'' = (A'', R'')$, where $A'' = args(G')\backslash(E \cup E^+)$ and $R'' = R' \cap (A'' \times A'')$. Then, G' is a complete subgraph of G^p w.r.t. E iff the following condition holds: $\forall \alpha \in A'', \alpha^- \neq \varnothing$.*

Proof. ($\Rightarrow$:) When G' is a complete subgraph of G^p w.r.t. E, assume that $\exists \alpha \in A''$ such that $\alpha^- = \varnothing$. It follows that α is acceptable w.r.t. E, and therefore E is not a complete extension, contradicting G' is a complete subgraph w.r.t. E.

($\Leftarrow$:) If E is a complete extension, according to Definition 5, G' is a complete subgraph of G^p w.r.t. E.

According to theorem 2, the function ρ^{co} is specified as follows:

$$\rho^{co}(E) = \{G' \in \rho^{ad}(E) \mid \forall \alpha \in args(G')\backslash(E \cup E^+) : \alpha^- \neq \varnothing\} \quad (8)$$

Example 5. Continue Example 4. Among the eight admissible subgraphs, except G_1^2 and G_1^6, others are complete subgraphs w.r.t. $\{a\}$ (as shown in the fourth column of Table 2), i.e., $\rho^{co}(\{a\}) = \{G_1^1, G_1^3, G_1^4, G_1^5, G_1^7, G_1^8\}$.

With regard to G_1^2, let $G_1'' = (A_1'', R_1'')$, where $A_1'' = \{a, b, c\}\backslash(\{a\} \cup \{b\}) = \{c\}$ and $R_1'' = \varnothing$. So, $c^- = \varnothing$, and therefore G_1^2 is not a complete subgraph w.r.t. $\{a\}$. Similarly, G_1^6 is not a complete subgraph w.r.t. $\{a\}$.

Table 2. σ-subgraphs of G_1^p w.r.t. $\{a\}$

subgraph	admissible subgraph w.r.t. $\{a\}$	comple subgraph w.r.t. $\{a\}$	stable subgraph w.r.t. $\{a\}$	preferred subgraph w.r.t. $\{a\}$	grounded subgraph w.r.t. $\{a\}$
G_1^1 $a \leftrightarrow b \rightarrow c \leftrightarrow d \circlearrowleft$	Yes	Yes	No	No	No
G_1^2 $a \leftrightarrow b \rightarrow c$	Yes	No	No	No	No
G_1^3 $a \leftrightarrow b \quad d \circlearrowleft$	Yes	Yes	No	Yes	No
G_1^4 $a \leftrightarrow b$	Yes	Yes	Yes	Yes	No
G_1^5 $a \quad c \leftrightarrow d \circlearrowleft$	Yes	Yes	No	No	Yes
G_1^6 $a \quad c$	Yes	No	No	No	No
G_1^7 $a \quad d \circlearrowleft$	Yes	Yes	No	Yes	Yes
G_1^8 a	Yes	Yes	Yes	Yes	Yes
G_1^9 $b \rightarrow c \leftrightarrow d \circlearrowleft$	No	No	No	No	No
G_1^{10} $b \rightarrow c$	No	No	No	No	No
G_1^{11} $b \quad d \circlearrowleft$	No	No	No	No	No
G_1^{12} b	No	No	No	No	No
G_1^{13} $c \leftrightarrow d \circlearrowleft$	No	No	No	No	No
G_1^{14} c	No	No	No	No	No
G_1^{15} $d \circlearrowleft$	No	No	No	No	No
G_1^{16}	No	No	No	No	No

Third, since every stable extension is a complete extension, and under stable semantics no argument is undecided, we may infer that a complete subgraph is a stable subgraph (w.r.t. a set of arguments E), iff all arguments in the subgraph are included in $E \cup E^+$. We directly have the following theorem.

Theorem 3. Let $G^p = (A, R, p)$ be a PrAG, $E \subseteq A$ be a conflict-free set of arguments, and $G' = (A', R')$ be a complete subgraph of G^p w.r.t. E. Then, G' is a stable subgraph of G^p w.r.t. E iff the following condition holds: $E \cup E^+ = args(G')$.

According to theorem 3, the function ρ^{st} is specified as follows:

$$\rho^{st}(E) = \{G' \in \rho^{co}(E) \mid E \cup E^+ = args(G')\} \qquad (9)$$

Example 6. Continue Example 5. According to Theorem 3, it is not difficult to verify that among the six admissible subgraphs, only G_1^4 and G_1^8 are stable subgraphs w.r.t. $\{a\}$ (as shown in the fifth column of Table 2), i.e., $\rho^{st}(\{a\}) = \{G_1^4, G_1^8\}$.

The above theorems and formulas show that under admissible, complete, and stable semantics, the set of subgraphs with respect to an extension can be identified without computing the extensions of subgraphs. However, under preferred and grounded semantics, partial computation of extensions is needed.

Theorem 4. *Let $G^p = (A, R, p)$ be a PrAG, $E \subseteq A$ be a conflict-free set of arguments, and $G' = (A', R')$ be a complete subgraph of G^p w.r.t. E. Let $G'' = (A'', R'')$, where $A'' = A' \backslash (E \cup E^+)$ and $R'' = R' \cap (A'' \times A'')$. Then, G' is a preferred subgraph of G^p w.r.t. E iff the following condition holds: G'' has only one empty admissible extension.*

Proof. ($\Rightarrow$): Assume the contrary, i.e., G'' has a non-empty admissible extension $E' \subseteq A''$. It follows that $E \cup E'$ is admissible, in that:

- $E \cup E'$ is conflict-free: both E and E' are conflict-free; E does not attack E' (otherwise, $E' \cap E^+ \neq \varnothing$, contradicting $E' \subseteq A'' = A' \backslash (E \cup E^+)$); E' does not attack E (otherwise, E attacks E', contradiction).
- $\forall \alpha \in E'$, α is acceptable w.r.t. $E \cup E'$.

So, $E \cup E'$ is an admissible extension of G'. So, E is not a preferred extension of G', contradicting "G' is a preferred subgraph of G^p w.r.t. E".

($\Leftarrow$): Since G'' has only one empty admissible extension, no argument in $A'' = A' \backslash (E \cup E^+)$ is acceptable w.r.t. E or any conflict-free superset of E. It turns out that E is a preferred extension of G', i.e., G' is a preferred subgraph of G^p w.r.t. E.

According to theorem 4, the function ρ^{pr} is specified as follows:

$$\rho^{pr}(E) = \{G' \in \rho^{co}(E) \mid \mathcal{E}_{ad}(G'') = \{\varnothing\}\} \tag{10}$$

Example 7. Continue Example 5. Among the six complete subgraphs, except G_1^1 and G_1^5, others are preferred subgraphs w.r.t. $\{a\}$ (as shown in the sixth column of Table 2). With regard to G_1^1, let $G_1'' = (A_1'', R_1'')$, where $A_1'' = \{a, b, c, d\} \backslash (\{a\} \cup \{b\}) = \{c, d\}$ and $R_1'' = \{(c, d), (d, c), (d, d)\}$. So, there is an admissible set of G_1'' (i.e., $\{c\}$) which is not empty, and therefore G_1^1 is not a preferred subgraph w.r.t. $\{a\}$. Similarly, G_1^5 is not a preferred subgraph w.r.t. $\{a\}$.

Theorem 5. *Let $G^p = (A, R, p)$ be a PrAG, $E \subseteq A$ be a conflict-free set of arguments, and $G' = (A', R')$ be a complete subgraph of G^p w.r.t. E. Let $G'' = (A'', R'')$, where $A'' = E \cup E^+$ and $R'' = R' \cap (A'' \times A'')$. Then, G' is a grounded subgraph of G^p w.r.t. E iff the following condition holds: E is a grounded extension of G''.*

Proof. ($\Rightarrow$): Since G' is a grounded subgraph of G^p w.r.t. E, it holds that E is the grounded extension of G'. Since for all $\alpha \in A' \backslash A''$, α does not attack E (otherwise, E attacks α, and therefore $\alpha \in E^+$, contradicting $\alpha \notin A''$), the status of arguments in $A' \backslash A''$ does not affected by the arguments in E. In other words, E is grounded extension of G'' where arguments in $A' \backslash A''$ are not considered.

($\Leftarrow$): Since E is a grounded extension of G'' and the status of arguments in $A' \backslash A''$ does not affected by the arguments in E, E is a grounded extension of G', i.e. G' is a grounded subgraph of G^p w.r.t. E.

Example 8. Continue Example 5. Among the six complete subgraphs, G_1^5 and G_1^7 and G_1^8 are grounded subgraphs w.r.t. $\{a\}$ (as shown in the last column of Table 2).

Theorems 4 and 5 show that under preferred and grounded semantics, when identifying the set of subgraphs with respect to an extension, only partial computation of extensions is needed, i.e., rather than computing the extensions of a whole subgraph G', the extensions of a part of G' (i.e., G'') are computed.

Finally, the relations between the characterized subgraphs under different semantics can be formulated as follows.

Theorem 6. *It hods that $\rho^{ad}(E) \supseteq \rho^{co}(E) \supseteq \rho^{pr}(E) \supseteq \rho^{st}(E)$, and $\rho^{co}(E) \supseteq \rho^{gr}(E)$.*

Proof. According to Theorems 2, 4 and 5, we directly have $\rho^{ad}(E) \supseteq \rho^{co}(E)$, $\rho^{co}(E) \supseteq \rho^{pr}(E)$, and $\rho^{co}(E) \supseteq \rho^{gr}(E)$. Now, let us verify that $\rho^{pr}(E) \supseteq \rho^{st}(E)$.

According to formulas (9) and (10), the condition $E \cup E^+ = args(G')$ implies $\mathcal{E}_{ad}(G'') = \{\varnothing\}$ where $G'' = \{\varnothing, \varnothing\}$. It follows that $\rho^{pr}(E) \supseteq \rho^{st}(E)$.

4   Semantics of Probabilistic Argumentation

According to the approach introduced in the previous section, given a PrAG $G^p = (A, R, p)$, a conflict-free set of arguments $E \subseteq A$ and a semantics $\sigma \in \{ad, co, pr, gr, st\}$, we get a set of σ-subgraphs w.r.t. E, i.e., $\rho^\sigma(E)$ without computing (or with less computation of) the extensions of subgraphs.

Then, according to formula (5), semantics of probabilistic argumentation, i.e., the probability of E being a σ-extension (denoted as $p(E^\sigma)$), is represented as follows.

$$p(E^\sigma) = \Sigma_{G' \in \rho^\sigma(E)} \, p(G') \qquad (11)$$

Note that $Q_\sigma(E)$ in formula (5) is replaced by $\rho^\sigma(E)$ in formula (11).

Furthermore, according to Theorem 1, each admissible subgraph w.r.t. an extension E is characterized by the conditions under which all arguments in E appear, while all arguments in $E^- \backslash E^+$ do not appear. In other words, the probability of a set of arguments being an admissible extension can be evaluated by the probabilities of arguments appearing or not appearing, without constructing the subgraphs and computing their extensions.

Formally, we have the following theorem.

Theorem 7. *Let $G^p = (A, R, p)$ be a PrAG, and $E \subseteq A$ be a conflict-free set of arguments. It holds that:*

$$p(E^{ad}) = \Pi_{\alpha \in E} p(\alpha) \times \Pi_{\beta \in E^- \backslash E^+} p(\bar{\beta}) \qquad (12)$$

Proof. Let $\Phi = A \backslash (E \cup (E^- \backslash E^+))$. For all $B \in 2^\Phi$, let $G^{E:B} = (E \cup B, R \cap ((E \cup B) \times (E \cup B)))$. According to Theorem 1 and formula (7), it holds that $G^{E:B} \in \rho^{ad}(E)$, and $\rho^{ad}(E) = \{G^{E:B} \mid B \in 2^\Phi\}$. Then, according to formula (11), $p(E^{ad}) = \Sigma_{G' \in \rho^{ad}(E)} \, p(G') = \Sigma_{B \in 2^\Phi} \, p(G^{E:B})$.

Since in $G^{E:B}$,

- every argument in E appears,
- every argument in $E^-\backslash E^+$ does not appear,
- every argument in B appears, and
- and every argument in $\Phi\backslash B$ do not appear,

it holds that $p(G^{E:B}) = \Pi_{\alpha\in E}p(\alpha) \times \Pi_{\beta\in E^-\backslash E^+}p(\bar\beta) \times \Pi_{\gamma\in B}p(\gamma) \times \Pi_{\eta\in\Phi\backslash B}p(\bar\eta)$. Since $\Sigma_{B\in 2^\Phi}(\Pi_{\gamma\in B}p(\gamma) \times \Pi_{\eta\in\Phi\backslash B}p(\bar\eta)) = 1$, we may conclude that:

$$
\begin{aligned}
p(E^{ad}) &= \Sigma_{B\in 2^\Phi}\, p(G^{E:B}) \\
&= \Sigma_{B\in 2^\Phi}(\Pi_{\alpha\in E}p(\alpha) \times \Pi_{\beta\in E^-\backslash E^+}p(\bar\beta) \times \Pi_{\gamma\in B}p(\gamma) \times \Pi_{\eta\in\Phi\backslash B}p(\bar\eta)) \\
&= (\Pi_{\alpha\in E}p(\alpha) \times \Pi_{\beta\in E^-\backslash E^+}p(\bar\beta)) \times \Sigma_{B\in 2^\Phi}(\Pi_{\gamma\in B}p(\gamma) \times \Pi_{\eta\in\Phi\backslash B}p(\bar\eta)) \\
&= \Pi_{\alpha\in E}p(\alpha) \times \Pi_{\beta\in E^-\backslash E^+}p(\bar\beta) \times 1 \\
&= \Pi_{\alpha\in E}p(\alpha) \times \Pi_{\beta\in E^-\backslash E^+}p(\bar\beta)
\end{aligned}
$$

5   Conclustions and Future Work

In this paper, we have proposed a new approach to formulate semantics of probabilistic argumentation. Given a PrAG, a set of subgraphs each of which has a certain extension under a given semantics is characterized by defining some conditions. As a result, semantics of probabilistic argumentation can be evaluated without computing (or with less computation of) the extensions of subgraphs. More specifically, under admissible semantics, it is neither necessary to construct subgraphs, nor to compute the extensions of the subgraphs. Under complete and stable semantics, it is not necessary to compute the extensions of any subgraphs. Under preferred and grounded semantics, for each subgraph, only a part of it is computed.

As to the best of our knowledge, our approach is the first attempt to formulate semantics of probabilistic argumentation by characterizing subgraphs. In existing literature, probability values are associated to arguments [3, 4, 8], or attacks [7], or both arguments and attacks [5, 9, 10]. Although there are some differences between the notions of PrAGs, to evaluate the probability of a set of arguments being an extension, for a PrAG with n nodes, it is necessary to construct 2^n subgraphs, and to compute their extensions.

Future work is as follows. First, based on the theory introduced in this paper, it is worth studying the computational complexity and developing efficient algorithms for computing the semantics of probabilistic argumentation. Second, the PrAG handled in this paper is based on Dung's abstract argumentation. In recent years, some extended argumentation frameworks have been proposed. Among them, the work on abstract dialectical frameworks (ADFs) [12] is increasingly active. So, it could be interesting to formulate the semantics of probabilistic abstract dialectical frameworks [13] by exploring the approach introduced in this paper.

Acknowledgment. The research reported in this paper was partially supported by the National Science Foundation of China (No.61175058, No.61203324), the National Research Fund Luxembourg (FNR), and Zhejiang Provincial Natural Science Foundation of China (No. LY14F030014).

References

1. Dung, P.M.: On the acceptability of arguments and its fundamental role in non-monotonic reasoning, logic programming and n-person games. Artificial Intelligence 77(2), 321–357 (1995)
2. Baroni, P., Giacomin, M.: On principle-based evaluation of extension-based argumentation semantics. Artificial Intelligence 171(10-15), 675–700 (2007)
3. Dung, P.M., Thang, P.M.: Towards (probabilistic) argumentation for jury-based dispute resolution. In: Proceedings of the COMMA 2010, pp. 171–182. IOS Press (2010)
4. Rienstra, T.: Towards a probabilistic dung-style argumentation system. In: Proceedings of the AT, pp. 138–152 (2012)
5. Li, H., Oren, N., Norman, T.J.: Probabilistic argumentation frameworks. In: Modgil, S., Oren, N., Toni, F. (eds.) TAFA 2011. LNCS, vol. 7132, pp. 1–16. Springer, Heidelberg (2012)
6. Dunne, P.E., Hunter, A., McBurney, P., Parsons, S., Wooldridge, M.: Weighted argument systems: Basic definitions, algorithms, and complexity results. Artificial Intelligence 175(2), 457–486 (2011)
7. Hunter, A.: Probabilistic qualification of attack in abstract argumentation. International Journal of Approximate Reasoning 55(2), 607–638 (2014)
8. Hunter, A.: Some foundations for probabilistic abstract argumentation. In: Proceedings of the 4th International Conference on Computational Models of Argument, pp. 117–128. IOS Press (2012)
9. Fazzinga, B., Flesca, S., Parisi, F.: Efficiently estimating the probability of extensions in abstract argumentation. In: Liu, W., Subrahmanian, V.S., Wijsen, J. (eds.) SUM 2013. LNCS, vol. 8078, pp. 106–119. Springer, Heidelberg (2013)
10. Fazzinga, B., Flesca, S., Parisi, F.: On the complexity of probabilistic abstract argumentation. In: Proceedings of the Twenty-Third International Joint Conference on Artificial Intelligence, pp. 898–904. AAAI Press (2013)
11. Li, H., Oren, N., Norman, T.J.: Relaxing independence assumptions in probabilistic argumentation. In: Proceedings of Argumentation in Multi-Agent Systems, ArgMAS (2013)
12. Brewka, G., Woltran, S.: Abstract dialectical frameworks. In: Proceedings of KR 2010, pp. 102–111. AAAI Press (2010)
13. Polberg, S., Doder, D.: Probabilistic abstract dialectical frameworks. In: Fermé, E., Leite, J. (eds.) JELIA 2014. LNCS, vol. 8761, pp. 591–599. Springer, Heidelberg (2014)

Algebraic Semantics for Dynamic Dynamic Logic

Minghui Ma[1] and Jeremy Seligman[1,2,*]

[1] Institute for Logic and Intelligence, Southwest University, Chongqing, China
mmh.thu@gmail.com
[2] Department of Philosophy, University of Auckland, Aukland, New Zealand
jeremy@seligman.info

Abstract. Dynamic dynamic logic (DDL) is a generalisation of propositional dynamic logic PDL and dynamic epistemic logic. In this paper, we develop algebraic semantics for DDL without the constant program. We introduce inductive and continuous modal Kleene algebras for PDL and show the validity of reduction axioms in algebraic models and hence the algebraic completeness of DDL.

1 Introduction

Dynamic epistemic logic (DEL) is a formalism extending epistemic modal logic by adding dynamic operations that change epistemic models (cf. e.g. [21]). Propositional dynamic logic (PDL) is the modal logic of programs. Girard, Seligman and Liu [16] introduced a general dynamic dynamic logic (GDDL) that generalises both DEL and PDL. In this logic, the update is done in two steps: first the multiple transformations of the original epistemic model are done for each action in the given action model, and then those multiple models are combined into a new epistemic model. The axiomatisation of GDDL is given by providing a computable translation of each GDDL-formula into a logically equivalent PDL-formula. The proof involves switching between representing programs as regular expressions and as automata.

Relational models are not unique as models for dynamic epistemic logics. Recently alternative approaches, using neighbourhood models [11,20] or algebraic models [12,8], have been proposed. The aim of this paper is to develop an algebraic semantics for the dynamic dynamic logic (DDL) which is also first presented in [16]. This logic is a restricted version of GDDL which excludes the product construction; DDL is to GDDL as public announcement logic (cf. [18]) is to DEL. Nonetheless, it captures the essential idea behind GDDL of using program expressions to define both modal operators within a model and also transformations of the model. One further restriction we make, for economy and elegance of presentation, is to exclude the constant atomic program ϵ, which corresponds to the universal modal operator. Our results can be extended to include this. Moreover, the algebraic semantics for dynamic logics, in particular, for dynamic operators

---

* The work of both authors was supported by China national funding of social sciences (grant no. 14ZDB016).

W. van der Hoek et al. (Eds.): LORI 2015, LNCS 9394, pp. 255–267, 2015.
DOI: 10.1007/978-3-662-48561-3_21

that represent ways of updating Kripke models, is significant for understanding the metamathematics of dynamic logics.

The algebraic models we will use for interpreting DDL are based on modal Kleene algebras with tests. The paper contains two parts. In the first part, we will show that PDL with tests can be characterised by both inductive and continuous modal Kleene algebras with tests. As far as we know, these modal Kleene algebras with tests are presented for the first time in this paper. Their non-modal parts, which are called Kleene algebras with tests, can be found, e.g. in Kozen [7]. A feature of such algebras is that each Kleene algebra has a Boolean subalgebra, and each formula stands for a Boolean term and each program for a term in the whole algebra. In the second part, we will introduce an operation of using a Boolean term to restrict a modal Kleene algebra with tests, as a way of representing updates. This enables us to give an interpretation of the "PDL transformations" described in [16]. We will give *explicit* reduction axioms for DDL, and show the algebraic completeness for it. This is in contrast to [16], which only proves the existence of such reduction axioms.

2   Preliminaries

Propositional dynamic logic with tests, PDL, is a logic of programs. The language of PDL consists of a set Prop of propositional variables (or atomic propositions) and a set Rel of relational variables (or atomic programs). The program constructors include unary ones $*$ (iteration) and $?\varphi$ (test), and binary ones ; (composition) and $\cup$ (choice).

Definition 1. *The set of* PDL*-programs* π *and the set of* PDL*-formulas* φ *are defined simultaneously by the following inductive rules:*

$$\pi ::= r \mid (\pi_1; \pi_2) \mid (\pi_1 \cup \pi_2) \mid \pi^* \mid ?\varphi$$
$$\varphi ::= p \mid \bot \mid \neg\varphi \mid (\varphi_1 \vee \varphi_2) \mid \langle\pi\rangle\varphi$$

where $r \in$ Rel$, p \in$ Prop$.$ *The program* $\pi_1; \pi_2$ *is the* composition *of* π_1 *and* π_2, $\pi_1 \cup \pi_2$ *is the* choice *of* π_1 *and* π_2, π^* *is the* iteration *of* π, *and* $?\varphi$ *is the* test *of* φ. *Other connectives* $\top, \wedge, \rightarrow$ *and* $\leftrightarrow$ *are defined as usual. In particular, the dual of* $\langle\pi\rangle$ *is defined by* $[\pi]\varphi := \neg\langle\pi\rangle\neg\varphi$.

We will introduce a Kripke-style semantics for the language of PDL, which is slightly different from the standard one given in [17]. It will make use of the binary operations of $\circ$ (relational composition) and $\cup$ (union), the unary operation $*$ (reflexive transitive closure), the constant relation I_W (identity), defined for binary relations R, R_1, R_2 on a non-empty set W as follows: $R_1 \circ R_2 = \{(x, y) \mid \exists z \in W(xR_1z \ \& \ zR_2y)\}; R_1 \cup R_2 = \{(x, y) \mid xR_1y \text{ or } xR_2y\};$ $I_W = \{(x, x) \mid x \in W\}; R^* = \bigcup_{n\in\omega} R^n$ where $R^0 = I_W$ and $R^{n+1} = R \circ R^n$. Also, for each $x \in W, X \subseteq W$, we define the images $R(x)$ and $R[X]$, and the unary operations $\langle R \rangle$ and $[R]$ on subsets of W by setting: $R(x) = \{y \in W \mid xRy\};$ $R[X] = \bigcup_{x\in X} R(x); \langle R \rangle X = \{x \in W \mid R(x) \cap X \neq \emptyset\}; [R]X = W \setminus \langle R \rangle(W \setminus X).$

Definition 2. *A Kripke model is a triple* $\mathfrak{M} = (W, V_R, V_P)$ *where W is a set,* $V_R : \mathsf{Rel} \to \mathcal{P}(W^2)$ *assigns a binary relation on W to each relational variable and $V_P : \mathsf{Prop} \to \mathcal{P}(W)$ assigns a subset of W to each propositional variable.*

Definition 3. *For any model* $\mathfrak{M} = (W, V_R, V_P)$, *the denotations* $[\![\pi]\!]^{\mathfrak{M}}$, *of a PDL-program π, and* $[\![\varphi]\!]^{\mathfrak{M}}$, *of a PDL-formula φ, are defined recursively by*

$$
\begin{aligned}
[\![r]\!]^{\mathfrak{M}} &= V_R(r), & [\![p]\!]^{\mathfrak{M}} &= V_P(p), \\
[\![\pi_1; \pi_2]\!]^{\mathfrak{M}} &= [\![\pi_1]\!]^{\mathfrak{M}} \circ [\![\pi_2]\!]^{\mathfrak{M}}, & [\![\bot]\!]^{\mathfrak{M}} &= \emptyset, \\
[\![\pi_1 \cup \pi_2]\!]^{\mathfrak{M}} &= [\![\pi_1]\!]^{\mathfrak{M}} \cup [\![\pi_2]\!]^{\mathfrak{M}}, & [\![\neg\varphi]\!]^{\mathfrak{M}} &= W \setminus [\![\varphi]\!]^{\mathfrak{M}}, \\
[\![\pi^*]\!]^{\mathfrak{M}} &= ([\![\pi]\!]^{\mathfrak{M}})^*, & [\![\varphi_1 \vee \varphi_2]\!]^{\mathfrak{M}} &= [\![\varphi_1]\!]^{\mathfrak{M}} \cup [\![\varphi_2]\!]^{\mathfrak{M}}, \\
[\![?\varphi]\!]^{\mathfrak{M}} &= I_W \cap ([\![\varphi]\!]^{\mathfrak{M}})^2, & [\![\langle\pi\rangle\varphi]\!]^{\mathfrak{M}} &= \langle[\![\pi]\!]^{\mathfrak{M}}\rangle [\![\varphi]\!]^{\mathfrak{M}}.
\end{aligned}
$$

We write $\mathfrak{M}, x \models \varphi$ if $x \in [\![\varphi]\!]_{\mathfrak{M}}$. By $\mathfrak{M} \models \varphi$ we mean that $[\![\varphi]\!]_{\mathfrak{M}} = W$. We say that φ is *valid* (notation: $\models \varphi$), if $\mathfrak{M} \models \varphi$ for all models $\mathfrak{M}$.

Remark 1. Definition 3 of semantics for PDL-formulas is essentially the same as the standard semantics (cf. [17]) using regular models. The only difference is that the signature we used consists of a set of program variables and a set of propositional variables. The validities under this semantics do not change.

PDL can be defined as the set of all valid formulas. Fischer and Ladner [3], and Segerberg [15] proposed Hilbert-style axiomatisations for PDL. Segerberg's system $\mathsf{H_{PDL}}$ (also see [17]) consists of the following axioms and rules:

(Tau)	All instances of propositional tautologies,
(K)	$[\pi](\varphi \to \psi) \to ([\pi]\varphi \to [\pi]\psi)$,
(Com)	$\langle\pi_1; \pi_2\rangle\varphi \leftrightarrow \langle\pi_2\rangle\langle\pi_1\rangle\varphi$,
(Choice)	$\langle\pi_1 \cup \pi_2\rangle\varphi \leftrightarrow \langle\pi_1\rangle\varphi \vee \langle\pi_2\rangle\varphi$,
(Iteration)	$\langle\pi^*\rangle\varphi \leftrightarrow (\varphi \vee \langle\pi\rangle\langle\pi^*\rangle\varphi)$,
(Ind)	$[\pi^*](\varphi \to [\pi]\varphi) \to (\varphi \to [\pi^*]\varphi)$,
(Test)	$\langle?\varphi\rangle\psi \leftrightarrow \varphi \wedge \psi$,
(MP)	from φ and $\varphi \to \psi$ infer ψ,
(Gen)	from φ infer $[\pi]\varphi$.

By $\vdash_{\mathsf{H_{PDL}}} \varphi$ we mean that φ is a theorem in $\mathsf{H_{PDL}}$. $\mathsf{H_{PDL}}$ is sound and complete, i.e., for any PDL-formula φ, $\vdash_{\mathsf{H_{PDL}}} \varphi$ iff $\models \varphi$.

In 1991, Pratt [13] provided an algebraic semantics using the concept of a *dynamic algebra with tests*, defined to be a tuple $\mathfrak{B} = ((B, \vee, -, 0), (R, \cup, ; , *), ?, \Diamond)$, where $(B, \vee, -, 0)$ is a Boolean algebra, and $? : B \to R$ and $\Diamond : R \times B \to B$ are operations satisfying the following axioms for all $a, b \in R$ and $x, y \in B$:

(A1) $\Diamond(a, 0) = 0$,
(A2) $\Diamond(a, x \vee y) = \Diamond(a, x) \vee \Diamond(a, y)$,
(A3) $\Diamond(a \cup b, x) = \Diamond(a, x) \vee \Diamond(b, x)$,
(A4) $\Diamond(a; b, x) = \Diamond(a, \Diamond(b, x))$,
(A5) $x \vee \Diamond(a, \Diamond(a^*, x)) \leq \Diamond(a^*, x) \leq x \vee \Diamond(a^*, -x \wedge \Diamond(a, x))$,
(A6) $\Diamond(?x, y) = x \wedge y$.

Since these axioms are obtained directly from those of $\mathsf{H_{PDL}}$, its completeness with respect to the class of dynamic algebras with tests can be easily obtained by the standard Lindenbaum-Tarski construction.

3   Varieties of Modal Kleene Algebras with Tests

In this section we define several varieties of modal Kleene algebras with tests, which will be shown to be variants of Pratt's dynamic algebras with tests. We will prove that PDL is sound and complete with respect to both the variety of continuous modal Kleene algebras with tests and the variety of inductive ones.

3.1   Modal Kleene Algebras with Tests

A Kleene algebra with tests is a combination of a regular algebra for actions and a Boolean algebra for propositions, introduced by Kozen in [7]. We first recall the basic concepts and then extend the definition with the addition of a new modal operator "$\downarrow$" that checks if a given action can be performed.

Definition 4. *An algebra* $(A, +, \cdot, 0, 1, *)$ *is a* Kleene algebra *if A is a set (of actions), $0, 1 \in A$, $*$ is a unary operation and $+$ and $\cdot$ are binary operations on A satisfying the following equations:*

$$a + (b + c) = (a + b) + c \quad (1)$$
$$a + b = b + a \quad (2)$$
$$a + 0 = a \quad (3)$$
$$a + a = a \quad (4)$$
$$a(bc) = (ab)c \quad (5)$$
$$1a = a \quad (6)$$
$$a1 = a \quad (7)$$

$$a(b + c) = ab + ac \quad (8)$$
$$(b + c)a = ba + ca \quad (9)$$
$$0a = 0 \quad (10)$$
$$a0 = 0 \quad (11)$$
$$1 + aa^* = a^* \quad (12)$$
$$1 + a^*a = a^* \quad (13)$$

where $ab = a \cdot b$ and the precedence order of operations is $\langle *, \cdot, + \rangle$; and the rules:

$$\text{if } ab \leq b, \text{ then } a^*b \leq b. \quad (14)$$
$$\text{if } ab \leq a, \text{ then } ab^* \leq a. \quad (15)$$

where $a \leq b$ is defined as $a + b = b$.

An algebra $(A, T, +, \cdot, 0, 1, *, -)$ is a *Kleene algebra with tests*, if $(A, +, \cdot, 0, 1, *)$ is a Kleene algebra and $(T, +, \cdot, 0, 1, -)$ is a Boolean algebra with $T \subseteq A$, where $-$ is a unary operator defined only on T. The elements of T are called *tests*. The class of all Kleene algebras with tests is denoted by KAT.

Definition 5. *An algebra* $\mathfrak{A} = (A, T, +, \cdot, 0, 1, *, -, \downarrow)$ *is a* modal Kleene algebra with tests *if $(A, T, +, \cdot, 0, 1, *, -)$ is a Kleene algebra with tests and $\downarrow : A \to T$ is a function that satisfies the following equations for all $t \in T$ and $a, b \in A$:*

$$t\downarrow = t \quad (16)$$
$$(ab)\downarrow = (a(b\downarrow))\downarrow \quad (17)$$
$$(a + b)\downarrow = a\downarrow + b\downarrow . \quad (18)$$

The class of all modal Kleene algebras with tests is denoted by MKAT. *We also write equations so that $\downarrow$ has precedence between $\cdot$ and $+$, i.e. $ab\downarrow = (ab)\downarrow$ and $a\downarrow + b\downarrow = (a\downarrow) + (b\downarrow)$.*

Remark 2. One should think of $a\downarrow$ as a test that the action a is possible. In that sense, we call $\downarrow$ a *modal operator*. In computability theory, $\pi\downarrow$ means that the program π is defined in the current state, i.e., that it terminates (cf. [2], p.6). Moreover, the axioms (16)-(18) for defining the modal operator $\downarrow$ has a natural interpretation in quantale theory. The operator $\downarrow$ is interpreted as the domain of relations projected on the identity relation I_W, i.e., for each relation a, $a\downarrow = \{(x,x) \mid x \in dom(a)\}$ where $dom(a)$ is the domain of a, i.e., it is the set $\{x \in W \mid \exists y(x,y) \in a\}$. Then one can easily check that axioms (16)-(18) hold. In fact, we get a supported quantale (cf. [14], p.48). In Definition 8, we will adopt this interpretation of the downarrow for defining Kripke models.

Lemma 1. *Given a modal Kleene algebra with tests* $\mathfrak{A} = (A, T, +, \cdot, 0, 1, *, -, \downarrow$ *), the following holds for all* $a, b \in A$ *and* $s, t \in T$:

(i) $0\downarrow = 0$.

(ii) $-(a0\downarrow) = 1$.

(iii) $(a(s+t))\downarrow = as\downarrow + at\downarrow$.

(iv) $-((a\cdot -t)\downarrow)\cdot -((a\cdot -s)\downarrow) = -((a\cdot -(ts))\downarrow)$.

(v) *if* $a \leq b$, *then* $a\downarrow \leq b\downarrow$.

(vi) $-((a\cdot -t)\downarrow)\cdot as\downarrow \leq a(ts)\downarrow$.

Proof. (i) is an instance of (16). For (ii), $-(a0\downarrow) = -(0\downarrow) = -0 = 1$. For (iii), $(a(s+t))\downarrow = (as+at)\downarrow = as\downarrow + at\downarrow$ by (18). For (iv), $-((a\cdot -t)\downarrow)\cdot -((a\cdot -s)\downarrow) = -((a\cdot -t)\downarrow + (a\cdot -s)\downarrow) = -(a(-t+-s)\downarrow) = -((a\cdot -(t\cdot s))\downarrow)$. For (v), assume $a \leq b$. Then $b\downarrow = (a+b)\downarrow = a\downarrow + b\downarrow$. Hence $a\downarrow \leq b\downarrow$. For (vi), first, we have $s \leq -t + ts$. Then $as \leq a(-t+ts)$. By (v), $as\downarrow \leq (a(-t+ts))\downarrow = (a\cdot -t + a(ts))\downarrow = (a\cdot -t)\downarrow + a(ts)\downarrow = --((a\cdot -t)\downarrow) + a(ts)\downarrow$. Hence $-(a\cdot -t)\downarrow \cdot a(ts)\downarrow \leq a(ts)\downarrow$. $\square$

Remark 3. Modal Kleene algebras with tests are straightforward variants of dynamic algebras with tests in which the two sorts B and R of the latter correspond to the two sorts T and A of the former, with functions $\Diamond : A \times T \to T$ and $? : T \to A$ given by $?t = t$ and $\Diamond(a,t) = (at)\downarrow$. For the converse, we can define $a\downarrow = \Diamond(a,1)$. Note also that since $T \subseteq A$ we can take $\vee = \cup = +$ and $\wedge = ; = \cdot$. The Kleene-Kozen algebra approach is therefore somewhat more economical in terms of operators than Pratt's.

Definition 6. *A* modal Kleene model *is a tuple* $M = (\mathfrak{A}, \theta_P, \theta_R)$ *in which* $\mathfrak{A} = (A, T, +, \cdot, 0, 1, *, -, \downarrow)$ *is in* MKAT, $\theta_P : $ Prop $\to T$ *assigns a test to each propositional variable and* $\theta_R : $ Rel $\to A$ *assigns an action to each relation variable.*

Definition 7. *For each* PDL-*program* π *and each* PDL-*formula* φ, *define denotations* $[\![\pi]\!]^M$ *and* $[\![\varphi]\!]^M$ *in a modal Kleene model* M *recursively as follows:*

$$[\![p]\!]^M = \theta_P(p),$$
$$[\![\bot]\!]^M = 0,$$
$$[\![\neg\varphi]\!]^M = -[\![\varphi]\!]^M,$$
$$[\![\varphi \vee \psi]\!]^M = [\![\varphi]\!]^M + [\![\psi]\!]^M,$$
$$[\![\langle\pi\rangle\varphi]\!]^M = ([\![\pi]\!]^M \cdot [\![\varphi]\!]^M)\downarrow,$$

$$[\![r]\!]^M = \theta_R(p),$$
$$[\![\pi_1;\pi_2]\!]^M = [\![\pi_1]\!]^M \cdot [\![\pi_2]\!]^M,$$
$$[\![\pi_1 \cup \pi_2]\!]^M = [\![\pi_1]\!]^M + [\![\pi_2]\!]_2,$$
$$[\![\pi^*]\!]^M = ([\![\pi]\!]^M)^*,$$
$$[\![?\varphi]\!]^M = [\![\varphi]\!]^M \downarrow .$$

We say that a PDL-formula φ is *valid* in $\mathfrak{A}$ (notation: $\mathfrak{A} \models \varphi$), if $[\![\varphi]\!]^M = 1$ for all models M based on $\mathfrak{A}$. For any $\mathbb{C} \subseteq \mathsf{MKAT}$, we say that a PDL-formula α is *valid* in $\mathbb{C}$ (notation: $\mathbb{C} \models \varphi$), if $\mathfrak{A} \models \varphi$ for all $\mathfrak{A} \in \mathbb{C}$.

3.2   Continuous Modal Kleene Algebras with Tests

A modal Kleene algebra $\mathfrak{A} = (A, T, +, \cdot, 0, 1, *, -, \downarrow)$ is said to be *continuous*, if it satisfies the following equation for all $a \in A$ and $t \in T$:

$$a^*t = \bigvee_{n \in \omega} a^n t \tag{19}$$

where $a^n = \underbrace{a \cdot \ldots \cdot a}_{n}$; and $(a^*t)\downarrow = \bigvee\{(a^nt)\downarrow \mid n \in \omega\}$. Let MKAT_C be the class of all continuous modal Kleene algebras with tests.

Lemma 2. *For every continuous modal Kleene algebra $\mathfrak{A} = (A, T, +, \cdot, 0, 1, *, -, \downarrow)$, for all $a \in A$ and $t \in T$,*

$$(a^*t)\downarrow \leq t + (a^*(-t \cdot (at\downarrow)))\downarrow . \tag{20}$$

Proof. We give the sketch of the proof which similar to the proof given by Kozen [5]. For all $n \geq 0$, $-(a^nt\downarrow) \cdot (a^{n+1}t\downarrow) = -(a^nt\downarrow) \cdot (a^nat\downarrow) \leq (a^n \cdot (-t \cdot at))\downarrow \leq (a^* \cdot (-t \cdot at))\downarrow = (a^* \cdot (-t \cdot (at\downarrow)))\downarrow$, by Lemma 1 (5) and the continuous condition. Then $1 = t + (-t \cdot (at\downarrow)) + (-(at\downarrow) \cdot (a^2t\downarrow)) + (-(a^2t\downarrow) \cdot (a^3t\downarrow)) + \ldots + (-(a^{n-1}t\downarrow) \cdot (a^nt\downarrow)) + -(a^nt\downarrow)$. Hence $1 = t + (a^* \cdot (-t \cdot (at\downarrow)))\downarrow + -(a^nt\downarrow)$. Hence $a^nt\downarrow \leq t + (a^* \cdot (-t \cdot (at\downarrow)))\downarrow$. Hence $(a^*t)\downarrow \leq t + (a^*(-t \cdot (at\downarrow)))\downarrow$. $\square$

For showing the completeness of PDL with respect to MKAT_C, we make use of continuous modal Kleene algebras with tests constructed from Kripke models.

Definition 8. *For any Kripke model $\mathfrak{M} = (W, V_R, V_P)$, the dual modal Kleene model of $\mathfrak{M}$ is defined as $\mathfrak{M}^+ = (\mathfrak{A}_{\mathfrak{M}}, \theta_P^{\mathfrak{M}}, \theta_R^{\mathfrak{M}})$, where $\mathfrak{A}_{\mathfrak{M}} = (A_{\mathfrak{M}}, T_{\mathfrak{M}}, +, \cdot, 0, 1, *, -, \downarrow)$ is the algebra defined as follows: (i) $A_{\mathfrak{M}} = \mathcal{P}(W^2)$, $T_{\mathfrak{M}} = \mathcal{P}(1)$, $0 = \emptyset$, $1 = I_W$; (ii) $a + b = a \cup b$, $-a = 1 \setminus a$, $a \cdot b = a \circ b$; (iii) $a^* = \bigcup_{n \in \mathbb{N}} a^n$ where $a^0 = 1$ and $a^{n+1} = a^n \cdot a$; and (iv) $a\downarrow = \{(x, x) \mid (x, y) \in a$ for some $y \in W\}$. Finally, let $\theta_P^{\mathfrak{M}}(p) = \{(x, x) \mid x \in V_P(p)\}$ for each $p \in \mathsf{Prop}$, and $\theta_R^{\mathfrak{M}} = V_R$.*

Fact 1 *For any Kripke model $\mathfrak{M}$, we have $\mathfrak{A}_{\mathfrak{M}} \in \mathsf{MKAT}_C$.*

Proof. The definition of a^* guarantees that $\mathfrak{A}_{\mathfrak{M}}$ is continuous. Other conditions for modal Kleene algebra with tests can be easily proved. $\square$

Proposition 1. *Given a Kripke model $\mathfrak{M} = (W, V_R, V_P)$ and $x \in W$, for any PDL-formula φ, $\mathfrak{M}, x \models \varphi$ iff $(x, x) \in [\![\varphi]\!]^{\mathfrak{M}^+}$.*

Proof. By induction on φ. Atomic and Boolean cases are easy. For the case $\varphi := \langle \pi \rangle \psi$, assume that $\mathfrak{M}, x \models \langle \pi \rangle \psi$. Then there exists $y \in W$ such that

$(x, y) \in [\![\pi]\!]^{\mathfrak{M}}$ and $\mathfrak{M}, y \models \psi$. By inductive hypothesis, $(y, y) \in [\![\psi]\!]^{\mathfrak{M}^+}$. By the definition of $\theta_R^{\mathfrak{M}}$, we have $(x, y) \in [\![\pi]\!]^{\mathfrak{M}^+}$. Thus $(x, y) \in [\![\pi]\!]^{\mathfrak{M}^+} \cdot [\![\psi]\!]^{\mathfrak{M}^+}$. Hence $(x, x) \in ([\![\pi]\!]^{\mathfrak{M}^+} \cdot [\![\psi]\!]^{\mathfrak{M}^+}) \downarrow$. Hence $(x, x) \in [\![\langle\pi\rangle\psi]\!]^{\mathfrak{M}^+}$. The other direction is shown by definitions similarly. $\qquad\square$

Theorem 2. $\mathsf{H_{PDL}}$ *is sound and complete with respect to* $\mathsf{MKAT_C}$.

Proof. The equation (20) guarantees the validity of inductive axioms in $\mathsf{H_{PDL}}$. The validity of other axioms are shown regularly. For the completeness, assume that $\nvdash_{\mathsf{H_{PDL}}} \varphi$. By relational completeness of $\mathsf{H_{PDL}}$, there exists a Kripke model $\mathfrak{M} = (W, V_R, V_P)$ such that $\mathfrak{M}, x \nvDash \varphi$ for some $x \in W$. Thus by Proposition 1, $(x, x) \notin [\![\varphi]\!]^{\mathfrak{M}^+}$. Hence $\mathfrak{M}^+ \nvDash \varphi$, while $\mathfrak{M}^+ \in \mathsf{MKAT_C}$. $\qquad\square$

3.3 Inductive Modal Kleene Algebras with Tests

Definition 9. *A modal Kleene algebra* $\mathfrak{A} = (A, T, +, \cdot, 0, 1, *, -, \downarrow)$ *is said to be* inductive, *if it satisfies the following equation for all* $t \in T$ *and* $a \in A$:

$$a^* t = t + a^* \cdot (-t) \cdot at \qquad (21)$$

The class of all inductive modal Kleene algebras is denoted by $\mathsf{MKAT_{Ind}}$.

Fact 3 *Every continuous modal Kleene algebra with tests is inductive.*

Proof. The induction principle follows from continuity. See [5,6]. $\qquad\square$

The axiom (21) makes the axiom (Ind) $[\pi^*](\varphi \to [\pi]\varphi) \to (\varphi \to [\pi^*]\varphi)$ valid. It suffices to observe that the axiom (Ind) is logically equivalent to $\langle\pi^*\rangle\neg\varphi \to \neg\varphi \vee \langle\pi^*\rangle(\varphi \wedge \langle\pi\rangle\varphi)$, the validity of which is granted by the axiom (21).

Proposition 2. *For* $\mathfrak{A} = (A, T, +, \cdot, 0, 1, *, -, \downarrow) \in \mathsf{MKAT_{Ind}}$, $t \in T$ *and* $a \in A$, *the following equations hold:*

$$t + (aa^* t)\downarrow = (a^* t)\downarrow = t + (a^* \cdot (-t \cdot (at\downarrow)))\downarrow . \qquad (22)$$

Hence the axiom (Iteration) $\langle\pi^*\rangle\varphi \leftrightarrow (\varphi \vee \langle\pi\rangle\langle\pi^*\rangle\varphi)$ *is valid in* $\mathsf{MKAT_{Ind}}$.

Proof. For the first equation, using axioms (12), (9), (6) and (18), one gets $a^* t \downarrow = (1 + aa^*)t \downarrow = (1t + aa^* t) \downarrow = (t + aa^* t) \downarrow = t\downarrow + (aa^* t)\downarrow = t + (aa^* t)\downarrow$. For the second one, using axioms (19), (18), (16), (5) and (17), one gets $a^* t\downarrow = (t + a^* \cdot (-t) \cdot at)\downarrow = t\downarrow + (a^* \cdot (-t) \cdot at)\downarrow = t + (a^* \cdot (-t) \cdot at)\downarrow = t + ((a^* \cdot -t) \cdot at)\downarrow = t + ((a^* \cdot -t) \cdot (at\downarrow))\downarrow = t + (a^* \cdot (-t \cdot (at\downarrow)))\downarrow.$ $\quad\square$

Theorem 4. $\mathsf{H_{PDL}}$ *is sound and complete with respect to* $\mathsf{MKAT_{Ind}}$.

Proof. The soundness is shown by induction on proofs in $\mathsf{H_{PDL}}$, using Proposition 2. The completeness follows from Fact 3, and Theorem 2. $\qquad\square$

4   Algebraic Semantics for Dynamic Dynamic Logic

In this section, we introduce algebraic semantics for dynamic dynamic logic (DDL). DDL was first suggested in [16]. The algebraic approach we make use of was first systematically presented in [12] for (intuitionistic) public announcement logics. However, the updated algebra defined below differs from that in [12].

Definition 10. *Define the sets of* DDL-*formulas* φ, DDL-*programs* π *and* DDL-*dynamic operators* λ *as follows:*

$$\varphi ::= p \mid \neg\varphi \mid (\varphi \wedge \varphi) \mid \langle\pi\rangle\varphi \mid \langle\lambda\rangle\varphi$$
$$\pi ::= r \mid (\pi;\pi) \mid (\pi \cup \pi) \mid \pi^* \mid \varphi? \mid \langle\lambda\rangle\pi$$
$$\lambda ::= \varphi \mid p := \varphi \mid r := \pi$$

where $p \in$ Prop *and* $r \in$ Rel. *Standard abbreviations are used, including the definition of* $\top$ *as some arbitrarily chosen tautology.*

A formula φ is said to be *static* if it contains no dynamic operators. Otherwise, it is said to be *dynamic*. There are three kinds of dynamic operator:

(1) *Domain restrictions* $\langle\varphi\rangle$. This is understood as restricting the evaluation of the following subformula to the case where φ holds. It is a kind of conditional: given that φ holds, evaluate . . . .

(2) *Propositional substitution* $\langle p := \varphi \rangle$. This reassigns propositional variable p to have the denotation of φ.

(3) *Relational substitution* $\langle r := \pi \rangle$. This reassigns relational variable r to have the denotation of π.

The effect of applying dynamic operators $\lambda_1, \ldots, \lambda_n$ simultaneously, written $\langle\lambda_1, \ldots, \lambda_n\rangle$, will be defined inductively. For example, we define $\langle p_1 := \varphi_1 \ldots p_n := \varphi_n\rangle$ to be $\langle q_1 := p_1\rangle \ldots \langle q_n := p_n\rangle\langle p_1 := \varphi_1'\rangle \ldots \langle p_n := \varphi_n'\rangle$, where $\varphi_i' = \varphi_i[p_j \mapsto q_j]_{1 \leq j \leq n}$ and $q_1, \ldots, q_n$ are any distinct variables not occuring in the formula.

Definition 11. *Let* $\mathfrak{A} \in$ MKAT. *Given* $s \in T$, *for each* $a \in A$, *let* $a_s = s{\cdot}a{\cdot}s$. *Define the restriction of* $\mathfrak{A}$ *to* s *to be the algebra* $\mathfrak{A}_s = (A_s, T_s, +_s, \cdot_s, 0_s, 1_s, *_s, -_s, \downarrow_s$ *) with actions* $A_s = \{a_s \mid a \in A\}$ *and tests* $T_s = \{t_s \mid t \in T\}$ *and operations defined by first performing the operation in* $\mathfrak{A}$ *and then taking the image under the map* $a \mapsto a_s$, *i.e., for all* $x, y \in A_s$, $0_s = s0s = 0$; $1_s = s1s = ss$; $x +_s y = (x+y)_s$; $x \cdot_s y = (x \cdot y)_s$; $-_s x = (-x)_s$; $*_s x = (x^*)_s$; $x\downarrow_s = (x\downarrow)_s$.

Although the map $a \mapsto a_s$ is a Boolean homomorphism, it is not in general a Kleene algebra homomophism. In particular, for $a, b \in A$, $(a \cdot b)_s = sabs$ but $a_s \cdot_s b_s = s(sas \cdot sbs)s = sasbs$ and it is possible that $ab \neq asb$.

It is easy to see that $x_s = sxs = x$, $sxs = sx = xs = x$ for any $x \in A_s$. Moreover, $ss = s$ since $s \in T$ and T forms a Boolean algebra.

Lemma 3. $\mathfrak{A}_s$ *is a modal Kleene algebra with tests.*

Proof. It is easy to check that T_s is a Boolean algebra. Now we check axioms for modal Kleene algebra with tests one by one. Given any $x, y, z \in A_s$, we have

(1) By definition, $x +_s (y +_s z) = (x + (y + z))_s = ((x + y) + z)_s = (x +_s y) +_s z$. Axioms (2) and (5) are shown quite similarly.

(3) $x +_s 0 = (x + 0)_s = x_s = x$, since $x \in A_s$ is a_s for some $a \in A$ and $(a_s)_s = ssass = sas = a_s$.

(4) $x +_s x = (x + x)_s = x_s = x$.

(6) $1_s \cdot_s x = (1_s \cdot x)_s = ss1sxs = s1sxs = ssxs = sxs = x$. The axiom (7) is shown quite similarly.

(8) $x \cdot_s (y +_s z) = (x(y + z))_s = (xy + xz)_s = x \cdot_s y +_s x \cdot_s z$. The axiom (9) is shown quite similarly.

(10) $0_s \cdot_s x = (0 \cdot x)_s = 0_s$. The axiom (11) is shown quite similarly.

(12) $1_s +_s x \cdot_s x^{*_s} = s(s1s + sxsx^*ss)s = s(s1s + ssxsx^*s)s = ss(1s + xx^*s)s = s(1 + xx^*)s = sx^*s = x^{*_s}$. The axiom (13) is shown similarly.

(14) Assume that $x \cdot_s y \le y$. Then $ssxyss = sxys = xys \le y = ys$. Hence $x^*ys \le ys$. Thus $ssx^*ys \le ssys$. Then $ssx^*sys \le y$, i.e., $x^{*_s} \cdot_s y \le y$. The axiom (15) is shown quite similarly.

(16) Let $t \in T_s$. Then $t{\downarrow}_s = (t{\downarrow})_s = t_s = t$.

(17) Firstly, $(x \cdot_s y){\downarrow}_s = (xy)_s {\downarrow}_s = (sxys) {\downarrow}_s = ((sxys){\downarrow})_s = ((xy){\downarrow})_s = ((x(y{\downarrow})){\downarrow})_s$. On the other hand, $(x(y{\downarrow}_s)){\downarrow}_s = ((x(y{\downarrow}_s)){\downarrow})_s = ((xs(y{\downarrow}s)){\downarrow})_s = ((x(y{\downarrow})){\downarrow})_s$. Hence, $(x \cdot_s y){\downarrow}_s = (x(y{\downarrow}_s)){\downarrow}_s$.

(18) Firstly, $(x +_s y){\downarrow}_s = ((s(x + y)s){\downarrow})_s = ((sxs + sys){\downarrow})_s = ((x + y){\downarrow})_s = (x{\downarrow} +_s y{\downarrow})_s$. On the other hand, $x{\downarrow}_s +_s y{\downarrow}_s = ((x{\downarrow})_s + (y{\downarrow})_s)_s = (s(x{\downarrow})s + s(y{\downarrow})s)_s = s(s(x{\downarrow})s + s(y{\downarrow})s)s = ss(x{\downarrow} + y{\downarrow})ss = s(x{\downarrow} + y{\downarrow})s = (x{\downarrow} + y{\downarrow})_s$. Then $(x +_s y){\downarrow}_s = x{\downarrow}_s +_s y{\downarrow}_s$.   □

Proposition 3. *If $\mathfrak{A}$ is inductive (continuous), then the restriction $\mathfrak{A}_s$ is also inductive (continuous).*

Proof. Assume that $\mathfrak{A}$ is inductive. Let $a_s \in A_s$ and $t_s \in T_s$. We calculate as follows: $(a_s^{*_s}) \cdot_s t_s = s(sa_s^* s \cdot t_s)s = s(a_s^* t_s)s = s[t_s + sa_s^*(-t_s)a_s t_s]s = s[t_s + ss(sa_s^* ss(-t_s)s \cdot sa_s t_s s)s]s = t_s +_s ss(sa_s^* ss(-t_s)s \cdot sa_s t_s s)s = t_s +_s [s(sa_s^* s \cdot s(-t_s)s \cdot s(a_s t_s)s] = t_s +_s a_s^{*_s} \cdot_s (-_s t_s) \cdot_s a_s \cdot_s t_s$. Assume that $\mathfrak{A}$ is continuous. Let $a_s \in A_s$ and $t_s \in T_s$. We calculate as follows: $(a_s^{*_s} \cdot_s t_s) = s(sa_s^* s \cdot t_s)s = s(sa_s^* t_s)s = s(a_s^* t_s)s = a_s^* t_s = \bigvee_{n \in \omega} a_s^n t_s$, as desired.   □

An *algebraic model* $M = (\mathfrak{A}, \theta_P, \theta_R)$ consists of a modal Kleene algebra with test $\mathfrak{A} = (A, T, +, \cdot, 0, 1, *, -, {\downarrow})$ and functions $\theta_P : \mathsf{Prop} \to T$ and $\theta_R : \mathsf{Rel} \to A$. Let $\mathcal{K}$ be the class of all such models.

Definition 12. *The denotations $[\![\varphi]\!]^M$ and $[\![\pi]\!]^M$ of each formula φ and program π, and the dynamic operation $[\![\lambda]\!]$ on algebraic models for each dynamic operator λ are defined recursively as follows:*

(1) *Formulas:*

$$[\![p]\!]^M = \theta_P(p); \quad [\![\neg\varphi]\!]^M = -[\![\varphi]\!]^M; \quad [\![\varphi \wedge \psi]\!]^M = [\![\varphi]\!]^M \cdot [\![\psi]\!]^M;$$
$$[\![\langle\pi\rangle\varphi]\!]^M = ([\![\pi]\!]^M \cdot [\![\varphi]\!]^M){\downarrow}; \quad [\![\langle\lambda\rangle\varphi]\!]^M = [\![\varphi]\!]^{[\![\lambda]\!]M}.$$

(2) *Programs:*

$$[\![r]\!]^M = \theta_R(r); \quad [\![\pi_1; \pi_2]\!]^M = [\![\pi_1]\!]^M \cdot [\![\pi_2]\!]^M; \quad [\![\pi_1 \cup \pi_2]\!]^M = [\![\pi_1]\!]^M + [\![\pi_2]\!]^M;$$

$$[\![\pi^*]\!]^M = [\![\pi]\!]^{M*}; \quad [\![\varphi?]\!]^M = [\![\varphi]\!]^M; \quad [\![\langle\lambda\rangle\pi]\!]^M = [\![\pi]\!]^{[\lambda]M}.$$

(3) *Dynamic operators:*

$$[\![\varphi]\!]M = (\mathfrak{A}_{[\![\varphi]\!]M}, \theta_P|_{[\![\varphi]\!]M}, \theta_R|_{[\![\varphi]\!]M}); [\![p := \varphi]\!]M = (\mathfrak{A}, \theta_P[p \mapsto [\![\varphi]\!]M], \theta_R);$$

$$[\![r := \pi]\!]M = (\mathfrak{A}, \theta_P, \theta_R[r \mapsto [\![\pi]\!]M]).$$

The notation $f|_s$ *is the function that maps* x *to* $s \cdot f(x) \cdot s$. *In particular,* $\theta_P|_{[\![\varphi]\!]M}(p) = [\![\varphi]\!]^M \cdot \theta_P(p) \cdot [\![\varphi]\!]^M$ *and* $\theta_R|_{[\![\varphi]\!]M}(r) = [\![\varphi]\!]^M \cdot \theta_R(r) \cdot [\![\varphi]\!]^M$.

A DDL-formula φ is *valid* in a modal Kleene algebra with tests $\mathfrak{A}$ (notation: $\mathfrak{A} \models \varphi$), if $[\![\varphi]\!]_M = 1$ for any model M based on $\mathfrak{A}$. A DDL-formula φ is valid in a class $\mathsf{C} \subseteq \mathsf{MKAT}$ (notation: $\mathsf{C} \models \varphi$), if $\mathfrak{A} \models \varphi$ for all $\mathfrak{A} \in \mathsf{C}$.

5 Reduction Axioms and Algebraic Completeness

Firstly, we will state two rules of replacement. Suppose that p and r are two propositional and relational variables, and that ψ is a formula or a program. Given a formula φ, let $\psi[\varphi]$ be the result of replacing p by φ in ψ. Likewise, given a program π, let $\psi[\pi]$ be the result of replacing r by π in ψ. Our two rules of replacement can be stated as:

REF: From $\varphi_1 \leftrightarrow \varphi_2$ and $\psi[\varphi_1]$ infer $\psi[\varphi_2]$.
RET: From $\langle\pi_1\rangle p \leftrightarrow \langle\pi_2\rangle p$ and $\psi[\pi_1]$ infer $\psi[\pi_2]$, if p is not in π_1, π_2.

Definition 13. *Let* $\mathsf{H_{DDL}}$ *be the Hilbert-style system consisting of the axioms and rules of* $\mathsf{H_{PDL}}$, *rules REF and RET, plus the following reduction axioms:*

(RA$_1$)	$\langle\psi\rangle p$	$\leftrightarrow \psi \wedge p$,	(RA$_{10}$)	$\langle\langle\psi\rangle r\rangle\varphi$	$\leftrightarrow \langle\psi?; r; \psi?\rangle\varphi$,
(RA$_2$)	$\langle p := \psi\rangle p$	$\leftrightarrow \psi$,	(RA$_{11}$)	$\langle\langle p := \psi\rangle r\rangle\varphi$	$\leftrightarrow \langle r\rangle\varphi$,
(RA$_3$)	$\langle q := \psi\rangle p$	$\leftrightarrow p \quad (p \neq q)$,	(RA$_{12}$)	$\langle\langle r := \pi\rangle r\rangle\varphi$	$\leftrightarrow \langle\pi\rangle\varphi$,
(RA$_4$)	$\langle r := \pi\rangle p$	$\leftrightarrow p$,	(RA$_{13}$)	$\langle\langle s := \pi\rangle r\rangle\varphi$	$\leftrightarrow \langle r\rangle\varphi \quad (r \neq s)$,
(RA$_5$)	$\langle\psi\rangle\neg\varphi$	$\leftrightarrow \psi \wedge \neg\langle\psi\rangle\varphi$,	(RA$_{14}$)	$\langle\langle\lambda\rangle\psi?\rangle\varphi$	$\leftrightarrow \langle(\langle\lambda\rangle\psi)?\rangle\varphi$,
(RA$_6$)	$\langle p := \psi\rangle\neg\varphi$	$\leftrightarrow \neg\langle p := \psi\rangle\varphi$,	(RA$_{15}$)	$\langle\langle\lambda\rangle(\pi_1; \pi_2)\rangle\varphi$	$\leftrightarrow \langle\langle\lambda\rangle\pi_1; \langle\lambda\rangle\pi_2\rangle\varphi$,
(RA$_7$)	$\langle r := \pi\rangle\neg\varphi$	$\leftrightarrow \neg\langle r := \pi\rangle\varphi$,	(RA$_{16}$)	$\langle\langle\lambda\rangle(\pi_1 \cup \pi_2)\rangle\varphi$	$\leftrightarrow \langle\langle\lambda\pi_1\rangle \cup \langle\lambda\rangle\pi_2\rangle\varphi$,
(RA$_8$)	$\langle\lambda\rangle(\varphi \wedge \psi)$	$\leftrightarrow \langle\lambda\rangle\varphi \wedge \langle\lambda\rangle\psi$,	(RA$_{17}$)	$\langle\langle\lambda\rangle\pi^*\rangle\varphi$	$\leftrightarrow \langle(\langle\lambda\rangle\pi)^*\rangle\varphi$.
(RA$_9$)	$\langle\lambda\rangle\langle\pi\rangle\varphi$	$\leftrightarrow \langle\langle\lambda\rangle\pi\rangle\langle\lambda\rangle\varphi$,			

By $\vdash_{\mathsf{H_{DDL}}} \varphi$ *we mean that* φ *is a theorem of* $\mathsf{H_{DDL}}$.

Remark 4. We could formulate an axiom system without these replacement rules but the number of axiom schemas increases significantly to cope with chains of dynamic operators and the * operation. The use of dynamic operators over programs is partly motivated to allow for the above axiomatisation, without it would have no simple reduction axiom for $\langle\langle\lambda\rangle\pi^*\rangle\varphi$. An alternative is to define the reduction directly, by induction but without using reduction axioms.

Theorem 5 (Soundness). *If $\vdash_{H_{DDL}} \varphi$, then $\mathsf{MKAT_C} \models \varphi$ and $\mathsf{MKAT_{Ind}} \models \varphi$.*

Proof. It suffices to show the validity of reduction axioms. Let $M = (\mathfrak{A}, \theta_P, \theta_R)$ be a model where $\mathfrak{A} = (A, T, +, \cdot, 0, 1, *, -, \downarrow) \in \mathsf{MKAT_C}$ or $\mathsf{MKAT_{Ind}}$. The proof is done as follows:

- (RA_1) $[\![\langle\psi\rangle p]\!]^M = [\![p]\!]^{[\![\psi]\!]^M} = \theta_P|_{[\![\psi]\!]^M}(p) = [\![\psi]\!]^M \cdot \theta(p) \cdot [\![\psi]\!]^M = [\![\psi \wedge p]\!]^M$.
- (RA_2) $[\![\langle p := \psi\rangle p]\!]^M = [\![p]\!]^{[\![p:=\psi]\!]M} = \theta_P[p \mapsto [\![\psi]\!]^M](p) = [\![\psi]\!]^M$.
- (RA_3) $[\![\langle q := \psi\rangle p]\!]^M = [\![p]\!]^{[\![q:=\psi]\!]M} = \theta_P[q \mapsto [\![\psi]\!]^M](p) = \theta_P(p) = [\![p]\!]^M$.
- (RA_4) $[\![\langle r := \pi\rangle p]\!]^M = [\![p]\!]^{[\![r:=\pi]\!]M} = \theta_P(p) = [\![p]\!]^M$.
- (RA_5) Let $[\![\psi]\!]^M = s$ and $[\![\varphi]\!]^M = t$. One proof is as follows: $[\![\langle\psi\rangle\neg\varphi]\!]^M = [\![\neg\varphi]\!]^{[\![\psi]\!]M} = -{}^s[\![\varphi]\!]^{[\![\psi]\!]M} = s \cdot -(sts) \cdot s = s \cdot -(sts) = s \cdot -([\![\varphi]\!]^{[\![\psi]\!]M}) = [\![\psi]\!]^M \cdot -([\![\varphi]\!]^{[\![\psi]\!]M}) = [\![\psi \wedge \neg\langle\psi\rangle\varphi]\!]^M$.
- (RA_6) $[\![\langle p := \psi\rangle\neg\varphi]\!]^M = -[\![\langle p : \psi\rangle\varphi]\!]^M = -[\![\varphi]\!]^{[\![p:=\psi]\!]M} = -[\![\langle p := \psi\rangle\varphi]\!]^M = [\![\neg\langle p := \psi\rangle\varphi]\!]^M$.
- (RA_7) $[\![\langle r := \pi\rangle\neg\varphi]\!]^M = -[\![\langle r := \pi\rangle\varphi]\!]^M = -[\![\varphi]\!]^{[\![r:=\pi]\!]M} = -[\![\langle r := \pi\rangle\varphi]\!]^M = [\![\neg\langle r := \pi\rangle\varphi]\!]^M$.
- (RA_8) For the cases when $\lambda = p := \xi$ or $\lambda = r := \pi$, it is easy to check the result. Let $\lambda = \xi$, $[\![\xi]\!]^M = s$, $[\![\varphi]\!]^M = t$ and $[\![\psi]\!]^M = u$. One proof is as follows: $[\![\langle\xi\rangle(\varphi \wedge \psi)]\!]^M = [\![\varphi \wedge \psi]\!]^{[\![\xi]\!]M} = [\![\varphi]\!]^{[\![\xi]\!]M} \cdot {}^s[\![\psi]\!]^{[\![\xi]\!]M} = ssts \cdot suss = sts \cdot sus = [\![\varphi]\!]^{[\![\xi]\!]M} \cdot [\![\psi]\!]^{[\![\xi]\!]M} = [\![\langle\xi\rangle\varphi \wedge \langle\xi\rangle\psi]\!]^M$.
- (RA_9) $[\![\langle\lambda\rangle\langle\pi\rangle\varphi]\!]^M = [\![\langle\pi\rangle\varphi]\!]^{[\![\lambda]\!]M} = ([\![\pi]\!]^{[\![\lambda]\!]M}[\![\varphi]\!]^{[\![\lambda]\!]M}) \downarrow = ([\![\langle\lambda\rangle\pi]\!]^M[\![\langle\lambda\rangle]\!]^M)\downarrow = [\![\langle\langle\lambda\rangle\pi\rangle\langle\lambda\rangle\varphi]\!]^M$.
- (RA_{10}) First, $[\![\langle\langle\psi\rangle r\rangle\varphi]\!]^M = [\![\varphi]\!]^{[\![\langle\psi\rangle r]\!]M}$, $[\![\langle\psi?; r := \psi?\rangle\varphi]\!]^M = [\![\varphi]\!]^{[\![\psi?;r:=\psi?]\!]M}$. It suffices to show $[\![\langle\psi\rangle r]\!]^M = [\![\psi?; r; \psi?]\!]^M$. One proof is as follows: $[\![\langle\psi\rangle r]\!]^M = [\![r]\!]^{[\![\psi]\!]M} = \theta_R|_{[\![\psi]\!]M}(r) = [\![\psi]\!]^M \cdot [\![r]\!]^M \cdot [\![\psi]\!]^M = [\![\psi?; r; \psi?]\!]^M$.
- (RA_{11}) It suffices to show that $[\![r]\!]^M = [\![\langle p := \psi\rangle r]\!]^M$. This is obtained by the definition $[\![p := \psi]\!]M$. Similarly, it is easy to show the validity of $\langle\langle p := \psi\rangle r\rangle\varphi \leftrightarrow \langle\pi\rangle\varphi$ and $\langle\langle s := \pi\rangle r\rangle\varphi \leftrightarrow \langle r\rangle\varphi$ ($r \neq s$).
- The case of (RA_{12}) and (RA_{13}) is similar to the case of (RA_{11}).
- (RA_{14}) It suffices to show $[\![\langle\lambda\rangle\psi?]\!]^M = [\![(\langle\lambda\rangle\psi)?]\!]^M$. This is done as follows: $[\![\langle\lambda\rangle\psi?]\!]^M = [\![\psi?]\!]^{[\![\lambda]\!]M} = [\![\psi]\!]^{[\![\lambda]\!]M} = [\![(\langle\lambda\rangle\psi)?]\!]^M$.
- (RA_{15}) It suffices to show that $[\![\langle\lambda\rangle(\pi_1; \pi_2)]\!]^M = [\![\langle\langle\lambda\rangle\pi_1; \langle\lambda\rangle\pi_2]\!]^M$. One proof is as follows: $[\![\langle\lambda\rangle(\pi_1; \pi_2)]\!]^M = [\![\pi_1; \pi_2]\!]^{[\![\lambda]\!]M} = [\![\pi_1]\!]^{[\![\lambda]\!]M} \cdot [\![\pi_2]\!]^{[\![\lambda]\!]M} = [\![\langle\lambda\rangle\pi_1]\!]^M \cdot [\![\langle\lambda\rangle\pi_2]\!]^M = [\![\langle\lambda\rangle\pi_1; \langle\lambda\rangle\pi_2]\!]^M$.
- (RA_{16}) It suffices to show $[\![\langle\lambda\rangle\pi_1 \cup \pi_2]\!]^M = [\![\langle\langle\lambda\rangle\pi_1 \cup \langle\lambda\rangle\pi_2]\!]^M$. One proof is as follows: $[\![\langle\lambda\rangle(\pi_1 \cup \pi_2)]\!]^M = [\![\pi_1 \cup \pi_2]\!]^{[\![\lambda]\!]M} = [\![\pi_1]\!]^{[\![\lambda]\!]M} + [\![\pi_2]\!]^{[\![\lambda]\!]M} = [\![\langle\lambda\rangle\pi_1]\!]^M + [\![\langle\lambda\rangle\pi_2]\!]^M = [\![\langle\lambda\rangle\pi_1 \cup \langle\lambda\rangle\pi_2]\!]^M$.
- (RA_{17}) $[\![\langle\lambda\rangle\pi^*]\!]^M = [\![\pi^*]\!]^{[\![\lambda]\!]M} = ([\![\pi]\!]^{[\![\lambda]\!]M})^* = ([\![\langle\lambda\rangle\pi]\!]^M)^* = [\![(\langle\lambda\rangle\pi)^*]\!]^M$.

This completes the proof.   □

Lemma 4 (Reduction). *For every DDL-formula φ, there exists a PDL-formula φ' that is computable from φ, such that $\vdash_{H_{DDL}} \varphi \leftrightarrow \varphi'$.*

Proof. By reduction axioms in Definition 13.   □

Theorem 6 (Completeness). *For any* DDL-*formula* φ, *the following hold:* (i) $\vdash_{\mathsf{H}_{\mathsf{DDL}}} \varphi$ *iff* $\mathsf{MKAT}_\mathsf{C} \models \varphi$; *and* (ii) $\vdash_{\mathsf{H}_{\mathsf{DDL}}} \varphi$ *iff* $\mathsf{MKAT}_{\mathrm{Ind}} \models \varphi$.

Proof. Suppose that φ is valid. By Lemma 4 there is a PDL-formula φ' such that $\vdash_{\mathsf{H}_{\mathsf{DDL}}} \varphi \leftrightarrow \varphi'$, and so by Theorem 5 $\varphi \leftrightarrow \varphi'$ is valid. Thus φ' is also valid, and by the completeness of $\mathsf{H}_{\mathsf{PDL}}$ with respect to MKAT_C and $\mathsf{MKAT}_{\mathrm{Ind}}$, φ' is provable in $\mathsf{H}_{\mathsf{PDL}}$, and so provable in $\vdash_{\mathsf{H}_{\mathsf{DDL}}}$. Hence φ is provable in $\vdash_{\mathsf{H}_{\mathsf{DDL}}}$. □

One can easily get the decidability of the satisfiability in Kripke models for DDL. As [3] proved, the satisfiability in Kripke models for PDL is decidable. The satisfiability for DDL follows from Lemma 4 immediately.

6 Conclusion and Further Directions

We developed an algebraic semantics for propositional dynamic logic (with tests), using continuous and inductive modal Kleene algebras with tests. Then we introduced an algebraic operation of restriction to give a semantics for the dynamic operators of dynamic dynamic logic, with a complete set of explicit reduction axioms. The result is a much cleaner formulation of the core of DDL than provided in [16] with a new, simplified and modular syntax that makes the two senses of 'dynamic', i.e., PDL-transformations and epistemic updates, clearer.

The algebraic approach we develop for DDL has two novel aspects. The first one is that the static part, PDL without the constant program, is shown to be complete with respect to both continuous and inductive modal Kleene algebras. The continuity or inductiveness is needed because of the property of the Kleene star. Such algebras are more economical than Pratt's in the sense that we save operations. The second aspect is that we make clear the epistemic update on modal Kleene algebras which is helpful for understanding the metamathematics of those dynamic operators.

There are several directions for extending the present approach. One is to consider additional programs or program constructors, such as the universal program ϵ and the converse program operator. These extensions of PDL are known to preserve decidability. For example, in [10], Lutz proved that PDL extended with intersection and converse is decidable. Extending with program intersection is also possible but would result in an undecidable logic with deterministic programs (cf. [4]) and a decidable logic with non-deterministic programs (cf. [1]). Likewise, adding the converse program to PDL results a decidable logic (cf. [19]). Adding the complement of programs to PDL will result an undecidable logic (cf. [4]), while adding only the complement of atomic programs gives a decidable logic (cf. [9]). So we would expect to get decidable dynamic dynamic versions. We expect all these results to be extendable to the dynamic dynamic case via suitable reduction axioms.

The second direction is to extend our algebraic approach to cover the general dynamic dynamic logic developed in [16], which involves a product construction similar to that of DEL. For each element in an action model, we can get a updated (restricted) model. The problem is to consider how to add the action structure to those updated algebraic models.

Acknowledgement . We would like to thank the reviewers for their constructive comments that are helpful for revising our paper.

References

1. Danecki, R.: Nondeterministic propositional dynamic logic with intersection is decidable. In: Skowron, A. (ed.) SCT 1984. LNCS, vol. 208, pp. 34–53. Springer, Heidelberg (1985)
2. Enderton, H.B.: Computability Theory: An Introduction to Recursion Theory. Elsevier (2011)
3. Fischer, M.J., Ladner, R.E.: Propositional dynamic logic of regular programs. Journal of Computer and System Sciences 18(2), 194–211 (1979)
4. Harel, D.: Dynamic logic. In: Handbook of Philosophical Logic, vol. II, pp. 496–604. D. Reidel Publishers (1984)
5. Kozen, D.: A representation theorem for models of *-free pdl. In: de Bakker, J.W., van Leeuwen, J. (eds.) ICALP 1980. LNCS, vol. 85, pp. 351–362. Springer, Heidelberg (1980)
6. Kozen, D.: On induction vs.*-continuity. In: Kozen, D. (ed.) Logic of Programs 1981. LNCS, vol. 131, pp. 167–176. Springer, Heidelberg (1982)
7. Kozen, D.: Kleene algebra with tests. ACM Transactions on Programming Languages and Systems 19(3), 427–443 (1997)
8. Kurz, A., Palmigiano, A.: Epistemic updates on algebras. Logical Methods in Computer Science 9(4:17), 1–28 (2013)
9. Lutz, C., Walther, D.: PDL with Negation of Atomic Programs. Journal of Applied Non-Classical Logic 15(2), 189–214 (2005)
10. Lutz, C.: PDL with intersection and converse is decidable. In: Ong, L. (ed.) CSL 2005. LNCS, vol. 3634, pp. 413–427. Springer, Heidelberg (2005)
11. Ma, M., Sano, K.: How to update neighborhood models. In: Grossi, D., Roy, O., Huang, H. (eds.) LORI. LNCS, vol. 8196, pp. 204–217. Springer, Heidelberg (2013)
12. Ma, M., Palmigiano, A., Sadrzadeh, M.: Algebraic semantics and model completeness for intuitionistic public announcement logic. Annals of Pure and Applied Logic 165(4), 963–995 (2014)
13. Pratt, V.: Dynamic algebras: examples, constructions, applications. Studia Logica 50(3-4), 571–605 (1991)
14. Resende, P.: Lectures on étale groupoids, inverse semigroups and quantales (2006), http://www.math.ist.utl.pt/~pmr/poci55958/gncg51gamap-version2.pdf
15. Segerberg, K.: A completeness theorem in the modal logic of programs. Banach Center Publications 9(1), 31–46 (1982)
16. Girard, P., Seligman, J., Liu, F.: General dynamic dynamic logic. In: Ghilardi, S., Bolander, T., Moss, L. (eds.) Advances in Modal Logic, vol. 9, pp. 239–260. Colledge Publications (2012)
17. Tiuryn, J., Harel, D., Kozen, D.: Dynamic Logic. MIT Press (2000)
18. Plaza, L.: Logics of public communications. Synthese 158(2), 165–179 (2007)
19. Vakarelov, D.: Filtration theorem for dynamic algebras with tests and inverse operator. In: Salwicki, A. (ed.) Logic of Programs 1980. LNCS, vol. 148, pp. 314–324. Springer, Heidelberg (1983)
20. van Benthem, J., Pacuit, E.: Dynamic logics of evidence-based beliefs. Studia Logica 99(1), 61–92 (2011)
21. van Ditmarsch, H., van der Hoek, W., Kooi, B.P.: Dynamic epistemic logic. Springer Science & Business Media (2007)

Logic and Ethics:
An Integrated Model for Norms,
Intentions and Actions

Alessandra Marra[1] and Dominik Klein[2,3]

[1] Tilburg Center for Logic, Ethics and Philosophy of Science,
Tilburg University, Tilburg, The Netherlands
[2] Political Theory Group, University of Bamberg, Bamberg, Germany
[3] Philosophy Department, University of Bayreuth, Bayreuth, Germany

Abstract. The paper investigates the way norms relate to and affect
agents' intentions and actions. Current work in deontic logic dealing with
agency mainly falls within two different groups: a variety of frameworks
which adopt a purely external approach and represent agency in terms of
possible outcomes of actions, and frameworks which instead endorse an
internal approach and focus exclusively on the agent's intentions. The
paper argues that neither of these models alone can produce a satis-
factory analysis. An integrated model which combines the internal and
external approaches is therefore put forward. The model is dynamic and
represents the change that accepting a goal norm triggers in an agent's
intentions (especially the so-called "prior-intentions") and actions.

1 Introduction

While there exists an extensive philosophical literature developing, analyzing
and connecting theories of norms, intentions and actions,[1] the interplay between
these notions has yet received little attention on the side of formal logic. Several
logical models have been proposed to reason about intentions,[2] and even more
to reason about actions,[3] and about norms concerning what an agent ought to
do or what ought to be the case.[4] To the best of our knowledge, however, just
very few logical models take all three notions into account jointly and, if they

---

[1] See the seminal works of Bratman [1], Gibbard [9] and Searle [20].
[2] See Cohen&Levesque [5], van Ditmarsch *et al.* [8], van der Hoek *et al.* [12], Icard *et al.* [15], Lorini&Herzig [17], Roy [19], Shoham [21] and van Zee *et al.* [23].
[3] See the development of dynamic logics in van Ditmarsch *et al.* [7].
[4] For an overview, see Hilpinen [10] and Hilpinen&McNamara [11].

W. van der Hoek et al. (Eds.): LORI 2015, LNCS 9394, pp. 268–281, 2015.
DOI: 10.1007/978-3-662-48561-3_22

do so, they are not concerned with explicitly representing how norms, intentions and actions relate to each other.[5]

It is especially from the perspective of deontic logic, i.e., the logic which deals with what ought to be done and what ought to be the case, that having an integrated model for norms, intentions and actions would be conceptually beneficial. When it comes to reasoning about norms concerning what an agent ought to do or about what ought to be the case, considerations about an agent's intentions and possible actions play indeed a central role. Some examples may illustrate the situations we have in mind:

1. Ann, a practicing Christian, invites her friend Julia to have dinner at a restaurant on Friday. Following the corresponding Christian norm, Ann refrains from eating meat on Fridays. She takes fish for dinner and Julia, finding the idea of eating fish compelling, does likewise.

2. A doctor tells Carla that, given her health conditions, she ought to drink milk or apple juice. Both are fine, so she is in principle free to pick any. However, once at home, Carla sees that there is no apple juice in her fridge, just milk. Therefore, she ought to drink milk.

As scenario 1 shows, an agent's intentions have a relevant role: both Ann and Julia happen to eat fish on Friday, but only Ann can be said to have accepted and fulfilled the Christian norm. Looking simply at the outcomes of an agent's actions would not be sufficient to discriminate between those two cases. Scenario 2, on the other hand, illustrates that focusing exclusively on an internal dimension (constituted by an agent's intentions) would not be enough to represent Carla's situation. The external properties of the world (or the agent's belief thereof) and the possible actions an agent can undertake play indeed a predominant role: Carla's initial possibility of choosing freely between milk and apple juice is lost given what the external world looks like.

The paper aims to provide a formal model of deontic logic which integrates those two dimensions: the internal one, constituted by an agent's intentions, and the external one, which is given by the external properties of the world, an agent's possible actions and their outcomes. In particular, such an integration is brought about by focusing on the role of norms in the formation of an agents' intentions (especially the so-called "prior intentions") and, consequently, on the effects on the agent's possible actions. The formal model, which makes use of tools from Stit-logics [13,14] combined with Veltman's [22] internal approach, not only can

---

[5] See, for instance, Dignum *et al.* [6] for an analysis of the notions of intentions, commitments and obligations in modal logic, Broersen *et al.* [2] for a BOID logic which deals with conflicts between beliefs, obligations, intentions and desires, and Broersen [4] for an analysis of the so-called "intentional actions" in Stit-logics. All those works take the notion of intention as primitive, and then impose external constraints or axioms to deal with the interplay between intentions, other mental states and actions. By doing so, however, the relation between norms, intentions and actions rests, at best, implicit. In particular, those works remain silent on the role of norms in the process of intentions' formation, and on how intentions affect (possible) actions. This is, indeed, the focus of the present paper.

be used to represent situations like 1 and 2 but can also be seen as a first step to fill in the gap between the philosophical work on norms, intentions and actions and their logical analysis as provided by current approaches in deontic logic.

The paper has the following structure. In §2 we have a closer look at other current works in deontic logic that deal with agency, and discuss in particular the external approach underlying Stit logics [13,14], and the internal one proposed by Veltman [22]. Moreover, some basic concepts such as "goal norms" and "prior intentions" are introduced. In §3 we present our integrated model, called NIA for *N*orms, *I*ntentions and *A*ctions. We introduce our formal language, discuss the main components of NIA (§3.1), show how the model can dynamically change every time a new norm is accepted by an agent (§3.2), and provide a formal representation of the examples 1 and 2 (§3.3). Finally, §4 concludes.

2   Conceptual Foundation

There are currently two main traditions in deontic logics dealing with agency: frameworks that adopt a purely external approach and exclusively model agents' actions and their outcomes, and frameworks which endorse an internal approach and represent only agents' intentions. Stit logics [13,14] are a prominent example of the first tradition,[6] while Veltman's model [22] is an example of the second. Before discussing how those two approaches can be bridged, we provide an overview of their main characteristics.

2.1   External and Internal Approaches

Stit logics concentrate on sentences of the form "agent α ought to see to it that Φ", where the basic construct "agent α sees to it that Φ" indicates that the agent's actions ensure that Φ is the case in the resulting state. Stit models provide an indeterministic, temporal representation of the external world: a tree-like structure indicates the temporal evolution of the world, its past, its present and the possible states of affairs that can be realized in the future. The main intuition behind the notion of agency as it is represented by Stit logics is that, by acting in the world, agents constrain the course of events: when an agent α sees to it that Φ, her actions have the effect of forcing the possible course of events to lie within the states of affairs in which Φ is realized.[7] For the purposes of this paper, it is worth mentioning some characteristics of Stit logics. First, the tree-like structure at the basis of Stit models is meant to exclusively represent the external properties of the world as shaped by nature and agents' actions. Second, as we have mentioned above, Stit logics endorse a purely external representation of agency: agency

---

[6] Another example is given, for instance, by Meyer's [18] dynamic logic. In this paper, however, we focus our attention only on the more recent Stit-logics.

[7] For a detailed presentation of Stit models, we refer the reader to Horty&Belnap [13] and Horty [14]. For the present paper, we limit ourselves to a deterministic fragment, where the agent's action uniquely determines the immediate future. We leave the treatment of the general case for future work.

is analyzed in terms of the outcomes of agents' actions, abstracting from any internal component. Finally, on a more technical note, given that Stit logics concentrate on the states of affairs that are brought about by agents' actions, rather than on the specific actions themselves, the formal language used by Stit logics does not contain any names for actions.

The deontic logic developed by Veltman [22] takes a completely different approach. While Stit models are based on an external representation of the world, Veltman's models represent exclusively an agent's internal, cognitive state. In particular, Veltman concentrates on the relation between norms and an agent's intentions, where the latter are represented by using "to-do lists" that indicate what an agent is committed to carry out in the future.[8] When an agent accepts a new norm, Veltman argues, that acceptance induces a change of intentions in the cognitive state of the agent in such a way that the agent intends to carry out what the norm requires.

Even though Veltman is not explicit about the nature of intentions considered in his models, it is clear that he is referring to what in the literature are usually called *prior, future-looking intentions*.[9] Prior intentions differ from the so-called "intentions in action" in that they *anticipate* and *cause* actions but are not parts of the actions themselves. More precisely, while prior intentions are intentions to do something, to perform a certain action (for instance, taking a picture with a camera), intentions in action are events responsible for the simple bodily movements (for instance, pressing the shutter bottom) which make up the action and therefore occur during the action, not prior to it.[10] Moreover, prior future-looking intentions are intentions to do something in the future,[11] and differ from prior present-directed intentions which, in turn, are intentions to perform an action here and now.

Another notion which is assumed but not explicitly mentioned by Veltman is that of *goal norm*.[12] Goal norms are norms about the results or outcomes of actions. They require that a certain state of affairs should obtain, but do not deal with the entire process of reaching that state of affairs. Different from goal norms are the so called "process norms" which also concern the single sequential actions whose execution is needed to reach a certain state of affairs. In the same line as Stit logics, the formal language used by Veltman does not contain names for actions, and the norms he considers in his framework are exclusively goal norms which require a certain state of affairs to be brought about.[13]

---

[8] We refer the reader to Veltman [22] for the formal details.

[9] See Searle [20], Lorini and Herzig [17] and Roy [19].

[10] For a detailed discussion on the difference between prior intentions and intentions in action, see Bratman [1] and Searle [20].

[11] Veltman [22], pp.29,33, indeed talks about "successor worlds", possible future evolutions of the current world in which an agent's intentions are realized.

[12] Here we are using Broersen's [3] terminology.

[13] Veltman's (*ibid.*) framework concentrates on norms in the form of imperatives.

2.2   The NIA Principle

In §1, we have seen that there exist situations which neither of the two approaches alone, the external one and the internal one, can satisfactorily deal with. The external approach fails to distinguish between Ann not eating meat with the intention to follow the Christian norm and Julia accidentally behaving in accordance with that norm without being aware of it. The internal approach of Veltman, on the the other hand, cannot deal with the case of Carla realizing that contingent factors about the content of her fridge limit her choice options, while being still committed to follow her doctor's prescription. These limitations, we have seen, call for an integration between the two frameworks. But how do Stit-logics' and Veltman's approaches relate to each other? More generally, which kind of relation between norms, intentions and actions is in place?

Bratman [1] extensively argues that prior future-looking intentions carry a distinctive element of *commitment*, which allows them to not only influence, but even guide and control actions. Forming a prior intention to do a certain action in the future motivates and engages the agent to act upon that intention and, normally, also moves her to act.[14] Generalizing Bratman's analysis a bit, we can say the following: If it is a norm the agent intends to follow, the agent's prior future-looking intentions control her actions at least in the sense that, in considering what to do in the future, the agent discriminates between those actions which permit her to fulfill the norm and those which do not.[15] Following Gibbard [9], we can then say that the agent's possible actions get divided between those which are *admissible*, or okay, and those which are not admissible, or not okay.[16] Then, if the agent intends to follow the norm, she is committed to those actions that are admissible.

We have seen that the relation between prior future-looking intentions and an agent's actions is spelled out in terms of commitment by Bratman, that is, in terms of an element which is intrinsic to those intentions themselves. However, when it comes to the relation between norms and the formation of prior future-looking intentions, Searle [20] notices there seems to be a gap. Norms, which Searle recognizes as reasons for actions, are indeed not causally sufficient to force the formation of those intentions in an agent.[17] For a norm to impact an agent's intentions, it is necessary that the agent recognizes the norm. In other words, the norm should be *accepted* by the agent.

---

[14] In particular, Bratman identifies two dimensions of commitment in prior intentions: (i) the so-called "volitional dimension", according to which prior intentions control an agent's conduct, and (ii) the so called "reasoning centered dimension" which concerns the role that prior intentions, thanks to their characteristic stability, play in the agent's reasoning in the period between their formation and their eventual execution. See Bratman [1], pp. 15-18, 107-110.

[15] More precisely: the agent discriminate between those actions *she believes* permit her to fulfill the norm, and those *she believes* do not. However, for the sake of the present paper, we abstract from issues related to ignorance and uncertainty. We leave them for future work.

[16] The "okay/not okay" is the original terminology used by Gibbard, see [9] pp.19,20.

[17] See Searle [20] pp. 40-41 and p.131.

From this discussion, it emerges that there is indeed a conceptual relation between norms, intentions and actions. And it is such a relation which permits the integration between Stit-logics' and Veltman's approaches. The model we present in the paper bridges the two approaches through the following principle:

NIA Principle: An agent's acceptance of a goal norm triggers a change in her prior future-looking intentions in such a way that the agent is committed to bring about what the norm requires. Consequently, the agent's admissible future actions get restricted to the ones permitting her to fulfill the goal norm, i.e., to reach the required state of affairs.

3   An Integrated Model: NIA

We start by introducing the formal languages on which NIA models are based. We make use of a descriptive language, to provide descriptions of the facts of the world, and a normative language, to talk about goal norms. As descriptive language we adopt the standard language of propositional logic, while we define the normative language as follows:

Definition 1 (Normative Language).   Let $At = \{p_1 \dots p_n\}$ be a set of atomic propositions. The normative language $\mathcal{L}_{norm}$ is given by the BNF:

$$\varphi := p^\exists | p^\forall | \neg p^\exists | \neg p^\forall | \varphi \wedge \varphi | \varphi \vee \varphi$$

In this definition, $p^\exists$ and $\neg p^\exists$ refer to *one-time* goal norms, requiring p (resp. $\neg p$) to be true once, while $p^\forall$ and $\neg p^\forall$ indicate *standing* goal norms, requiring p (resp. $\neg p$) to be always true.[18] The difference is crucial, especially when it comes to fulfillment: one-time goal norms expire once fulfilled (like: write your PhD thesis), while standing goal norms are always active and cannot be completely discharged (like: respect your parents).

3.1   NIA Models: Trees, Obligations and To-Do Lists

As mentioned above, the NIA model integrates external and internal approaches to agency by making use of tools from Stit logics' and Veltman's framework. Just like in Stit logics, we adopt a tree-like structure to represent the temporal evolution of the world; moreover, as in Veltman's framework, we make use of to-do lists to represent an agent's prior future-looking intentions. Finally, just as Stit logics and Veltman, we talk about actions in terms of their outcomes, and formalize actions as propositions, i.e., as sets of states of affairs in which those actions are realized (see [13], p.558).

We introduce the NIA models in a step to step approach. To begin with, we introduce the underlying temporal tree, as taken from Stit logic. But first, let us clarify some notation about trees.

---

[18] The terminology is taken from Lindström&Segerberg [16], p.1208.

In this paper, we take a *tree* to be a finite set W together with a tree-order $\prec$.[19] If x comes before y in the tree, i.e., $x \prec y$, we say that x is a *predecessor* of y and y is a *successor* of x. The *root* r of a tree $\langle W, \prec \rangle$ is the element r that lies below all other elements, i.e., $r \prec x$ for all $x \neq r \in W$. The *leaves* of a tree are nodes without successors, i.e., l is a leaf if $l \not\prec y$ for all $y \in W$. Finally, a *history* is a maximal *branch* in a tree, i.e., a sequence of immediate successors $r \prec x_1, \ldots, \prec x_n \prec l$ where r is the root of the tree and l a leaf.

Definition 2 (Tree model). A tree model is a 4-tuple $\mathcal{T} = \langle W, w_0, \prec, V \rangle$ where:

- W is a set of worlds with $w_0 \in W$
- $\prec$ is a tree-order on W with root w_0
- $V : At \to \mathcal{P}(W)$ is the atomic valuation.

Next, we add norms to our model. To be precise, we want to incorporate the obligations arising upon accepting a norm. For a given tree $\mathcal{T}$ and each possible norm $\varphi \in \mathcal{L}_{norm}$, we want to identify the subtree $\mathcal{O}_\varphi$ (called "obligation set") of $\mathcal{T}$ of possible histories compatible with satisfying φ.[20] The subtree $\mathcal{O}_\varphi$ is the formal representation of the admissible actions described in Sec.2.2.

Crucially, we impose some consistency conditions on the concept of admissibility. For a possible state $w \in W$ to be admissible for an agent, it is not enough to ensure that, at w, she does not violate any of the norms she has accepted. We furthermore demand that w is compatible with the agent not running into a violation in the future. That is, we want to exclude those states where the agent has not yet violated any norm, but will inevitably violate such a norm in the future, no matter what she does. To state these conditions formally, we introduce the following notation: For $S \subseteq W$, we define:

$$\overline{S} = \bigcup \{ h \subseteq S \mid h \text{ is a history of } \mathcal{T} \}$$

With other words, $\overline{S}$ is the set of all $s \in S$ that are part of a branch that lies completely in S.

Now, we can give the inductive definition of obligations sets.

Definition 3 (Obligation Sets, $\mathcal{O}_\varphi$). Let $\mathcal{T}$ be a tree-model. The obligation set $\mathcal{O}_\varphi$ for $\varphi \in \mathcal{L}_{norm}$ is inductively defined as follows:

- $\mathcal{O}_{p^\exists} = \bigcup \{ h \mid h \text{ is a history of } \mathcal{T} \text{ and some world in } h \text{ satisfies } p \}$
- $\mathcal{O}_{\neg p^\exists} = \bigcup \{ h \mid h \text{ is a history of } \mathcal{T} \text{ and some world in } h \text{ satisfies } \neg p \}$
- $\mathcal{O}_{p^\forall} = \bigcup \{ h \mid h \text{ is a history of } \mathcal{T} \text{ and every world in } h \text{ satisfies } p \}$
- $\mathcal{O}_{\neg p^\forall} = \bigcup \{ h \mid h \text{ is a history of } \mathcal{T} \text{ and every world in } h \text{ satisfies } \neg p \}$
- $\mathcal{O}_{\varphi \lor \psi} = \mathcal{O}_\varphi \cup \mathcal{O}_\psi$
- $\mathcal{O}_{\varphi \land \psi} = \overline{\mathcal{O}_\varphi \cap \mathcal{O}_\psi}$

Finally, as a third element, we introduce our representation of an agent's intentions. In our approach, we follow Veltman [22] in using consistent to do-lists

---

[19] That is, for any $w_1 \in W$, the set $\{ w \mid w \prec w_1 \}$ is linearly ordered by $\prec$.

[20] Norms and obligations are therefore distinct: while a norm is a linguistic item (part of the $\mathcal{L}_{norm}$ language), an obligation is a semantic item (subset of the tree $\mathcal{T}$).

and plans, sets of to-do lists, to model an agent's prior future-looking intentions. Intuitively, a to-do list is a list of all basic commitments of an agent, and contains elements of the form $\langle p^\exists, true \rangle$ (read as: "make p true once") or $\langle p^\forall, false \rangle$ ("make q false always"). The relation between prior-future looking intentions and commitments is the one discussed in Sec.2.2. In listing what the agent is committed to realize, to-do lists then serve to represent and keep track of what the agent intends to bring about.

Definition 4 (To-Do List, Plan). [21]

- A to-do list is a set $D \subseteq At^{\{\forall, \exists\}} \times \{true, false\}$.
- A to-do list is *consistent* if it does not contain contradicting commitments, that is, pairs of commitments of the form:[22]
  - $\langle p^\forall, true \rangle$ and $\langle p^\forall, false \rangle$
  - $\langle p^\forall, true \rangle$ and $\langle p^\exists, false \rangle$
  - $\langle p^\forall, false \rangle$ and $\langle p^\exists, true \rangle$
- A plan P is a set of consistent to-do lists such that for all $D, D' \in P$ it holds that $D \nsubseteq D'$.

A to-do list is the set of basic commitments the agent aims to jointly realize. For instance, the to-do list of Ann in our first example is $\{\langle eat\ meat\ on\ friday^\forall, false \rangle\}$. Now, it can happen that an agent accepts a norm that allows her to choose between different courses of action. This is indeed the case of Carla in our second example. This freedom of choice is reflected in plans: A plan contains several to-do lists, and indicates the agent's alternative possible commitments.[23] Thus, Carla's plan consists of two to-do lists, one saying $\langle drink\ milk^\exists, true \rangle$, the other one stating $\langle drink\ apple\ juice^\exists, true \rangle$.

Before proceeding, we should emphasize the crucial difference between an *empty to-do list* (or, equivalently, the plan containing only the empty to-do list, $P = \{\emptyset\}$) and an *empty plan* ($P = \emptyset$).[24] The first case, the empty to-do list, describes an agent that has not accepted any norms yet. The agent has no commitments to carry out, and, in this sense, any action she chooses is admissible. In the second case, the empty plan, the set of admissible courses of actions she could choose from is empty – whatever she does she will be in a state of violation. Consequentially, we call the case $P = \emptyset$ a *state of violation*.

As a next step, we show how the norms an agent accepts translate into her to-do lists. To be a bit more precise, we give a formal definition of how newly accepting a norm *changes* the agent's to-do lists.

---

[21] We adapt Veltman's definition to the normative language of NIA, which contains both standing and one-time norms. Cf. Veltman [22], pp.12-13.

[22] Notably, pairs of commitments like $\langle p^\exists, true \rangle$ ("make p true once") and $\langle p^\exists, false \rangle$ ("make p false once") *are* consistent.

[23] It is worth pointing to the fact that the notion of plan here differs from the one adopted by Bratman [1]. While in Bratman plans are typically partial and hierarchical, here plans are simply meant to illustrate the alternative actions an agent is committed to perform. Cf. Bratman [1], pp. 28-32, and Veltman [22], pp. 13,29.

[24] See Veltman [22] p.13

Definition 5 (Updating a plan, $P \uparrow \varphi$). Let P be a plan. The update of P by accepting some formula $\varphi \in \mathcal{L}_{norm}$ (written $P \uparrow \varphi$) is defined inductively as follows:[25]

- For φ of the form $p^\exists$ (resp. $p^\forall$, $\neg p^\exists$, $\neg p^\forall$)
  $P \uparrow \varphi = min\{D' \mid D' \text{ consistent } D' = D \cup \{\langle p^\exists, true\rangle\} \text{ for some } D \in P\}$
  (resp. $\langle p^\forall, true\rangle / \langle p^\exists, false\rangle / \langle p^\forall, false\rangle$)
- φ of the form $\psi \vee \chi$
  $P \uparrow \varphi = min(P \uparrow \psi \cup P \uparrow \chi)$
- φ of the form $\psi \wedge \chi$
  $P \uparrow \varphi = min\{D \cup D' \mid D \in P \uparrow \psi, D' \in P \uparrow \chi, D \cup D' \text{ consistent}\}$

Where $min(P)$ denotes the $\subseteq$-minimal elements of P. In particular, upon updating with some φ that is incompatible with the current plans, we have $P \uparrow \varphi = \emptyset$, the state of violation, not to be confused with the empty to-do list $P = \{\emptyset\}$.

Now, we have introduced all necessary components for defining a NIA model. For convenience, we start by first introducing a pre-NIA model. We later expand this to a NIA-model by adding some further coherence constraints.

Definition 6 (pre-NIA model). A pre-NIA model is a 6-tuple $M = \langle W, w_0, \prec, V, \mathcal{O}, F\rangle$ where

- $\langle W, w_0 \prec, V\rangle$ is a tree model
- $\mathcal{O} \subseteq W$ with $\overline{\mathcal{O}} = \mathcal{O}$ is the obligation set
- $F : W \to \mathcal{P}(\mathcal{P}(At^{\{\exists, \forall\}} \times \{true, false\}))$ is the planning function which attaches a plan to each world.

Of course, the intended interpretation is that w_0 is the current moment in time. In particular, the agent decides to accept or reject new norms and forms her to-do list at moment w_0. To-do lists and plans at all subsequent nodes are meant to keep track of which commitments have already been fulfilled, and which are still open. Some coherence constraints concerning individual plans, and the relation between different plans are in place. Before we can define these, we need to fix one piece of notation: If v is the immediate predecessor of w and $D \in F(w)$, then we call $D' \in F(v)$ a *source* of D if $D \subseteq D'$ and every item in $D' - D$ is of the form $\langle p^\exists, true\rangle$ for $w \in V(p)$ or $\langle p^\exists, false\rangle$ for $w \notin V(p)$.

Definition 7 (Coherence of F). Let M be a NIA model with planning function F. Then we call F coherent iff:

i) *Success:* If w is a leaf of NIA then no $D \in F(w)$ contains commitments of the form "$\langle p^\exists, true/false\rangle$"
ii) *Non-redundancy:* If $w \in W$ with $w \in V(p)$, then no $D \in F(w)$ contains $\langle p^\exists, true\rangle$. Similarly for $w \notin V(p)$ and $\langle p^\exists, false\rangle$.
iii) *Non-violation of standing norms:* If $w \in W$ with $w \in V(p)$, then no $D \in F(w)$ contains $\langle p^\forall, false\rangle$. Similarly, for $w \notin V(p)$ and $\langle p^\forall, true\rangle$.

---

[25] These definitions are adaptions of Veltman's [22], p.15.

iv) *Conservativity:* Let v be the immediate predecessor of w. Then for every $D \in F(w)$, there is a source D' of D in $F(v)$.

v) *Free Choice:* For every $D \in F(w)$ there is some immediate successor v of w and some $D' \in F(v)$ such that D is a source of D'.

Before we proceed, we should elaborate a bit on the above definition. The first item, *success*, expresses the fact that the tree in M constitutes the agent's time horizon, and every one-time norm has to be satisfied within that horizon. Next, the *non-redundancy* condition states that one-time norms get discharged once satisfied. Third, the *non-violation* condition expresses that the plan $F(w)$ cannot contain any to-do lists that are incompatible with w. Fourth, the *conservativity* condition expresses that no new commitments are introduced along the way. The agent can only form new commitments in her starting state w_0. All to-do lists at subsequent nodes can only track how these commitments are gradually satisfied. Finally, the last condition, *free choice*, states that every to-do list in $F(w)$ is compatible with some future state of affairs. That is, no matter which to-do list the agent picks at some node, that list is guaranteed to be satisfiable in some successor of the current world. Thus, taken together, the five conditions express that the different $F(w)$ cohere and that every to-do list in every plan is satisfiable. Now, we can finally define a NIA model.

Definition 8 (NIA-model). A NIA-model is a pre-NIA model in which the planing function F is coherent.

With those definitions at hand, we can now turn to the dynamics of the NIA model, and give a formal representation of the **NIA Principle**.

3.2 Dynamics

In this section we define the update operation induced by an agent's acceptance of a goal norm.[26] The update operation formalizes the **NIA Principle**: accepting a goal norm ϕ triggers a change in the agent's prior future-looking intentions (i.e, ϕ is added, as a commitment, to the agent's to do-lists) and, consequently,

---

[26] In the current paper, we focus exclusively on *updates*, and do not deal with the *revision* of norms or intentions. In particular, we are interested in the process of intention formation, as triggered by the acceptance of a goal norm, and its coherence requirements, rather than with reconsidering about norms/intentions that are already in place. This will, however, be important for future work. For works dealing with intentions' update and revision, see van der Hoek *et al.* [12], van Ditmarsch *et al.* [8] and Icard *et al.* [15]. Similar to our approach on intentions' update are van der Hoek *et al.* [12], also inspired by Bratman [1], and Icard *et al.* [15]. However, Icard *et al.* [15]'s approach differs from ours in two crucial ways. For once, their approach cannot deal with intentions triggered by what we call one-time norms. Also, our construction of temporal trees is more general than theirs. Finally, it should be noted that all those works on intentions' update and revision do not treat the relation between norms and intentions which is, on the other hand, at the core of the present paper.

it also affects the agent's actions by restricting the set $\mathcal{O}$ to that subtree $\mathcal{O}_\phi$ which is compatible with fulfilling the norm ϕ.

In Def.5, we have seen how newly accepted norms change the agent's to-do lists. We now define an *Upward-Downward Algorithm* that will be used for updating the *entire* planning function F, following an update of the agent's current plan at w_0. Basically, this algorithm is needed to ensure that, after a new norm is accepted, the F-function is coherent.[27]

- **do** $F(w_0) \uparrow \varphi$
- **do** $upward(w_0)$
- For all $w \in W$, **do**: $F(w) = min(F(w))$
- For all leaves $l \in W$, **do**: $downward(l)$
 end

def $upward(x)$:

- if $x \neq w_0$ and y ImPred(x)
  let $F(x) = F(y)$
- For all $D \in F(x)$ and $p \in At$:
  If $x \in V(p)$:
  do $D - \{\langle p^{\exists}, true \rangle\}$
  Else:
  do $D - \{\langle p^{\exists}, false \rangle\}$
- For all $D \in F(x)$ and $p \in At$:
  If $x \in V(p)$ and $\langle p^{\forall}, false \rangle \in$
  D:
  do $F(x) - D$
  If $x \notin V(p)$ and $\langle p^{\forall}, true \rangle \in D$:
  do $F(x) - D$
- For all z ImSucc(x):
  do $upward(z)$
 end

def $downward(x)$:

- If x is a leaf:
  do $F(x) - \{D \in F(x) \mid$
  $\exists \varphi \in D$ **with** $\varphi \in At^{\{\exists\}} \times$
  $\{true, false\}\}$
- for y ImPred(x):
  do $F(y) = \{D \in F(y) \mid \exists j$
  ImSucc(y) $\exists D' \in F(j)$:
  $D - D' \subseteq \{\langle p^{\exists}, true, \rangle | j \in$
  $V(p)\} \cup \{\langle p^{\exists}, false \rangle | j \notin$
  $V(p)\}$

  do $downward(y)$
 end

Intuitively, the working of the algorithm is the following: First, the agent's initial plan at w_0 is updated. In the upward part of the algorithm, the gradual fulfillment of this plan is successively traced throughout the tree: At each world, fulfilled commitments generated by one-time norms get removed. Also, the upward algorithm tracks the non-violation of standing norms by removing all to-do lists that violate standing norms. The downward part of the algorithm then checks whether all one-time norms will be satisfied eventually. For each leaf, we remove those to-do lists which contain open commitments generated by one-time norms, backtrack and remove the corresponding to-do lists down to the root.

The following lemmas show that the upward-downward algorithm functions as desired, that is, it terminates in finite time and guarantees that the F-function obeys to the constraints described in Def.7:[28]

---

[27] In what follows, ImPred(x) and ImSucc(x) stand for the the immediate Predecessors/Successors of x.

[28] Due to the limited space, we omit all proofs.

Lemma 1. *The upward-downward algorithm terminates in finite time.*

Lemma 2. *Let M be a pre-NIA model and let $\varphi \in \mathcal{L}_{norm}$. Then applying the upward-downward algorithm to M and φ makes the function F coherent. It even does so with the minimal necessary changes to F.*

Now, we can finally define the update with a norm:

Definition 9 (NIA Model Update). Let $M = \langle W, w_0, \prec, V, \mathcal{O}, F \rangle$ be a NIA model and $\varphi \in \mathcal{L}_{norm}$. Accepting the norm φ updates M to a new model $M' = \langle W, w_0, \prec, V, \mathcal{O}', F' \rangle$ such that:

- F' is obtained by upward-downward algorithm starting with F and φ at M
- $\mathcal{O}' = \overline{\mathcal{O} \cap \mathcal{O}_\varphi}$

The definition of update operation provides then a formal characterization of the NIA Principle we have introduced and discussed in §2. Notably, our approach generates the following agreement between an agent's prior future-looking intentions and her obligation set, that is, the set of admissible actions.

Theorem 1. *Let M be a NIA-model and let $\varphi \in \mathcal{L}_{norm}$. Then the updated model M' satisfies the following:*

i) If $\{x \in W | F(x) \neq \emptyset\} = \mathcal{O}$, then also $\{x \in W | F'(x) \neq \emptyset\} = \mathcal{O}'$

ii) Every world in $\mathcal{O}'$ is compatible with satisfying φ

Notably, *i)* implies that, if an agent's commitments are derived by starting with an empty to-do list and $\mathcal{O} = W$ and repeatedly accepting new norms, we are guaranteed that her obligation set is exactly the set of all nodes compatible with her fulfilling all accepted commitments. Point *ii)* follows from *i)*.

3.3   Examples

We can now return to our examples from section 1. Example 1: Given that NIA models take into account both an internal (an agent's intentions) and an external dimension (the world as shaped by nature and an agent's actions), it is possible to distinguish between Ann's and Julia's cases. While the state of affairs resulting from Ann's and Julia's actions is the same (having fish for dinner), only in Ann's case that state of affairs is reached through incorporating the Christian norm "¬ eat-meat-on-Friday$^\forall$" in her to-do lists. In particular, Ann and Julia differ in the content of their respective to-do lists and obligation sets.

Example 2: The norm "drink milk or apple juice" generates a set containing two to-do lists, $D = \{\langle milk^\exists, true\rangle\}$ and $D' = \{\langle apple\ juice^\exists, true\rangle\}$. Thus, Carla is free to choose between bringing about D or D'. However, since there is no apple juice available (i.e., all the nodes in the NIA model are such that "*apple juice*" is false), the to-do list D' gets eliminated by the Upward-Downward Algorithm. In particular, D' gets eliminated in the downward procedure.

4   Conclusion

The present paper aimed to propose a formal model in which the interplay between goal norms, prior future-looking intentions and possible actions could be represented. Motivated by the philosophical works on those topics, the paper proposed a so-called NIA Principle and formalized it in a dynamic model which combined tools from Stit logics and Veltman's models. It was shown that such an integrated model allowed to treat some relevant examples which would have remained problematic with external and internal approaches kept apart.

References

1. Bratman, M.: Intention, Plans and Practical Reason. Harvard University Press (1987)
2. Broersen, J., Dastani, M., van der Torre, L.W.N.: Resolving Conflicts Between Beliefs, Obligations, Intentions, and Desires. In: Benferhat, S., Besnard, P. (eds.) ECSQARU 2001. LNCS (LNAI), vol. 2143, p. 568. Springer, Heidelberg (2001)
3. Broersen, J.: Modal Action Logic for Reasoning about Reactive Systems. PhD Thesis, Vrije Universiteit, Amsterdam (2003)
4. Broersen, J.: Logics for (Artificial) Agency, Unpublished Manuscript, ESSLLI, Düsseldorf (2013)
5. Cohen, P.R., Levesque, H.J.: Intention is Choice with Commitment. Artificial Intelligence 42, 213–261 (1990)
6. Dignum et al.: A Modal Approach to Intentions, Commitments and Obligations. In: Deontic Logic, Agency and Normative Systems. Springer (1996)
7. van Ditmarsch, H., et al.: Dynamic Epistemic Logic. Springer (2008)
8. van Ditmarsch, H., de Lima, T., Lorini, E.: Intention Change via Local Assignments. In: Dastani, M., El Fallah Seghrouchni, A., Hübner, J., Leite, J. (eds.) LADS 2010. LNCS, vol. 6822, pp. 136–151. Springer, Heidelberg (2011)
9. Gibbard, A.: Reconciling our Aims. Oxford University Press (2011)
10. Hilpinen, R.: Deontic Logic. In: Goble, L. (ed.) Blackwell Guide to Philosophical Logic, pp. 159–182. Blackwell (2001)
11. Hilpinen, R., McNamara, P.: Deontic Logic: A Historical Survey and Introduction. In: Gabbay, D., et al. (eds.) Handbook of Deontic Logic and Normative Systems. College Publications (2013)
12. van der Hoek, W., et al.: Towards a theory of intention revision. Synthese 155(2), 265–290 (2007)
13. Horty J.F., Belnap N.: The Deliberative Stit: A Study of Action, Omission, Ability, and Obligation. Journal of Philosophical Logic 24(6) (1995)
14. Horty, J.F.: Agency and Deontic Logic. Oxford University Press (2001)
15. Icard, T., et al.: Joint Revision of Belief and Intention. In: Proceedings of the Twelfth International Conference on the Principles of Knowledge Representation and Reasoning, KR (2010)
16. Lindström, S., Segerberg, K.: Modal Logic and Philosophy. In: Blackburn, P., van Benthem, J., Wolter, F. (eds.) Handbook of Modal Logic, Studies in Logic and Practical Reasoning, vol. 3, pp. 1153–1218 (2006)
17. Lorini, E., Herzig, A.: A Logic of Intention and Attempt. Synthese 163, 45–77 (2008)

18. Meyer, J.-J.C.: A Different Approach to Deontic Logic: Deontic Logic Viewed as a Variant of Dynamic Logic. Notre Dame Journal of Formal Logic 29(1) (1988)
19. Roy, O.: Thinking Before Acting. PhD Thesis, Institute for Logic, Language and Computation, Amsterdam (2008)
20. Searle, J.: Making the Social World. The Structure of Human Civilization. Oxford University Press (2010)
21. Shoham, Y.: Logical Theories of Intention and the Database Perspective. Journal of Philosophical Logic 38(6) (2009)
22. Veltman, F.: Or else, what? Imperatives on the borderline of semantics and pragmatics. Institute for Logic, Language and Computation, Amsterdam (2011), https://staff.fnwi.uva.nl/u.endriss/teaching/lolaco/2011/slides/veltman.pdf
23. van Zee, M., et al.: AGM Revision of Beliefs about Action and Time. In: Proceedings of the International Joint Conference on Artificial Intelligence, IJCAI 2015 (2015)

A General Framework
for Modal Correspondence
in Dynamic Epistemic Logic

Shota Motoura

Research Institute for Mathematical Sciences, Kyoto University, Japan
motoura@kurims.kyoto-u.ac.jp

Abstract. We introduce a unified framework for dynamic epistemic logics, which in particular encompasses Public Announcement Logic (**PAL**), Epistemic Action (**EA**) and Preference Upgrade (**PU**). Our framework consists of a generic language, in which some of the known reduction axioms are expressible, together with relational and algebraic semantics. We then establish correspondences between generic reduction axioms and semantic properties, in both relational and algebraic settings. This leads to alternative proofs of the completeness of **PAL**, **EA**, **PU** with respect to their relational semantics and algebraic semantics (for the former two).

1 Introduction

Dynamic Epistemic Logic (DEL) is a branch of modal logic for reasoning about knowledge changes or belief revisions caused by communication. This is technically materialised by adding, to a static epistemic logic, dynamic operators that express actions of communication. These operators are interpreted as transformations of Kripke models (*model transformations*). The pioneer study on DEL is *Public Announcement Logic* (**PAL**) [5]. Then *Epistemic Action* (**EA**) [2] was proposed for reasoning about a greater variety of communication, including public announcements, and this has made the research area much more active. Until now, many DELs for various kinds of actions of communications have been proposed and studied: *Update Model* [11], *Command Logic* [14], *Belief Change* [6], *Preference Upgrade* (**PU**) [8], *Evidence Dynamics* [9] and *Manipulative Update* [12]. **PAL** and **EA**, among others, have been studied well: recently, algebraic counterparts of model transformations were proposed as the algebraic semantics of **PAL** [4] and **EA** [3].

Modal Correspondence in DEL: We aim at developing a modal correspondence theory for DEL in general, which establishes a link between axioms and properties of model transformations. There are some precedents in the literature [6–8]. Among these, van Benthem [7] gives a quite comprehensive account based on the concept of *update universe* to **PAL**, **EA**, Belief Change and Evidence Dynamics. In this paper we further this line of research, proposing a more general framework. We consider a wider class of dynamic operators than those studied in [7], which are generally expressed by formulas. In addition, we give not only

W. van der Hoek et al. (Eds.): LORI 2015, LNCS 9394, pp. 282–294, 2015.
DOI: 10.1007/978-3-662-48561-3_23

frame/model correspondences but also soundness-completeness-type correspondences for model transformations in general.

Organisation and Novel Contribution: The main contribution of our paper is in proposing a general framework for modal correspondence in DEL. We proceed as follows:

- Section 2: Language. We first propose a generic DEL language that is defined by using abstract action expressions. The languages of **PAL**, **EA** and **PU** (without the auxiliary universal modality) can be obtained by substituting their action expressions for abstract ones.
- Section 3: Relational Semantics. We then propose the notion of a two-layered relational model in which model transformations are expressed by abstract update relations instead of ordinary operational ways. The language and the two-layered models allow us to develop a general modal correspondence theory for the generic fragment. We then give correspondence results specific to **PAL**, **EA** and **PU**. As corollaries, we obtain alternative proofs of completeness of these logics. While the ordinary proofs are based on translation of dynamic formulas into purely static ones by reduction axioms, our new proofs consist in matching each reduction axiom with a corresponding semantic property. Thus our proofs are modular.
- Section 4: Algebraic Semantics. We undertake a similar analysis in an algebraic setting: we propose an algebraic notion of model; give general correspondence results and ones specific to **PAL** and **EA**; and also obtain alternative modular proofs of the completeness of these two logics.
- Section 5: Duality. To conclude the paper, we give several results on the duality between our relational and algebraic models.

2 Generic DEL Language

Let us begin by proposing a generic language for DEL.

Definition 1 (Generic DEL Language). *Let $\mathcal{P}$ be a set of* atomic propositions, *E a set of* epistemic expressions *and $\mathfrak{A}$ a set of* action expressions. *We define a generic DEL language $\mathcal{L}(E, \mathfrak{A})$ by the following rule:*

$$\varphi ::= \top \mid p \mid \neg\varphi \mid \varphi \vee \psi \mid \langle e \rangle \varphi \mid \langle\!\langle \alpha \rangle\!\rangle \varphi$$

where p ranges over $\mathcal{P}$, e over E and α over $\mathfrak{A}$.

In other words, the language $\mathcal{L}(E, \mathfrak{A})$ is the multimodal language with modalities $\langle e \rangle$ and $\langle\!\langle \alpha \rangle\!\rangle$ ($e \in E$, $\alpha \in \mathfrak{A}$). The individual languages of *Public Announcement Logic* (**PAL**), *Epistemic Action* (**EA**) and *Preference Upgrade* (excluding the universal modality) (**PU**) can be seen as special cases of $\mathcal{L}(E, \mathfrak{A})$:

Example 1 (Specific DEL Languages). Let Ag be a given set of agents.
- The language of **PAL** can be expressed as $\mathcal{L}_{\mathsf{PAL}} = \mathcal{L}(Ag, \mathfrak{A}_{\mathsf{PAL}})$, where $\mathfrak{A}_{\mathsf{PAL}} = \{!\varphi \mid \varphi \in \mathcal{L}_{\mathsf{PAL}}\}$. As $\mathfrak{A}_{\mathsf{PAL}}$ depends on $\mathcal{L}_{\mathsf{PAL}}$, these two sets are actually defined by simultaneous induction; however, the resulting language fits

the pattern of $\mathcal{L}(E, \mathfrak{A})$. The same remark applies to the other two examples below. The intended meaning of $[n]\varphi := \neg\langle n\rangle\neg\varphi$ is 'agent n knows φ', while $[[!\varphi]]\psi := \neg\langle\!\langle !\varphi\rangle\!\rangle\neg\psi$ means 'ψ holds after a truthful public announcement of φ'.

- The language of **EA** is $\mathcal{L}_{EA} = \mathcal{L}(Ag, \mathfrak{A}_{EA})$ where $\mathfrak{A}_{EA}$ is the set of *action models* (U, s) [2]. An action model (U, s) consists of a finite Kripke frame $(U, \{\to_n\}_{n\in Ag})$ together with a precondition function $\text{Pre} : U \to \mathcal{L}_{EA}$ and s is a state of U: $[[(U, s)]]\varphi := \neg\langle\!\langle(U, s)\rangle\!\rangle\neg\psi$ is read as 'φ holds after an epistemic action (U, s)', and $[n]\varphi$ is as in **PAL**.

- The language $\mathcal{L}_{PU}$ of **PU** can also be expressed as $\mathcal{L}_{PU} = \mathcal{L}(E_{PU}, \mathfrak{A}_{PU})$ where $E_{PU} = \{n, \bar{n} \mid n \in Ag\}$ and $\mathfrak{A}_{PU} = \{\varphi!, \sharp\varphi \mid \varphi \in \mathcal{L}_{PU}\}$. $[n]\varphi$ and $[[\varphi!]]\psi := \neg\langle\!\langle\varphi!\rangle\!\rangle\neg\varphi$ are as $[n]\varphi$ and $[[!\varphi]]\psi$ in **PAL**, while $[\bar{n}]\varphi := \neg\langle\bar{n}\rangle\neg\varphi$ and $[[\sharp\varphi]]\psi := \neg\langle\!\langle\sharp\varphi\rangle\!\rangle\neg\psi$ express 'all the worlds which agent n considers at least as good as the current one satisfy φ' ([8]) and 'ψ holds after suggestion of φ', respectively.

A common feature of dynamic epistemic logics is the use of *reduction axioms*, which are intended to transform any dynamic formula (involving dynamic modalities $\langle\!\langle\alpha\rangle\!\rangle$) into a purely static one. Some reduction axioms are already expressible in the generic DEL language $\mathcal{L}(E, \mathfrak{A})$:

Definition 2 (Generic Reduction Axioms). *We call the following axioms* generic reduction axioms:

$$\mathsf{R_N} : \langle\!\langle\alpha\rangle\!\rangle\neg\varphi \leftrightarrow \langle\!\langle\alpha\rangle\!\rangle\top \wedge \neg\langle\!\langle\alpha\rangle\!\rangle\varphi \qquad \mathsf{R_P} : \langle\!\langle\alpha\rangle\!\rangle p \leftrightarrow \langle\!\langle\alpha\rangle\!\rangle\top \wedge p$$
$$\mathsf{R_K} : \langle\!\langle\alpha\rangle\!\rangle\langle e\rangle\varphi \leftrightarrow \langle\!\langle\alpha\rangle\!\rangle\top \wedge \langle e\rangle\langle\!\langle\alpha\rangle\!\rangle\varphi \qquad \mathsf{R_A} : \langle\!\langle\alpha\rangle\!\rangle\top \leftrightarrow \top$$

Notice that $\mathsf{R_P}$ refers to atomic propositions p, thus logics involving $\mathsf{R_P}$ are not closed under uniform substitution.

We can give proof systems $\mathsf{PAL}, \mathsf{EA}$ and PU to the above three logics—**PAL**, **EA**, **PU**—by choosing a suitable set of generic reduction axioms and adding some extra ones: here we consider the multimodal logic K (without the substitution rule) in the language $\mathcal{L}(E, \mathfrak{A})$ as the base logic and use the symbol $\oplus$ for the addition of axiom schemata.

Example 2
- $\mathsf{PAL} = \mathsf{K} \oplus \mathsf{R_N}\mathsf{R_K}\mathsf{R_P} \oplus \mathsf{R_T} : \langle\!\langle !\varphi\rangle\!\rangle\top \leftrightarrow \varphi$.
- $\mathsf{EA} = \mathsf{K} \oplus \mathsf{R_N}\mathsf{R_P} \oplus \text{Pre} : \langle\!\langle(U, s)\rangle\!\rangle\top \leftrightarrow \text{Pre}(s) \oplus \mathsf{A_{EA}} : \langle\!\langle(U, s)\rangle\!\rangle\langle n\rangle\varphi \leftrightarrow \langle\!\langle(U, s)\rangle\!\rangle\top \wedge \bigvee\{\langle n\rangle\langle\!\langle(U, t)\rangle\!\rangle\varphi \mid s \to_n t\}$.
- $\mathsf{PU} = \mathsf{K} \oplus \mathsf{R_N}$ (for $\varphi!$ and $\sharp\varphi$) $\oplus \mathsf{R_P}$ (for $\varphi!$ and $\sharp\varphi$) $\oplus \mathsf{R_K}$ (for $(\varphi!, n)$, $(\varphi!, \bar{n})$, $(\sharp\varphi, n)$) $\oplus \mathsf{R_T}$ (for $\varphi!$) $\oplus \mathsf{R_A}$ (for $\sharp\varphi$) $\oplus \mathsf{A_{PU}} : \langle\!\langle\sharp\varphi\rangle\!\rangle\langle\bar{n}\rangle\psi \leftrightarrow (\neg\varphi \wedge \langle\bar{n}\rangle\langle\!\langle\sharp\varphi\rangle\!\rangle\psi) \vee (\langle\bar{n}\rangle(\varphi \wedge \langle\!\langle\sharp\varphi\rangle\!\rangle\psi))$.

We can easily see that these proof systems are equivalent to the original ones in [2, 5, 8].

3   Relational Semantics

3.1   Model Transition System

Usually, the language of a DEL is interpreted by using a Kripke model $\mathbf{M} = (S, \{R_e\}_{e\in E}, V)$. The effect of an action α is explained in terms of *model transformation*: $\mathbf{M}$ is transformed into another model $\mathbf{M}^\alpha$ and a state v in $\mathbf{M}$ is sent to a state w in $\mathbf{M}^\alpha$ (cf. Baltag [1]). Since we want to treat a family (property) of model transformations in one general framework, it is convenient to consider a family of Kripke models, linked to each other by dynamic action relations. Hence, we consider the following novel system, which modifies the update universe (see Remark 1 infra) in [7]:

Definition 3 (Model Transition System). *A* model transition system (MTS) *for $\mathcal{L}(E, \mathfrak{A})$ is a triple $\mathfrak{M} = (\mathfrak{M}_I, \Phi, \mathfrak{R})$ such that*
1. *$\mathfrak{M}_I$ is a family of Kripke models $\mathbf{M}_i = (S, \{R_e\}_{e \in E}, V)$ indexed by $i \in I$ ($\mathbf{M}_i$ is allowed to be an empty structure),*
2. *$\Phi : I \times \mathfrak{A} \to I$ is a function (notation: $\mathbf{M}_i^\alpha := \mathbf{M}_{\Phi(i,\alpha)}$),*
3. *$\mathfrak{R}$ assigns a binary relation $\mathfrak{R}_i^\alpha \subseteq \mathbf{M}_i \times \mathbf{M}_i^\alpha$ to each $(i, \alpha) \in I \times \mathfrak{A}$.*

Analogously, a *frame transition system (FTS)* $\mathfrak{F} = (\mathfrak{F}_I, \Phi, \mathfrak{R})$ is defined by using indexed Kripke frames instead of indexed Kripke models. We say that $\mathfrak{F} = (\mathfrak{F}_I, \Phi, \mathfrak{R})$ is the *underlying FTS* of an MTS $\mathfrak{M} = (\mathfrak{M}_I, \Phi, \mathfrak{R})$ and write $\mathfrak{F} = \mathbf{U}(\mathfrak{M})$ if $\mathbf{F}_i$ is the underlying frame of $\mathbf{M}_i$ for each $i \in I$. An MTS expresses model transformations: a model $\mathbf{M}_i$ is transformed into $\mathbf{M}_i^\alpha$ by action α, and the state v in $\mathbf{M}_i$ is sent to w in $\mathbf{M}_i^\alpha$ if $v \mathfrak{R}_i^\alpha w$. As a result, a pointed model $(\mathbf{M}_i, v)$ is transformed into $(\mathbf{M}_i^\alpha, w)$ that satisfies $v \mathfrak{R}_i^\alpha w$, if such a w exists.

The generic DEL language $\mathcal{L}(E, \mathfrak{A})$ is interpreted by an MTS:

Definition 4. *Suppose that $\mathbf{M}_i = (S, \{R_e\}_{e \in E}, V)$ is a Kripke model in an MTS $\mathfrak{M} = (\mathfrak{M}_I, \Phi, \mathfrak{R})$ and v is a state in $\mathbf{M}_i$. We inductively define the notion of a formula φ being* satisfied *at state v in $\mathbf{M}_i \in \mathfrak{M}_I$ (notation: $\mathfrak{M}, \mathbf{M}_i, v \models \varphi$) as follows:*

$\mathfrak{M}, \mathbf{M}_i, v \models \top$	*iff*	*always*
$\mathfrak{M}, \mathbf{M}_i, v \models p$	*iff*	$v \in V(p)$
$\mathfrak{M}, \mathbf{M}_i, v \models \neg\varphi$	*iff*	$\mathfrak{M}, \mathbf{M}_i, v \not\models \varphi$
$\mathfrak{M}, \mathbf{M}_i, v \models \varphi \vee \psi$	*iff*	$\mathfrak{M}, \mathbf{M}_i, v \models \varphi$ *or* $\mathfrak{M}, \mathbf{M}_i, v \models \psi$
$\mathfrak{M}, \mathbf{M}_i, v \models \langle e \rangle \varphi$	*iff*	*for some* $w \in S$, $v R_e w$ *and* $\mathfrak{M}, \mathbf{M}_i, w \models \varphi$
$\mathfrak{M}, \mathbf{M}_i, v \models \langle\!\langle \alpha \rangle\!\rangle \varphi$	*iff*	*for some* $w \in \mathbf{M}_i^\alpha$, $v \mathfrak{R}_i^\alpha w$ *and* $\mathfrak{M}, \mathbf{M}_i^\alpha, w \models \varphi$

We say that $\mathfrak{M}$ *validates* φ if $\mathfrak{M}, \mathbf{M}_i, v \models \varphi$ for any Kripke model $\mathbf{M}_i$ in $\mathfrak{M}$ and state v in $\mathbf{M}_i$. Validity in an FTS is defined analogously.

The model transformations of the three logics—**PAL, EA, PU**—are expressed by MTSs $(\mathfrak{M}_I, \Phi, \mathfrak{R})$ by defining Φ and $\mathfrak{R}$ as follows (the families $\mathfrak{M}_I$ have to be given so that Φ and $\mathfrak{R}$ are well-defined):

Example 3 (Specific Model Transition Systems)

- *PAL-MTS*: for every $!\varphi \in \mathfrak{A}_{\mathsf{PAL}}$ and $\mathbf{M}_i = (S, \{R_n\}_{n \in Ag}, V) \in \mathfrak{M}_I$,
  - $\mathbf{M}_i^{!\varphi} = (S', \{R_n'\}_{n \in Ag}, V')$ is the submodel of $\mathbf{M}_i$ whose carrier set is $S' = \{v \in S \mid \mathfrak{M}, \mathbf{M}_i, v \models \varphi\}$, and
  - $\mathfrak{R}_i^{!\varphi} = \{(v, v) \in \mathbf{M}_i \times \mathbf{M}_i^{!\varphi} \mid v \in \mathbf{M}_i^{!\varphi}\}$.
- *EA-MTS*: for every action model $(U, s) \in \mathfrak{A}_{\mathsf{EA}}$ with $U = (U, \{\to_n\}_{n \in Ag})$ and $\mathbf{M}_i = (S, \{R_n\}_{n \in Ag}, V) \in \mathfrak{M}_I$,

- $\mathbf{M}_i^{(U,s)} = (S', \{R'_n\}_{n\in Ag}, V')$ is given by
  * $S' = \{(v,t) \mid v \in \mathbf{M}_i, t \in U$ and $\mathfrak{M}, \mathbf{M}_i, v \models \mathrm{Pre}(t)\}$,
  * $(v,t)R'_n(w,u)$ iff vR_nw and $t{\to}_n u$ for any $n \in Ag$,
  * $(v,t) \in V'(p)$ iff $v \in V(p)$,
- $\Phi(i,(U,s)) = \Phi(i,(U,t))$ for any $i \in I$ and $(U,s),(U,t) \in \mathfrak{A}_{\mathsf{EA}}$, and
- $\mathfrak{R}_i^{(U,s)} = \{(v,(v,s)) \in \mathbf{M}_i \times \mathbf{M}_i^{(U,s)} \mid (v,s) \in \mathbf{M}_i^{(U,s)}\}$.

- PU-MTS: for every $\varphi!, \sharp\varphi \in \mathfrak{A}_{\mathsf{PU}}$, and $\mathbf{M}_i = (S, \{R_n, R_{\bar{n}}\}_{\bar{n}\in Ag}, V) \in \mathfrak{M}_I$,
  - $\mathbf{M}_i^{\varphi!} = (S, \{R'_n, R_{\bar{n}}\}_{n\in Ag}, V)$ is given by
    * $R'_n = \{(v,w) \in R_n \mid \mathfrak{M}, \mathbf{M}_i, v \models \varphi$ iff $\mathfrak{M}, \mathbf{M}_i, w \models \varphi\}$,
  - $\mathbf{M}_i^{\sharp\varphi} = (S, \{R_n, R'_{\bar{n}}\}_{n\in Ag}, V)$ is given by
    * $R'_{\bar{n}} = \{(v,w) \in R_{\bar{n}} \mid \mathfrak{M}, \mathbf{M}_i, v \models \neg\varphi$ or $\mathfrak{M}, \mathbf{M}_i, w \models \varphi\}$,
  - $\mathfrak{R}_i^{\varphi!} = \{(v,v) \in \mathbf{M}_i \times \mathbf{M}_i^{\varphi!} \mid \mathfrak{M}, \mathbf{M}_i, v \models \varphi\}$, and $\mathfrak{R}_i^{\sharp\varphi} = \{(v,v) \in \mathbf{M}_i \times \mathbf{M}_i^{\sharp\varphi} \mid v \in \mathbf{M}_i^{\sharp\varphi}\}$.

Usually bounded morphisms are defined between Kripke models. We extend them to morphisms between MTSs as follows:

Definition 5. *Let* $\mathfrak{M} = (\mathfrak{M}_I, \Phi, \mathfrak{R})$ *and* $\mathfrak{N} = (\mathfrak{N}_J, \Psi, \mathfrak{Q})$ *be MTSs. A bounded morphism* $\mathfrak{f} : \mathfrak{M} \to \mathfrak{N}$ *is a pair* $(f, \{f_i\}_{i\in I})$ *of a function* $f : I \to J$ *and bounded morphisms (in the ordinary sense)* $f_i : \mathbf{M}_i \to \mathbf{N}_{f(i)}$ *that satisfies the following conditions for any* $i \in I$ *and action expression* $\alpha \in \mathfrak{A}$: (*Here* f_i^α *denotes* $f_{\Phi(i,\alpha)}$.)
  1. $f(\Phi(i,\alpha)) = \Psi(f(i),\alpha)$,   2. *if* $v\mathfrak{R}_i^\alpha w$ *then* $f_i(v)\, \mathfrak{Q}_{f(i)}^\alpha\, f_i^\alpha(w)$,
  3. *if* $f_i(v)\, \mathfrak{Q}_{f(i)}^\alpha\, w'$ *then* $v\mathfrak{R}_i^\alpha w$ *and* $f_i^\alpha(w) = w'$ *for some* $w \in \mathbf{M}_i^\alpha$.

Items 2 and 3 in the definition correspond to the homomorphic condition and the back condition in the definition of ordinary bounded morphisms. Item 1 is their precondition. As expected, we have:

Proposition 1. *Let* $(f, \{f_i\}_{i\in I}) : \mathfrak{M} \to \mathfrak{N}$ *be a bounded morphism between MTSs. Then, for any* $\mathbf{M}_i$ *in* $\mathfrak{M}$ *and state* v *in* $\mathbf{M}_i$, $(\mathfrak{M}, \mathbf{M}_i, v)$ *and* $(\mathfrak{N}, \mathbf{N}_{f(i)}, f_i(v))$ *satisfy exactly the same formulas.*

We call a bounded morphism $(f, \{f_i\}_{i\in I}) : (\mathfrak{M}_I, \Phi, \mathfrak{R}) \to (\mathfrak{N}_J, \Psi, \mathfrak{Q})$ *surjective* if for any $j \in J$ there is an $i \in I$ such that $f(i) = j$ and $f_i : \mathbf{M}_i \to \mathbf{N}_j$ is surjective, and we say that $\mathfrak{N}$ is a *bounded morphic image* of $\mathfrak{M}$ if there is a surjective bounded morphism from $\mathfrak{M}$ to $\mathfrak{N}$. Similar notions are defined for FTSs.

Remark 1. Our MTSs generalise the idea of *update universe* [7] to the generic DEL language $\mathcal{L}(E, \mathfrak{A})$. In particular, PAL-MTSs for the specific language $\mathcal{L}_{\mathsf{PAL}}$ (Example 3) roughly correspond to the original. The difference is that [7] considers relations $(\mathbf{M}, s)R_P(\mathbf{N}, t)$ with P a subset of the carrier set of $\mathbf{M}$. These relations, for example, interpret an announcement $!\varphi$ as $R_{[\![\varphi]\!]}$, which may be called 'extensional' in the sense that $R_{[\![\varphi]\!]} = R_{[\![\psi]\!]}$ whenever $[\![\varphi]\!] = [\![\psi]\!]$. In comparison with this interpretation, our interpretation is 'intensional', since $\mathfrak{R}_i^{!\varphi}$ does not necessarily coincide with $\mathfrak{R}_i^{!\psi}$, even if φ and ψ are logically equivalent.

3.2   General Correspondence Results

We now give a correspondence between classes of MTSs (or FTSs) and the generic reduction axioms. The results below extend some of the observations made in [7].

Definition 6 (Deterministic FTS). *An FTS* $(\mathfrak{F}_I, \Phi, \mathfrak{R})$ *[or an MTS* $(\mathfrak{M}_I, \Phi, \mathfrak{R})$*] is deterministic if for each* $(i, \alpha) \in I \times \mathfrak{A}$, $\mathfrak{R}_i^\alpha$ *is a partial function.*

This means that the result of each action is completely determined by the current state.

Proposition 2. *An FTS validates* $\mathsf{R_N}$ *iff it is deterministic.*

Definition 7 (Epistemic MTS). *An MTS* $\mathfrak{M} = (\mathfrak{M}_I, \Phi, \mathfrak{R})$ *is epistemic if for each* $(i, \alpha) \in I \times \mathfrak{A}$, $v\mathfrak{R}_i^\alpha w$ *implies that* v *and* w *satisfy exactly the same atomic propositions.*

Proposition 3. *An MTS* $\mathfrak{M}$ *validates* $\mathsf{R_P}$ *and* $\mathbf{U}(\mathfrak{M})$ *validates* $\mathsf{R_N}$ *iff* $\mathfrak{M}$ *is deterministic and epistemic.*

Proposition 3 indicates that $\mathsf{R_N}\mathsf{R_P}$ corresponds to the model transformations that are deterministic and preserve the facts (the valuations). Examples of this kind of action include suggestion [8], lying [12] and commanding [14].

Definition 8 (Eliminative FTS). *An FTS* $\mathfrak{F} = (\mathfrak{F}_I, \Phi, \mathfrak{R})$ *is called* eliminative *if for any* $(i, \alpha) \in I \times \mathfrak{A}$, $\mathbf{F}_i^\alpha$ *is a subframe of* $\mathbf{F}_i$ *and the inverse relation* $(\mathfrak{R}_i^\alpha)^{-1}$ *embeds* $\mathbf{F}_i^\alpha$ *into* $\mathbf{F}_i$.

Proposition 4. *An FTS validates* $\mathsf{R_N}$ *and* $\mathsf{R_K}$ *iff it is a bounded morphic image of an eliminative FTS.*

Proof. For convenience, we assume that E is a singleton. Suppose that FTS $\mathfrak{F} = (\mathfrak{F}_I, \Phi, \mathfrak{R})$ validates $\mathsf{R_N}$ and $\mathsf{R_K}$. It is easy to see that for each $(i, \alpha) \in I \times \mathfrak{A}$, $\mathfrak{R}_i^\alpha$ of $\mathfrak{F}$ is a partial bounded morphism between Kripke frames.

However, $\mathfrak{F}$ may not be an eliminative FTS. We construct an eliminative FTS $\mathfrak{F}' = (\mathfrak{F}_{I \times \mathfrak{A}^*}, \Phi', \mathfrak{R}')$ from $\mathfrak{F}$ thus: let us denote each Kripke frame $\mathbf{F}_i$ in $\mathfrak{F}_I$ by (S_i, R_i). We first extend the notation $\Phi(i, \alpha)$ and $\mathfrak{R}_i^\alpha$ ($\alpha \in \mathfrak{A}$) (cf. Definition 3) to $\Phi(i, \gamma)$ and $\mathfrak{R}_i^\gamma$ for each string $\gamma \in \mathfrak{A}^*$; $\Phi(i, \epsilon) = i$ and $\Phi(i, \gamma\alpha) = \Phi(\Phi(i, \gamma), \alpha)$ for $\gamma \in \mathfrak{A}^*$ and $\alpha \in \mathfrak{A}$, and $\mathfrak{R}_i^\epsilon = \{(x, x) \mid x \in \mathbf{F}_i\}$ and $\mathfrak{R}_i^{\gamma\alpha} = \mathfrak{R}_i^\gamma \circ \mathfrak{R}_{\Phi(i,\gamma)}^\alpha$. Then, each Kripke frame $\mathbf{F}_{(i,\gamma)} = (S_{(i,\gamma)}, R_{(i,\gamma)})$ in $\mathfrak{F}_{I \times \mathfrak{A}^*}$ is defined as follows: $S_{(i,\gamma)} := \{v \in \mathbf{F}_i \mid$ there exists a state $w \in \mathbf{F}_{\Phi(i,\gamma)}$ such that $v\mathfrak{R}_i^\gamma w\}$; $R_{(i,\gamma)} := R_i \cap (S_{(i,\gamma)} \times S_{(i,\gamma)})$. Lastly, Φ' and $\mathfrak{R}'$ are defined to be $\Phi'((i, \gamma), \alpha) = \Phi(i, \gamma\alpha)$ and $\mathfrak{R}'^\alpha_{(i,\gamma)} = \{(v, v) \mid v \in \mathbf{F}_{(i,\gamma\alpha)}\}$.

A surjective bounded morphism $\mathfrak{f} = (f, \{f_{(i,\gamma)}\}_{(i,\gamma) \in I \times \mathfrak{A}^*})$ from $\mathfrak{F}'$ to $\mathfrak{F}$ can be defined as follows: $f : I \times \mathfrak{A}^* \to I$ maps (i, γ) to $\Phi(i, \gamma)$; $f_{(i,\gamma)} : \mathbf{F}_{(i,\gamma)} \to \mathbf{F}_{\Phi(i,\gamma)}$ maps $v \in S_{(i,\gamma)} \subseteq S_i$ to $w \in S_{\Phi(i,\gamma)}$ such that $v\mathfrak{R}_i^\gamma w$. Since each $\mathfrak{R}_i^\alpha$ is a partial bounded morphism, $\mathfrak{f}$ is indeed a bounded morphism. $\qquad\square$

Proposition 4 means that $R_N R_K$ corresponds to the actions that eliminate several possible states as **PAL** actions do, but may cause change of truth values of atomic propositions.

In the context of DEL, we often restrict our attention to epistemic actions, which do not change the truth values of atomic propositions. Thus here we name an epistemic MTS $\mathfrak{M}$ whose underlying FTS $\mathbf{U}(\mathfrak{M})$ is eliminative, an *eliminative MTS*. Eliminative updates appear in the literature not only in this *state-eliminating* style [5, 6] but in the *link-cutting* style as in PU-MTSs [6, 8]. By Proposition 3 and the construction of the proof in Proposition 4, this class is captured by R_N, R_K and R_P:

Proposition 5. *An MTS $\mathfrak{M}$ validates R_P and $\mathbf{U}(\mathfrak{M})$ validates R_N and R_K iff $\mathfrak{M}$ is a bounded morphic image of an eliminative MTS.*

Proposition 5 means that $R_N R_K R_P$ corresponds to the actions that eliminate possible states, i.e. increase agents' knowledge in the epistemic case. Public announcements are a typical example of such actions, but these do not always have to be expressed by a formula. For example, think of the computer game Minesweeper. Each time the player clicks on a cell on the board, the player's knowledge about the mines' locations is updated. Thus $\langle\!\langle \text{click} \rangle\!\rangle$ can be equally considered an action. The logic $K \oplus R_N R_K R_P$ could be useful to model such a situation.

Definition 9 (Unconditional FTS). *An FTS $\mathfrak{F} = (\mathfrak{F}_I, \Phi, \mathfrak{R})$ [or an MTS $(\mathfrak{M}_I, \Phi, \mathfrak{R})$] is called* unconditional *if for each $(i, \alpha) \in I \times \mathfrak{A}$, $\mathfrak{R}_i^\alpha$ satisfies the condition that for any $v \in \mathbf{F}_i$, there exists $w \in \mathbf{F}_i^\alpha$ such that $v \mathfrak{R}_i^\alpha w$.*

If we were to express this differently, there is no precondition to the action α and therefore α is always possible. R_A expresses this property:

Proposition 6. *An FTS validates R_A iff it is unconditional.*

Note that Propositions 2, 3 and 6 do not involve the construction of a bounded morphism, thus they can be freely combined.

We know that all the above combinations of generic reduction axioms are *canonical* in the ordinary sense, that is, their canonical models of the form $(S, \{R_e\}_{e \in E}, \{R_\alpha\}_{\alpha \in \mathfrak{A}}, V)$ satisfy their corresponding frame/model properties. Given this and all the correspondence results above, soundness and completeness hold for each of the following pairs:

1. $K \oplus R_N$ and the class of deterministic FTSs
2. $K \oplus R_N R_P$ and the class of deterministic and epistemic MTSs
3. $K \oplus R_N R_K$ and the class of eliminative FTSs
4. $K \oplus R_N R_K R_P$ and the class of eliminative MTSs
5. $K \oplus R_A$ and the class of unconditional FTSs
6. $K \oplus R_A R_N$ and the class of unconditional and deterministic FTSs
7. $K \oplus R_A R_N R_P$ and the class of unconditional, deterministic and epistemic MTSs

Remark 2. R_N plays an important role; for example, without R_N we cannot even prove that R_P is linked to the class of epistemic MTSs.

3.3 Specific Correspondence Results

So far we have been concerned with the generic DEL language $\mathcal{L}(E, \mathfrak{A})$ and MTSs/FTSs in general. We now proceed to the extra reduction axioms in Example 2, which are specific to the languages of **PAL**, **EA** and **PU**. Our modular approach leads to an alternative proof of completeness for each of the three logics.

Proposition 7

1. *Let $\mathfrak{M}$ be a deterministic and epistemic MTS for $\mathcal{L}_{\mathsf{PAL}}$. $\mathfrak{M}$ validates $\mathsf{R_T}$ and $\mathbf{U}(\mathfrak{M})$ validates $\mathsf{R_K}$ iff $\mathfrak{M}$ is a bounded morphic image of an eliminative MTS $\mathfrak{N}$ where each transformed model $\mathbf{N}_i^{!\varphi}$ is the submodel of $\mathbf{N}_i$ whose carrier set is $\{v \in \mathbf{N}_i \mid \mathfrak{N}, \mathbf{N}_i, v \models \varphi\}$, i.e. $\mathfrak{M}$ is a bounded morphic image of a PAL-MTS.*
2. *PAL is sound and complete with respect to the class of PAL-MTSs.*

As a PAL-MTS precisely expresses the intended model transformation of **PAL**, we obtain an alternative proof of the completeness of PAL with respect to the original semantics in [5].

Remark 3. An alternative and modular proof of the completeness of PAL has already been given in [13]. It also uses a canonical Kripke model of the form $(S, \{R_n\}_{n \in Ag}, \{R_{!\varphi}\}_{!\varphi \in \mathfrak{A}_{\mathsf{PAL}}}, V)$. However, our approach stresses the modular nature of the argument by starting from the generic framework.

For **EA**, we have the following result:

Proposition 8

1. *Let $\mathfrak{M} = (\mathfrak{M}, \Phi, \mathfrak{R})$ be an MTS for $\mathcal{L}_{\mathsf{EA}}$ that satisfies $\Phi(i, (U, s)) = \Phi(i, (U, t))$ for any $i \in I$ and $(U, s), (U, t) \in \mathfrak{A}_{\mathsf{EA}}$. Then, $\mathfrak{M}$ validates $\mathsf{R_P}$ and Pre while $\mathbf{U}(\mathfrak{M})$ validates $\mathsf{R_N}$ and $\mathsf{A_{EA}}$ iff $\mathfrak{M}$ is a bounded morphic image of an EA-MTS.*
2. *EA is complete with respect to the class of EA-MTSs.*

From these two results, we obtain an alternative proof of the completeness with respect to the semantics in [2], in a way analogous to the case of **PAL**.

For PU, we do not yet have an adequate characterisation of $\mathsf{A_{PU}}$. Nevertheless, the previous generic results turn out to be useful when proving the following result:

Proposition 9. *PU is complete with respect to the class of PU-MTSs.*

This also leads to an alternative proof of the completeness of PU with respect to the semantics in [8].

4 Algebraic Semantics

We discuss algebraic semantics of DEL and develop a similar correspondence theory as above.

4.1 Algebraic Model Transition System

We first introduce an algebraic counterpart of the notion of MTS (Definition 3). By *algebraic model*, we mean $\mathcal{M} = (\mathcal{A}, \theta)$, where $\mathcal{A} = (A, +, -, 1, \{f_e\}_{e \in E})$ is a *Boolean algebra with operators (BAO)* and $\theta : \mathcal{P} \to A$ is an *assignment*.

Definition 10 (Algebraic Model Transition System). *An* algebraic model transition system (AMTS) *for* $\mathcal{L}(E, \mathfrak{A})$ *is a triple* $\mathsf{M} = (\mathsf{M}_I, \Phi, \mathsf{F})$ *such that*
1. M_I *is a family of algebraic models* $\mathcal{M}_i$ *indexed by* $i \in I$,
2. $\Phi : I \times \mathfrak{A} \to I$ *is a function (notation:* $\mathcal{M}_i^\alpha := \mathcal{M}_{\Phi(i,\alpha)}$),
3. F *assigns a 0-preserving additive function*[1] F_i^α *from the carrier set of* $\mathcal{M}_i^\alpha$ *to that of* $\mathcal{M}_i$ *for each* $(i, \alpha) \in I \times \mathfrak{A}$ *(notation:* $\mathsf{F}_i^\alpha : \mathcal{M}_i^\alpha \to \mathcal{M}_i$).

Analogously, an *algebra transition system (ATS)* $\mathsf{A} = (\mathsf{A}_I, \Phi, \mathsf{F})$ is also defined by using indexed BAOs instead of indexed algebraic models. We say that $\mathsf{A} = (\mathsf{A}_I, \Phi, \mathsf{F})$ is the *underlying* ATS of an AMTS $\mathsf{M} = (\mathsf{M}_I, \Phi, \mathsf{F})$ and write $\mathsf{A} = \mathsf{U}(\mathsf{M})$ if $\mathcal{A}_i \in \mathsf{A}_I$ is the underlying BAO of algebraic model $\mathcal{M}_i$ for each $i \in I$.

Definition 11. *Let* $\mathsf{M} = (\mathsf{M}_I, \Phi, F)$ *be an AMTS for* $\mathcal{L}(E, \mathfrak{A})$ *and* $\mathcal{M}_i = (A, +, -, 1, \{f_e\}_{e \in E}, \theta)$ *in* M_I. *The meaning* $[\varphi]_{\mathcal{M}_i, \mathsf{M}}$ *of an* $\mathcal{L}(E, \mathfrak{A})$-*formula* φ *is inductively defined as follows:*

$$[\top]_{\mathcal{M}_i, \mathsf{M}} = 1 \qquad\qquad [p]_{\mathcal{M}_i, \mathsf{M}} = \theta(p)$$
$$[\neg \varphi]_{\mathcal{M}_i, \mathsf{M}} = -[\varphi]_{\mathcal{M}_i, \mathsf{M}} \qquad [\varphi \vee \psi]_{\mathcal{M}_i \mathsf{M}} = [\varphi]_{\mathcal{M}_i, \mathsf{M}} + [\psi]_{\mathcal{M}_i, \mathsf{M}}$$
$$[\langle e \rangle \varphi]_{\mathcal{M}_i, \mathsf{M}} = f_e([\varphi]_{\mathcal{M}_i, \mathsf{M}}) \quad [\langle\!\langle \alpha \rangle\!\rangle \varphi]_{\mathcal{M}_i, \mathsf{M}} = \mathsf{F}_i^\alpha([\varphi]_{\mathcal{M}_i^\alpha, \mathsf{M}})$$

We say that an AMTS M *validates* φ if $[\varphi]_{\mathcal{M}_i, \mathsf{M}} = 1$ for any $\mathcal{M}_i$ in M. Validity in an ATS is defined analogously.

The algebraic semantics of **PAL** [4] and of **EA** [3] can be rephrased in terms of our AMTSs as follows. An algebraic semantics of **PU** has not been established. This shall be dealt with in our forthcoming work.

Example 4 (Specific Algebraic Model Transition Systems)
   PAL: Let $\mathcal{A} = (A, +, -, 1, \{f_e\}_{e \in E})$ be a BAO and $\mathcal{M} = (\mathcal{A}, \theta)$ an algebraic model. For each $a \in A$, we define $\mathcal{A} \!\downarrow\! a = (A \!\downarrow\! a, +', -', 1', \{f_e'\}_{e \in E})$ and $\mathcal{M} \!\downarrow\! a = (\mathcal{A} \!\downarrow\! a, \theta_a)$ as follows:

$$A \!\downarrow\! a = \{x \in A \mid x \leq a\} \quad x +' y = a \cdot (x + y) = x + y$$
$$-'(x) = a \cdot (-x) \qquad\qquad 1' = a \cdot 1 = a$$
$$f_e'(x) = a \cdot f_e(x) \qquad\qquad \theta_a(p) = a \cdot \theta(p)$$

A *PAL-AMTS* (for the language $\mathcal{L}_{\mathsf{PAL}}$) is then given as follows: for every $\mathcal{M}_i = (A, +, -, 1, \{f_n\}_{n \in Ag})$ and $!\varphi \in \mathfrak{A}_{\mathsf{PAL}}$, let $\mathcal{M}_i^{!\varphi} = \mathcal{M} \!\downarrow\! [\varphi]_{\mathcal{M}_i, \mathsf{M}}$, and $\mathsf{F}_i^{!\varphi}$ be the set inclusion function from $A \!\downarrow\! [\varphi]_{\mathcal{M}_i, \mathsf{M}}$ to A. It is easy to see that $\mathsf{F}_i^{!\varphi}$ is indeed 0-preserving and additive.

   EA: Suppose that $\mathcal{M} = (A, +, -, 1, \{f_n\}_{n \in Ag}, \theta)$ is an algebraic model and that (U, s) is an action model with $U = (U, \{\to_n\}_{n \in Ag})$. We define $\prod_U \mathcal{M} = (\prod_U A, +', -', 1', \{f_n'\}_{n \in Ag}, \theta')$: $\prod_U A$ is the $|U|$-colored product (i.e. the power of A with each coordinate indexed by $u \in U$); $+', -', 1'$ and θ' are defined

---
[1] That is, F_i^α such that $\mathsf{F}_i^\alpha(0) = 0$ and $\mathsf{F}_i^\alpha(x + y) = \mathsf{F}_i^\alpha(x) + \mathsf{F}_i^\alpha(y)$.

coordinatewise; and $f'_n : \prod_U A \to \prod_U A$ is defined by $f'_n(k)(s) = \bigvee\{f_n(k(t)) \mid s\to_n t\}$ for any $k \in \prod_U A$ and $s \in U$. An *EA-AMTS* M is then given as follows (here $[\mathrm{Pre}]_{\mathcal{M}_i,\mathsf{M}}$ denotes the element $\langle [\mathrm{Pre}(s)]_{\mathcal{M}_i,\mathsf{M}}\rangle_{s\in U}$ of $\prod_U \mathcal{M}$): each transformed model $\mathcal{M}_i^{(U,s)}$ is given by $(\prod_U \mathcal{M}_i)\downarrow [\mathrm{Pre}]_{\mathcal{M}_i,\mathsf{M}}$; $\mathsf{F}_i^{(U,s)} : \mathcal{M}_i^{(U,s)} \to \mathcal{M}_i$ is defined to be the composition of the set inclusion $\mathcal{M}_i^{(U,s)} \hookrightarrow \prod_U \mathcal{M}_i$ and the s-th projection $\prod_U \mathcal{M}_i \to \mathcal{M}_i$; and we impose the condition that $\Phi(i,(U,s)) = \Phi(i,(U,t))$ for any $i \in I$ and $(U,s), (U,t) \in \mathfrak{A}_{\mathsf{EA}}$.

Homomorphisms for AMTSs are defined as follows:

Definition 12. *Let* $\mathsf{M} = (\mathsf{M}_I, \Phi, \mathsf{F})$ *and* $\mathsf{N} = (\mathsf{N}_J, \Psi, \mathsf{G})$ *be AMTSs. A homomorphism* h *from* M *to* N *is a pair* $(h, \{h_j\}_{j\in J})$ *of a function* $h : J \to I$ *and algebraic model homomorphisms*[2] $h_j : \mathcal{M}_{h(j)} \to \mathcal{N}_j$*, such that for any* $j \in J$ *and action expression* $\alpha \in \mathfrak{A}$*, 1.* $\Phi(h(j), \alpha) = h(\Psi(j, \alpha))$ *and 2.* $h_j \circ \mathsf{F}_{h(j)}^\alpha = \mathsf{G}_j^\alpha \circ h_{\Psi(j,\alpha)}$*.*

Proposition 10. *Let* $(h, \{h_j\}_{j\in J}) : \mathsf{M} \to \mathsf{N}$ *be a homomorphism between AMTSs. Then, for any* $\mathcal{N}_j$ *in* N *and* $\mathcal{L}(E, \mathfrak{A})$*-formula* φ*,* $h_j([\varphi]_{\mathcal{M}_{h(j)},\mathsf{M}}) = [\varphi]_{\mathcal{N}_j,\mathsf{N}}$*.*

We call a homomorphism $(h, \{h_j\}_{j\in J}) : (\mathsf{M}_I, \Phi, \mathsf{F}) \to (\mathsf{N}_J, \Psi, \mathsf{G})$ *injective* if for any $i \in I$ there is a $j \in J$ such that $h(j) = i$ and $h_j : \mathcal{M}_i \to \mathcal{N}_j$ is an injective homomorphism, and we say that M can be *embedded* into N if there is an injective homomorphism from M to N. Similar notions are defined for ATSs.

4.2   General Algebraic Correspondence Results

Let us now discuss correspondences between reduction axioms and algebraic properties.

Definition 13. *An ATS* $\mathsf{A} = (\mathsf{A}_I, \Phi, \mathsf{F})$ *is* deterministic *if each* $\mathsf{F}_i^\alpha : \mathcal{A}_i^\alpha \to \mathcal{A}_i$ *preserves all meets (i.e.* $\mathsf{F}_i^\alpha(x \cdot y) = \mathsf{F}_i^\alpha(x) \cdot \mathsf{F}_i^\alpha(y)$*).*

Proposition 11. *An ATS validates* R_{N} *iff it is deterministic.*

Correspondence results concerning R_{K} are expressed by the following notions:

Definition 14. *An ATS* $\mathsf{A} = (\mathsf{A}_I, \Phi, \mathsf{F})$ *is* eliminative *if for any algebraic model* $\mathcal{A}_i \in \mathsf{A}_I$ *and* $\alpha \in \mathfrak{A}$*, the transformed model* $\mathcal{A}_i^\alpha$ *is given by* $\mathcal{A}_i\downarrow a$ *for some* $a \in \mathcal{A}_i$*, and* $\mathsf{F}_i^\alpha : \mathcal{A}_i^\alpha \to \mathcal{A}_i$ *the set inclusion function. An* eliminative AMTS *is analogously defined by using* $\mathcal{M}_i\downarrow a$ *instead of* $\mathcal{A}_i\downarrow a$*.*

Proposition 12. *Let* A *be an ATS and* M *an AMTS.*
1. A *validates* R_{N} *and* R_{K} *iff it can be embedded into an eliminative ATS.*
2. M *validates* R_{P} *and* $\mathbf{U}(\mathsf{M})$ *validates* R_{N} *and* R_{K} *iff* M *can be embedded into an eliminative AMTS.*

---

[2] These are BAO homomorphisms that preserve assignments.

As a corollary, soundness and completeness hold for each of the following pairs:

1. $K \oplus R_N$ and the class of deterministic ATSs
2. $K \oplus R_N R_K$ and the class of eliminative ATSs
3. $K \oplus R_N R_K R_P$ and the class of eliminative AMTSs

4.3 Specific Correspondence Results

We next turn to correspondence results specific to **PAL** and **EA**.

Proposition 13
1. *An AMTS* M *for* $\mathcal{L}_{PAL}$ *validates* R_P *and* R_T *while* $U(M)$ *validates* R_N *and* R_K *iff* M *can be embedded into a PAL-AMTS.*
2. PAL *is sound and complete with respect to the class of PAL-AMTSs.*

The above results lead to an alternative and modular proof of the completeness of PAL with respect to the algebraic semantics in [4] since a PAL-AMTS expresses the intended algebraic model transformations of **PAL**.

Analogously, in the case of **EA**, the following results give its completeness with respect to the algebraic semantics in [3]:

Proposition 14
1. *Let* $M = (M_I, \Phi, F)$ *be an AMTS for* $\mathcal{L}_{EA}$ *that satisfies* $\Phi(i, (U, s)) = \Phi(i, (U, t))$ *for any* $i \in I$ *and* $(U, s), (U, t) \in \mathfrak{A}_{EA}$. *Then,* M *validates* R_P *and* Pre *while* $U(M)$ *validates* R_N *and* A_{EA} *iff* M *can be embedded into an EA-AMTS.*
2. EA *is sound and complete with respect to the class of EA-AMTSs.*

5 On Duality between MTSs and AMTSs

The correspondence results in Section 4 were generated by the duality between MTSs and AMTSs. This can be summarised as follows.

First of all, all MTSs and all bounded morphisms, and all AMTSs and all homomorphisms constitute categories $\mathcal{MTS}$ and $\mathcal{AMTS}$. Here, the composition of morphisms is defined as follows: for bounded morphisms $\mathfrak{f} = (f, \{f_i\}_{i \in I}) : (\mathfrak{L}_I, \Phi, \mathfrak{P}) \to (\mathfrak{M}_J, \Psi, \mathfrak{Q})$ and $\mathfrak{g} = (g, \{g_j\}_{j \in J}) : (\mathfrak{M}_J, \Psi, \mathfrak{Q}) \to (\mathfrak{N}_K, X, \mathfrak{R})$, their composition $\mathfrak{g} \circ \mathfrak{f}$ is given by $(g \circ f, \{g_{f(i)} \circ f_i\}_{i \in I})$; and for homomorphisms $f = (f, \{f_j\}_{j \in J}) : (L_I, \Phi, F) \to (M_J, \Psi, G)$ and $g = (g, \{g_k\}_{k \in K}) : (M_J, \Psi, G) \to (N_K, X, H)$, their composition $g \circ f$ is defined by $(f \circ g, \{g_k, \circ f_{g(k)}\}_{k \in K})$.

Between these two categories, there are contravariant functors as follows: (Here, $\mathbf{M}^+$ and $\mathcal{M}_+$ denote the full complex algebra with the assignment of an ordinary Kripke model $\mathbf{M}$ and the ultrafilter model of an algebraic model $\mathcal{M}$.)

Definition 15

1. *A contravariant functor* $(-)^+ : \mathcal{MTS} \to \mathcal{AMTS}$ *is given as follows: its object function assigns to each MTS* $\mathfrak{M} = (\{\mathbf{M}_i\}_{i \in I}, \Phi, \mathfrak{R})$ *the AMTS* $\mathfrak{M}^+ = (\{\mathbf{M}_i^+\}_{i \in I}, \Phi, \mathfrak{R}^+)$ *where* $\mathfrak{R}^+(i, \alpha)$ *is given by* $(\mathfrak{R}_i^\alpha)^{-1} : (\mathbf{M}_i^\alpha)^+ \to \mathbf{M}_i^+$; *its arrow function assigns to each bounded morphism* $(f, \{f_i\}_{i \in I}) : \mathfrak{M} \to \mathfrak{N}$ *the homomorphisms* $(f, \{f_i^{-1} : \mathbf{M}_{f(i)}^+ \to \mathbf{M}_i^+\}_{i \in I}) : \mathfrak{N}^+ \to \mathfrak{M}^+$.

2. *A contravariant functor* $(-)_+ : \mathcal{AMTS} \to \mathcal{MTS}$ *is given as follows: its object function assigns to each AMTS* $\mathsf{M} = (\{\mathcal{M}_i\}_{i \in I}, \Phi, \mathsf{F})$ *the MTS* $\mathsf{M}_+ = (\{\mathcal{M}_{i+}\}_{i \in I}, \Phi, \mathsf{F}_+)$ *where* $\mathsf{F}_+(i, \alpha)$ *is defined by* $v \mathsf{F}_{+i}^\alpha w \Leftrightarrow \mathsf{F}_i^\alpha[w] \subseteq v$; *its arrow function assigns to each homomorphism* $(f, \{f_j : \mathcal{M}_{f(j)} \to \mathcal{N}_j\}_{j \in J}) : \mathsf{M} \to \mathsf{N}$ *the bounded morphism* $(f, \{f_j^{-1} : \mathcal{N}_{j+} \to \mathcal{M}_{f(j)+}\}_{j \in J}) : \mathsf{N}_+ \to \mathsf{M}_+$.

In particular, surjective bounded morphisms of MTSs and injective homomorphisms of AMTSs are 'dual' via the above contravariant functors.

On the relationship between these two functors $(-)^+$ and $(-)_+$, the following result is immediate:

1. An MTS $\mathfrak{M} = (\mathfrak{M}_I, \Phi, \mathfrak{R})$ is 'embedded' into $(\mathfrak{M}_+)^+$ by $\epsilon_{\mathfrak{M}} = (Id_I, \{\Pi_i\}_{i \in I}) : \mathfrak{M} \to (\mathfrak{M}^+)_+$ where each embedding $\Pi_i : \mathbf{M}_i \to (\mathbf{M}_i^+)_+$ assigns the principal ultrafilter π_x to $x \in \mathbf{M}_i$.
2. An AMTS $\mathsf{M} = (\mathsf{M}_I, \Phi, \mathsf{F})$ is 'embedded' into $(\mathsf{M}_+)^+$ by $\eta_{\mathsf{M}} = (Id_I, \{r_i\}_{i \in I}) : \mathsf{M} \to (\mathsf{M}_+)^+$ where $r_i : \mathcal{M}_i \to (\mathcal{M}_{i+})^+$ is the canonical embedding.

All the $\epsilon_{\mathfrak{M}}$ meet the condition of a natural transformation from $Id_{\mathcal{MTS}}$ to $(-)_+ \circ (-)^+$ and all the η_{M} meet that of a natural transformation from $Id_{\mathcal{AMTS}}$ to $(-)^+ \circ (-)_+$. However, placing a condition on $\mathfrak{M}$ and M is necessary for $\epsilon_{\mathfrak{M}}$ and $\eta_{\mathcal{M}}$ to be arrows in the categories $\mathcal{MTS}$ and $\mathcal{AMTS}$, and to obtain natural transformations ϵ and η. For instance, as an easy example, let us take the condition that all $\mathbf{M}_i \in \mathfrak{M}$ and all $\mathcal{M}_i \in \mathsf{M}$ are finite. Those objects satisfying this condition constitute full subcategories $Fin\mathcal{MTS}$ and $Fin\mathcal{AMTS}$ of $\mathcal{MTS}$ and $\mathcal{AMTS}$, which are equivalent via the restricted functors of $(-)^+$ and $(-)_+$ as η and ϵ become natural isomorphisms in this case. It is this duality that underlies our algebraic development: for example, the algebraic characterisation of the axiom $\mathsf{R_N}$ (Proposition 4.2) is obtained by the fact that $\mathfrak{R}_i^\alpha$ is a partial function (i.e. deterministic) iff $(\mathfrak{R}_i^\alpha)^{-1}$ preserves intersections (i.e. meets).

6   Conclusion

We have proposed a general framework for modal correspondence in Dynamic Epistemic Logic (DEL) in both relational and algebraic semantics. (i) We first introduced a generic DEL language and (ii) accordingly introduced model transition systems (MTSs) and algebraic model transition systems (AMTSs) as 'static' formalisations of model transformations. Using our framework, (iii) we gave general correspondence results for generic reduction axioms and (iv) extended them to specific reduction axioms defining **PAL**, **EA** and **PU**. (v) All these constitute modular proofs to the completeness of the three logics with respect to both relational and algebraic semantics.

An exception is the algebraic study of **PU**, which shall be addressed in our future work. It would be also interesting to study other DELs, such as **LCC** [10], and other operators, like common knowledge operators. In this paper we have only considered reduction axioms that already exist in the literature. However, since our language is generic, it is perhaps possible to treat a more general class of axioms and develop a 'dynamic' Sahlqvist theory for them. This too shall form the object of our future studies.

Acknowledgments. I would like to thank Kazushige Terui, Ichiro Hasuo, Shin-ya Katsumata, Masahito Hasegawa, Toru Takisaka, Manuela Antoniu, Johan van Benthem, Alexandru Baltag, Katsuhiko Sano, Yanjing Wang, Minghui Ma, Nobu-yuki Suzuki, Mamoru Kaneko and the anonymous referees for many helpful comments.

References

1. Baltag, A., Moss, L.S.: Logics for epistemic programs. Synthese 139(2), 165–224 (2004)
2. Baltag, A., Moss, L.S., Solecki, S.: The logic of public announcements, common knowledge, and private suspicions. In: Proceedings of the 7th Conference on Theoretical Aspects of Rationality and Knowledge, pp. 43–56. Morgan Kaufmann Publishers Inc. (1998)
3. Kurz, A., Palmigiano, A.: Epistemic updates on algebras. Logical Methods in Computer Science 9(4) (2013)
4. Ma, M.: Mathematics of public announcements. In: van Ditmarsch, H., Lang, J., Ju, S. (eds.) LORI 2011. LNCS, vol. 6953, pp. 193–205. Springer, Heidelberg (2011)
5. Plaza, J.: Logics of public communications. In: Proceedings of the Fourth International Symposium on Methodologies for Intelligent Systems: Poster Session Program, pp. 201–216. Oak Ridge National Laboratory (1989)
6. van Benthem, J.: Dynamic logic for belief revision. Journal of Applied Non-Classical Logics 17(2), 129–155 (2007)
7. van Benthem, J.: Two logical faces of belief revision. In: Krister Segerberg on Logic of Actions, pp. 281–300. Springer (2014)
8. van Benthem, J., Liu, F.: Dynamic logic of preference upgrade. Journal of Applied Non-Classical Logics 17(2), 157–182 (2007)
9. van Benthem, J., Pacuit, E.: Dynamic logics of evidence-based beliefs. Studia Logica 99(1-3), 61–92 (2011)
10. van Benthem, J., van Eijck, J., Kooi, B.: Logics of communication and change. Information and Computation 204(11), 1620–1662 (2006)
11. van Ditmarsch, H., Kooi, B.: Semantic results for ontic and epistemic change. In: Logic and the Foundations of Game and Decision Theory (LOFT 7), pp. 87–117 (2008)
12. van Ditmarsch, H., van Eijck, J., Sietsma, F., Wang, Y.: On the logic of lying. In: van Eijck, J., Verbrugge, R. (eds.) Games, Actions and Social Software 2010. LNCS, vol. 7010, pp. 41–72. Springer, Heidelberg (2012)
13. Wang, Y., Cao, Q.: On axiomatizations of public announcement logic. Synthese 190, 103–134 (2013)
14. Yamada, T.: Acts of commanding and changing obligations. In: Inoue, K., Satoh, K., Toni, F. (eds.) CLIMA 2006. LNCS (LNAI), vol. 4371, pp. 1–19. Springer, Heidelberg (2007)

Intuitionistic Epistemology and Modal Logics of Verification

Tudor Protopopescu*

The Graduate Center, City University of New York
tprotopopescu@gradcenter.cuny.edu

Abstract. The language of intuitionistic epistemic logic, IEL [3], captures basic reasoning about intuitionistic knowledge and belief, but its language has expressive limitations. Following Gödel's explication of IPC as a fragment of the more expressive system of classical modal logic S4 we present a faithful embedding of IEL into S4V – S4 extended with a verification modality. The classical modal framework is finer-grained and more flexible, allowing us to make explicit various properties of verification.

1 Introduction

Intuitionistic epistemic logic, IEL, was introduced in [3]. The systems developed there provide a formal foundation for intuitionistic epistemology based on the Brouwer-Heyting-Kolmogorov (BHK) semantics. Our purpose here is to study intuitionistic knowledge and belief from a classical modal perspective. The classical modal language is more expressive, enabling us to make explicit assumptions which the intuitionistic epistemic language cannot express, thereby gaining us a more nuanced understanding of intuitionistic knowledge and belief.

Intuitionistic knowledge is the product of verification, not necessarily of proof. IEL extends intuitionistic propositional logic, IPC, by adding an epistemic modality **K** asserting a proposition is known on the basis of verification. Just as Gödel's translation [12] faithfully embeds IPC into S4, likewise IEL faithfully embeds into S4V, the result of extending S4 with a verification modality **V**.

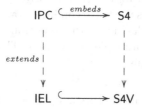

Fig. 1. Relationship between IPC–S4 IEL–S4V

* Many thanks to Sergei Artemov for helpful suggestions and inspiring discussions. Thanks also to anonymous referees for helpful suggestions.

© Springer-Verlag Berlin Heidelberg 2015
W. van der Hoek et al. (Eds.): LORI 2015, LNCS 9394, pp. 295–307, 2015.
DOI: 10.1007/978-3-662-48561-3_24

The same relationships hold of the logics of intuitionistic belief and strong knowledge presented below.

2   Intuitionistic Epistemic Logic

According to the BHK semantics a proposition, A, is true if there is a proof of it and false if the assumption that there is a proof of A yields a contradiction. This is extended to complex propositions by the following clauses:

- a proof of $A \wedge B$ consists in a proof of A and a proof of B;
- a proof of $A \vee B$ consists in giving either a proof of A or a proof of B;
- a proof of $A \to B$ consists in a construction which given a proof of A returns a proof of B;
- $\neg A$ is an abbreviation for $A \to \bot$, and $\bot$ is a proposition that has no proof.

The fundamental principle of verification-based intuitionistic knowledge, and belief, is that

$$A \to \mathbf{K}A \qquad\qquad \text{(Co-Reflection)}$$

is valid on a BHK reading. Intuitionistic truth is based on proof; since any proof is a verification, the intuitionistic truth of a proposition yields a verification and hence knowledge/belief.

By similar reasoning the converse principle,

$$\mathbf{K}A \to A, \qquad\qquad \text{(Reflection)}$$

is not valid on a BHK reading. A verification may warrant knowledge, but need not be, or yield a method for obtaining, a proof.[1] Co-reflection, along with the distributivity of $\mathbf{K}$ over implication $\mathbf{K}(A \to B) \to (\mathbf{K}A \to \mathbf{K}B)$, forms the basic logic of intuitionistic belief, IEL^-.

Definition 1 (IEL^-). *The list of axioms and rules of* IEL^- *consists of:*

IA0.   Axioms of propositional intuitionistic logic;
IA1. $\mathbf{K}(A \to B) \to (\mathbf{K}A \to \mathbf{K}B)$;
IA2. $A \to \mathbf{K}A$;

IR0.   Modus Ponens.

The difference between intuitionistic knowledge and belief, as in the classical case, is that knowledge obeys the truth condition: falsehoods cannot be known, or only truths can be known. Classically the reflection principle expresses this idea; intuitionistically a weaker principle is required. The minimal intuitionistically acceptable formulation of this is

$$\neg \mathbf{K}\bot, \qquad\qquad \text{(Truth Condition)}$$

adding this yields the basic intuitionistic logic of knowledge, IEL.[2]

---

[1] For example, interpreting $\mathbf{K}A$ as a 'truncated' or 'squash' type of Intuitionistic Type Theory, [19,6] yields the invalidity of reflection.

[2] Other acceptable formulations of the truth condition are $\neg A \to \neg \mathbf{K}A$ and $\neg(\mathbf{K}A \wedge \neg A)$. Adding these as axioms to IEL^- yields equivalent systems, see [3].

Definition 2 (IEL). IEL *is the system* IEL$^-$ *with the additional axiom:*

IA3. $\neg \mathbf{K}\bot.$

One might argue that a weak form of reflection is acceptable, namely that a verification of a verification of A yields a verification of A, though not necessarily a proof of A:[3]

$$\mathbf{KK}A \rightarrow \mathbf{K}A. \qquad \text{(Weak Reflection)}$$

Since positive introspection $\mathbf{K}A \rightarrow \mathbf{KK}A$ (and negative introspection $\neg\mathbf{K}A \rightarrow \mathbf{K}\neg\mathbf{K}A$) is simply an instance of co-reflection this immediately yields that knowledge or belief is idempotent, $\mathbf{KK}A \leftrightarrow \mathbf{K}A$. Adding weak reflection yields the logic of strong intuitionistic knowledge, IEL$^+$.

Definition 3 (IEL$^+$). IEL$^+$ *is the system* IEL *with the additional axiom:*

IA4. $\mathbf{KK}A \rightarrow \mathbf{K}A.$

Definition 4 (Semantics for $\mathcal{L} \in \{$IEL$^-$, IEL, IEL$^+\}$). *Models for $\mathcal{L}$ are intuitionistic Kripke models, $\langle W, R, \Vdash \rangle$, with an additional accessibility relation E.*

IEL$^-$: *An* IEL$^-$ *model satisfies the following conditions on E, for states u, v, w*

IM1.   *uEv yields uRv;*
IM2.   *uRv and vEw yield uEw;*
IM3.   *$u \Vdash \mathbf{K}A$ iff $v \Vdash A$ for all v such that uEv.*

IEL: *An* IEL *model is an* IEL$^-$ *model with the additional condition on E that:*

IM4.   *E is serial, for all u, there is a v such that uEv.*

IEL$^+$: *An* IEL$^+$ *model is an* IEL *model with the additional condition on E that:*

IM5.   *E is dense, uEv implies there is a w such that uEw and wEv.*

IEL$^-$, IEL, IEL$^+$ are each sound and complete, satisfy monotonicity, have the disjunction property, and the rule of $\mathbf{K}$-necessitation is derivable, see [3].

For other formulations of an intuitionistic epistemic logic, though not necessarily from a BHK perspective, see [20,17,14]. All these endorse reflection and are arguably too classical in their view of knowledge as a result.

As intuitionistic modal logics IEL$^-$, IEL and IEL$^+$ (e.g. [7,8,21,22]) are similar to Došen's [9] in that reflection fails and co-reflection holds, but his $\Box$ is rather a simulation of classical logic inside intuitionistic logic rather than an epistemic modality.

---

[3] E.g. this holds in the intuitionistic type theoretical interpretation of $\mathbf{K}$, see Footnote 1, and [3].

2.1   Two Readings of Intuitionistic Knowledge

Within a BHK context $\mathbf{K}A$ can be read in two ways. On the first reading $\mathbf{K}A$ asserts that:

> *it is verified that A holds intuitionistically, i.e. that A has a proof, not necessarily specified in the process of verification.*

This kind of knowledge amounts to checking for the existence of a proof of A, not necessarily by an explicit BHK-proof. For instance, the existence of a proof of A may be checked by a zero-knowledge protocol, or a probabilistic procedure known to yield correct results with a very small probability of error.

However, the intuitionistic epistemic language supports another reading, which considers non-proof verification as another means of constructively establishing the truth of a proposition, along-side proof. On this readings $\mathbf{K}A$ asserts that:

> *it is verified in some 'non-proof' constructive sense that A holds.*

On such a reading a proposition may be either true in the sense of having a proof, or true in the weaker sense of being constructively verified. The intuitionistic epistemic language does not distinguish these readings; a classical modal framework does, and so enables us to choose which reading to work with.

2.2   Stability of Truth and Knowledge

Intuitionistic knowledge and belief are monotonic with respect to truth, this means that both are indefeasible, once $\mathbf{K}A$ is true it can never become false. This is due to the stability of intuitionistic truth, i.e. proof; once a proposition is proved it can never become 'unproved'. The stability of truth is encoded by the definition of $\Vdash$. This stability is extended to $\mathbf{K}$ by Condition IM2, and accounts for the indefeasibility of knowledge and belief, as well as positive and negative introspection. These are essential properties of intuitionistic knowledge and belief precisely because they are aspects of the intuitionistic notion of truth.

In a classical modal framework truth and knowledge are not stable by default, offering the flexibility to assume explicitly the stability of knowledge.[4]

3   Modal Logics of Verification and Proof

The well-known Gödel translation yields a faithful embedding of the intuitionistic propositional calculus, IPC, into the classical modal logic S4 (see [12,16,5,18]).

Following Gödel [12] we interpret the $\Box$ of S4 as provability; this reading has been made precise within the framework of the Logic of Proofs [1]. On this reading appending a $\Box$ to a proposition is a way of expressing in a classical language that it is constructively true. The translation takes a formula, A, of IPC and returns a formula of S4, $tr(A)$, according to the rule

---

[4] For the role of stability in a constructive resolution of the 'knowability paradox' see [2].

box every subformula of A.

By extending S4 with a verification modality $\mathbf{V}$, the translation can be extended to each of the logics IEL$^-$, IEL, IEL$^+$. We will define the systems S4V$^-$, S4V, S4V$^+$ and show that the Gödel translation yields a faithful embedding of each intuitionistic system into its classical modal companion. In this way we interpret intuitionistic truth in a setting where we can make explicit when (and if) a proposition is intuitionistically true, or verified, or some combination of them.

Intuitionistic $\mathbf{K}$ represents verifications which are not necessarily proofs, which is why intuitionistic reflection can fail. Similarly, $\mathbf{V}$ represents a verification procedure which is not necessarily factive (unlike $\square$, which represents proof). This is a realistic assumption given many, if not most, of our justifications are fallible, and hence so is the knowledge based on them.[5] The systems S4V$^-$, S4V, and S4V$^+$ may be regarded as systems of proof and verification-based belief or fallible knowledge. $\mathbf{V}A \rightarrow A$ could be added to the systems in question to yield systems of verification-based infallible, i.e. factive, knowledge and proof. The embedding results below do not require reflection for $\mathbf{V}$, nor would adding reflection alter them.

3.1   Modal Logics S4V$^-$, S4V, S4V$^+$

Definition 5 (S4V$^-$). *The list of axioms and rules of* S4V$^-$ *consists of*

    A0.   *Axioms of* S4;
    A1. $\mathbf{V}(A \rightarrow B) \rightarrow (\mathbf{V}A \rightarrow \mathbf{V}B)$;
    A2. $\square A \rightarrow \mathbf{V}A$;

    R0.   *Modus Ponens;*
    R1. $\square$*-Necessitation.*

S4V$^-$ represents basic, not necessarily consistent, verification, the only requirement of which is that anything which is proved be regarded as verified.

Definition 6 (S4V). S4V *is* S4V$^-$ *with the additional axiom:*

    A3. $\neg\mathbf{V}\bot$.

S4V represents consistent verification, which does not guarantee the truth of the proposition verified.

Definition 7 (S4V$^+$). S4V$^+$ *is* S4V *with the additional axiom:*

    A4. $\mathbf{V}\mathbf{V}A \rightarrow \mathbf{V}A$.

S4V$^+$ represents verifications which can correctly evaluate the fact of verification.

---

[5] Fallibilism is a position which "...contemporary [mainstream] epistemologists almost universally agree in endorsing" [15]. See e.g. [13] for an opposing view.

Proposition 1. *The rule of* $\mathbf{V}$*-Necessitation is derivable in* $\mathcal{L}_\square$.

Proof. Assume $\vdash A$, by $\square$-necessitation $\vdash \square A$ follows, hence by Axiom A2 $\vdash \mathbf{V}A$.

Definition 8 (Semantics for $\mathcal{L}_\square \in \{S4V^-, S4V, S4V^+\}$). *Models for* $\mathcal{L}_\square$ *are* S4 *Kripke models,* $\langle W, R_\square, \Vdash \rangle$*, with an additional accessibility relation* $R_{\mathbf{V}}$.

$S4V^-$: *An* $S4V^-$*-model satisfies the following conditions on* $R_{\mathbf{V}}$*, for states* x, y, z

    M1. $xR_{\mathbf{V}}y$ *yields* $xR_\square y$;
    M2. $x \Vdash \mathbf{V}A$ *iff* $y \Vdash A$ *for all* y *such that* $xR_{\mathbf{V}}y$.

$S4V$: *An* $S4V$*-model is an* $S4V^-$*-model with the additional condition on* $R_{\mathbf{V}}$ *that:*

    M3. $R_{\mathbf{V}}$ *is serial, for all* x *there is a* y *such that* $xR_{\mathbf{V}}y$.

$S4V^+$: *An* $S4V^+$*-model is an* $S4V$*-model with the additional condition on* $R_{\mathbf{V}}$ *that:*

    M4. $R_{\mathbf{V}}$ *is dense,* $xR_{\mathbf{V}}y$ *implies there is a* z *such that* $xR_{\mathbf{V}}z$ *and* $zR_{\mathbf{V}}y$.

Proposition 2. *The inclusions* $S4V^- \subset S4V \subset S4V^+$ *are strict.*

Proof. See [3, Theorem 3], the models there can be regarded, respectively, as an $S4V^-$-model in which Axiom A3 is not valid, and an $S4V$-model in which Axiom A4 in not valid.

Theorem 1 ($\mathcal{L}_\square$ Soundness and Completeness). *For* $\mathcal{L}_\square \in \{S4V^-, S4V, S4V^+\}$,

$$\mathcal{L}_\square \vdash A \Leftrightarrow \mathcal{L}_\square \Vdash A.$$

Proof. Soundness is shown by induction on derivations in $\mathcal{L}_\square$, with respect to the appropriate class of models. As an example let us check that Axiom A2, $\square A \rightarrow \mathbf{V}A$, holds in any $S4V^-$-model. Let $x \Vdash \square A$ for some x in an $S4V^-$-model. Hence for all y such that $xR_\square y$ $y \Vdash A$ holds. By Condition M1 for any z such that $xR_\square z$ $xR_{\mathbf{V}}z$ also holds, hence $z \Vdash A$, in which case $x \Vdash \mathbf{V}A$ also.

Completeness is proved by the standard maximal consistent set/canonical model/truth lemma construction (see e.g. [4,5,11]). The canonical relations $R_\square^c$ and $R_{\mathbf{V}}^c$ are defined as follows: for maximal consistent sets Γ and Δ, $\Gamma R_\square^c \Delta$ iff $\Gamma_\square = \{X | \square X \in \Gamma\} \subseteq \Delta$, and $\Gamma R_{\mathbf{V}}^c \Delta$ iff $\Gamma_{\mathbf{V}} = \{X | \mathbf{V}X \in \Gamma\} \subseteq \Delta$.

The key thing to show is that the canonical $\mathcal{L}_\square$-model is an $\mathcal{L}_\square$-model, which comes down to showing that $R_\square^c$ and $R_{\mathbf{V}}^c$ have the right properties. As an example let us show Condition M1, $R_{\mathbf{V}}^c$ yields $R_\square^c$, which holds for each $\mathcal{L}_\square$. Let Γ and Δ be maximal consistent sets of formulas. Assume $\Gamma R_{\mathbf{V}}^c \Delta$ and that $\square X \in \Gamma$. Since $\square X \rightarrow \mathbf{V}X \in \Gamma$ by maximal consistency, $\mathbf{V}X \in \Gamma$ also, hence $X \in \Delta$. So $\{X | \square X \in \Gamma\} \subseteq \Delta$, i.e. $\Gamma R_\square^c \Delta$.

Theorem 2 (Conservativity).

1. *For each* $\mathcal{L}_\square \in \{\text{S4V}^-, \text{S4V}, \text{S4V}^+\}$ *its* $\square$-*fragment is* S4.
2. *The* **V**-*fragment of* S4V$^-$ *is* K.
3. *The* **V**-*fragment of* S4V *is* KD.
4. *The* **V**-*fragment of* S4V$^+$ *is* KDWR.[6]

Proof.

1) Suppose S4 $\nvdash A$, hence there is an S4-model, $\mathcal{M} = \langle W, R_\square, \Vdash \rangle$, such that $\mathcal{M} \nVdash A$. $\mathcal{M}$ can be turned into an $\mathcal{L}_\square$-model $\mathcal{M}' = \langle W, R_\square, R_\mathbf{V} \Vdash \rangle$ by defining $R_\mathbf{V}$ so that for all $x \in W$ $xR_\mathbf{V}x$. So $R_\mathbf{V}$ is serial and dense, hence in each case $\mathcal{M}'$ is an $\mathcal{L}_\square$-model, and $\mathcal{M}' \nVdash A$, hence $\mathcal{L}_\square \nvdash A$.

2) Suppose K $\nvdash A$, so there is a K-model, $\mathcal{M} = \langle W, R_\mathbf{V}, \Vdash \rangle$, such that $\mathcal{M} \nVdash A$. $\mathcal{M}$ can be turned into an $\mathcal{L}_\square$-model $\mathcal{M}' = \langle W, R_\square, R_\mathbf{V} \Vdash \rangle$ where for all $x, y \in W$ $xR_\square y$. $R_\square$ is transitive and reflexive, hence $\mathcal{M}'$ is an $\mathcal{L}_\square$-model, and $\mathcal{M}' \nVdash A$, hence S4V$^- \nvdash A$

3) and 4) We can define S4V and S4V$^+$ models, respectively, on the basis of KD and KDWR models in the same fashion as 2) above. Hence if KD $\nvdash A$ then S4V $\nvdash A$, and if KDWR $\nvdash A$ then S4V$^+ \nvdash A$.

3.2 Embedding Intuitionistic Epistemic Logics into Modal Logics of Provability and Verification

For $\mathcal{L} \in \{\text{IEL}^-, \text{IEL}, \text{IEL}^+\}$ and $\mathcal{L}_\square \in \{\text{S4V}^-, \text{S4V}, \text{S4V}^+\}$, respectively, we will show that

$$\mathcal{L} \vdash F \Leftrightarrow \mathcal{L}_\square \vdash tr(F)$$

where for each F of the appropriate $\mathcal{L}$ $tr(F)$ is the result of prefixing each sub-formula of F with $\square$.

Lemma 1
$$\mathcal{L} \vdash F \Rightarrow \mathcal{L}_\square \vdash tr(F).$$

Proof By induction on derivations in $\mathcal{L}$.

The case of the propositional intuitionistic axioms IA0 and modus ponens is the embedding of IPC into S4. The cases for each of Axioms IA1 to IA4 are all quite similar involving repeated use of necessitation and distribution; as an example let us check S4V$^- \vdash \square(\square A \to \square\mathbf{V}\square A) = tr(A \to \mathbf{K}A)$. To keep notation simple we assume that A is an atomic formula.

1. $\square\square A \to \square\square\square A$, S4 Axiom A0;
2. $\square\square A \to \mathbf{V}\square A$, Axiom A2;
3. $\square\square\square A \to \square\mathbf{V}\square A$, from 2 $\square$-necessitation and distribution;
4. $\square\square A \to \square\mathbf{V}\square A$, from 1,3 by propositional reasoning;
5. $\square A \to \square\square A$, S4 Axiom A0;

---

[6] I.e. weak reflection, also known as converse-4 C4.

6. $\Box A \to \Box\mathbf{V}\Box A$, from 4,5 by propositional reasoning;

7. $\Box(\Box A \to \Box\mathbf{V}\Box A)$, from 6 by necessitation.

To show the converse, consider an $\mathcal{L}$-model $\mathcal{M} = \langle W, R, E, \Vdash \rangle$. We can consider $\mathcal{M}$ to be an $\mathcal{L}_\Box$-model $\mathcal{M}' = \langle W, R_\Box, R_\mathbf{V}, \Vdash' \rangle$ by taking $R = R_\Box$, $E = R_\mathbf{V}$ and treating $\Vdash$ as a classical forcing $\Vdash'$.

Clearly for all $\mathcal{L}_\Box$ $R_\Box$ is transitive and reflexive and $R_\mathbf{V}$ yields $R_\Box$, hence all axioms of S4V$^-$ hold in $\mathcal{M}'$. Where $\mathcal{M}$ is an IEL-model it is additionally the case that $R_\mathbf{V}$ is serial, hence all axioms of S4V hold in $\mathcal{M}'$. Where $\mathcal{M}$ is an IEL$^+$-model furthermore $R_\mathbf{V}$ is dense, hence all axioms of S4V$^+$ hold in $\mathcal{M}'$.

Lemma 2. *For each formula F of $\mathcal{L}$ and each $u \in W$,*

$$\mathcal{M}', u \Vdash' tr(F) \Leftrightarrow \mathcal{M}, u \Vdash F$$

Proof. By induction on F.

Case 1 (F is atomic p). Assume $\mathcal{M}, u \Vdash p$, then for all v such that uRv $\mathcal{M}, v \Vdash p$, hence for all v such that $uR_\Box v$ $v \Vdash p$, so $u \Vdash \Box p$, i.e. $tr(p)$.

Conversely, assume $\mathcal{M}, u \nVdash p$, then $\mathcal{M}', u \nVdash' \Box p$ since $R_\Box$ is reflexive, hence $\mathcal{M}', u \nVdash' tr(p)$.

Case 2 (Boolean cases $F = A \wedge B$ and $F = A \vee B$ are standard).

Case 3 ($F = A \to B$). Assume $\mathcal{M}, u \Vdash A \to B$, hence for all v such that uRv either $\mathcal{M}, v \nVdash A$ or $v \Vdash B$. By the induction hypothesis $\mathcal{M}', v \nVdash' tr(A)$ or $\mathcal{M}', v, \Vdash' tr(B)$, hence $\mathcal{M}', u \Vdash' \Box(tr(A) \to tr(B))$, and $\mathcal{M}', u \Vdash' tr(A \to B)$.

Conversely, assume $\mathcal{M}, u \nVdash A \to B$, hence there is a v such that uRv in which $\mathcal{M}, v \Vdash A$ and $v \nVdash B$. By the induction hypothesis $\mathcal{M}', v \Vdash' tr(A)$ and $\mathcal{M}', v, \nVdash' tr(B)$, hence $\mathcal{M}', v \nVdash' tr(A) \to tr(B)$. Since $R = R_\Box$ $\mathcal{M}', u \nVdash' \Box(tr(A) \to tr(B))$, hence $\mathcal{M}', u \nVdash' tr(A \to B)$.

Case 4 ($F = \mathbf{K}A$). Assume $\mathcal{M}, u \Vdash \mathbf{K}A$; for any u such that uRv and any w such that vEw uEw holds by Condition IM2, hence $\mathcal{M}, w \Vdash A$. By the induction hypothesis $\mathcal{M}', w \Vdash' tr(A)$, hence $v \Vdash' \mathbf{V}tr(A)$ and $\mathcal{M}', u \Vdash' \Box\mathbf{V}tr(A)$, hence $\mathcal{M}', u \Vdash' tr(\mathbf{K}A)$

Conversely, assume $\mathcal{M}, u \nVdash \mathbf{K}A$ so there is a v such that uEv in which $v \nVdash A$. By induction hypothesis $\mathcal{M}', v \nVdash' tr(A)$. Since $E = R_\mathbf{V}$ $\mathcal{M}', u \nVdash' \mathbf{V}tr(A)$. Since $R_\Box$ is reflexive $\mathcal{M}', u \nVdash' \Box\mathbf{V}tr(A)$, hence $\mathcal{M}', u \nVdash' tr(\mathbf{K}A)$.

Lemma 3

$$\mathcal{L}_\Box \vdash tr(F) \implies \mathcal{L} \vdash F.$$

Proof Assume $\mathcal{L} \nvdash F$. By $\mathcal{L}$-completeness, there is an $\mathcal{L}$-model $\mathcal{M} = \langle W, R, E, \Vdash \rangle$ and a state $u \in W$ such that $u \nVdash F$. By Lemma 2, $u \nVdash' tr(F)$ in an $\mathcal{L}_\Box$-model $\mathcal{M}'$. By $\mathcal{L}_\Box$-soundness, $\mathcal{L}_\Box \nvdash tr(F)$.

Hence for each of IEL$^-$, IEL, and IEL$^+$, their embedding into S4V$^-$, S4V, and S4V$^+$ respectively, is faithful. Lemma 1 and Lemma 3 yield:

Theorem 3 (Embedding). *The Gödel translation faithfully embeds each $\mathcal{L} \in \{$IEL$^-$, IEL, IEL$^+\}$ into each $\mathcal{L}_\Box \in \{$S4V$^-$, S4V, S4V$^+\}$ respectively:*

$$\mathcal{L} \vdash F \Leftrightarrow \mathcal{L}_\Box \vdash tr(F).$$

4   Making Explicit Properties of Intuitionistic Knowledge

There are several assumptions about verification-based knowledge implicit in the intuitionistic epistemic framework; verification is only of provability; truth, hence knowledge and belief, is stable, and consequently both positive and negative introspection hold. These properties do not necessarily hold for the classical modal counterparts in $\mathcal{L}_\square$. If we wish to model a view of knowledge or belief which has any one of these properties then we must assume each explicitly. The greater expressive strength of the modal language gives us more control over our assumptions.

4.1   The Two Readings of Intuitionistic Knowledge

Section 2.1 outlined two ways in which intuitionistic verification, $\mathbf{K}$, can be understood. According to the first verification amounts to a kind of proof-checking – a verification of A is a verification of a proof of A. This reading is reflected by the Gödel translation of $\mathbf{K}A$ which is $\square\mathbf{V}\square A$. According to the second reading of verification a proposition can be verified directly by a non-proof justificatory procedure. In accepting a proposition as known such evidence may be perfectly adequate, or the only kind practically available.

The modal framework can accommodate this latter understanding of verification by extending $\mathcal{L}_\square$ with the additional principle

$$\mathbf{V}A \to \mathbf{V}\square A \tag{P}$$

which states that a non-proof verification is sufficiently robust to guarantee the existence of a proof. We often accept informal arguments based on general theoretical reasons or clear examples in place of specific proofs when it is clear that such proofs can be obtained. For instance, we might justify that IPC $\vdash \neg\neg(A \vee \neg A)$ by reasoning informally on the basis of the BHK interpretation, rather than exhibiting a derivation in IPC (see e.g. [10, Section 1.3] for examples).

Definition 9 ($\mathcal{L}_\square$+ P). $\mathcal{L}_\square$+P is $\mathcal{L}_\square \in \{S4V^-, S4V, S4V^+\}$ with the additional axiom:

$A5.$ $\mathbf{V}A \to \mathbf{V}\square A.$

Definition 10. A model for $\mathcal{L}_\square$+P is an $\mathcal{L}_\square$-model with the additional condition

$M5.$ For states x, y, z in a model $xR_\mathbf{V}y$ and $yR_\square z$ yield $xR_\mathbf{V}z$.

Proposition 3. Let $\mathfrak{F} = \langle W, R_\square, R_\mathbf{V} \rangle$ be a frame. P holds at all states of a model based on $\mathfrak{F}$ iff $\mathfrak{F}$ satisfies Condition M5.

Proof. $\Leftarrow$: Assume $\mathfrak{F}$ satisfies Condition M5 and there is some state $a \in W$ s.t. $a \Vdash \mathbf{V}A$, in which case for all b s.t. $aR_\mathbf{V} b$ $b \Vdash A$. Assume further that $bR_\square c$ for an arbitrary c; by M5 $aR_\mathbf{V}c$, so $c \Vdash A$ also. Hence $a \Vdash \mathbf{V}\square A$.

$\Rightarrow$: By contrapositive. Assume $\mathfrak{F}$ does not satisfy Condition M5. Hence there are states $a, b, c \in W$ such that $a R_{\mathbf{V}} b$ and $b R_\square c$ but $\neg a R_{\mathbf{V}} c$. Define a valuation $V(p) = \{x \in W \mid a R_{\mathbf{V}} x\}$. In the resulting model $a \Vdash \mathbf{V} p$ but $c \nVdash p$, hence $b \nVdash \square p$ and so $a \nVdash \mathbf{V} \square p$.

Theorem 4 ($\mathcal{L}_\square + \mathsf{P}$ Soundness and Completeness).

$$\mathcal{L}_\square + \mathsf{P} \Vdash A \Leftrightarrow \mathcal{L}_\square + \mathsf{P} \vdash A.$$

Proof. Soundness follows from Proposition 3. For completeness we verify that the $\mathcal{L}_\square + \mathsf{P}$ canonical model satisfies Condition M5.

Assume $\Gamma R_{\mathbf{V}}^c \Delta$ and $\Delta R_\square^c \Omega$, and that $\mathbf{V} X \in \Gamma$. By maximal consistency $\mathbf{V} X \to \mathbf{V} \square X \in \Gamma$, hence $\mathbf{V} \square X \in \Gamma$. Since $\Gamma R_{\mathbf{V}}^c \Delta$ it follows that $\square X \in \Delta$ and hence $X \in \Omega$. So $\mathbf{V} X \in \Gamma$ yields that $X \in \Omega$, i.e. $\Gamma R_{\mathbf{V}}^c \Omega$.

Given the equivalence of $\mathbf{V} A$ and $\mathbf{V} \square A$ in $\mathcal{L}_\square + \mathsf{P}$ we can simplify the translations of IEL formulas by substituting $\mathbf{V} A$ for $\mathbf{V} \square A$, hence for example $tr(A \to \mathbf{K} A) = \square(\square A \to \square \mathbf{V} A)$. With this observation and Lemma 1 it is clear that this modified translation holds in the respective systems $\mathcal{L}_\square + \mathsf{P}$.

4.2 Stability of Knowledge

Intuitionistic truth, hence intuitionistic $\mathbf{K}$, is stable, but $\mathbf{V}$ in $\mathcal{L}_\square$ is not.

Theorem 5. *Neither truth nor $\mathbf{V}$ are monotonic with respect to $R_\square$ for any $\mathcal{L}_\square \in \{\mathsf{S4V}^-, \mathsf{S4V}, \mathsf{S4V}^+\}$, i.e. if $x R_\square y$ then 1) $x \Vdash A$ does not necessarily yield $y \Vdash A$ and 2) $x \Vdash \mathbf{V} A$ does not necessarily yield $y \Vdash \mathbf{V} A$.*

Proof. Consider the $\mathsf{S4V}^+$-model (hence $\mathsf{S4V}^-$- and $\mathsf{S4V}$-) 2:

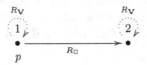

Fig. 2. $\mathsf{S4V}^+$-model $\mathcal{M}_2$

1) holds by definition of 2. For 2) since $1 \Vdash p$ then $1 \Vdash \mathbf{V} p$, and since $2 \nVdash p$ $2 \nVdash \mathbf{V} p$. Hence $\mathbf{V} p$ does not hold at all the $R_\square$-successors of 1 where $\mathbf{V} p$ holds.

To ensure $\mathbf{V}$ is monotonic we can adopt the principle

$$\mathbf{V} A \to \square \mathbf{V} A, \tag{M}$$

which says that whenever we have a verification we can prove it to be correct, but such a proof guarantees the verification can never be defeated, so can never be lost. Adding M to a system in $\mathcal{L}_\square$ yields a logic in which $\mathbf{V}$ is monotonic with respect to $R_\square$.

Definition 11 ($\mathcal{L}_\Box + M$). $\mathcal{L}_\Box + M$ *is any system* $\mathcal{L}_\Box$ *with the additional axiom:*

A6. $\mathbf{V}A \to \Box\mathbf{V}A.$

Definition 12 (V-Monotonic Models). *A* **V***-Monotonic model is an* $\mathcal{L}_\Box$*-model with the additional condition:*

M6.   *For states* x, y, z *in a model* $xR_\Box y$ *and* $yR_\mathbf{V}z$ *yield* $xR_\mathbf{V}z$.

Proposition 4. *Let* $\mathfrak{F} = \langle W, R_\Box, R_\mathbf{V}\rangle$ *be a frame. M holds at all states of a model based on* $\mathfrak{F}$ *iff* $\mathfrak{F}$ *satisfies Condition M6.*

Proof. Virtually identical to the proof of Proposition 3.

Theorem 6 (Monotonicity). *If a model satisfies Condition M6 then* $x \Vdash \mathbf{V}A$ *yields that for any* y *such that* $xR_\Box y$ $y \Vdash \mathbf{V}A$ *holds.*

Proof. Assume there is a state $a \in W$ such that $a \Vdash \mathbf{V}A$. Take an arbitrary b such that $aR_\Box b$, and an arbitrary c such that $bR_\mathbf{V}c$; by M6 $aR_\mathbf{V}c$, hence $c \Vdash A$. Hence $b \Vdash \mathbf{V}A$, since c is arbitrary.

Theorem 7 ($\mathcal{L}_\Box + M$ Soundness and Completeness).

$$\mathcal{L}_\Box + M \vdash A \Leftrightarrow \mathcal{L}_\Box + M \Vdash A.$$

Proof. Soundness follows from Proposition 4 and the soundness of $\mathcal{L}_\Box$. The canonicity of Condition M6 is shown in an identical manner to that of Theorem 4.

4.3   Positive Introspection and Negative Introspection

In $\mathcal{L}$ positive and negative introspection are instances of the 'proof yields verification' co-reflection principle IA2.

For positive introspection in $\mathcal{L}_\Box$ the principle (M) suffices for the stability of positive verification statements $\mathbf{V}A$, hence yields positive introspection.

Theorem 8. $\mathcal{L}_\Box + M \vdash \mathbf{V}A \to \mathbf{V}\mathbf{V}A.$

Proof. Argue in $\mathsf{S4V}^- + M$:

1. $\Box\mathbf{V}A \to \mathbf{V}\mathbf{V}A$, Axiom A2;
2. $\mathbf{V}A \to \Box\mathbf{V}A$, Axiom A6;
3. $\mathbf{V}A \to \mathbf{V}\mathbf{V}A$, propositional reasoning.

We note in passing that positive introspection also holds in $\mathcal{L}_\Box + P$, consequently adding M or P to $\mathsf{S4V}^+$ yields idempotency of $\mathbf{V}$.

(M) asserts only that positive verification statements, $\mathbf{V}A$, are stable. To ensure that negative verification statements, $\neg\mathbf{V}A$, are also stable we can adopt the principle

$$\neg\mathbf{V}A \to \Box\neg\mathbf{V}A. \tag{N}$$

which says that the failure of verification is provable, hence where a verification has not succeeded it can never succeed.

Definition 13 ($\mathcal{L}_\square + \mathsf{N}$). *$\mathcal{L}_\square + \mathsf{N}$ is any system $\mathcal{L}_\square$ with the additional axiom:*

A7. $\neg \mathbf{V}A \rightarrow \square \neg \mathbf{V}A$

$\mathcal{L}_\square + \mathsf{N}$ yields negative introspection by an obvious modification of Theorem 8.

Definition 14. *A model for $\mathcal{L}_\square + \mathsf{N}$ is an $\mathcal{L}_\square$ model with the additional condition*

M7.   *For states x, y, z in a model $xR_\square y$ and $xR_\mathbf{V} z$ yield $yR_\mathbf{V} z$.*

Proposition 5. *Let $\mathfrak{F} = \langle W, R_\square, R_\mathbf{V} \rangle$ be a frame. N holds at all states of a model based on $\mathfrak{F}$ iff $\mathfrak{F}$ satisfies Condition M7.*

Proof. $\Leftarrow$: Assume $\mathfrak{F}$ satisfies Condition M7 and there is some $a \in W$ such that $a \Vdash \neg \mathbf{V}A$ holds, hence there is a $c \in W$ such that $aR_\mathbf{V} c$ and $c \nVdash A$. Let b be an arbitrary state such that $aR_\square b$, by M7 $bR_\mathbf{V} c$ holds, hence $b \nVdash \mathbf{V}A$, i.e. $b \Vdash \neg \mathbf{V}A$. Since b is arbitrary, $a \Vdash \square \neg \mathbf{V}A$.

$\Rightarrow$: Assume $\mathfrak{F}$ does not satisfy Condition M7, hence there is a model based on $\mathfrak{F}$ with states a, b and c such that $aR_\square b$ and $aR_\mathbf{V} c$, but $\neg bR_\mathbf{V} c$. Define a valuation such that $V(p) = \{x \in W | x \neq c\}$; hence $b \Vdash \mathbf{V}p$, hence $b \nVdash \neg \mathbf{V}p$, in which case $a \nVdash \square \neg \mathbf{V}p$. Since $c \nVdash p$ then $a \nVdash \mathbf{V}p$, hence $a \Vdash \neg \mathbf{V}p$, so $a \nVdash \neg \mathbf{V}p \rightarrow \square \neg \mathbf{V}p$.

Theorem 9 ($\mathcal{L}_\square + \mathsf{N}$ **Soundness and Completeness**).

$$\mathcal{L}_\square + \mathsf{N} \vdash A \Leftrightarrow \mathcal{L}_\square + \mathsf{N} \Vdash A$$

Proof. Soundness follows from Proposition 5 and the soundness of $\mathcal{L}_\square$. For completeness we check that the $\mathcal{L}_\square + \mathsf{N}$ canonical model satisfies Condition M7.

Assume $\Gamma R_\square^c \Delta$ and $\Gamma R_\mathbf{V}^c \Omega$. Suppose $\mathbf{V}A \in \Delta$ but $A \notin \Omega$. Hence $\mathbf{V}A \notin \Gamma$, so $\neg \mathbf{V}A \in \Gamma$; by maximal consistency $\neg \mathbf{V}A \rightarrow \square \neg \mathbf{V}A \in \Gamma$ so $\square \neg \mathbf{V}A \in \Gamma$, and so $\neg \mathbf{V}A \in \Delta$, which is a contradiction. Hence if $\mathbf{V}A \in \Delta$ then $A \in \Omega$, i.e. $\Delta R_\mathbf{V}^c \Omega$.

5   Conclusion – Further Applications

The logics in $\mathcal{L}_\square$ and their extensions offer a more nuanced way of understanding verification-based epistemic-doxastic states than do the logics in $\mathcal{L}$. The logic S4V$^-$ can be regarded as a logic of verification-based belief, like IEL$^-$, but without the assumption that belief is indefeasible, allowing for the possibility that one's beliefs may change. The logic S4V$^-$ with (M), on the other hand, can be regarded as the logic of provably correct, hence indefeasible, beliefs – though reflection does not hold such beliefs might still qualify as a form of fallible knowledge.

This points to a more general application for the logics outlined above as calculi of conclusive vs. non-conclusive justification. Many epistemologists hold some form of fallibilism to be true, according to which it is possible to know, or at least have rational belief, on the basis of non-conclusive justification. The logics in $\mathcal{L}_\square$ can model different versions of the distinction between conclusive and non-conclusive justifications – $\mathbf{V}A$ is not factive, hence it does not guarantee truth, whereas $\square A$ does.

References

1. Artemov, S.: Explicit Provability and Constructive Semantics. Bulletin of Symbolic Logic 7(1), 1–36 (2001)
2. Artemov, S., Protopopescu, T.: Discovering Knowability: A Semantical Analysis. Synthese 190(16), 3349–3376 (2013)
3. Artemov, S., Protopopescu, T.: Intuitionistic Epistemic Logic. Tech. rep. (December 2014), http://arxiv.org/abs/1406.1582v2
4. Blackburn, P., de Rijke, M., Vedema, Y.: Modal Logic. Cambridge University Press (2002)
5. Chagrov, A., Zakharyaschev, M.: Modal Logic. Clarendon Press (1997)
6. Constable, R.: Types in Logic, Mathematics and Programming. In: Buss, S. (ed.) Handbook of Proof Theory, pp. 683–786. Elsevier (1998)
7. Došen, K., Božić, M.: Models for Normal Intuitionistic Modal Logics. Studia Logica 43(3), 217–245 (1984)
8. Došen, K.: Models for Stronger Normal Intuitionistic Modal Logics 44(1)
9. Došen, K.: Intuitionistic Double Negation as a Necessity Operator. Publications de L'Institute Mathématique (Beograd)(NS) 35(49), 15–20 (1984)
10. Dummett, M.A.E.: Elements of Intuitionism. Clarendon Press (1977)
11. Fitting, M., Mendelsohn, R.: First-Order Modal Logic. Kluwer Academic Publishers (1998)
12. Gödel, K.: An Interpretation of the Intuitionistic Propositional Calculus. In: Feferman, S., Dawson, J.W., Goldfarb, W., Parsons, C., Solovay, R.M. (eds.) Collected Works, vol. 1, pp. 301–303. Oxford Univeristy Press (1933)
13. Hendricks, V.F.: Formal and Mainstream Epistemology. Cambridge University Press (2006)
14. Hirai, Y.: An Intuitionistic Epistemic Logic for Sequential Consistency on Shared Memory. In: Clarke, E.M., Voronkov, A. (eds.) LPAR-16 2010. LNCS, vol. 6355, pp. 272–289. Springer, Heidelberg (2010)
15. Leite, A.: Fallibilism. In: Dancy, J., Sosa, E., Steup, M. (eds.) A Companion to Epistemology, 2nd edn., pp. 370–375. Blackwell (2010)
16. McKinsey, J.C.C., Tarski, A.: Some Theorems About the Sentential Calculi of Lewis and Heyting 13(1), 1–15 (1948)
17. Proietti, C.: Intuitionistic Epistemic Logic, Kripke Models and Fitch's Paradox. Journal of Philosophical Logic 41(5), 877–900 (2012)
18. Troelstra, A., Schwichtenberg, H.: Basic Proof Theory. Cambridge University Press (2000)
19. Univalent Foundations Program: Homotopy Type Theory. Univalent Foundations Program (2013)
20. Williamson, T.: On Intuitionistic Modal Epistemic Logic. Journal of Philosophical Logic 21(1), 63–89 (1992)
21. Wolter, F., Zakharyaschev, M.: Intuitionistic Modal Logics as Fragments of Classical Bimodal Logics. In: Orlowska, E. (ed.) Logic at Work, pp. 168–186. Springer (1999)
22. Wolter, F., Zakharyaschev, M.: Intuitionistic Modal Logic. In: Casari, E., Cantini, A., Minari, P. (eds.) Logic and Foundations of Mathematics, pp. 227–238. Kluwer Academic Publishers (1999)

An Argument for Permissivism from Safespots

Thomas Raleigh

Department of Philosophy and Religious Studies,
Norwegian University of Science and Technology,
N.T.N.U. Dragvoll,
7491 Trondheim,
Norway
thomas.raleigh@ntnu.no

Abstract. I present an argument against the thesis of Uniqueness and in favour of Permissivism. Counterexamples to Uniqueness are provided, based on 'Safespot' propositions – i.e. a proposition that is guaranteed to be true provided the subject adopts a certain attitude towards it. The argument relies on the following plausible principle: If S knows that her believing p would be a true belief, then it is rationally permitted for S to believe p. One motivation for denying this principle – viz. opposition to 'epistemic consequentialism' – is briefly discussed.

Keywords: Uniqueness, Permissivism, Blindspots, Rationality, Belief, Epistemic Consequentialism, Evidence, Epistemology, Philosophy.

Take UNIQUENESS[1] to be the following thesis:

- For any subject S, proposition p and set of evidence E, *exactly one* of the 3 doxastic attitudes to p – Belief, Disbelief[2] or Suspension – is rationally permitted for S on the basis of E.

Take A.B.U. (At Best Unique) to be the following thesis:

- For any subject S, proposition p and set of evidence E, *at most one* of the 3 doxastic attitudes to p is rationally permitted for S on the basis of E.

The negation of A.B.U. then is PERMISSIVISM[3]:

- It is possible that there could be some subject S, proposition p and set of evidence E such that *more than one* of the 3 doxastic attitudes to p is rationally permitted for S on the basis of E.

---

[1] Recent advocates of Uniqueness include: White [1], Christensen [2], Feldman [3], Sosa [4].

[2] I assume, as I take to be standard, that disbelieving a proposition is equivalent to believing its negation. I.e. DBp = B¬p. However, see Sturgeon (forthcoming) for a denial of this equivalence.

[3] Recent advocates of Permissivism include: Douven [5], Kelly [6], Schoenfield [7].

© Springer-Verlag Berlin Heidelberg 2015
W. van der Hoek et al. (Eds.): LORI 2015, LNCS 9394, pp. 308–315, 2015.
DOI: 10.1007/978-3-662-48561-3_25

As they stand, these might be thought somewhat imprecise formulations in (at least) a couple of respects. Firstly, in addition to belief, disbelief and suspension, there is also, you might think, a fourth possible option of forming no opinion whatsoever – or 'withdrawing'[4]. Secondly, one might wonder whether it is best to formulate these theses in an inter-subjective or intra-subjective way[5] – i.e. if UNIQUENESS is true, might it nevertheless be that different subjects with the same evidence could each be permitted to have a *different* uniquely permitted attitude to p? However, these complications will make no difference in what follows. The argument I will present, if sound, shows that a single subject can be permitted on the basis of her evidence both to believe or to disbelieve a proposition. No matter how advocates of UNIQUENESS or A.B.U. want to incorporate the possibility of 'withholding' into their preferred thesis, this is still going to be a problem. And it looks extremely plausible that if a single subject is permitted both to believe or to disbelieve a proposition, then *a fortiori* 2 different subjects with the same evidence could be permitted to each adopt different doxastic attitudes to it. For ease of presentation then, we can safely leave the three theses above as they stand.

At the risk of stating the obvious, for PERMISSIVISM to be true, all that is required is one counterexample to A.B.U. I will suggest that, given a plausible assumption, counterexamples can be formed, based on the kind of Moorean propositions that Roy Sorensen [10] labelled 'Blindspots'[6] or on their opposite kind, which I will label 'Safespots'[7].

An attitude A-blindspot for some subject S is a proposition that can be true and that S can take some propositional attitude, A, towards; but not both – i.e. it is bound to be false if S adopts attitude A to it.

An attitude A-safespot for some subject S is a proposition that can be false and that S can take some propositional attitude, A, towards; but not both – i.e. it is bound to be true if S adopts attitude A to it.

Some examples:

The proposition: 'It is raining & S does not believe that it is raining' is a belief-blindspot for S. (I assume here that it is possible to both Believe that p and Believe

---

[4] The term 'withdrawing' is used by Turri [8]. In fact, it is not clear to me that refusing to adopt any of the 3 doxastic attitudes, so not even suspending judgement, is a state of mind that is subject to the demands of theoretical/epistemic rationality (as opposed to practical rationality). It may be a more or less prudent option to 'withdraw', but just refusing to think any further about a certain proposition is not obviously evaluable at all as an intellectual or theoretical move that could be rationally correct or incorrect in light of the available evidence.

[5] For discussion of the distinction between inter-subjective and intra-subjective versions of these theses, see Kelly [9].

[6] Bykvist & Hattiangadi [11] argue that blindspot propositions provide a counter-example to the following 'truth-norm' for beliefs: For any S, *p*: if S considers *p*, then S ought to (believe that *p*) iff *p* is true.

[7] In Raleigh [12], I labelled such propositions 'true-turns'. I now prefer the label 'safespot', which makes clearer their relation to blindspot propositions.

that one does not Believe that p.) But it is not a hope-blindspot nor a desire-blindspot. And, on the assumption that it is possible for a subject to both disbelieve a proposition and disbelieve that they disbelieve it, nor is this a disbelief-blindspot.

Conversely, '2+2=4 & S believes that 2+2 = 4' is a belief-safespot for S, but not a disbelief safespot for S. (Nor, of course, a hope-safespot or a desire-safespot etc.)

Whereas, the proposition: 'S is dead' is both a belief-blindspot and a disbelief-blindspot for S (and a hope-blindspot and a desire-blindspot etc.) And conversely, 'S is alive' is both a belief-safespot and a disbelief-safespot – and so on for other attitudes – for S. (I assume here that only the living can have propositional attitudes.)

Notice also that there can be more specific *kinds* of belief-blindspots and belief-safespots. E.g. a proposition of the form: 'p & S cannot justifiably believe that p' is not a belief-blindspot for S, but it is a *justified-belief-blindspot* for S. (It can be truly but unjustifiedly believed by S, but it cannot be truly and justifiably believed by S.)

The existence of Safespots and Blindspots becomes a problem for A.B.U. once we grant the following principle:

- If S knows that her believing p would be a true belief, then it is rationally permitted for S to believe p.

Or in semi-formal terms:

- PRINCPLE: [SK (SBp → p)] → Rationally Permitted: SBp

I will not provide any further support for this principle other than to simply state that it has, I take it, a very large measure of prima facie plausibility. After all, if you *know* that it is impossible that your forming a belief in a particular proposition could result in a false belief, it looks like you have a pretty good rational basis to hold that belief. (Of course, that you *have* a rational basis to believe some proposition, does not guarantee that your actual belief in that proposition *is* a rational belief, for you might have formed the belief on some other irrational basis[8].)

A slightly over-simple formulation of a special case of PRINCIPLE is the following:

- If S knows that a belief *of some specific type* that p (by S) is bound to be a true belief, then S is rationally permitted to have a belief that p *of that specific type*.

E.g. If S knows that a belief in p *which has been formed on a Tuesday*, is bound to be a true belief, then S is rationally permitted to have a belief formed on a Tuesday that p.

Taking 'SB_tp' to mean that S has a belief that p is of type t, we could put this in semi-formal terms:

- [SK (SB_tp → p)] → Rationally Permitted: SB_tp

---

[8] Compare the familiar distinction in epistemology between doxastic and propositional justification – a subject's *having* justification for some belief, does not entail that the subject's belief is actually justified.

As it stands this is not quite right, for S may not know whether she is currently in a position to form beliefs of the specific type in question. E.g. when the type of belief in question is: *formed on a Tuesday*, S may not know which day of the week it is. Even if it *is* in fact Tuesday, if S has no idea what day of the week it is, then it would be rationally amiss of her to go ahead and form a belief that p solely on the basis of her knowledge that a belief formed on a Tuesday would be a true belief. What is required, in addition, is that the subject knows that she is in a position to form the specific type of belief in question – i.e. she must know in addition that it is indeed a Tuesday[9]. This additional requirement was not needed in the original, general formulation of PRINCIPLE, as when the belief that p in question need not be of any specific type, we assume that there are no specific circumstances, which the subject might be ignorant of, that need to obtain in order for her to be able to form such a belief that p (of no specific type). I.e. we assume that a subject always knows that she is in a position to form a belief that p where this belief need not be of any further specific type.

So a better formulation of this special case of PRINCIPLE is:

- If S knows that a belief *of some specific type* that p (by S) is bound to be a true belief, AND S knows that she is in a position to form a belief that p of that specific type, then S is rationally permitted to have a belief that p *of that specific type in these given circumstances*.

Or in semi-formal terms:

- SPECIAL CASE: [SK (SB$_t$p → p) & SK (S is in a position to B$_t$p)] → Rationally Permitted: SB$_t$p

The problem for A.B.U. and UNIQUENESS now arises as there can be a proposition that one can know to be a kind of belief-safespot for oneself without actually (yet) believing it to be true – indeed, whilst knowing that it is (currently) false. For example, consider the following proposition:

q: S has at least one non-innate belief.

Notice that this is *not* a belief-safespot for S, as if all of S's beliefs are innate, including her belief in q, then q would be both false and believed by S. But q is a *non-innate-belief*-safespot for S. For q can be false and q can be non-innately believed by S, but it cannot be falsely and non-innately believed by S.

---

[9] Here's another way of seeing the shortcomings of the over-simple formulation. For any proposition whatever, we know a priori the tautology that: a true belief that p is bound to be a true belief. But we don't want to say that a subject is thereby rationally permitted to believe any proposition that happens to be true, even those for which the subject has no evidence. I.e. we need to rule out that it is rationally permitted for a subject to form a belief that happens to be true by sheer lucky guess. This is ruled out by the additional requirement, as a subject who forms a true belief that p by sheer luck does not *know* that she is in a position to form a belief of the type in question – i.e. a true belief. Thanks to an anonymous referee for pressing me to consider this issue.

Now, it seems that S could innately know that q is a non-innate-belief-safespot for her – i.e. $SK_i(SB_{ni}q \to q)$, where K_i = innate knowledge and B_{ni} = non-innate belief. Moreover, S can know innately that any new belief she forms will be non-innate – so she knows that she is in a position to form a belief that q of that specific type. And so, by SPECIAL CASE, S is rationally permitted to $B_{ni}q$ – i.e. to have a non-innate belief that q.

But it seems possible that S could have excellent overall evidence[10] that q is (currently) false. S might have excellent evidence that all her current beliefs are innate (including her belief that q is a non-innate-belief-safespot for her). Indeed, S might innately *know* that q is false. E.g. perhaps S is an android, furnished with many innate beliefs, who has the ability to form new non-innate beliefs but who has only just been turned on for the first time innately *knowing* both that q is (currently) false and that q is a non-innate-belief-safespot for her. And so it seems that S is also rationally permitted to *innately* disbelieve that q (i.e. to $B_i\neg q$, i.e. to DB_iq). So S is rationally permitted to adopt 2 different doxastic attitudes to p – an innate disbelief or a non-innate belief. Hence A.B.U. and UNIQUENESS are both false, PERMISSIVISM is true.

Nothing in the foregoing hinges on the specifics of the innate vs. non-innate distinction. For it seems clear that further examples of this kind could be manufactured by appeal to other pairs of contrasting belief types. E.g. S could know *since before she turned 20 years old* that: S has formed no beliefs *after coming to know that she is over 20 years old*. But S could also know, again since before the age of 20, that the negation of this proposition – S has formed at least one belief after coming to know that she is 20 years old – would be true if first believed by S after learning that she is over 20 years old. And so, assuming that S knows that she has turned 20 and so knows that she is in a position to form beliefs of the type: believed-only-since-learning-I-am-over-twenty, it seems that S is rationally permitted (at a single time) both to disbelieve-since-before-the-age-of-twenty that p, and also to believe-only-since-learning-I-am-over-twenty that p.

I presented the argument above in terms of the safespot proposition q. But it could equally have been presented in terms of a blindspot. The negation of q is a non-innate-belief blindspot for S

¬q: S has no non-innate beliefs.

S could have excellent (innate) evidence that ¬q is true, and indeed innately know that ¬q. So it seems S is rationally permitted to innately believe ¬q. But S could also innately know that a non-innate disbelief in ¬q – i.e. a non-innate belief that ¬¬q – would bound to be true. So, by SPECIAL CASE, S is rationally permitted to non-innately believe that ¬¬q – i.e. disbelieve that ¬q.

---

[10] I do not wish to assume the truth of 'evidentialism' here. If you think that the rational permissibility of belief can be partially determined by non-evidential factors, then we can just stipulate that these extra factors also obtain in this case.

In general and semi-formal terms then, the issue is that safespots and blindspots allow for the possibility that both of the following conditions obtain:

$SK_{t1}\neg p$
So, Rationally Permitted: $SDB_{t1}p$

$SK_{t1} (SB_{t2}p \rightarrow p)$ & $SK_{t1} (S$ is in a position to $B_{t2}p)$
So, Rationally Permitted: $SB_{t2}p$ (by SPECIAL CASE)

We can have a situation then where a subject is both rationally permitted to believe a proposition and is rationally permitted to believe its negation.

Of course, this does not mean that the subject is rationally permitted to believe a contradiction. Permissivists who think that both belief or disbelief in some proposition, p, can be rationally permitted, will presumably wish to deny that this entails it is rationally permitted to believe (p & ¬p), and so they will presumably want to deny that the rational permissibility of belief is closed under conjunction. I.e. they will need to deny that:

Rationally Permitted (SBp) & Rationally Permitted (SBq) → Rationally Permitted (SBp & SBq)

But of course the following sort of closure principle for permissibility in general is clearly invalid:

It is permitted to: (do X) & it is permitted to: (do Y) → it is permitted to: (do X & do Y)

E.g. that S is allowed to marry Jack and S is allowed to marry Jill does not entail that S is allowed to be married to both Jack and to Jill! So there seems to be no obvious theoretical cost for denying this kind of closure principle in the specific case of rationally permitted doxastic attitudes.

Advocates of A.B.U. or of UNIQUENESS then apparently need to deny PRINCIPLE. I will now briefly consider one possible motivation for such a denial.

It might be objected that by tying rational permissibility to a subject's knowledge that a belief *would* be true, PRINCIPLE assumes, or at least is motivated and made plausible by, a form of *epistemic consequentialism* that advocates of UNIQUENESS (or ABU) might want to reject.

(EC) EPISTEMIC CONSEQUENTIALISM: The ultimate epistemic goals/values, in virtue of which the epistemic rationality of holding any particular belief, on the basis of any particular set of evidence, is to be determined are: (i) acquiring true beliefs, (ii) avoiding false beliefs.

An anti-consequentialist about epistemic rationality will insist, against (EC), that the norms imposed by evidence are fundamentally concerned simply that the subject believes what the evidence indicates *is* (now, actually) true. Such norms are not means to some further end concerning one's set of beliefs; they are not concerned with the *results* of respecting one's evidence in this way. So in particular the rationality of forming a belief in accord with one's evidence is not fundamentally to be explained in terms of its promoting the goals of gaining accurate beliefs and avoiding inaccurate ones; rather, the epistemic rationality of forming some belief *just is* determined simply by whether (and the extent to which) the evidence indicates that the proposition in question *is true*.

(AC) EPISTEMIC ANTI-CONSEQUENTIALISM: The epistemic rationality of holding any particular belief, on the basis of any particular set of evidence, is determined solely by which propositions the evidence indicates (more or less strongly) to be (currently/actually) true or false.

[Hence rationality is not fundamentally determined by reference to the promotion of the doxastic goals/values (i) or (ii), in the statement of (EC) above.]

This is not the place to mount a full discussion of epistemic consequentialism. But I will note that blindspots seem to yield a particularly unintuitive result for the sort of anti-consequentialism just sketched. For your evidence could very strongly indicate that some blindspot proposition is (now, actually) true. But this same evidence could also clearly indicate that a belief in the proposition is bound to be false. It sounds very strange then to say that your evidence here rationally permits belief in the blindspot, when it manifestly indicates to you that the proposition is bound to be false if you believe it.

In other words, if one rejects PRINCIPLE on general anti-consequentialist grounds, it would seem that one should also be committed to rejecting the following:

CONVERSE PRINCIPLE: [SK (SBp → ¬p)] → ¬Rationally Permitted: SBp

If anything, this CONVERSE PRINCIPLE seems, to me, even more plausible than the original PRINCIPLE. When you *know* that your believing something would bound to be a false belief, then it is not rationally permitted for you to form that belief. But if, in adherence to (AC), one insists that one's evidence rationalises one's beliefs *solely* in virtue of what it indicates *is actually* true/false, then so long as your evidence indicates strongly enough that a blindspot is currently true, you should be permitted to believe it even though you know this belief would be false.

We have seen then how PRINCIPLE is in conflict with (AC) – safespots can provide examples where the evidence can indicate both that a proposition is false and that it would be true if believed. In these cases PRINCIPLE insists, against (AC), that what is rationally permitted to believe is *not* determined solely by what the evidence indicates to be true/false. And likewise, blindspots can provide examples where the evidence can indicate both that a proposition is true and that it would be false if believed. In these cases CONVERSE PRINCIPLE insists, against (AC), that what is rational to believe is not determined solely by what the evidence indicates to be true/false.

But notice, accepting PRINCIPLE (or CONVERSE PRINCIPLE) does not obviously require going so far as endorsing (EC). I.e. accepting that there are *some* cases in which the consequences of forming a belief are *relevant* to assessing epsitemic rationality, does not obviously require accepting that in *every* case such consequences are the *sole* or *ultimate* determinants of epistemic rationality. And this may be just as well for the plausibility of PRINCIPLE and CONVERSE PRINCIPLE, because consequentialism may well have its own problems – see e.g. Berker [13] for arguments against (EC).

Finally, advocates of UNIQUENESS have, of course, given their own arguments in favour of the thesis and against PERMISSIVISM (see [1], [2], [3], [4], below); nothing I have said is supposed to indicate what flaws there are, if any, in those arguments. And so if one were sufficiently strongly convinced on independent grounds that UNIQUENESS must be correct, one could treat the foregoing argument of this paper as providing a reason to reject PRINCIPLE. But in any case, whatever the ultimate theoretical costs or benefits of denying PRINCIPLE, I hope that the need to deny it is at least an interestingly non-obvious and prima facie implausible consequence of UNIQUENESS and of A.B.U.

Acknowledgments. An earlier version of this paper was presented at the 89th Joint Session of the Aristotelian Society and the Mind Association, held at the University of Warwick. I am very grateful to members of the audience for helpful questions and comments on that occasion. This paper was written whilst part of the project 'Representationalism or Anti-Representationalism? Perspectives on Intentionality from Philosophy and Cognitive Science', which was supported by the Norwegian Research Council ISP-FIDE initiative.

References

1. White, R.: Epistemic Permissiveness. Philosophical Perspectives 19(1), 445–459 (2005)
2. Christensen, D.: Epistemology of Disagreement: The Good News. Philosophical Review 116(2), 187–217 (2007)
3. Feldman, R.: Reasonable Religious Disagreement. In: Antony, L. (ed.) Philosophers Without God, pp. 197–214. Oxford University Press, Oxford (2007)
4. Sosa, E.: The Epistemology of Disagreement. In: Armchair Philosophy. Princeton University Press, Princeton (2010)
5. Douven, I.: Uniqueness Revisited. American Philosophical Quarterly 46(4), 347–362 (2009)
6. Kelly, T.: How to Be an Epistemic Permissivist. In: Steup, M., Turri, J. (eds.) Contemporary Debates in Epistemology, 2nd edn., Blackwell Publishing, Malden (2014)
7. Schoenfield, M.: Permission to Believe: Why Permissivism Is True and What It Tells Us About Irrelevant Influences on Belief, Noûs (forthcoming)
8. Turri, J.: A Puzzle about withholding. Philosophical Quarterly 62(247), 355–364 (2012)
9. Kelly, T.: How to Be an Epistemic Permissivist. In: Steup, M., Turri, J. (eds.) Contemporary Debates in Epistemology, 2nd edn., Blackwell Publishing, Malden (2014)
10. Sorensen, R.: Blindspots. Clarendon Press, Oxford (1988)
11. Bykvist, K., Hattiangadi, A.: Does thought imply ought? Analysis 67, 277–285 (2007)
12. Raleigh, T.: Belief-Norms and Blindspots. The Southern Journal of Philosophy 51(2), 243–269 (2013)
13. Berker, S.: Epistemic Teleology and the Separateness of Propositions. Philosophical Review 122(3), 337–393 (2013)

Model Transformers for Dynamical Systems of Dynamic Epistemic Logic

Rasmus K. Rendsvig[1,2]

[1] Information Quality Research Group, Lund University, Lund, Sweden
[2] Center for Information and Bubble Studies, University of Copenhagen,
Copenhagen, Denmark
rendsvig@gmail.com

Abstract. This paper takes a dynamical systems perspective on the semantic structures of dynamic epistemic logic (DEL) and asks the question which orbits DEL-based dynamical systems may produce. The class of dynamical systems based directly on action models produce very limited orbits. Three types of more complex model transformers are equivalent and may produce a large class of orbits, suitable for most modeling purposes.

Keywords: dynamic epistemic logic, dynamical systems, model transformers, protocols, modeling.

1 Introduction

When modeling socio-epistemic phenomena, working with the temporally local models of dynamic epistemic logic (DEL) is both a blessing and a bane. It is a blessing as both epistemic state models and their updates are small relative to a fully explicated epistemic temporal structure. This eases both model construction and comprehension. It is a bane as the small models are incomplete: each is an individual time-step while we seek to model temporally extended dynamics. To form a 'complete model', we must specify the 'temporal glue' that ties individual epistemic states together to dynamics.

This 'temporal glue' is often presented informally in the DEL literature by way of a natural language problem description, typically involving conditional tests to determine which update to apply. Methodologically, this leaves modelers with a small gap: when modeling information dynamics using the semantic tools of DEL, what *mathematical object* shall we identify as *the model* of our target phenomenon?

It is an advantage of the DEL approach that a full sequential model need not be specified from the outset, but a drawback that a complete formalization of the problem under investigation is missing. Ideally, such 'complete models' should be both

1. Computably tractable (for each step), and
2. Informative (model the problem, not just describe the solution).

© Springer-Verlag Berlin Heidelberg 2015
W. van der Hoek et al. (Eds.): LORI 2015, LNCS 9394, pp. 316–327, 2015.
DOI: 10.1007/978-3-662-48561-3_26

The first *desideratum* is for implementation purposes. By the second, it is sought that eventual implementations are interesting: models that formalize problems without requiring they be solved first, allows one to draw informative conclusions about the modeled phenomena. The informal approach is typically informative.

This paper suggests a *dynamical systems* approach to specifying 'complete models' of information dynamics and provides some preliminary results.[1] As a (discrete time) dynamical system consists of only a state space X and a map $\tau : X \longrightarrow X$ iteratively applied, the future development of the dynamics depend only on the current state and the map τ. Dynamical systems thus provide a formal container for dynamical models in the local spirit of DEL. This stands in contrast to the only formal alternative, *DEL protocols* [3], which define dynamics globally. This approach is discussed in Section 3.

Dynamical systems are simple but may therefore also be limiting. E.g., if one's chosen model transformer class contains only action models, then the set of scenarios that can be modeled is very narrow: the same action model will be reapplied by the dynamical system, scenarios such as the well-known Muddy Children example [10] are among the unrepresentable phenomena. This provides a motivation for seeking broader classes of model transformers, the topic of Section 5. Three methods for defining complex model transformers are defined, being *multi-pointed action models*, *programs* and *problems*. The main technical results compare these approaches with respect to the orbits they can produce when used in dynamical systems.

2   DEL Preliminaries

Let be given a finite, non-empty set of propositional atoms Φ and a finite, non-empty set of agents, $\mathcal{A}$.

Definition 1 (Kripke Model). *A Kripke model is a tuple $M = (\llbracket M \rrbracket, R, \llbracket \cdot \rrbracket)$ where*

   $\llbracket M \rrbracket$ *is a non-empty set of states;*

   $R : \mathcal{A} \longrightarrow \mathcal{P}(S \times S)$ *is an accessibility function;*

   $\llbracket \cdot \rrbracket : \Phi \longrightarrow \mathcal{P}(S)$ *is a valuation function.*

A pair (M, s) with $s \in \llbracket M \rrbracket$ is called an epistemic state.

Definition 2 (Language, Semantics). *Where $p \in \Phi$ and $i \in \mathcal{A}$, define a language $\mathcal{L}$ by*

$$\varphi := \top \mid p \mid \neg\varphi \mid \varphi \wedge \varphi \mid K_i\varphi$$

with non-propositional formulas evaluated over epistemic state (M, s) by

$$(M, s) \models K_i\varphi \text{ iff } \forall t \in R_i(s), (M, t) \models \varphi.$$

---

[1] The approach to dynamical systems taken here thus differs from that [14], which mainly seeks modal logical descriptions of dynamical system concepts.

With a normal modal logical language like $\mathcal{L}$, the natural notion of equality of epistemic states is *bisimulation*:

Theorem 1 (Hennessy-Milner, [4], Thm.2.24). *Let M and M' be image-finite, i.e., $\forall s \in [\![M]\!], \forall i \in \mathcal{A}$, the set $\{t : (s,t) \in R_i\}$ is finite. Then for all $s \in [\![M]\!], s' \in [\![M']\!]$, s and s' are modally equivalent iff (M,s) and (M',s') are bisimilar.*

When working with finite models, $\mathcal{L}$ is strong enough to distinguish any two non-bisimilar models:

Theorem 2 ([11], Thm.32). *Let (M,s) and (M',s') be finite epistemic states that are not n-bisimilar. Then there exists $\delta \in \mathcal{L}$ such that $(M,s) \models \delta$ and $(M',s') \not\models \delta$.*

Dynamics are introduced by transitioning from one epistemic state to the next:

Definition 3 (Model Transformer). *Let $\mathcal{M}$ be the set of epistemic states based on $\mathcal{A}$. A* model transformer *is a (possibly partial) function $\tau : \mathcal{M} \longrightarrow \mathcal{M}$.*

Several model transformers have been suggested in the literature, the most well-known being public announcement, $!\varphi$ [12]. Primary to this paper is the rich class of action models [2] with postconditions [8].

Definition 4 (Action Model). *An* action model *is a tuple $\Sigma = ([\![\Sigma]\!], \mathsf{R}, \mathsf{pre}, \mathsf{post})$ where*

$[\![\Sigma]\!]$ *is a finite, non-empty set of* actions;

$\mathsf{R} : \mathcal{A} \longrightarrow \mathcal{P}([\![\Sigma]\!] \times [\![\Sigma]\!])$ *is an* accessibility *function;*

$\mathsf{pre} : [\![\Sigma]\!] \longrightarrow \mathcal{L}$ *is a* precondition *function;*

$\mathsf{post} : [\![\Sigma]\!] \longrightarrow \{\bigwedge_{i=0}^{n} \varphi_i \not\models \bot : \varphi_i \in \{\top, p, \neg p : p \in \Phi\}\}$ *is a* postcondition *function.*

A pair (Σ, σ) with $\sigma \in [\![\Sigma]\!]$ is called an epistemic action.

The precondition of an action σ specifies the conditions under which σ is executable; the postconditions specify how σ sets the values of select atoms. If $\mathsf{post}(\sigma) = \top$, then σ changes nothing.

An epistemic state is informationally updated with an epistemic action by taking their product:

Definition 5 (Product Update). *The* product update *of epistemic state $(M,s) = ([\![M]\!], R, [\![\cdot]\!], s)$ with epistemic action $(\Sigma, \sigma) = ([\![\Sigma]\!], \mathsf{R}, \mathsf{pre}, \mathsf{post}, \sigma)$ is the epistemic state*

$$(M \otimes \Sigma, (s, \sigma)) = ([\![M \otimes \Sigma]\!], R', [\![\cdot]\!]', (s, \sigma))$$

where

$[\![M \otimes \Sigma]\!] = \{(s, \sigma) \in [\![M]\!] \times [\![\Sigma]\!] : (M,s) \models \mathsf{pre}(\sigma)\}$

$R'_i = \{((s, \sigma), (t, \tau)) : (s,t) \in R_i \text{ and } (\sigma, \tau) \in \mathsf{R}_i\}$

$[\![p]\!]' = \{(s, \sigma) : s \in [\![p]\!], \mathsf{post}(\sigma) \not\models \neg p\} \cup \{(s, \sigma) : \mathsf{post}(\sigma) \models p\}.$

In combination, an epistemic action (Σ, σ) and product update $\otimes$ thus define a model transformer. Denote the class of such transformers by $\boldsymbol{\Sigma}$. Each $\tau \in \boldsymbol{\Sigma}$ has the following pleasant property:

Fact (Bisimulation Preservation). $\forall \tau \in \boldsymbol{\Sigma}$, if (M, s) and (M', s') are bisimilar, then so are $\tau(M, s)$ and $\tau(M', s')$.

$\boldsymbol{\Sigma}$ is a very powerful class: for any finite epistemic state (M, s), it contains a transformer that will map (M, s) to any other finite epistemic state (M', s'), as long as no agents with empty access in M has non-empty access in M' and as long as M and M' differ only in the truth value of a finite number of atoms. The restrictions are due to the 'and'-condition used in defining R_i' in product update and the finite conjunction used in defining postcondition maps. If the directed relation given by these restrictions holds from (M, s) to (M', s'), then call the transition from the first to the second *reasonable*:

Definition 6 (Reasonable Transition). *Let* $(M, s) = (\llbracket M \rrbracket, R, V, s)$ *and* $(M', s') = (\llbracket M' \rrbracket, R', V', s')$ *be two epistemic states. Then the transition from* (M, s) *to* (M', s') *is* reasonable *iff*

1. *it* preserves insanity*: there exists a submodel M^s of M such that $s \in \llbracket M^s \rrbracket$ and $\forall i \in \mathcal{A}$, if $R_i' \neq \emptyset$, then R_i is serial in M^s, and*
2. *it invokes* finite ontic change*:*

$$\{p : \llbracket p \rrbracket \neq \emptyset \text{ and } \llbracket p \rrbracket \neq \llbracket M \rrbracket\}$$
$$\cup \{p : \llbracket p \rrbracket = \emptyset\} \setminus \{p : \llbracket p \rrbracket' = \emptyset\}$$
$$\cup \{p : \llbracket p \rrbracket = \llbracket M \rrbracket\} \setminus \{p : \llbracket p \rrbracket' = \llbracket M' \rrbracket\}$$

is finite.

Theorem 3 (Arbitrary Change, [8], Prop.3.2). *Let the transition from finite (M, s) to finite (M', s') be reasonable. Then there exists a $(\Sigma, \sigma) \in \boldsymbol{\Sigma}$ such that $(M, s) \otimes (\Sigma, \sigma)$ and (M', s') are bisimilar.*

3 DEL Protocols

One framework which could be used to construct 'complete models' is *DEL protocols* [3,7,13,15].

Definition 7 (DEL Protocol). *Let $\boldsymbol{\Sigma}^*$ be the set of all finite sequences of transformers $\tau \in \boldsymbol{\Sigma}$. A set $\mathsf{P} \subseteq \boldsymbol{\Sigma}^*$ is a (uniform)* DEL protocol *iff P is closed under non-empty prefixes.*

A DEL protocol specifies which model transformers *may* be executed at a given time—whether they *can* be executed depends on the model transformers, e.g. their preconditions.

Where P is a DEL protocol and $\sigma = (\tau_1, ..., \tau_n) \in \mathsf{P}$, set $(M, s)^\sigma := \tau_n \circ \cdots \circ \tau_1(M)$. From an initial model (M, s) and time 0, a DEL protocol P produces a set of possible evolutions to each time n, namely

$\{(M, s)^{\sigma} : \text{len}(\sigma) = n\}$. Notice that $\text{len}(\sigma) = n$ does not imply that $(M, s)^{\sigma}$ exists: one of the transformers from σ may have been unexecutable at some earlier stage.

DEL protocols are dismissed as suitable for constructing 'complete models' as the results will be unexecutable, incorrect or uninformative. To see this, assume that some phenomenon that involves multiple model transformers $T = \{\tau_1, ..., \tau_n\}$, as e.g. Muddy Children does.

If the DEL protocol used is T^* (the set of all finite strings sequences of transformers from T) a very nice model is obtained: it is applicable to multiple initial states with varying mud distributions, and it may accordingly be used to obtain answers to questions about e.g. how the scenario unfolds as a function of the number of muddy children. Alas, T^* is infinite and as a model therefore unexecutable: given some initial state (M, s) it will not be possible to run T^* on (M, s) in finite time as the input to any function that is to determine the set $\{(M, s)^{\sigma} : \text{len}(\sigma) = 1\}$ will be infinite.

To obtain an executable model, T^* could be pruned to obtain a finite DEL protocol $\mathsf{T} \subseteq T^*$, e.g. by setting some upper bound on the length of $\sigma \in \mathsf{T}$. The risk associated with this move (pruning) is that the model becomes useless or uninformative: if the upper bound is set too low, the model will terminate too soon and not provide a correct output; to ensure the upper bound high enough, the problem must have been solved beforehand, leading to an uninformative model. In the extreme case where the only included maximal σ is 'the correct one' given some natural language protocol and initial state, a descriptive model is produced, but such a 'gold in, gold out' model is of little interest from an investigative perspective.

4   DEL and Dynamical Systems

Given Theorem 3, one might expect that dynamical systems based on the class of action models Σ would allow modeling of a plethora of phenomena. Surprisingly, not even even simple and well-known epistemic puzzles such as Muddy Children can be modeled by this class. To see this, let us first clarify the notion of dynamical system.

As standardly defined [6], a dynamical system is a tuple $D = (X, T, \mathcal{E})$ where X is set, called the *state space*, $T \subseteq \mathbb{R}$ is a *time set* which forms an additive semi-group $(t_1, t_2 \in T \Rightarrow t_1 + t_2 \in T)$ and $\mathcal{E} : X \times T \to X$ is an *evolution map* satisfying that $\mathcal{E}(x, 0) = 0$ and $\mathcal{E}(\mathcal{E}(x, t_1), t_2) = \mathcal{E}(x, t_1 + t_2)$.

To obtain a state space for DEL-based dynamical systems, it is natural, given Theorem 1, to equate bisimilar epistemic states, and let the state space consist of each bisimulation type's smallest representative. For an epistemic state (M, s), this representative is given by (M, s)'s *generated submodel rooted at s's bisimulation quotient* $(M[s]/\rho^M, [s]_{\rho}^M)$, see [11], Sec. 3.6. Setting

$$\mathbf{M} := \{(M[s]/\rho^M, [s]_{\rho}^M) : (M, s) \text{ is an epistemic state}\},$$

a class is obtained that contains a canonical representative of each epistemic state, each unique up to isomorphism.

As DEL updates are discrete and non-invertible, the suitable time set for a DEL-based dynamical system is $\mathbb{Z}_+$. The evolution function of any dynamical system $D = (X, \mathbb{Z}_+, \mathcal{E})$ with time set $\mathbb{Z}_+$ may be defined by the iterations of a function $e : X \to X$ by $\mathcal{E}(x, n) = e^n(x)$. Given the chosen state space, the suitable class of such functions e is the set of model transformers $\tau : \mathbf{M} \to \mathbf{M}$, denoted by $\mathbf{T}$.

Given these considerations, the following definition of DEL-based dynamical systems is obtained:

Definition 8 (DEL-based Dynamical System). *A DEL-based dynamical system is a pair $D = (\mathbf{X}, \tau)$ where $\mathbf{X} \subseteq \mathbf{M}$ and $\tau : \mathbf{X} \to \mathbf{X}$.*
The orbit of D from initial state $x_0 \in \mathbf{X}$ is the sequence $o(D, x_o) = (\tau^n(x_0))_{n \in \mathbb{Z}_+}$.

Remark. Given an epistemic action $\tau \in \Sigma$, $x \in \mathbf{M}$ does not imply that $\tau(x) \in \mathbf{M}$. There will however be a $x' \in \mathbf{M}$ that is bisimilar to $\tau(x)$. Given Fact 1, each $\tau \in \Sigma$ may be identified with a $\tau' \in \mathbf{T}$ by if $\tau(x) = (M, s)$, then $\tau'(x) = (M[s]/\rho^M, [s]_\rho^M)$. Henceforth, when executing an epistemic action (Σ, σ) in $x \in \mathbf{M}$, it is thus assumed that $x \otimes (\Sigma, \sigma) \in \mathbf{M}$.

It is immediately clear that any dynamical system $D = (\mathbf{X}, \tau)$ with $\tau \in \Sigma$ will be limited in its orbits. In particular, where s_0 is the actual state in the initial epistemic state x_0 and σ_0 is the actual state of τ, then for any n, the actual state of $\tau^n(x_0)$ will be of the form $(...(s_0, \sigma_0), ..., \sigma_0)$. Consequently, any phenomenon that involves the occurrence of more than one actual action is unmodelable. As most phenomena do involve shift in the performed action, e.g. by a shift in the announcement made, there is a motivation for seeking out a more general class of model transformers.

5 Complex Model Transformers

The limitation of DEL-based dynamical systems does not stem from action models, but rather from the fact that their usage is not controlled. This problem is solved by DEL protocols or update streams; simply specify at which time which action model should be executed. However, this requires a description of the evolution before execution, leaving little of the local DEL spirit intact.

A natural way to specify which transformer should be applied next that still remains local in spirit is by using a map $\pi : \mathbf{M} \longrightarrow \mathbf{T}$. Composing such a π with the model transformers it picks at each epistemic state is then again a model transformer $\tau_\pi : \mathbf{M} \longrightarrow \mathbf{M}$ given by $\tau_\pi(x) = \pi(x)(x)$.

To be interesting from modeling and implementation perspectives, such π must be finitely representable. This puts constraints on the dynamical systems definable, but, as will be shown, the restriction is still to a vast class of such systems.

We focus on three ways of specifying maps π, each picking model transformers from Σ. The choice to restrict attention to maps picking transformers from

Σ is warranted by Theorem 3: As basic transformers, this class has sufficient transformational power to construct a rich class of dynamical systems.

The first type is closely related to the (knowledge-based) programs known from interpreted systems [10], though defined to specify transformers based on the global, epistemic state rather than specifying sub-actions based on agents' local states:[2]

Definition 9 (Program). *A (finite, deterministic,$(\mathcal{L}, \Sigma)$) program is a finite set of formula-transformer pairs*

$$P = \{(\varphi_i, \tau_i) : \varphi_i \in \mathcal{L}, \tau_i \in \Sigma\}$$

where $\forall i, j$ if $\varphi_i \neq \varphi_j$ and $(\varphi_i, \tau_j), (\varphi_j, \tau_j) \in P$, then $\mathbf{M} \models \varphi_i \wedge \varphi_j \rightarrow \perp$.
  Each program P gives rise to a model transformer τ_P given by $\tau_P(x) = \tau_i(x)$ if $x \models \varphi_i$ and $(\varphi_i, \tau_i) \in P$. Denote this class by $\mathbf{P}$.

Each program may be read as a set of conditional tests of the form **if** φ_i, **do** τ_i, in form similar to the informal specifications often used in DEL literature.

The explicit specification of programs stands in contrast with the implicit specification of the second transformer type, *problems*, where each instruction may be read **if** φ_i, **obtain** ψ_i. Problems as defined here are related to *epistemic planning problems*, also know from the DEL literature [5].

Definition 10 (Problem). *A (finite $(\mathcal{L}, \Sigma)$) problem is a pair*

$$\Pi = (Q, \Sigma_\Pi)$$

where $Q = \{(\varphi_i, \psi_i) : \varphi_i, \psi_i \in \mathcal{L}\}$ is a finite set of formula-formula pairs and $\Sigma_\Pi \subset \Sigma$ is a finite set of model transformers with an associated strict order $<$.

A solution to $\Pi = (Q, T)$ at epistemic state x is a model transformer $\tau \in T$ such that $\forall (\varphi_i, \psi_i) \in Q$, if $x \models \varphi_i$, then $\tau(x) \models \psi_i$. Denote the set of solution to Π at x by $\Pi(x)$.

  Each problem Π gives rise to a model transformer τ_Π given by $\tau_\Pi(x) = \min_< \Pi(x)$. Denote this class by $\mathbf{\Pi}$.

The model transformer τ_Π is defined using the strict order $<$ on Σ_Π to ensure that τ_Π is a function: nothing in the definition ensures that $|\Pi(x)| \leq 1$.

The last model transformer type to be considered is a slight generalization of action models [1], where each such may have multiple actual states. In the definition it is required, non-standardly, that the preconditions of the actual states must be mutually exclusive. This is to ensure that executing a multi-pointed action model using product update remains a single-pointed epistemic state.

---

[2] Programs based on agents' local states is also at least to some degree feasible in a DEL setting, using *parallel action model composition* [9].

Definition 11 (Multi-Pointed Epistemic Actions). *A (finite, deterministic) multi-pointed epistemic action is an epistemic action* (Σ, σ) *with* σ *replaced by a finite, non-empty set* $S \subseteq [\![\Sigma]\!]$, *where for each* $\sigma, \sigma' \in S$, *if* $\sigma \neq \sigma'$, *then* $\mathbf{M} \models pre(\sigma) \wedge pre(\sigma') \rightarrow \bot$.

Applied using product update, each (Σ, S) *is a model transformer* $\tau : (M \otimes \Sigma, (s, S)) \mapsto ([\![M \otimes \Sigma]\!], R', [\![\cdot]\!]', (s, \sigma_i))$ *where* $(M, s) \models pre(\sigma_i)$. *Denote this class by* Σ^+.

With mutually exclusive preconditions, a multi-pointed action model (Σ, S) encodes a map $\pi : \mathbf{M} \longrightarrow \mathbf{T}$ with image $\{(\Sigma, \sigma) : \sigma \in S\}$ by $\pi(x) = (\Sigma, \sigma)$, $x \models pre(\sigma)$.

6 Results

Note initially that DEL-based dynamical systems fair better than DEL protocols in regard to executability and informativity. DEL-based dynamical systems resting on either a program or a multi-pointed action model are step-wise computable, as both transformer types are finite and therefore require only check of a finite set of formulas at each (M, s). The case for problems must be checked against [5]. Moreover, DEL-based dynamical systems will provide informative models: once a system is defined, one may start investigating how its orbits behave as a function of initial state without having pre-solved the encoded problem.

The first main result shows that dynamical systems based on the class Π of problem-based model transformers can model any reasonable, deterministic, finite or cyclic sequence of finite epistemic states. Problem-based dynamical systems can thus model a large class of phenomena.

The proof of Proposition 1 is by brute force. The construction results in a large, cumbersome problem fully pre-encoding the target orbit. For many modeling purposes, far more economical complex model transformers will do.

Definition 12 (Finite Variation, Deterministic). *Let* $\bar{x} = (x_0, x_1, ...)$ *be a sequence of epistemic states from* $\mathbf{M}$. $\bar{x}$ *has* finite variation *iff*

1. $\bar{x}$ *is finite, or*
2. $\exists n, m, k \in \mathbb{Z}_+ \backslash \{0\} : x_k = x_{k+m}$ *for all* $k \geq n$.

$\bar{x}$ *is* deterministic *iff if* $x_k, x_{k+1}, x_m \in \bar{x}$ *and* $x_k = x_m$, *then* $x_{m+1} \in \bar{x}$ *and* $x_{k+1} = x_{m+1}$.

Proposition 1 (Arbitrary Orbits). *Let the sequence* $\bar{x} = (x_0, x_1, ...)$ *of finite epistemic states be deterministic, with finite variation and where the transition between each* x_i *and* x_{i+1} *is reasonable. Then there exists a dynamical system* $D = (\mathbf{M}, \tau_\Pi)$ *with* $\tau_\Pi \in \Pi$ *such that* $o(D, x_0) = \bar{x}$.

Proof. By constructing a problem $\Pi = (Q, \Sigma_\Pi)$ that gives rise to the sought τ_Π.

For each $x_i, x_j \in \bar{x}$, $x_i \neq x_j$, let $\delta_{i,j}$ be a formula that distinguishes x_i from x_j such that $x_i \models \delta_{i,j}$ and $x_j \not\models \delta_{i,j}$; this $\delta_{i,j}$ exists by Theorem 2. As $\bar{x}$ has finite variation, $\delta_i := \bigwedge_{j:x_j \in \bar{x} \setminus \{x_i\}} \delta_{i,j}$ is a formula that distinguishes x_i from all other $x_j \in \bar{x}$. For each $x_i, x_{i+1} \in \bar{x}$, let $\tau_i \in \Sigma$ be a model transformer such that $\tau_i(x_i) = x_{i+1}$; this exists by Theorem 3.

Let Q be the smallest set that for each $x_i, x_{i+1} \in \bar{x}$ contains (δ_i, δ_{i+1}). Let Σ_Π be the smallest set that for each $x_i, x_{i+1} \in \bar{x}$ contains τ_i. Both Q and Σ_Π are finite by the assumption of finite variation, so $\Pi = (Q, \Sigma_\Pi)$ is a finite program, so τ_Π is a model transformer.

That $o(D, x_0) = \bar{x}$ when $D = (\mathbf{M}, \tau_\Pi)$ is shown by induction on x_n:
Base: $\tau_\Pi^0(x_0) = x_0$. *Step:* Assume $\tau_\Pi^n(x_0) = x_n$. If $\bar{x} = (x_0, ..., x_n)$, then $o(D, x_0) = \bar{x}$ as $(\delta_n, \varphi) \notin Q$ for any φ, by determinism of $\bar{x}$, so $\tau_\Pi(x_n)$ is undefined. If $x_{n+1} \in \bar{x}$, then $(\delta_n, \delta_{n+1}) \in Q$ and $\tau_n \in \Sigma_\Pi$. By construction, $\Pi(x_n) = \tau_n$, so $\tau_\Pi(x) = x_{n+1}$. □

Proposition 2 (Problem Orbit Properties). *Let* $o(D, x_0) = \bar{x}$ *with* $D = (\mathbf{M}, \tau_\Pi)$, $\tau_\Pi \in \mathbf{\Pi}$. *Then* $\bar{x}$ *is deterministic and for each* $x_i, x_{i+1} \in \bar{x}$, *the transition from* x_i *to* x_{i+1} *is reasonable.*

Proof. $\bar{x}$ is deterministic as τ_Π is a function; each transition is reasonable as $x_{i+1} = \tau(x_i)$ for some $\tau \in \mathbf{\Sigma}$.

Propositions 1 and 2 cannot be strengthened to a characterization result as not all problem-based dynamical system have finite variation:

Proposition 3 (Infinite Variation). *There exists a dynamical system* $D = (\mathbf{M}, \tau_\Pi)$ *with* $\tau_\Pi \in \mathbf{\Pi}$ *such that* $o(D, x_0)$ *does not have finite variation.*

Proof. Let $D = (\mathbf{M}, \tau_\Pi)$ with problem $\Pi = (\{(\top, \top)\}, \{(\Sigma, \sigma_1)\})$. This trivial problem has unique solution (Σ, σ_1) for all $(M, s) \in \mathbf{M}$. Hence, for all $x \in \mathbf{M}$, $\tau_\Pi(x) = (M, s) \otimes (\Sigma, \sigma_1)$.

Let M and Σ given by

$$M: \quad s: \overset{}{\circlearrowleft}\bigcirc \leftarrow \boxed{p} \; t: \qquad \Sigma: \quad \sigma_1: \circlearrowleft\bigcirc\!(\neg p, \top) \to \boxed{(\neg p, p)}\overset{\sigma_2:}{\to} \boxed{(p, \top)}\circlearrowright \; \sigma_3:$$

Then $o(D, (M, s))$ does not have finite variation: for each iteration of τ_Π, the state not satisfying p will split, inserting a new p state as it's child with σ_2:

$$(M, s) \otimes (\Sigma, \sigma_1): \quad s\sigma_1: \circlearrowleft\bigcirc \overset{s\sigma_2:}{\to} \boxed{p} \overset{t\sigma_3:}{\to} \boxed{p}$$

All other states have only one child, with σ_3.

In all further applications of (Σ, σ_1), the circular structure seen in $(M, s) \otimes (\Sigma, \sigma_1)$ is preserved, only with an additional p state. No two such models are bisimilar, and hence the orbit does not have finite variation. □

The second main result shows that also program-based dynamical systems and dynamical systems based on multi-pointed action models can produce a vast class of orbits.

Proposition 4 (Equivalence). *Let* $\bar{x} = (x_0, x_1, ...)$ *be a sequence of epistemic states. Then*

1. $\exists \tau_\Pi \in \Pi$ *such that for* $D = (\mathbf{M}, \tau_\Pi)$, $o(D, x_0) = \bar{x}$.

   $\Uparrow$

2. $\exists \tau_P \in \mathbf{P}$ *such that for* $D = (\mathbf{M}, \tau_P)$, $o(D, x_0) = \bar{x}$.

   $\Updownarrow$

3. $\exists \tau_{\Sigma^+} \in \Sigma^+$ *such that for* $D = (\mathbf{M}, \tau_{\Sigma^+})$, $o(D, x_0) = \bar{x}$.

If $\bar{x} = (x_0, x_1, ...)$ *has finite variation and* x_0 *is finite, then the three statements are equivalent.*

Proof.
Case: 2. $\Rightarrow$ 1. Let $D = (\mathbf{M}, \tau_P)$, $\tau_P \in \mathbf{P}$ with $o(D, x_0) = \bar{x} = (x_0, x_1, ...)$ be given.

Construct a problem $\Pi = (Q, \Sigma_\Pi)$ as follows: Let Q be the smallest set that for each $(\varphi_i, \tau_i) \in P$ contains $(\varphi_i, \top)$. Let Σ_Π be the smallest set that for each $(\Sigma, \sigma) \in \Sigma_P$ contains (Σ, σ^*) identical to (Σ, σ) in all respects except that $pre(\sigma^*) = pre(\sigma) \wedge \varphi_i$. As P is finite, $\Pi = (Q, \Sigma_\Pi)$ is a finite problem; τ_Π is a model transformer as the φ_i's of P are mutually exclusive.

Then $o((\mathbf{M}, \tau_\Pi), x_0) = o((\mathbf{M}, \tau_P), x_0)$: Assume $x_i, x_{i+1} \in \bar{x}$. Then $x_{i+1} = \tau(x_i)$ for some $\tau = (\Sigma, \sigma)$ such that for some φ, $(\tau, \varphi) \in P$. Hence for some $\varphi, (\tau, \varphi) \in P$, it holds that $x_i \models \varphi$. Given the preconditions and that $(\varphi, \top) \in Q$, $\tau^* = (\Sigma, \sigma^*) \in \Sigma_\Pi$ will be the only solution to Π at x_i. As $x_i \models \varphi$, $\tau^*(x_i) = \tau(x_i)$.

Assume $\bar{x} = (x_0, ..., x_n)$ is finite. Then either $x_n \not\models \varphi_i$ for all $(\varphi_i, \tau_i) \in P$ or if $x_n \models \varphi_i$ for $(\varphi_i, (\Sigma, \sigma)) \in P$, then $x_n \not\models pre(\sigma)$. In the first case, $x_n \not\models \varphi_i$ for all $(\varphi_i, \top) \in Q$; in the second, $x_n \not\models pre(\sigma^*)$. In either case, $\tau_\Pi(x_n)$ is undefined.

Case: 2. $\Rightarrow$ 3. Let $D = (\mathbf{M}, \tau_P)$, $\tau_P \in \mathbf{P}$ with $o(D, x_0) = \bar{x} = (x_0, x_1, ...)$ be given. Let Σ_Π be as in the case 2. $\Rightarrow$ 1. Define a multi-pointed action model (Σ^+, S) by $\Sigma^+ = \biguplus \{\Sigma : (\Sigma, \sigma^*) \in \Sigma_\Pi\}$ and $S = \{\sigma^* : (\Sigma, \sigma^*) \in \Sigma_\Pi\}$. Let τ_{Σ^+} be the associated model transformer.

Then $o((\mathbf{M}, \tau_{\Sigma^+}), x_0) = o((\mathbf{M}, \tau_P), x_0)$: Assume $x_i, x_{i+1} \in \bar{x}$. Then $x_{i+1} = \tau(x_i)$ for some $\tau = (\Sigma, \sigma)$ such that for some φ, $(\tau, \varphi) \in P$. Hence for some $\varphi, (\tau, \varphi) \in P$, it holds that $x_i \models \varphi \wedge pre(\sigma)$, so by construction, $x_i \models pre(\sigma^*)$. Hence only the submodel (Σ, σ^*) of Σ^+ is executable at x_i, so $\tau_{\Sigma^+}(x_i) = \tau_P(x_i)$.

If $\bar{x} = (x_0, ..., x_n)$ is finite, then either $x_n \not\models \varphi_i$ for all $(\varphi_i, \tau_i) \in P$ or if $x_n \models \varphi_i$ for $(\varphi_i, (\Sigma, \sigma)) \in P$, then $x_n \not\models pre(\sigma)$. In the first case, $x_n \not\models pre(\sigma^*)$ for all $(\Sigma, \sigma^*) \in \Sigma^+$; in the second, $x_n \not\models pre(\sigma^*)$. In either case, $\tau_{\Sigma^+}(x_n)$ is undefined.

Case: 3. $\Rightarrow$ 2. Let $D = (\mathbf{M}, \tau_{\Sigma^+})$, $\tau_{\Sigma^+} \in \Sigma^+$ with $o(D, x_0) = \bar{x} = (x_0, x_1, ...)$ be given. Let the Σ^+ of τ_{Σ^+} be $\Sigma^+ = (\Sigma, S)$ and create from it a set of $|S|$ single-pointed action models $A = \{(\Sigma, \sigma) : \sigma \in S\}$. Create a program $P = \{(pre(\sigma), (\Sigma, \sigma)) : (\Sigma, \sigma) \in A\}$. P is both finite and deterministic.

Then $o((\mathbf{M}, \tau_P), x_0) = o((\mathbf{M}, \tau_{\Sigma^+}), x_0)$: Assume $x_i, x_{i+1} \in \bar{x}$. Then $x_i \models pre(\sigma)$ for exactly one $\sigma \in S$. As $(pre(\sigma), (\Sigma, \sigma)) \in P$, $\tau_P(x_i) = \tau_{\Sigma^+}(x_i)$.

If If $\overline{x} = (x_0, ..., x_n)$ is finite, then $x_n \not\models pre(\sigma)$ for all $\sigma \in S$. Hence for all $(\varphi, \tau) \in P$, $x_n \not\models \varphi$, so $\tau_P(x_n)$ is undefined.

Case: 1. $\Rightarrow$ 2., *if $\overline{x} = (x_0, x_1, ...)$ has finite variation and x_0 is finite*: Let $D = (\mathbf{M}, \tau_\Pi)$, $\tau_\Pi \in \mathbf{\Pi} = (Q, \Sigma_\Pi)$ with $o(D, x_0) = \overline{x} = (x_0, x_1, ...)$ having finite variation. Brute force construct a program using characteristic formulas: let δ_i be the characteristic formula of $x_i \in \overline{x}$. For each pair $x_i, x_{i+1} \in \overline{x}$, there is a unique $\tau_i \in \Sigma_\Pi$ such that $\tau_i(x_i) = x_{i+1}$. Let $P = \{(\delta_i, \tau_i) : x_i \in \overline{x}\}$. As $\overline{x}$ has finite variation, P is finite and gives rise to a model transformer τ_P.

Then $o((\mathbf{M}, \tau_P), x_0) = o((\mathbf{M}, \tau_\Pi), x_0)$: Assume $x_i, x_{i+1} \in \overline{x}$. Then $(\delta_i, \tau_i) \in P$, so $\tau_P(x_i) = x_{i+1}$. If $\overline{x} = (x_0, ..., x_n)$ is finite, then by Proposition 2, for no $x_i, i < n$ is $x_i = x_n$. Hence $(\delta_n, \tau) \notin P$, for any τ. Hence $\tau_P(x_n)$ is undefined. $\qquad\square$

Corollary 1 (Orbit Properties). *For any dynamical system $D = (\mathbf{M}, \tau)$ with $\tau \in \mathbf{P} \cup \Sigma^+$ and any $x_0 \in \mathbf{M}$, $o(D, x_0)$ is deterministic and for each $x_i, x_{i+1} \in \overline{x}$, the transition from x_i to x_{i+1} is reasonable.*

Proof. Let D be as described. By Proposition 4 there exists a $D' = (\mathbf{M}, \tau_\Pi)$, $\tau_\Pi \in \mathbf{\Pi}$, that recreates $o(D, x_0)$. The corollary then follows from Proposition 2.

7   Conclusion

The main contributions are

▷ that although dynamical systems defined using epistemic action models can produce only very limited orbits, dynamical systems that control when particular action models are used may produce orbits sufficient for most modeling purposes, and

▷ that the three methods for controlling which action models are applied are equivalent under the presented conditions.

The first result shows that DEL-based dynamical systems provide a rich framework for producing mathematically specified models of information dynamics. The latter shows that there are multiple ways of extending the DEL toolbox compatible with modeling using dynamical systems.

It would be interesting to make an in-depth comparison between DEL protocols and DEL-based dynamical systems, comparing the orbits they may produce and under which conditions such might be equivalent. Two considerations here involve the finite nature of DEL protocols, guaranteeing finite variation not guaranteed by DEL-based dynamical systems, and the 'bisimulation respecting' behavior of DEL-based dynamical systems, which is not necessarily followed by DEL protocols. Obtaining such results could be used to link DEL-based dynamical systems with Epistemic Temporal Logic via the results in [3].

Moreover, it would be interesting to investigate any deeper relationship between dynamic epistemic logic and dynamical systems; the latter field is well-developed, and one could envision that methods and results may be transferable.

References

1. Baltag, A., Moss, L.S.: Logics for Epistemic Programs. Synthese 139(2), 165–224 (2004)
2. Baltag, A., Moss, L.S., Solecki, S.: The Logic of Public Announcements, Common Knowledge, and Private Suspicions (extended abstract). In: Proc. of the Intl. Conf. TARK 1998, pp. 43–56. Morgan Kaufmann Publishers (1998)
3. van Benthem, J., Gerbrandy, J., Hoshi, T., Pacuit, E.: Merging Frameworks for Interaction. Journal of Philosophical Logic 38(5), 491–526 (2009)
4. Blackburn, P., de Rijke, M., Venema, Y.: Modal Logic. Cambridge University Press (2001)
5. Bolander, T., Birkegaard, M.: Epistemic planning for single- and multi-agent systems. Journal of Applied Non-Classical Logics 21(1), 9–34 (2011)
6. Broer, H.W., Takens, F.: Preliminaries of Dynamical Systems Theory. In: Hasselblatt, B., Broer, H.W., Takens, F. (eds.) Handbook of Dynamical Systems, North-Holland, vol. 3 (2010)
7. Dégremont, C.: The Temporal Mind: Observations on the logic of belief change in interactive systems. PhD thesis, University of Amsterdam (2010)
8. van Ditmarsch, H., Kooi, B.: Semantic Results for Ontic and Epistemic Change. In: Bonanno, G., van der Hoek, W., Wooldridge, M. (eds.) Logic and the Foundations of Game and Decision Theory (LOFT 7). Texts in Logic and Games, vol. 3, pp. 87–117. Amsterdam University Press (2008)
9. van Eijck, J., Sietsma, F., Wang, Y.: Composing models. Journal of Applied Non-Classical Logics 21(3-4), 397–425 (2011)
10. Fagin, R., Halpern, J.Y., Moses, Y., Vardi, M.Y.: Reasoning About Knowledge. The MIT Press (1995)
11. Goranko, V., Otto, M.: Model Theory of Modal Logic. In: Blackburn, P., van Benthem, J., Wolter, F. (eds.) Handbook of Modal Logic, Elsevier (2008)
12. Plaza, J.A.: Logics of public communications. In: Emrich, M.L., Pfeifer, M.S., Hadzikadic, M., Ras, Z.W. (eds.) Proceedings of the 4th International Symposium on Methodologies for Intelligent Systems, pp. 201–216 (1989)
13. Rodenhäuser, B.: A logic for extensional protocols. Journal of Applied Non-Classical Logics 21(3-4), 477–502 (2011)
14. Sarenac, D.: Modal Logic for Qualitative Dynamics. In: Roy, O., Girard, P., Marion, M. (eds.) Dynamic Formal Epistemology. Synthese Library, vol. 351, pp. 75–101. Springer (2011)
15. Wang, Y.: Epistemic Modelling and Protocol Dynamics. Doctoral thesis, Universiteit van Amsterdam (2010)

'Transitivity' of Consequence Relations

David Ripley

University of Connecticut, Storrs CT 06268, USA

Abstract. A binary relation R on a set S is *transitive* iff for all $a, b, c \in S$, if aRb and bRc, then aRc. This almost never applies to the relations logicians tend to think of as *consequence relations*; where such relations are relations on a set at all, they are rarely transitive. Yet it is common to hear consequence relations described as 'transitive', and to see rules imposed to ensure 'transitivity' of these relations. This paper attempts to clarify the situation.

1 Introduction

After briefly substantiating the claims in the abstract, this paper focuses on exploring a number of different properties of consequence relations that have traveled under the name 'transitivity', mapping the implications among them. From here forward, I will use 'transitive' and 'transitivity' very little, and only in their standard relation-theoretic sense. To reiterate: to be transitive, a relation R must be a binary relation on a set S, and it must be such that for any $a, b, c \in S$, if aRb and bRc, then aRc.

Many familiar consequence relations are not relations on a set at all, but instead relate *sets* of formulas (collections of premises) to *single* formulas (conclusions). That is, where $\mathcal{F}$ is the set of formulas under consideration, such a relation is a relation between $\wp(\mathcal{F})$ and $\mathcal{F}$. Following [6], I'll say these relations work in the 'SET-FORM framework'. Such a relation is not the right kind of thing to be transitive. Of course, these relations can, and frequently do, exhibit a number of properties more and less closely connected to transitivity. But I will not explore this here; I mention SET-FORM relations to set them aside.

In what follows, I work entirely in the SET-SET framework. In this framework, consequence relations really are binary relations on a single set: the set $\wp(\mathcal{F})$. That is, they relate sets of formulas to sets of formulas. So they are at least the right *kind* of relation to be transitive.

Much research into SET-SET consequence relations (see eg [4, 13, 7, 11, 14, 6]) interprets the members of the set of conclusions as (in some sense) *different possibilities*. On this interpretation, arguments with *fewer* conclusions are stronger than those with more, since they *narrow down* more finely on a result. This is the interpretation I'll focus on in what follows.

These relations, too, are almost never transitive. Consider, for example, the SET-SET consequence relation $\vdash$ determined by classical logic, explored and defended in [7], among other places. This relation relates $\{A \vee B\}$ to $\{A, B\}$, and relates $\{A, B\}$ to $\{A \wedge B\}$, but does not relate $\{A \vee B\}$ to $\{A \wedge B\}$; it is thus

© Springer-Verlag Berlin Heidelberg 2015
W. van der Hoek et al. (Eds.): LORI 2015, LNCS 9394, pp. 328–340, 2015.
DOI: 10.1007/978-3-662-48561-3_27

not transitive. The reason is nothing particularly to do with classical logic; it is instead to do with how *sets* of formulas are interpreted. As premises, they are meant *conjunctively*: as all available to be drawn on together in establishing conclusions. As conclusions, they are meant *disjunctively*: as jointly exhausting the space where the truth must lie, given the premises. This difference in interpretation prevents linking valid SET-SET arguments together in the simple way guaranteed by transitivity.

2   A Catalog of Linking Properties

In this section, I lay out the assumptions that will frame the paper, and then present a catalog of ten properties that a SET-SET consequence relation might exhibit, all of which, I think, are recognizable as related to what logicians often mean by 'transitivity'. These ten properties form the basis of the paper, which fully maps the implications among arbitrary conjunctions of these properties.

Some notational preliminaries: I use capital Roman letters for formulas, and capital Greek letters (that are not also capital Romans) for sets of formulas. $\mathcal{F}$ is the set of formulas in the language under consideration; each SET-SET consequence relation, then, is a binary relation on $\wp(\mathcal{F})$. (As above, I restrict attention entirely to SET-SET relations.) I abbreviate freely in usual sequent-calculus ways, so, for example, '$\Gamma, A, \Sigma \vdash$' abbreviates '$\Gamma \cup \{A\} \cup \Sigma \vdash \emptyset$'. When I talk of 'partitions' of a set, this should be understood to *include* partitions with an empty entry; for example, $\langle \emptyset, \Sigma \rangle$ *is* a partition of Σ, on this usage.

2.1   Assumptions

I assume in places that the language $\mathcal{F}$ contains infinitely many formulas; its cardinality does not otherwise matter. I make no assumptions about the nature or structure of formulas; $\mathcal{F}$ can be any infinite set.

Consequence relations are often defined as relations that are '*reflexive, monotonic*, and *transitive*'. The final condition, of course, is the subject of this paper, so I am certainly not assuming it. Nor will I assume reflexivity, although this turns out not to matter; all the results of the paper remain unchanged with such an assumption in place.[1]

---

[1] 'Reflexive' here is like 'transitive'; it does not have, in its usual application to SET-SET consequence relations, its usual relation-theoretic sense. In the usual sense, a relation R on a set S is *reflexive* iff for all $x \in S$, xRx. For consequence relations, this would require that for every set Γ of formulas, $\Gamma \vdash \Gamma$. As it happens, this is almost never the case; at the very least, the empty set does not entail itself in any familiar setting. There are two usual things one might mean by 'reflexivity' here: that $\Gamma \vdash \Gamma$ for all *singleton* Γ, or all *nonempty* Γ; these are the assumptions that would not change anything in what follows. To show this, I take care to make sure that all the examples I discuss are reflexive (in both of these senses), and that no proof of any claim depends on reflexivity (in any sense).

I will, however, assume throughout the paper that all consequence relations are *monotonic*: that whenever $\Gamma \vdash \Delta$, then $\Gamma, \Gamma' \vdash \Delta, \Delta'$.[2] This matters a great deal; the situation is very different if this assumption is not imposed, and many of the results to follow would not hold without it.

A consequence relation $\vdash$ is *compact* iff whenever $\Gamma \vdash \Delta$, then there are finite $\Gamma_{\text{fin}} \subseteq \Gamma$ and $\Delta_{\text{fin}} \subseteq \Delta$ such that $\Gamma_{\text{fin}} \vdash \Delta_{\text{fin}}$. In what follows, I will not require compactness in general, but I will keep track of compactness, and show what the effects of requiring compactness are.

2.2 The Catalog

Table 1 gives ten properties that a consequence relation $\vdash$ may or may not exhibit. Each of the properties is a *closure* property: they are all of the form 'if these things stand in the relation, then those things must also stand in the relation'. These should be understood as universally quantified; for example, $\vdash$ has the property KS iff whenever $\Gamma \vdash A$ and $A \vdash \Delta$, then $\Gamma \vdash \Delta$, for all choices of Γ, Δ, and A. The properties to be considered in this paper are the ten in Table 1, and arbitrary conjunctions of these.

Table 1. Linking properties

Name:	If	and	then
S	$C \vdash A$	$A \vdash D$	$C \vdash D$
KS	$\Gamma \vdash A$	$A \vdash \Delta$	$\Gamma \vdash \Delta$
/F	$\Gamma \vdash A$	$A, \Gamma \vdash \Delta$	$\Gamma \vdash \Delta$
F/	$\Gamma \vdash \Delta, A$	$A \vdash \Delta$	$\Gamma \vdash \Delta$
FG	$\Gamma \vdash \Delta, A$	$A, \Gamma \vdash \Delta$	$\Gamma \vdash \Delta$
/C	$\Gamma \vdash A$ for all $A \in \Sigma$	$\Sigma, \Gamma \vdash \Delta$	$\Gamma \vdash \Delta$
C/	$\Gamma \vdash \Delta, \Sigma$	$A \vdash \Delta$ for all $A \in \Sigma$	$\Gamma \vdash \Delta$
/C$^+$	$\Gamma \vdash \Delta, A$ for all $A \in \Sigma$	$\Sigma, \Gamma \vdash \Delta$	$\Gamma \vdash \Delta$
C$^+$/	$\Gamma \vdash \Delta, \Sigma$	$A, \Gamma \vdash \Delta$ for all $A \in \Sigma$	$\Gamma \vdash \Delta$
CG	$\Sigma^+, \Gamma \vdash \Delta, \Sigma^-$ for all partitions $\langle \Sigma^+, \Sigma^- \rangle$ of Σ		$\Gamma \vdash \Delta$

Each allows valid arguments to be *linked* in a specific way; in the antecedent of these properties, the formula A and/or the set Σ of formulas figures among the conclusions of the left conjunct and the premises of the right conjunct, but does not appear in the consequent at all. (CG is the only exception to this, as its antecedent does not have left and right conjuncts.) Two of these properties—S and KS—are special cases of transitivity. The others, however, are not.

The abbreviations for the properties are intended to be (at least somewhat) mnemonic without taking up too much space. The properties that have received the most attention are S for 'simple', FG for 'finite generalized', and CG for 'complete generalized'.[3] The remaining properties are lopsided; each focusses in on

---

[2] Unlike 'reflexive' and 'transitive', 'monotonic' here *does* have its usual relation-theoretic sense, w/r/t the order $\subseteq$ on sets of formulas.

[3] I take the terms 'simple' and 'generalized' from [16]. Weir's 'simple transitivity' is my S; his 'generalized transitivity' is my FG. (He does not consider CG.)

either the premise or conclusion side of the relation in question. The abbreviations for these properties include a '/'; where the property focusses on the premise side, a letter appears before '/', and where it focusses on the conclusion side, a letter appears after '/'. The 'F' and 'C' are for 'finite' and 'complete'.

Each property on the list has a *dual* also on the list. Properties P and P' are duals, in the sense relevant here, iff: for a consequence relation $\vdash$ to have P is for its converse $\dashv$ to have P'. The properties S, KS, FG, and CG are all self-dual. For the remaining properties, the names indicate duality; for example, /F and F/ are duals. Also, the assumptions in play about consequence relations (that they are monotonic SET-SET relations) are self-dual; a relation $\vdash$ meets them iff its converse $\dashv$ does. So too is compactness self-dual, in this sense. Noting these symmetries will allow for some of the following proofs to get away with only half the work they would otherwise take. For example, once we see that FG implies /F, we can immediately conclude that it implies F/ as well; and once we see that /c^+ does not imply c^+/, we can immediately conclude that c^+/ does not imply c^+/ either. I will use this style of reasoning frequently in what follows.

3 Previous Work

3.1 FG and Cut

In the present setting, FG is equivalent to the following property: if $\Gamma \vdash \Delta, A$ and $A, \Gamma' \vdash \Delta$, then $\Gamma, \Gamma' \vdash \Delta, \Delta'$. This property, in turn, is closely connected to [4]'s rule of *cut* in the sequent calculus LK. (Just like 'transitivity', 'cut' means many different things in different contexts. Most of them, however, are related to Gentzen's use of 'cut'.)

Cut looms large in many proof-theoretic investigations; FG, then, has real proof-theoretic import. But it also, at times, has philsophical import. For example, [7, 8] understand FG (as a condition on a particular consequence relation) as encoding the following constraint on certain conversational norms: if a certain combination of assertions and denials is within the norms, then for any formula A, either adding an assertion of A to that combination remains within the norms, or else adding a denial of A to that combination remains within the norms. [7, 8] endorse this constraint; [9, 10] dispute it.

3.2 CG and Bivaluations

One way to present a consequence relation on a language $\mathcal{F}$ is via *bivaluations*: binary partitions $\langle T, F \rangle$ of $\mathcal{F}$. By specifying a set $\mathfrak{M}$ of such partitions, one specifies a consequence relation $\vdash_{\mathfrak{M}}$ in the following way: $\Gamma \vdash_{\mathfrak{M}} \Delta$ iff there is no $\langle T, F \rangle \in \mathfrak{M}$ such that $\Gamma \subseteq T$ and $\Delta \subseteq F$. (Informally, you might think: the argument is valid iff there is no model on which all the premises are true and all the conclusions false.) This way of thinking is stressed in [13, 6], but even where it is not stressed it is often applicable. For example, any way of presenting a consequence relation using models with *designated values* in the usual way fits this mould directly: we can understand each model as partitioning the language into those formulas that receive a designated value and those that do not.

Any consequence relation arrived at in this way will have certain structural properties: it will be *reflexive* (in the senses of footnote 1), *monotonic*, and it will have the property CG. (For proof, see [13, p. 30].) As we will shortly see, CG in fact implies all the other properties in Table 1. This means that bivaluations will not prove useful in what follows; they obscure the relations between the linking properties under consideration, by forcing them all to hold.[4]

Many monotonic SET-SET consequence relations encountered in the wild can be presented in terms of bivaluations, and so exhibit CG and thus all the linking properties to be considered here. (Note, however, that [11, p. 83] complains that CG is overstrong, claiming that it requires "much more than the transitivity of consequence".) It is only in cases where CG fails that the distinctions explored here are revealed.

3.3 Quantum Logic

[3, p. 44] and [1] both consider forms of quantum logic, and attribute to it the conjunction of /F and F/, which I will call F/F. In quantum logic, distribution of conjunction over disjunction fails; as it happens, there are important connections between distribution and FG, which I do not have space to explore here (but see [6, p. 10], particularly Exercise 0.13.7(i)). In these authors' settings, quantum logic does not obey FG, which they take to be a default expression of transitivity; F/F is substituted to "reflect the transitivity of implication" [1, p. 247].

In both cases, the authors restrict their attention to compact relations, for which the conjunction of /C and C/, which I will call C/C, is equivalent to F/F.[5] (More on compactness presently.) Neither source discusses /F or F/ on their own.

3.4 Neo-Classical Logic

The 'neo-classical' logic explored in [15, 16], among other places, is another consequence relation that exhibits some of these properties but not others. As [16, p. 100] points out, this consequence relation obeys S. In fact it also obeys KS; as we will see, this is stronger. However, it does not exhibit any of the other properties in Table 1. Weir claims that S "should be incorporated in any genuine notion of logical consequence", but does not elaborate.

3.5 Cut₃

There is one other property not listed in Table 1 I'm aware of that has been considered a form of 'transitivity' for SET-SET consequence relations. This is

---

[4] Related techniques from [5], however, can avoid imposing CG.

[5] In fact, Dummett (but not Cutland & Gibbins) only considers finite sequents. Note as well that the discussion in [3] in support of F/F, if cogent, in fact supports the full strength of C/C, even for noncompact relations.

the property called 'Cut$_3$' in [13, p. 32]. A consequence relation $\vdash$ has Cut$_3$ iff whenever $\Gamma \vdash \Delta, A$ for all $A \in \Sigma_1$, and $B, \Gamma \vdash \Delta$ for all $B \in \Sigma_2$, and $\Sigma_1, \Gamma \vdash \Delta, \Sigma_2$, then $\Gamma \vdash \Delta$. But as Shoesmith and Smiley immediately show, Cut$_3$ is equivalent to the conjunction of /c$^+$ and c$^+$/; I will later call this conjunction c$^+$/c$^+$. (Their proof depends on monotonicity.)[6]

[12, p. 37], oddly, calls this property (there defined directly as the conjunction of /c$^+$ and c$^+$/) 'Cut', and takes it to be of some import. In particular, Segerberg points to FG, claims that it is not sufficient when infinite sets of premises and conclusions are considered, and then offers this property as the appropriate replacement. (He also points out that s, which he calls 'transitivity', is a 'very special case' of this property (p. 38).) I know of no other sources that have attended to this property.

4 Implications

There are ten properties listed in Table 1, and this paper will consider arbitrary conjunctions of these. Our exploration begins, then, with $2^{10} = 1024$ property-specifications to consider. Fortunately, there are many fewer distinct properties actually in play. In this section, I explore implications among these properties, and show that from our 1024, there are at most 21 distinct properties, and at most 7 if compactness is assumed. (I identify properties iff they imply each other.) In fact, these counts are exact, but the 'at least' part of the claim will not be proved until §5. First, I will lay out these implications in three categories: implications by special case, implications by monotonicity, and implications by semilattice properties. Then, I will consider the effects of compactness, and show additional implications among our properties that hold when compactness is assumed.

4.1 Three Kinds of Implications

Some implications from one property to another happen in the easiest possible way: when one property covers only certain special cases of another. These implications can be verified directly by inspection. In this way, five implications are secured: KS implies S; /c implies /F; c/ implies F/; and each of /c$^+$ and c$^+$/ implies FG.

Other implications are not so direct; these require some appeal to monotonicity. The needed appeals to monotonicity, however, are quite formulaic: when one property's antecedent follows by monotonicity from another property's antecedent, then the first property implies the second. This gives eight more implications: each of /F and F/ implies KS; FG implies both /F and F/; /c$^+$ implies /c; c$^+$/ implies c/; and CG implies both /c$^+$ and c$^+$/.

---

[6] [13, p. 30ff.] considers FG, /c$^+$, c$^+$/, Cut$_3$, and CG; the implications and nonimplications among these properties shown there are among what is shown in the present paper.

Finally, implication among properties forms a semilattice with conjunction as the meet.[7] That is, implication is transitive, and the conjunction of two properties is their greatest lower bound w/r/t the implication order. Together with the implications recorded above, this secures a large range of additional implications among the properties under consideration. For example, since FG implies both /F and F/, it follows that it implies their conjunction. Since /c$^+$ implies /c, and /c implies /F, then /c$^+$ implies /F. And so on.

4.2 Twenty-One Properties

These implications narrow the space of properties under consideration to twenty-one: the ten properties that appear in Table 1, plus the eleven additional properties given in Table 2, generated from the original ten by conjunction.

Table 2. Additional linking properties formed by conjunction

Name:	Definition:		Name:	Definition:
F/F	F/ and /F.		c/c	c/ and /c.
F/c	F/ and /c.		c/F	c/ and /F.
/FG/c	/c and FG.		c/FG/	c/ and FG.
c/FG/c	c/, /c, and FG.		c$^+$/c$^+$	c$^+$/ and /c$^+$.
c/c$^+$	c/ and /c$^+$.		c$^+$/c	c$^+$/ and /c.
⊤	The empty conjunction, exhibited by every consequence relation.			

Given the implications already recorded, each of the $2^{10} = 1024$ property-specifications we can generate from Table 1 by conjunction specifies one of these twenty-one properties. For example, for a consequence relation to exhibit the properties FG, F/, and /c is just for it to exhibit /FG/c, since FG already implies F/. Similarly, for a consequence relation to exhibit KS, /c$^+$, and F/ is just for it to exhibit /c$^+$, which implies the other two properties. And so on, for every combination.

4.3 Compactness

For compact relations, there are more implications to take account of among the properties in play; this section records these and takes account of their impact.

Proposition 1. *If ⊢ is compact and has FG, then it has CG.*

Proof. See [13, p. 37] for proof. (Their 'cut for formulae' is exactly FG, and their 'cut for sets' is exactly CG.)

Proposition 2. *If ⊢ is compact and has /F, then it has /c.*

---

[7] For semilattices (and lattices), see [2].

Proof. Suppose ⊢ is compact and has /F, that $\Gamma \vdash A$ for all $A \in \Sigma$, and that $\Sigma, \Gamma \vdash \Delta$. Since ⊢ is compact, this gives $\Sigma_{\text{fin}}, \Gamma_{\text{fin}} \vdash \Delta_{\text{fin}}$ for some finite $\Sigma_{\text{fin}} \subseteq \Sigma$, $\Gamma_{\text{fin}} \subseteq \Gamma$, and $\Delta_{\text{fin}} \subseteq \Delta$. By monotonicity, $\Sigma_{\text{fin}}, \Gamma \vdash \Delta$. Since $\Sigma_{\text{fin}} \subseteq \Sigma$, we have $\Gamma \vdash A$ for all $A \in \Sigma_{\text{fin}}$. Now, where n is the cardinality of Σ_{fin}, let $\Sigma_{\text{fin}} = \{\sigma_0, \ldots, \sigma_{n-1}\}$, and for $m \leq n$, let $\Sigma_{\text{fin}}^m = \{\sigma_m, \ldots, \sigma_{n-1}\}$. Thus, $\Sigma_{\text{fin}}^0 = \Sigma_{\text{fin}}$, and $\Sigma_{\text{fin}}^n = \emptyset$.

I claim that for any i from 0 to n (inclusive), $\Sigma_{\text{fin}}^i, \Gamma \vdash \Delta$; when $i = n$, this is $\Gamma \vdash \Delta$, and the proposition follows. This can be shown by induction. The case where $i = 0$ is already shown. So suppose the claim is true for $i < n$; then $\Sigma_i, \Gamma \vdash \Delta$, which is to say $\sigma_i, \Sigma_{i+1}, \Gamma \vdash \Delta$. By assumption, $\Gamma \vdash \sigma_i$; monotonicity gives $\Sigma_{i+1}, \Gamma \vdash \sigma_i$. Now, applying /F, $\Sigma_{i+1}, \Gamma \vdash \Delta$.

Proposition 3. *If* ⊢ *is compact and has* F/, *then it has* C/.

Proof. From Proposition 2, by duality.

For compact relations, then, /F and /C are equivalent to each other, as are F/ and C/. This also means that F/F, F/C, C/F, and C/C are all equivalent to each other for such relations. In addition, Since FG implies every property under consideration for compact relations, all of FG, /C⁺, C⁺/, CG, /FG/C, C/FG/, C/FG/C, C⁺/C, C/C⁺, and C⁺/C⁺ are equivalent to each other for these relations. This leaves (at most) seven distinct properties: T, s, KS, /F (= /C), F/ (= C/), F/F (= C/C), and FG (= CG).

The situation so far is recorded in Figure 1. In this figure, each arrow is an implication already recorded; the double-thickness arrows are implications that we have seen become equivalences in the presence of compactness. (For now, you can ignore the letters that label the arrows.) When compactness is assumed, only FG and the six other nodes implied by it remain distinct; each of the other fourteen nodes is connected to one of these seven by a path containing only double-thickness arrows.

5   Nonimplications

So far, only implications have been recorded. So while we know there are *at most* twenty-one distinct properties in play here, and at most seven if compactness is assumed, it's still possible, for all I've said so far, that there are fewer. In fact, there are not; the implications so far recorded exhaust the implications among these properties. This section shows that the remaining potential implications do not hold. In each case, I will show this by counterexample.

5.1   Presenting Consequence Relations

I will present consequence relations using a very simple kind of 'proof system'. I work with *sequents*; a sequent for a language $\mathcal{F}$ is a pair $\langle \Gamma, \Delta \rangle$ of subsets of $\mathcal{F}$;

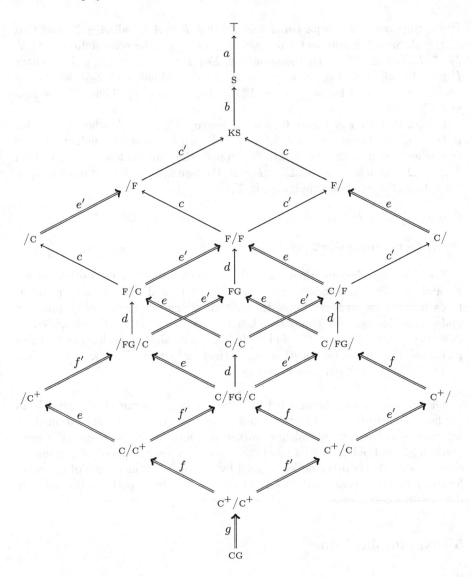

Fig. 1. Implications

I will write such a pair $[\Gamma \therefore \Delta]$. It is handy to consider the *subsequent* relation $\sqsubseteq$, defined: $[\Gamma' \therefore \Delta'] \sqsubseteq [\Gamma \therefore \Delta]$ iff $\Gamma' \subseteq \Gamma$ and $\Delta' \subseteq \Delta$.

A sequent-based proof system involves two components: some set of *initial sequents*, which are simply given as valid, and some *rules* that allow new validities to be generated from old. The proof systems I will draw on here are all quite simple. For each of them, I will specify a set $\mathfrak{P}$ of sequents; the initial sequents of the system are then all those sequents in $\mathfrak{P}$, together with all sequents of the form $[A \therefore A]$, for any $A \in \mathcal{F}$. There is only a single rule in any of these systems: the rule of *infinitary weakening*, which allows us to derive $[\Gamma, \Gamma' \therefore \Delta, \Delta']$ from $[\Gamma \therefore \Delta]$, for any $\Gamma, \Gamma', \Delta, \Delta' \subseteq \mathcal{F}$.

So for any set $\mathfrak{P}$ of sequents, we have a consequence relation $\vdash_{\mathfrak{P}}$ determined as follows: $\Gamma \vdash_{\mathfrak{P}} \Delta$ iff either 1) $\Gamma \cap \Delta \neq \emptyset$, or 2) there is some $[\Gamma' \therefore \Delta'] \in \mathfrak{P}$ such that $[\Gamma' \therefore \Delta'] \sqsubseteq [\Gamma \therefore \Delta]$. A binary relation on $\wp(\mathcal{F})$ is $\vdash_{\mathfrak{P}}$ for some $\mathfrak{P}$ iff it is monotonic and reflexive (in the senses of footnote 1, which are equivalent given monotonicity), so this approach works at the right level of generality for present purposes. It also gives a tractable way to explore compactness: note that $\vdash_{\mathfrak{P}}$ is compact iff every infinite sequent in $\mathfrak{P}$ has a finite subsequent in $\mathfrak{P}$.

5.2   A Menagerie of Consequence Relations

Table 3 presents seven distinct consequence relations, a–g. For each, it notes two of the twenty-one properties: one that the consequence relation has and one that it lacks. These two properties are chosen so that the implications already recorded suffice to settle the situation as regards the remaining nineteen: each other property is either implied by the property the relation has, or else implies the property the relation lacks. (I find this easiest to see by referring to Figure 1.) Let B, C, D, E, F be five distinct formulas, and let $\Theta \subseteq \mathcal{F}$ be infinite. For each of these, Table 4 gives a counterexample to the property that it is listed in Table 3 as lacking; these are easy to check.[8]

Table 3. Seven consequence relations

Name:	$\mathfrak{P}$	Has:	Lacks:
a	$\{[B \therefore C], [C \therefore D]\}$	$\top$	S
b	$\{[\Gamma \therefore \Delta] : \max(|\Gamma|, |\Delta|) > 2 \text{ and } B \in \Gamma \cup \Delta\}$	S	KS
c	$\{[E \therefore B, C, D], [B \therefore C, D]\}$	/C	F/
d	$\{[C \therefore D, B], [B, C \therefore D]\}$	C/C	FG
e	$\{[\Gamma \therefore \Delta] : \Delta \text{ is infinite or } \Gamma \cap \Theta \neq \emptyset\}$	/C$^+$	C/
f	$\{[\Gamma \therefore \Delta] : \Delta \text{ is infinite or } |\Gamma| \geq 2\}$	C/C$^+$	C$^+$/
g	$\{[\Gamma \therefore \Delta] : \Gamma \cup \Delta \text{ is infinite}\}$	C$^+$/C$^+$	CG

For space reasons, I do not prove here that every relation in Table 3 has the property it is there claimed to have; none of the needed proofs is particularly devious. Here are two examples to give the flavour.

Proposition 4. *Relation d has* C/C.

Proof. Suppose it lacks /C; then there are Γ, Δ, Σ such that $\Gamma \nvdash_d \Delta$ while $\Gamma \vdash_d A$ for every $A \in \Sigma$ and $\Sigma, \Gamma \vdash_d \Delta$. If $\Sigma \subseteq \Gamma$, then $\Gamma \vdash_d \Delta$, contrary

---

[8] [13, p. 31] gives the relation here called g, for the same purpose: to show that C$^+$/C$^+$ and CG are distinct. See their Theorem 2.7 (p. 32).

Table 4. Counterexamples

Name:	Lacks:	Validates:	And:	But:
a	S	$B \vdash C$	$C \vdash D$	$B \nvdash D$
b	KS	$C, D \vdash B$	$B \vdash E, F$	$C, D \nvdash E, F$
c	F/	$E \vdash B, C, D$	$B \vdash C, D$	$E \nvdash C, D$
d	FG	$C \vdash D, B$	$B, C \vdash D$	$C \nvdash D$
e	C/	$\vdash \Theta$	$A \vdash$ for all $A \in \Theta$	$\nvdash$
f	C$^+$/	$B \vdash \mathcal{F} \setminus \{B\}$	$B, A \vdash$ for all $A \in \mathcal{F} \setminus \{B\}$	$B \nvdash$
g	CG	$\mathcal{F}^+ \vdash \mathcal{F}^-$ for every partition $\langle \mathcal{F}^+, \mathcal{F}^- \rangle$ of $\mathcal{F}$		$\nvdash$

to supposition. So there must be some $A \in \Sigma$ with $A \notin \Gamma$. Since $\Gamma \vdash_d A$ and $A \notin \Gamma$, it must be that $B, C \in \Gamma$ and $A = D$. Since $\Gamma \nvdash_d \Delta$ while $B, C \in \Gamma$, it must be that $D \notin \Delta$. Now, suppose $E \in \Sigma \cap \Delta$; since $\Gamma \vdash_d E$ and E is not D, we must have $E \in \Gamma$. Then $\Gamma \vdash_d \Delta$, contrary to supposition. So $\Sigma \cap \Delta$ is empty. But then $(\Sigma \cup \Gamma) \cap \Delta$ is empty, and since $D \notin \Delta$, it follows that $\Sigma, \Gamma \nvdash_d \Delta$. Contradiction.

For C/, the argument is dual, reversing the roles of C and D.

Proposition 5. *Relation f has* C/C$^+$.

Proof. First, that it has C/. If $A \vdash_f \Delta$ for each $A \in \Sigma$, then either Δ is infinite, in which case $\Gamma \vdash_f \Delta$ directly, or else $\Sigma \subseteq \Delta$; the only valid arguments with finitely many conclusions and a single premise are those where the premise is among the conclusions. But if $\Sigma \subseteq \Delta$, then if $\Gamma \vdash_f \Delta, \Sigma$, this is already $\Gamma \vdash_f \Delta$.

Second, that it has /C$^+$. Suppose $\Sigma, \Gamma \vdash_f \Delta$ and $\Gamma \vdash_f \Delta, A$ for each $A \in \Sigma$, to show $\Gamma \vdash_f \Delta$. If Δ is infinite, we're done; if $|\Gamma| \geq 2$ we're done; if $\Gamma \cap \Delta \neq \emptyset$ we're done. So suppose Δ is finite, $|\Gamma| < 2$, and $\Gamma \cap \Delta = \emptyset$. Since $\Sigma, \Gamma \vdash_f \Delta$, either $\Sigma \cap \Delta \neq \emptyset$ or $|\Sigma \cup \Gamma| \geq 2$. In the first case, take some $A \in \Sigma \cap \Delta$; since $\Gamma \vdash_f \Delta, A$ and $\Delta \cup \{A\} = \Delta$, we're done. In the second case, there must be some $A \in \Sigma$ but $A \notin \Gamma$; we then have $\Gamma \vdash_f \Delta, A$. But $|\Gamma| < 2$ and $\Gamma \cap (\Delta \cup \{A\}) = \emptyset$, so this is impossible.

5.3 No Further Implications

To see that there are no further implications, return to Figure 1, now attending to the letters that label the arrows. These letters correspond to consequence relations from §5.2; letters with ' pick out converse relations. The indicated consequence relation, in each case, is a counterexample to the claim that the implication in question is an equivalence. Moreover, where the implication is a single-line arrow—that is, where it is not already known to become an equivalence in the presence of compactness—the indicated counterexample is compact; this shows that no additional implications collapse to equivalences in the presence of compactness.[9]

---

[9] a–d are compact. For a, c, and d, $\mathfrak{P}$ contains only finite sequents. For b, $\mathfrak{P}$ contains a finite subsequent of each infinite sequent it contains. (e–g are not compact, and could not be, given the combinations of properties they exhibit.)

Because the properties under consideration are closed under conjunction, this suffices to rule out any additional implications. Any additional implication would bring with it an additional equivalence (if P implies Q, then P is equivalent to the conjunction of P and Q; this conjunction is already known to imply P); so the fact that there are no additional equivalences suffices to show that there are no additional implications.

As a result, the implications between these properties are now completely characterized. By taking arbitrary conjunctions of the ten properties in Table 1, there are exactly twenty-one distinct properties we can reach, twenty of which (all but ⊤) are linking properties, properties of the sort that can plausibly travel under the name 'transitivity'. For compact relations, these twenty-one collapse to seven, six of which (again, all but ⊤) are such linking properties.

6 Conclusion

'Transitivity', as applied to consequence relations, can conceal more than it reveals. When someone says a consequence relation is 'transitive', then, it is worth asking just what is meant. It is almost never the case that they mean that it is *transitive*, in the usual relation-theoretic sense. But then what can they mean?

This paper has explored some possible answers. It's a safe bet that nobody means ⊤ by 'transitivity', but the remaining twenty properties (in the general case) or six properties (in the presence of compactness) are all possible ways to fill in the idea. When *we* call consequence relations 'transitive', then, it behooves us to make clear exactly what we are saying; there is no single thing we must obviously mean.

Acknowledgements. For helpful discussion and comments, thanks to the Melbourne Logic Group (especially Rohan French and Lloyd Humberstone), audiences at the University of Groningen (GroLog) and the Australasian Association for Logic 2015 meeting, and three anonymous referees for LORI. This research was partially supported by the grant "Non-Transitive Logics", number FFI2013-46451-P, from the Ministerio de Economía y Competitividad, Government of Spain.

References

[1] Cutland, N.J., Gibbins, P.F.: A regular sequent calculus for quantum logic in which ∧ and ∨ are dual. Logique et Analyse 25(99), 221–248 (1982)
[2] Davey, B.A., Priestley, H.A.: Introduction to Lattices and Order. Cambridge University Press, Cambridge (2002)
[3] Dummett, M.: The Logical Basis of Metaphysics. Duckworth, London (1991)
[4] Gentzen, G.: Investigations into logical deduction. In: Szabo, M.E. (ed.) The Collected Papers of Gerhard Gentzen, pp. 68–131. North-Holland Publishing Company, Amsterdam (1969)

[5] Humberstone, L.: Heterogeneous logic. Erkenntnis 29, 395–435 (1988)

[6] Humberstone, L.: The Connectives. MIT Press, Cambridge (2012)

[7] Restall, G.: Multiple conclusions. In: Hajek, P., Valdes-Villanueva, L., Wester-ståhl, D. (eds.) Logic, Methodology, and Philosophy of Science: Proceedings of the Twelfth International Congress, pp. 189–205. Kings' College Publications, London (2005)

[8] Restall, G.: Truth values and proof theory. Studia Logica 92(2), 241–264 (2009)

[9] Ripley, D.: Paradoxes and failures of cut. Australasian Journal of Philosophy 91(1), 139–164 (2013)

[10] Ripley, D.: Anything goes. Topoi 34(1), 25–36 (2015)

[11] Rumfitt, I.: Knowledge by deduction. Grazer Philosophische Studien 77(1), 61–84 (2008)

[12] Segerberg, K.: Classical Propositional Operators: An Exercise in the Foundations of Logic. Clarendon Press, Oxford (1982)

[13] Shoesmith, D.J., Smiley, T.J.: Multiple-conclusion Logic. Cambridge University Press, Cambridge (1978)

[14] Steinberger, F.: Why conclusions should remain single. Journal of Philosophical Logic 40(3), 333–355 (2011)

[15] Weir, A.: Naive truth and sophisticated logic. In: Beall, J.C., Armour-Garb, B. (eds.) Deflationism and Paradox, pp. 218–249. Oxford University Press, Oxford (2005)

[16] Weir, A.: A robust non-transitive logic. Topoi 34(1), 99–107 (2015)

Boolean Game with Prioritized Norms

Xin Sun

Faculty of Science, Technology and Communication, University of Luxembourg
xin.sun@uni.lu

Abstract. In this paper we study boolean game with prioritized norms. Norms distinguish illegal strategies from legal strategies. Notions like legal strategy and legal Nash equilibrium are introduced. Our formal model is a combination of (weighted) boolean game and so called (prioritized) input/output logic. After formally presenting the model, we use examples to show that non-optimal Nash equilibrium can be avoided by making use of norms. We study various complexity issues related to legal strategy and legal Nash equilibrium.

Keywords: Boolean game, norm, input/output logic.

1   Introduction

The study of the interplay of games and norms can be divided into two main branches: the first, mostly originating from economics and game theory [11,19,20], treats norms as mechanisms that enforce desirable properties of social interactions; the second, that has its roots in social sciences and evolutionary game theory [29,12] views norms as (Nash or correlated) equilibrium that results from the interaction of rational agents. A survey of the interaction between games and norms can be found in Grossi *et al* [15]. This paper belongs to the first branch.

In this paper we study the combination of boolean games and norms. Boolean game is a class of games based on propositional logic. It was firstly introduced by Harrenstein et al. [17] and further developed by several researchers [16,23,13,9,7,26]. In a boolean game, each agent i is assumed to have a goal, represented by a propositional formula ϕ_i over some set of propositional variables $\mathbb{P}$. Each agent i is associated with some subset $\mathbb{P}_i$ of the variables, which are under the unique control of agent i. The choices, or strategies, available to i correspond to all the possible assignment of truth or falsity to the variables in $\mathbb{P}_i$. An agent will try to choose an assignment so as to satisfy his goal ϕ_i. Strategic concerns arise because whether i's goal is in fact satisfied will depend on the choices made by other agents.

Norms are social rules regulating agents' behavior by prescribing which actions are obligatory, forbidden or permitted. In the game theoretical setting, norms distinguish illegal strategies form legal strategies. By designing norms appropriately, non-optimal equilibrium might be avoided. To represent norms in boolean games, we need a logic of norms, which has been extensively studied in the deontic logic community.

© Springer-Verlag Berlin Heidelberg 2015
W. van der Hoek et al. (Eds.): LORI 2015, LNCS 9394, pp. 341–352, 2015.
DOI: 10.1007/978-3-662-48561-3_28

Various deontic logic has been developed since von Wright's first paper [30] in this area. In the first volume of the handbook of deontic logic [14], input/output logic [21,22] appears as one of the new achievement in deontic logic in recent years. Input/output logic takes its origin in the study of conditional norms. The basic idea is: norms are conceived as a deductive machine, like a black box which produces normative statements as output, when we feed it factual statements as input.

In this paper we use a simplification of Parent's prioritized input/output logic [25] as the logic of norms. Given a normative multi-agent system, which contains a boolean game, a set of prioritized norms and certain environment. Every strategy of every agent is classified as legal or illegal. Notions like legal Nash equilibrium are then naturally defined.

The structure of this paper is the following: We present some background knowledge, including boolean game, input/output logic and complexity theory in Section 2. Normative multi-agent system are introduced and its complexity issues are studied in Section 3. We conclude this paper in Section 4.

2   Background

2.1   Propositional Logic

Let $\mathbb{P} = \{p_0, p_1, \ldots\}$ be a finite set of propositional variables and let $L_{\mathbb{P}}$ be the propositional language built from $\mathbb{P}$ and boolean constants $\top$ (true) and $\bot$ (false) with the usual connectives $\neg, \vee, \wedge, \rightarrow$ and $\leftrightarrow$. Formulas of $L_{\mathbb{P}}$ are denoted by ϕ, ψ etc. A literal is a variable $p \in \mathbb{P}$ or its negation. $2^{\mathbb{P}}$ is the set of the valuations for $\mathbb{P}$, with the usual convention that for $V \in 2^{\mathbb{P}}$ and $p \in V$, V gives the value true to p if $p \in V$ and false otherwise. $\vDash$ denotes the classical logical consequence relation.

Let $X \subseteq \mathbb{P}$, 2^X is the set of X-valuations. A partial valuation (for $\mathbb{P}$) is an X-valuation for some $X \subseteq \mathbb{P}$. Partial valuations are denoted by listing all variables of X, with a "$+$" symbol when the variable is set to be true and a "$-$" symbol when the variable is set to be false: for instance, let $X = \{p, q, r\}$, then the X-valuation $V = \{p, r\}$ is denoted $\{+p, -q, +r\}$. If $\{\mathbb{P}_1, \ldots, \mathbb{P}_n\}$ is a partition of $\mathbb{P}$ and $V_1, \ldots, V_n$ are partial valuations, where $V_i \in 2^{\mathbb{P}_i}$, $(V_1, \ldots, V_n)$ denotes the valuation $V_1 \cup \ldots \cup V_n$.

2.2   Boolean Game

Boolean games introduced by Harrenstein *et al* [17] are zero-sum games with two players, where the strategies available to each player consist in assigning a truth value to each variable in a given subset of $\mathbb{P}$. Bonzon *et al* [8] give a more general definition of a boolean game with any number of players and not necessarily zero-sum. In this paper we further generalizes boolean games such that the utility of each agent is not necessarily in $\{0, 1\}$. Such generalization is reached by representing the goals of each agent as a set of weighted formulas.

We call such boolean game weighted boolean game. The idea of using weighted formulas to define utility can be found in many work among which we mention satisfiability game [5] and weighted boolean formula game [23].

Definition 1 (boolean game). *A weighted boolean game is a 4-tuple* $(Agent, \mathbb{P}, \pi, Goal)$, *where*

1. *Agent* = $\{1, \ldots, n\}$ *is a set of agents.*
2. $\mathbb{P}$ *is a finite set of propositional variables.*
3. $\pi : Agent \mapsto 2^{\mathbb{P}}$ *is a control assignment function such that* $\{\pi(1), \ldots, \pi(n)\}$ *forms a partition of* $\mathbb{P}$. *For each agent* i, $2^{\pi(i)}$ *is the strategy space of* i.
4. *Goal* = $\{Goal_1, \ldots, Goal_n\}$ *is a set of weighted formulas of* $L_{\mathbb{P}}$. *That is, each* $Goal_i$ *is a finite set* $\{\langle \phi_1, m_1 \rangle, \ldots, \langle \phi_k, m_k \rangle\}$ *where* $\phi_j \in L_{\mathbb{P}}$ *and* m_j *is a real number.*

A strategy for agent i is a partial valuation for all the variables i controls. Note that since $\{\pi(1), \ldots, \pi(n)\}$ forms a partition of $\mathbb{P}$, a strategy profile S is a valuation for $\mathbb{P}$. In the rest of the paper we make use of the following notation, which is standard in game theory. Let $G = (Agent, \mathbb{P}, \pi, Goal)$ be a weighted boolean game with $Agent = \{1, \ldots, n\}$, $S = (s_1, \ldots, s_n)$ be a strategy profile. s_{-i} denotes the projection of S on $Agent - \{i\}$: $s_{-i} = (s_1, \ldots, s_{i-1}, s_{i+1}, \ldots, s_n)$.

Agents' utilities in weighted boolean games are induced by their goals. For every agent i and every strategy profiles S, $u_i(S) = \Sigma\{m_j : \langle \phi_j, m_j \rangle \in Goal_i, S \models \phi_j\}$. Dominating strategies and pure-strategy Nash equilibria are defined as usual in game theory [24].

Example 1. *Let* $G = (Agent, \mathbb{P}, \pi, Goal)$ *where* $Agent = \{1, 2\}$, $\mathbb{P} = \{p, q, s\}$, $\pi(1) = \{p\}$, $\pi(2) = \{q, s\}$, $Goal_1 = \{\langle p \leftrightarrow q, 1 \rangle, \langle s, 2 \rangle\}$, $Goal_2 = \{\langle p \wedge q, 2 \rangle, \langle \neg s, 1 \rangle, \}$. *This boolean game is depicted as follows:*

	$+q, +s$	$+q, -s$	$-q, +s$	$-q, -s$
$+p$	$(3, 2)$	$(1, 3)$	$(2, 0)$	$(0, 1)$
$-p$	$(2, 0)$	$(0, 1)$	$(3, 0)$	$(1, 1)$

2.3   Input/Output Logic

In input/output logic, a norm is an ordered pair of formulas $(\phi, \psi) \in L_{\mathbb{P}} \times L_{\mathbb{P}}$, which is read as "given ϕ, it is obligatory to be ψ". A set of norm N can be viewed as a function from $2^{L_{\mathbb{P}}}$ to $2^{L_{\mathbb{P}}}$ such that for a set Φ of formulas, $N(\Phi) = \{\psi \in L_{\mathbb{P}} : (\phi, \psi) \in N \text{ for some } \phi \in \Phi\}$. A finite set of norms is called a (plain) normative system.

Definition 2 (Semantics of input/output logic [21]). *Given a normative system* N *and a finite set of formulas* Φ, $out(N, \Phi) = Cn(N(Cn(\Phi)))$, *where* Cn *is the consequence relation of propositional logic.*[1]

---

[1] In Makinson and van der Torre [21], this logic is called simple-minded input/output logic. Different input/output logics are developed in Makinson and van der Torre [21] as well. A technical introduction of input/output logic can be found in Sun [28].

Intuitively, the procedure of the semantics is as follows: We first have in hand a set of formulas Φ (call it the input) as a description of the current state. We then close it by logical consequence $Cn(\Phi)$. The set of norms, like a deductive machine, accepts this logically closed set and produces a set of formulas $N(Cn(\Phi))$. We finally get the output $Cn(N(Cn(\Phi)))$ by applying the logical closure again. $\psi \in out(N, \Phi)$ is understood as "ψ is obligatory given facts Φ and norms N".

Example 2. *Let p, q, r are propositional variables. Let $N = \{(p, q), (p \vee q, r), (r, p)\}$. Then $out(N, \{p\}) = Cn(N(Cn(\{p\}))) = Cn(\{q, r\})$.*

Input/output logic is given a proof theoretic characterization. We say that an ordered pair of formulas is derivable from a set N iff (a, x) is in the least set that extends N and is closed under a number of derivation rules. The following are the rules we need:

- SI (strengthening the input): from (ϕ, ψ) to (χ, ψ) whenever $\chi \vDash \phi$.
- WO (weakening the output): from (ϕ, ψ) to (ϕ, χ) whenever $\psi \vDash \chi$.
- AND (conjunction of output): from (ϕ, ψ) and (ϕ, χ) to $(\phi, \psi \wedge \chi)$.

The derivation system based on the rules SI, WO and AND is denoted as $deriv(N)$.

Example 3. *Let $N = \{(p \vee q, r), (q, r \to s)\}$, then $(q, s) \in deriv(N)$ because we have the following derivation*

1.	$(p \vee q, r)$	*Assumption*
2.	(q, r)	*1, SI*
3.	$(q, r \to s)$	*Assumption*
4.	$(q, r \wedge (r \to s))$	*2,3, AND*
5.	(q, s)	*4, WO*

In Makinson and van der Torre [21], the following soundness and completeness theorem is proved:

Theorem 1 ([21]). *Given a set of norms N,*

$$\psi \in out(N, \{\phi\}) \text{ iff } (\phi, \psi) \in deriv(N).$$

Prioritized Input/Output Logic. A prioritized normative system $N^{\geq} = (N, \geq)$ is a finite set of norms together with a priority relation over norms. We assume $\geq$ to be reflexive and transitive and understand $(\phi, \psi) \geq (\phi', \psi')$ as (ϕ, ψ) has higher priority than (ϕ', ψ'). The priority relation is further lifted to priority over sets of norms. Following Parent [25], we define the lifting as follows: $N_1 \succeq N_2$ iff for all $(\phi_2, \psi_2) \in N_2 - N_1$ there is $(\phi_1, \psi_1) \in N_1 - N_2$ such that $(\phi_1, \psi_1) \geq (\phi_2, \psi_2)$.

Definition 3 (output with priorities[2]). *Let $N^{\geq}$ be a prioritized normative system and Φ be a set of formulas.*

---

[2] Here our prioritized input/output logic is a simplification of the original version of Parent [25].

$\psi \in out_p(N^{\geq}, \Phi)$ iff $\psi \in \bigcap\{out(N', \Phi) : N' \in preffamily(N^{\geq}, \Phi)\}$.

Here $preffamily(N^{\geq}, \Phi)$ is defined via the following steps:

1. $maxfamily(N^{\geq}, \Phi)$ is the set of $\subseteq$-maximal subsets N' of N such that $out(N', \Phi)$ is consistent. That is, $out(N', \Phi)$ is consistent and for all N'' such that $N' \subset N''$, $out(N'', \Phi)$ is not consistent
2. $filterfamily(N^{\geq}, \Phi)$ is the set of norms $N' \in maxfamily(N^{\geq}, \Phi)$ that maximize the output, i.e., that are such that $out(N', \Phi) \subset out(N'', \Phi)$ for no $N'' \in maxfamily(N^{\geq}, \Phi)$.
3. $preffamily(N^{\geq}, \Phi)$ is the set of $\succeq$-maximal elements of $filterfamily(N^{\geq}, \Phi)$.

Permission in Input/Output Logic. Philosophically, it is common to distinguish between two kinds of permission: negative permission and positive permission. Negative permission is straightforward to describe: something is negatively permitted according to certain norms iff it is not prohibited by those norms. That is, iff there is no obligation to the contrary. Positive permission is more elusive. For the sake of simplicity, in this paper when only discuss negative permission and leave other types of permission as future work.

Definition 4 (permission). *Given a prioritized normative system $N^{\geq}$ and a finite set of formulas Φ, $Perm(N^{\geq}, \Phi) = \{\psi \in L_{\mathbb{P}} : \neg\psi \notin out_p(N^{\geq}, \Phi)\}$.*

Intuitively, ϕ is permitted iff ϕ is not forbidden. Since a formula is forbidden iff its negation is obligatory, ϕ is not forbidden is equivalent to $\neg\phi$ is not obligatory.

2.4 Complexity Theory

Complexity theory is the theory to investigate the time, memory, or other resources required for solving computational problems. In this subsection we briefly review those concepts and results from complexity theory which will be used in this paper. More comprehensive introduction of complexity theory can be found in [4]

We assume the readers are familiar with notions like Turing machine and the complexity class P, NP and coNP. Oracle Turing machine and two complexity classes related to oracle Turing machine will be used in this paper.

Definition 5 (oracle Turing machine [4]). *An oracle for a language L is a device that is capable of reporting whether any string w is a member of L. An oracle Truing machine M^L is a modified Turing machine that has the additional capability of querying an oracle. Whenever M^L writes a string on a special oracle tape it is informed whether that string is a member of L, in a single computation step.*

P^{NP} is the class of problems solvable by a deterministic polynomial time Turing machine with an NP oracle. $\mathrm{P}^{NP[O(\log n)]}$ only allows $O(\log n)$ oracle queries instead of polynomially-many. $P_{\parallel}^{NP}$ is the class of problems which can be solved by using the NP oracle only in parallel. Buss and Hay [10] show that $P_{\parallel}^{NP}$ coincide with $P_{\parallel O(1)}^{NP}$, where a fixed number of parallel rounds is allowed.

NP^{NP} is the class of problems solvable by a non-deterministic polynomial time Turing machine with an NP oracle. Another name for the class NP^{NP} is Σ_2^p. Σ_{i+1}^p is the class of problems solvable by a non-deterministic polynomial time Turing machine with a Σ_i^p oracle. Π_i^p is the class of problems of which the complement is in Σ_i^p.

3   From Boolean Game to Normative Multi-agent System

In recent years, normative multi-agent system [6,3] arises as a new interdisciplinary academic area bringing together researchers from multi-agent system [27,32,31], deontic logic [14] and normative system [1,18,2]. By combining boolean games and norms, we here develop a new approach to normative multi-agent system.

Definition 6 (normative multi-agent system). *A normative multi-agent system is a triple* $(G, N^{\geq}, E)$ *where*

- $G = (Agent, \mathbb{P}, \pi, Goal)$ *is a weighted boolean game.*
- $N^{\geq}$ *is a prioritized normative system.*
- $E \subseteq L_{\mathbb{P}}$ *is a finite set of formulas representing the environment.*

3.1   Legal Strategy

In a normative multi-agent system, agent's strategies are classified as either legal or illegal. The basic idea is viewing strategies as formulas and using the mechanism of input/output logic to decide whether a formula is permitted.

Definition 7 (legal   strategy). *Given a normative multi-agent system* $(G, N^{\geq}, E)$, *for each agent* i, *a strategy* $(+p_1, \ldots, +p_m, -q_1, \ldots, -q_n)$ *is legal if*

$$p_1 \wedge \ldots \wedge p_m \wedge \neg q_1 \wedge \ldots \wedge \neg q_n \in Perm(N^{\geq}, E).$$

Example 4. *Consider the prisoner's dilemma augmented with norms. Let* $(G, N^{\geq}, E)$ *be a normative multi-agent system as following:*

- $G = (Agent, \mathbb{P}, \pi, Goal)$ *is a weighted boolean game with*
  - $Agent = \{1, 2\}$,
  - $\mathbb{P} = \{p, q\}$,
  - $\pi(1) = \{p\}$, $\pi(2) = \{q\}$,
  - $Goal_1 = \{\langle p, 2 \rangle, \langle \neg q, 3 \rangle\}$, $Goal_2 = \{\langle q, 2 \rangle, \langle \neg p, 3 \rangle\}$.
- $N^{\geq} = (N, \geq)$ *where* $N = \{(\top, \neg p), (\top, \neg q), (\top, q)\}$, $(\top, \neg q) \geq (\top, q)$.
- $E = \emptyset$.

	$+q$	$-p$
$+p$	$(2,2)$	$(5,0)$
$-p$	$(0,5)$	$(3,3)$

Then $out(N,E)$ = $Cn(\{\neg p, \neg q, q\})$, $maxfamily(N^{\geq}, E)$ =
$\{\{(\top, \neg p), (\top, \neg q)\}$, $\{(\top, \neg p), (\top, q)\}\}$, $filterfamily(N^{\geq}, E)$ =
$maxfamily(N^{\geq}, E)$, $preffamily(N^{\geq},\ E)$ = $\{\{(\top, \neg p), (\top, \neg q)\}\}$.   There-
fore $out_p(N^{\geq}, E) = out(\{(\top, \neg p),\ (\top, \neg q)\}, E) = Cn(\{\neg p, \neg q\})$.
Therefore $\{-p\}$ and $\{-q\}$ are legal while $\{+p\}$ and $\{+q\}$ are not. $\dashv$

Having defined the notion legal strategy, a natural question to ask is how complex is it to decide whether a strategy is legal. Theorem 2 gives a first answer to this question. To prove Theorem 2, we need the following lemmas.

Lemma 1. *Given a normative system N, a finite set of formulas Φ and a formula ϕ, deciding whether $\phi \in out(N, \Phi)$ is coNP hard and in $P_{\parallel}^{NP}$.*

Proof. Concerning the coNP hardness, we prove by reducing the validity problem of propositional logic to our problem. Let ϕ be an arbitrary formula. Let $N = \emptyset$ and $\Phi = \emptyset$, then ϕ is a tautology iff $\phi \in Cn(\top)$ iff $\phi \in Cn(N(Cn(\Phi)))$ iff $\phi \in out(N, \Phi)$

Concerning the $P_{\parallel}^{NP}$ membership, we prove by giving an oracle Turing machine with oracle SAT, the set of all satisfiable propositional formulas, to solve this problem.

Let $N = \{(\phi_1, \psi_1), \ldots, (\phi_n, \psi_n)\}$.

1. for each $\phi_i \in \{\phi_1, \ldots, \phi_n\}$, use the oracle to test if $\Phi \models \phi_i$.
   (a) If yes, then mark ψ_i,
   (b) Otherwise do nothing.
2. Let $\psi_{i_1}, \ldots \psi_{i_k}$ be all those ψ_i which are marked in step 1.
3. Use the oracle to test if $\{\psi_{i_1}, \ldots, \psi_{i_k}\} \models \phi$.
   (a) If yes, then return "accept"
   (b) Otherwise return "reject".

It can be verified that $\phi \in Cn(N(Cn(\Phi)))$ iff the Turing machine returns "accept" and the time complexity of the oracle Turing machine runs in polynomial time and calls the oracle in parallel for 2 rounds. Therefore the problem is in $P_{\parallel O(1)}^{NP}$, which coincides with $P_{\parallel}^{NP}$. $\dashv$

Lemma 2. *Given a prioritized normative system $N^{\geq}$, a finite set of norms $N' \subseteq N$, a finite set of formulas Φ, deciding whether $N' \in maxfamily(N^{\geq}, \Phi)$ is coNP hard and in P^{NP}.*

Proof. The coNP hardness is easy to prove. Here we focuses on the P^{NP} membership. We prove by giving an oracle Turing machine with oracle SAT to solve this problem.

Let $N - N' = \{(\phi_1, \psi_1), \ldots, (\phi_n, \psi_n)\}$.

1. Test if $\perp \in out(N', \Phi)$.
2. If yes, return "reject". Otherwise continue.
3. For all $i \in \{1, \ldots, n\}$, test if $\perp \in out(N' \cup \{(\phi_i, \psi_i)\}, \Phi)$.
4. Return "accept" if $\perp \in out(N' \cup \{(\phi_i, \psi_i)\}, \Phi)$ for all $i \in \{1, \ldots, n\}$. Otherwise return "reject".

It can be verified that $N' \in maxfamily(N^{\geq}, \Phi)$ iff the Turing machine returns "accept" and the time complexity of the oracle Turing machine is polynomial. ⊣

Lemma 3. *Given a prioritized normative system $N^{\geq}$, a finite set of norms $N' \subseteq N$, a finite set of formulas Φ, deciding whether $N' \in filterfamily(N^{\geq}, \Phi)$ is coNP hard and in $coNP^{NP} = \Pi_2^p$.*

Proof. The coNP hardness is easy to prove. Here we focuses on the $coNP^{NP}$ membership.

We prove by giving a non-deterministic oracle Turing machine with oracle SAT to solve the complement of this problem.

1. Test if $N' \in maxfamily(N^{\geq}, \Phi)$. If no, return "accept". Otherwise continue.
2. Guess a set of norms $N'' \subseteq N$.
3. Test if $N'' \in maxfamily(N^{\geq}, \Phi)$. If no, return "reject" on this branch. Otherwise continue.
4. Test if $\models \bigwedge N''(Cn(\Phi)) \rightarrow \bigwedge N'(Cn(\Phi))$ meanwhile $\not\models \bigwedge N'(Cn(\Phi)) \rightarrow \bigwedge N''(Cn(\Phi))$. If yes, return "accept" on this branch. Otherwise return "reject" on this branch.

It can be verified that $N' \notin filterfamily(N^{\geq}, \Phi)$ iff the non-deterministic Turing machine returns "accept" on some branches. The time complexity of the non-deterministic Turing machine is polynomial because with the help of an NP oracle SAT, the test in step 1,3 and 4 can be done in polynomial time.   ⊣

Lemma 4. *Given a prioritized normative system $N^{\geq}$, a finite set of norms $N' \subseteq N$, a finite set of formulas Φ, deciding whether $N' \in preffamily(N^{\geq}, \Phi)$ is coNP hard and in Π_3^p.*

Proof. The hardness is easy to prove. Here we focus on the membership. We prove by giving a non-deterministic oracle Turing machine with a Σ_2^p oracle to solve the complement of this problem.

1. Test if $N' \in filterfamily(N^{\geq}, \Phi)$. If no, return "accept" on this branch. Otherwise continue.
2. Guess a set of norms $N'' \subseteq N$.
3. Test if $N'' \in filterfamily(N^{\geq}, \Phi)$. If no, return "reject" on this branch. Otherwise continue.
4. Test if $N'' \succeq N'$. If yes, return "accept" on this branch. Otherwise return "reject" on this branch.

It can be verified that $N' \notin preffamily(N^{\geq}, \Phi)$ iff the non-deterministic Turing machine returns "accept" on some branch. The time complexity of the non-deterministic Turing machine is polynomial because with the help of an Σ_2^p oracle, the test in step 1 and 3 can be done in polynomial time.

$\dashv$

Theorem 2. *Given a normative multi-agent system* $(G, N^{\geq}, E)$ *and a strategy* $(+p_1, \ldots, +p_m, -q_1, \ldots, -q_n)$, *deciding whether this strategy is legal is NP hard and in* Π_4^p.

Proof. To show that this problem is NP hard, we provide a reduction from the satisfiability problem of propositional logic to the problem of deciding whether a strategy is legal.

Let ϕ be a formula. Let $s = \{+p\}$ be a strategy, $N = \{(\neg\phi, \neg p)\}$, $E = \emptyset$. We will show ϕ is satisfiable iff s is legal.

Recall that $p \in Perm(N, E)$ iff $\neg p \notin out_p(N, E)$. In this case we have $out_p(N, E) = out(N, E) = Cn(N(Cn(E)))$.

From $E = \emptyset$ we know that $Cn(N(Cn(E))) = Cn(N(Cn(\top)))$. Therefore if ϕ is satisfiable, then $\neg\phi$ is not a tautology. Therefore $N(Cn(E)) = \emptyset$. Hence $\neg p \notin Cn(N(Cn(E))) = out(O, E)$, $p \in Perm(N, E)$. If ϕ is not satisfiable, then $\neg\phi$ is a tautology. Hence $\neg p \in N(Cn(E)) \subseteq Cn(N(Cn(E))) = out(N, E)$, which means $p \notin Perm(N, E)$.

For the Π_4^p membership, we prove by giving a non-deterministic oracle Turing machine with an Σ_3^p oracle to solve the complement of this problem.

1. Guess a set of norms $N' \subseteq N$.
2. Test if $N' \in preffamily(N^{\geq}, \Phi)$. If no, return "reject" on this branch. Otherwise continue.
3. Test if $\neg(p_1 \wedge \ldots \wedge p_m \wedge \neg q_1 \wedge \ldots \wedge \neg q_n) \in out(N', E)$. If yes, return "accept" on this branch.

It can be verified that $p_1 \wedge \ldots \wedge p_m \wedge \neg q_1 \wedge \ldots \wedge \neg q_n \notin out_p(N, E)$ iff the non-deterministic Turing machine returns "accept" on some branch and the time complexity of the Turing machine is polynomial. $\dashv$

3.2   Legal Nash Equilibrium

A (pure-strategy) legal Nash equilibrium is a strategy profile which contains only legal strategies and no agent can improve his utility by choosing another legal strategy, given others do not change their strategies.

Definition 8 (Legal Nash equilibrium). *Given a normative multi-agent system* (G, N, E), *A strategy profile* $S = (s_1, \ldots, s_n)$ *is a legal Nash equilibrium if*

- *for every agent* i, s_i *is a legal strategy*
- *for every agent* i, *for every legal strategy* $s_i' \in S_i$, $u_i(S) \geq u_i(s_i', s_{-i})$.

Example 5. *In the normative multi-agent system presented in Example 4,* $(-p, -q)$ *is the unique legal Nash equilibria.*

Example 6. *Let* $(G, N^{\geq}, E)$ *be a normative system as following:*

- $G = (Agent, \mathbb{P}, \pi, Goal)$ *is a weighted boolean game with*
  - $Agent = \{1, 2\}$,
  - $\mathbb{P} = \{p, q\}$,
  - $\pi(1) = \{p\}$, $\pi(2) = \{q\}$,
  - $Goal_1 = Goal_2 = \{\langle p \wedge q, 2 \rangle, \langle \neg p \wedge \neg q, 3 \rangle\}$.
- $N^{\geq} = (N, \geq)$, $N = \{(\top, \neg p), (\top, \neg q)\}$, $(\top, \neg p) \geq (\top, \neg q)$, $(\top, \neg q) \geq (\top, \neg p)$.
- $E = \emptyset$.

	$+q$	$-p$
$+p$	(2, 2)	(0, 0)
$-p$	(0, 0)	(3, 3)

Without normative system there are two Nash equilibrium: $(+p, +q)$ *and* $(-p, -q)$. *There is only one legal Nash equilibria:* $(-p, -q)$. *From the perspective of social welfare,* $(+p, +q)$ *is not an optimal equilibria because its social welfare is* $2 + 2 = 4$, *while the social welfare of* $(-p, -q)$ *is* $3 + 3 = 6$. *Therefore this example shows that by designing norms appropriately, non-optimal equilibrium might be avoided*

Theorem 3. *Given a normative multi-agent system* $(G, N^{\geq}, E)$ *and a strategy profile* $S = (s_1, \ldots, s_n)$. *Deciding whether* S *is a legal Nash equilibrium is NP hard and in* Π_5^p.

Proof. The NP hardness is trivial. For the Π_5^p membership, we prove by giving a non-deterministic oracle Turing machine with a Σ_4^p oracle to solve the complement of this problem.

1. Test if S is legal. If no, return "accept". Otherwise continue.
2. Guess a strategy profile S'
3. Test if S' is legal. If no, return "reject" on this branch. Otherwise continue.
4. For each agent i, test if $u_i(S) < u_i(S')$. Return "accept" on this branch if for some i, $u_i(S) < u_i(S')$. Otherwise return "reject" on this branch.

It can be verified that S is not a legal Nash equilibrium iff the non-deterministic Turing machine returns "accept" on some branch and the time complexity of the Turing machine is polynomial. ⊣

Theorem 4. *Given a normative multi-agent system* (G, N, E). *Deciding whether there is a legal Nash equilibrium of* G *is* Σ_2^P *hard and in* Σ_6^P.

Proof. The lower bound follows from the fact that deciding whether there is a Nash equilibria for boolean games without norms is Σ_2^P complete [8]. Concerning the upper bound, recall that $\Sigma_6^P = NP^{\Sigma_5^P}$. The problem can be solved by a polynomial time non-deterministic Turing machine with an Σ_5^P oracle. ⊣

4   Conclusion

In the present paper we introduce weighted boolean game with prioritized norms. Norms distinguish illegal strategies from legal strategies. Using ideas from (prioritized) input/output logic, legal strategies and legal Nash equilibrium are discussed. After formally presenting the model, we use examples to show that non-optimal Nash equilibrium can be avoided by making use of norms. We study the complexity issues related to legal strategy and legal Nash equilibrium. Our complexity results are not complete, which leaves rooms for future work. Other natural future work includes using a different input/output logic to reason about norms and using positive permission to define legal strategy.

References

1. Ågotnes, T., van der Hoek, W., Rodríguez-Aguilar, J.A., Sierra, C., Wooldridge, M.: On the logic of normative systems. In: Veloso, M.M. (ed.) Proceedings of the 20th International Joint Conference on Artificial Intelligence, IJCAI 2007, Hyderabad, India, January 6-12, pp. 1175–1180 (2007)
2. Alechina, N., Dastani, M., Logan, B.: Reasoning about normative update. In: Rossi, F. (ed.) Proceedings of the 23rd International Joint Conference on Artificial Intelligence, IJCAI 2013, Beijing, China, August 3-9. IJCAI/AAAI (2013)
3. Andrighetto, G., Governatori, G., Noriega, P., van der Torre, L.W.N. (eds.): Normative Multi-Agent Systems. Dagstuhl Follow-Ups, vol. 4. Schloss Dagstuhl - Leibniz-Zentrum fuer Informatik (2013)
4. Arora, S., Barak, B.: Computational Complexity: A Modern Approach. Cambridge University Press, New York (2009)
5. Bilò, V.: On satisfiability games and the power of congestion games. In: Kao, M.-Y., Li, X.-Y. (eds.) AAIM 2007. LNCS, vol. 4508, pp. 231–240. Springer, Heidelberg (2007)
6. Boella, G., van der Torre, L., Verhagen, H.: Introduction to the special issue on normative multiagent systems. Autonomous Agents and Multi-Agent Systems 17(1), 1–10 (2008)
7. Bonzon, E., Lagasquie-Schiex, M.-C., Lang, J.: Dependencies between players in boolean games. Int. J. Approx. Reasoning 50(6), 899–914 (2009)
8. Bonzon, E., Lagasquie-Schiex, M.-C., Lang, J., Zanuttini, B.: Boolean games revisited. In: Brewka, G., Coradeschi, S., Perini, A., Traverso, P. (eds.) 17th European Conference on Artificial Intelligence, ECAI 2006, Proceedings of th Including Prestigious Applications of Intelligent Systems (PAIS 2006), Riva del Garda, Italy, August 29 - September 1. Frontiers in Artificial Intelligence and Applications, vol. 141, pp. 265–269. IOS Press (2006)
9. Bonzon, E., Lagasquie-Schiex, M.-C., Lang, J., Zanuttini, B.: Compact preference representation and boolean games. Autonomous Agents and Multi-Agent Systems 18(1), 1–35 (2009)
10. Buss, S.R., Hay, L.: On truth-table reducibility to SAT. Inf. Comput. 91(1), 86–102 (1991)
11. Coase, R.: The problem of social cost. Journal of Law and Economics 1 (1960)
12. Coleman, J.: Foundations of Social Theory. Belnap Press (1998)

13. Dunne, P.E., van der Hoek, W., Kraus, S., Wooldridge, M.: Cooperative boolean games. In: Padgham, L., Parkes, D.C., Müller, J.P., Parsons, S. (eds.) 7th International Joint Conference on Autonomous Agents and Multiagent Systems (AAMAS 2008), Estoril, Portugal, May 12-16, vol. 2, pp. 1015–1022. IFAAMAS (2008)
14. Gabbay, D., Horty, J., Parent, X., van der Meyden, R., van der Torre, L. (eds.): Handbook of Deontic Logic and Normative Systems. College Publications, London (2014)
15. Grossi, D., Tummolini, L., Turrini, P.: Norms in game theory. In: Handbook of Agreement Technologies (2012)
16. Harrenstein, P.: Logic in conflict. PhD thesis, Utrecht University (2004)
17. Harrenstein, P., van der Hoek, W., Meyer, J.-J., Witteveen, C.: Boolean games. In: Proceedings of the 8th Conference on Theoretical Aspects of Rationality and Knowledge, TARK 2001, San Francisco, CA, USA, pp. 287–298. Morgan Kaufmann Publishers Inc. (2001)
18. Herzig, A., Lorini, E., Moisan, F., Troquard, N.: A dynamic logic of normative systems. In: Proceedings of the 22nd International Joint Conference on Artificial Intelligence, IJCAI 2011, Barcelona, Catalonia, Spain, July 16-22, pp. 228–233. IJCAI/AAAI (2011)
19. Hurwicz, L.: Institutions as families of game forms. Japanese Economic Review 47(2), 113–132 (1996)
20. Hurwicz, L.: But who will guard the guardians? American Economic Review 98(3), 577–585 (2008)
21. Makinson, D., van der Torre, L.: Input-output logics. Journal of Philosophical Logic 29, 383–408 (2000)
22. Makinson, D., van der Torre, L.: Permission from an input/output perspective. Journal of Philosophical Logic 32, 391–416 (2003)
23. Mavronicolas, M., Monien, B., Wagner, K.W.: Weighted boolean formula games. In: Deng, X., Graham, F.C. (eds.) WINE 2007. LNCS, vol. 4858, pp. 469–481. Springer, Heidelberg (2007)
24. Osborne, M., Rubinstein, A.: A Course in Game Theory. The MIT Press, Cambridge (1994)
25. Parent, X.: Moral particularism in the light of deontic logic. Artif. Intell. Law 19(2-3), 75–98 (2011)
26. Sauro, L., Villata, S.: Dependency in cooperative boolean games. J. Log. Comput. 23(2), 425–444 (2013)
27. Shoham, Y., Leyton-Brown, K.: Multiagent Systems - Algorithmic, Game-Theoretic, and Logical Foundations. Cambridge University Press (2009)
28. Sun, X.: How to build input/output logic. In: Bulling, N., van der Torre, L., Villata, S., Jamroga, W., Vasconcelos, W. (eds.) CLIMA XV 2014. LNCS (LNAI), vol. 8624, pp. 123–137. Springer, Heidelberg (2014)
29. Ulmann-Margalit, E.: The Emergence of Norms. Clarendon Press, Oxford (1977)
30. von Wright, G.: Deontic logic. Mind 60, 1–15 (1952)
31. Weiss, G. (ed.): Multiagent systems, 2nd edn. MIT Press (2013)
32. Wooldridge, M.: An Introduction to MultiAgent Systems, 2nd edn. Wiley (2009)

Boolean Network Games
and Iterated Boolean Games

Jeremy Seligman and Declan Thompson

University of Auckland, New Zealand
j.seligman@auckland.ac.nz, declanthompsonnz@gmail.com

Abstract. A Boolean Network Game is a game played on a network structure. Players choose actions depending on the actions of those in their neighbourhood and attempt to achieve some goal expressed in a modification of Linear Temporal Logic over an infinite run. Iterated Boolean Games are similar, but lack network structure. We define and give translations between these models, and give some complexity results.

1 Introduction

Network games are used to model numerous phenomena in social psychology and economics, such as voting, distribution of public goods, negotiation, etc. [6] is a recent survey. Players are represented as nodes on a graph, with payoffs that depend on their connections to other player, typically only their nearest neighbours. Boolean games, introduced by Harrenstein et al [5] and extended to an iterated version in [3,4], provide a nice interface between game theory and logic. Players get to control the value of disjoint sets of propositional variables, with the goal of satisfying a Boolean formula which may depend on variables controlled by other players.

We introduce the concept of a Boolean network game which combines these ideas. Players are represented as nodes of a graph, each of which may or may not satisfy each propositional variable. They can control only the value of variables at their own nodes, with the goal of satisfying a modal formula, describing their position in the network. This allows us to use logic to describe many network games and thereby to reason about which strategies are rational, the properties of equilibria, etc.

As a simple example, consider the Colouring Game of [7], which was studied experimentally as a model of social coordination problems. In this game, players get to choose repeatedly one of a number of colours with the goal of having a colour that differs from every neighbour. This can be modelled as a Boolean network game in which colours are represented by propositional variables $p_1, p_2, \ldots p_n$, with the constraint that only one of these can be satisfied by each player. Each then has the goal $\bigvee_i (p_i \wedge \Box \neg p_i)$, where $\Box$ is the modality over the network relation. One motivation for the present work is to develop a logical theory that can be applied to reasoning about Nash equilibria in such games. The present paper constitutes an initial investigation, exploring the relationship between Boolean network games and the existing literature on iterated Boolean games.

© Springer-Verlag Berlin Heidelberg 2015
W. van der Hoek et al. (Eds.): LORI 2015, LNCS 9394, pp. 353–365, 2015.
DOI: 10.1007/978-3-662-48561-3_29

2   Boolean Games

2.1   Iterated Boolean Games

Iterated Boolean Games (IBGs) were introduced in [3] where they were used to show how Nash equilibria can be affected by repeated plays of certain games, where each player's strategy can depend upon the choices of other players in the past. Here we summarise the notion of an iterated Boolean game.

Language. IBGs use the language of Linear Temporal Logic (LTL):

$$\varphi ::= p \mid \neg\varphi \mid (\varphi \vee \varphi) \mid \mathbf{X}\varphi \mid \varphi\mathbf{U}\varphi$$

where $p \in \Phi$, a finite set of Boolean variables. We call this language $\mathcal{L}_{IBG}$.

A *run* is a function[1] $\rho : \mathbb{N} \to \mathcal{P}(\Phi)$ that assigns a valuation $\rho[i]$ to every timestep i. $\mathcal{L}_{IBG}$ formulas are interpreted with respect to pairs (ρ, i) where ρ is a run and $i \in \mathbb{N}$. Satisfaction for formulas is defined as in standard LTL: $(\rho, i) \vDash \mathbf{X}\varphi$ iff $(\rho, i+1) \vDash \varphi$; and $(\rho, i) \vDash \varphi\mathbf{U}\psi$ iff $(\rho, k) \vDash \psi$ for some $i \le k$ and $(\rho, j) \vDash \varphi$ for all $i \le j < k$. We say $\rho \vDash \varphi$ iff $(\rho, 0) \vDash \varphi$.

Games. An *Iterated Boolean Game (IBG)* is a structure

$$G = (A, \Phi, \Phi_1, \dots \Phi_n, \gamma_1, \dots \gamma_n)$$

where $A = \{1, \dots n\}$ is a set of *agents*, Φ is a finite set of Boolean variables, $\Phi_a \subseteq \Phi$ is the set of Boolean variables controlled by agent a and $\gamma_a \in \mathcal{L}_{IBG}$ is the goal of player a. We require that the sets $\Phi_1, \dots \Phi_n$ partition Φ.

Strategies. Given an IBG $G = (A, \Phi, \Phi_1, \dots \Phi_n, \gamma_1, \dots \gamma_n)$, a *machine strategy* σ_a for player a is an automaton $\sigma_a = (Q_a, q_a^0, \delta_a, \tau_a)$ where Q_a is a finite non-empty set of *nodes*, q_a^0 is the *start node*, $\delta_a : Q_a \times \mathcal{P}(\Phi) \to Q_a$ is a *transition function* and $\tau_a : Q_a \to \mathcal{P}(\Phi_a)$ is a *choice function*.

A *strategy profile* is an n-tuple of strategies, one for each player. We denote strategy profiles as $\boldsymbol{\sigma} = (\sigma_1, \dots \sigma_n)$, where σ_a is the strategy for player a.

Strategy Induced Runs. A *node vector* of $\boldsymbol{\sigma}$ is an n-tuple $\boldsymbol{q} = (q_1, \dots q_n)$ where $q_a \in Q_a$ for every $a \in A$. We denote the node vector at timestep i by $\boldsymbol{q}[i] = (q_1[i], \dots q_n[i])$. Associated with each node vector $\boldsymbol{q}[i]$ is a valuation vector $\boldsymbol{v}[i] = (v_1[i], \dots v_n[i])$. These vectors are defined for all timesteps i as follows:

$$\boldsymbol{q}[0] = (q_1^0, \dots q_n^0) \qquad\qquad \boldsymbol{v}[0] = (\tau_1(q_1^0), \dots \tau_n(q_n^0))$$
$$\boldsymbol{q}[i+1] = (\delta_1(q_1[i], \boldsymbol{v}[i]), \dots \delta_n(q_n[i], \boldsymbol{v}[i])) \quad \boldsymbol{v}[i+1] = (\tau_1(q_1[i]), \dots \tau_n(q_n[i]))$$

The *run induced by* $\boldsymbol{\sigma}$ is defined as $\rho(\boldsymbol{\sigma})[i] = \bigcup_{1 \le a \le n} v_a[i]$, the set of Boolean variables chosen by all the players at each timestep.

---

[1] We take $\mathcal{P}(A)$ to denote the powerset of A.

Preferences and Nash Equilibrium. For each player a we have a preference relation between possible runs given by

$$\rho \succsim_a \rho' \quad \text{iff} \quad \rho' \vDash \gamma_a \text{ implies } \rho \vDash \gamma_a$$

If $\sigma = (\sigma_1, \ldots \sigma_a, \ldots \sigma_n)$ and σ'_a is an alternative strategy for a then let (σ_{-a}, σ'_a) denote the strategy profile $(\sigma_{-a}, \sigma'_a) = (\sigma_1, \ldots \sigma'_a, \ldots \sigma_n)$.

A strategy profile σ is a *Nash Equilibrium* if for every player a and every possible strategy $\sigma'_a \in \Sigma_a$ we have $\rho(\sigma) \succsim_a \rho(\sigma_{-a}, \sigma'_a)$. Informally, a cannot do better by changing strategy (assuming all other players' strategies are held constant). In this case, we write $\sigma \in NE(G)$.

2.2   Boolean Network Games

A Boolean Network Game (BNG) is a similar model to an IBG but with a network structure on the agents. BNGs are useful for modelling situations in which agents are attempting to find responses to situations with restricted information. For example, agents may be choosing times for a party. They want as many of their friends to attend as possible, but they have no knowledge of their friends' friends party times. Is there a good strategy for determining a party time?

Networks. A *network* $\langle A, R \rangle$ is a set A of agents and a binary accessibility relation R on A. For each $a \in A$, the *social neighbourhood* of a is the set $R_a = \{b \mid Rab\}$. We take a finite set of properties PROP. We require that R_a is finite for all $a \in A$. A *local state* is a subset of PROP. A *global state* is a function $g : A \to \mathcal{P}(\text{PROP})$. The *environment of a*, g_a, is the restriction of g to R_a. Intuitively, a local state is the variables a player sets true, and the environment is the variables a player's neighbours have chosen.

Strategies. A *strategy* for $a \in A$ is a Moore automaton $\langle N, T, I, O \rangle$ where N is a finite set of *nodes*, $I \in N$ is the *start node*, T is a *transition function* mapping nodes and environments of a to nodes and $O : N \to \mathcal{P}(\text{PROP})$ is an *output function* mapping nodes to local states of a (subsets of PROP).

A *strategy profile* s is a function mapping each agent a to a strategy $\langle N_{sa}, T_{sa}, I_{sa}, O_{sa} \rangle$. A *node profile* ξ for s is a function mapping each agent a to a node of N_{sa}. The *initial node profile* ξ_{Is} for s is the node profile mapping each agent a to I_{sa}. The *initial global state* g_{Is} is the global state mapping each agent a to $O_{sa}(I_{sa})$.

Suppose s is a strategy profile and $\mathfrak{s}$ is a strategy for a. The *modification* of s with $\mathfrak{s}$ for a is the function

$$s_{a:\mathfrak{s}}(b) = \begin{cases} \mathfrak{s} & b = a \\ s(b) & b \neq a \end{cases}$$

Outcomes. Given a global state g and a state profile ξ for s, the *next node profile* $\xi_{s,g}$ and the *next global state* are given by

$$\xi_{s,g}(a) = T_{sa}(\xi(a), g_a) \qquad g_{s,\xi}(a) = O_{sa}(T_{sa}(\xi(a), g_a))$$

These are the profiles after a single round of interaction between the agents. The *outcome behaviour* of s is the infinite sequence $\{\langle g^i, \xi^i \rangle\}_{i \in \mathbb{N}}$ defined by

$$\langle g^0, \xi^0 \rangle = \langle g_{Is}, \xi_{Is} \rangle \qquad \langle g^{i+1}, \xi^{i+1} \rangle = \langle g^i_{s,\xi^i}, \xi^i_{s,g^i} \rangle$$

The sequence $g^0, g^1, g^2 \ldots$ describes the evolution of the agents' properties over time. The sequence $\xi^0, \xi^1, \xi^2, \ldots$ describes the evolution of the agents' internal nodes over time.

Language. We use an extension of Linear Temporal Logic (LTL) called $\mathcal{L}_{BNG}$ to allow us to describe the network relation over time.

$$\varphi ::= p \mid \neg\varphi \mid (\varphi \vee \varphi) \mid \Box\varphi \mid X\varphi \mid \varphi U \varphi$$

where $p \in \mathsf{PROP}$. These propositions express properties at each agent. For example, $\Box p$ says that all my neighbours have property p.

A *network model* $M = \langle A, R, g \rangle$ is a network $\langle A, R \rangle$ with a global state g. Formulas are evaluated with respect to a strategy profile s for the network, an agent $a \in A$ and a timestep i as follows:

$$
\begin{aligned}
M, s, a, i &\models p & &\text{iff } p \in g^i(a) \\
M, s, a, i &\models \neg\varphi & &\text{iff } M, s, a, i \not\models \varphi \\
M, s, a, i &\models (\varphi \vee \psi) & &\text{iff } M, s, a, i \models \varphi \text{ or } M, s, a, i \models \psi \\
M, s, a, i &\models \Box\varphi & &\text{iff } M, s, b, i \models \varphi \text{ for all } b \in R_a \\
M, s, a, i &\models X\varphi & &\text{iff } M, s, a, i+1 \models \varphi \\
M, s, a, i &\models \varphi U \psi & &\text{iff } M, s, a, k \models \psi \text{ for some } i \leq k \\
& & &\text{and } M, s, a, j \models \varphi \text{ for all } i \leq j < k.
\end{aligned}
$$

We say that $M, s, a \models \varphi$ iff $M, s, a, 0 \models \varphi$.

Games. Given a network model $M = \langle A, R, g \rangle$, a *goal profile* is a function $\gamma : A \to \mathcal{L}_{BNG}$. A *Boolean network game* (BNG) is a pair $G = \langle M, \gamma \rangle$. For any player $a \in A$, a strategy for a $\langle N, T, I, O \rangle$ is *available* to a iff $O(I) = g(a)$.

The *utility* of a strategy profile s for a is given by

$$u_a(s) = \begin{cases} 1 & \text{if } M, s, a \models \gamma(a) \\ 0 & \text{otherwise} \end{cases}$$

A strategy profile s is a *Nash Equilibrium* if there is no player a and strategy $\mathfrak{s}$ for a such that $u_a(s_{a:\mathfrak{s}}) > u_a(s)$. That is, if no player can do better by choosing a different strategy (while all other players' strategies are kept constant). In this case, we write $s \in NE(G)$.

3   Expressivity of Boolean Network Games

Boolean network games and iterated Boolean games are similar structures with a differing basis. Where IBGs add a temporal structure to what is essentially a propositional base, BNGs add this temporality to a *modal* base.

Standard results allow us to model basic modal logics inside predicate logic (see, for example, the discussion on the Standard Translation in [2, pp83-90]). Propositional logic is ill-suited to this task however, with its lack of a relational structure. Even so, the similarity of BNGs to IBGs presents a natural question: Can BNGs be modelled by IBGs? That is, can we translate any BNG into an IBG in a way which preserves the impact the accessibility relation has on the interaction of the agents? In practice, the accessibility relation imposes a restriction on the *transition functions* of players' strategies. While a propositional setting cannot encode modal relations on its own, perhaps a restriction of transition functions can achieve the same ends.

We can also ask the converse question. Given an IBG, can we model it as a BNG? At first this seems an easy prospect. Take a complete graph for the relation, ensuring every player can see every other, and proceed as normal. But we quickly encounter problems. In a BNG every player has control over all the propositional variables; in an IBG, each player has control over only a subset, and different players may control different numbers of propositional variables. Perhaps a player can do more by controlling variables it shouldn't be able to?

In this section we propose two translations, first from BNGs to IBGs and second from IBGs to BNGs. For each translation, we consider what properties of games are preserved. Finally, we give some results on the complexity of certain decision problems related to BNGs.

3.1   Translation from BNGs to IBGs

In this section, we give a translation from Boolean network games to iterated Boolean games. By abuse of notation, we use $\mathcal{T}$ for all functions related to the translation; which function is intended will be clear from context.

Game Translation. Suppose we have a BNG $G = \langle M, \gamma \rangle$ where $M = \langle A, R, g \rangle$ and $A = \{1, 2, \ldots n\}$, and that $\mathsf{PROP} = \{p, p', \ldots p^{(k)}\}$.

- Define $\Phi = \{p_a : p \in \mathsf{PROP}, 1 \leq a \leq n\} = \{p_1, p_1', \ldots p_1^{(k)}, p_2, p_2', \ldots p_n^{(k)}\}$.
- For each $a \in A$ define $\Phi_a = \{p_a : p \in \mathsf{PROP}\} \subseteq \Phi$. It is easy to see that this will give a partition of Φ.
- For each agent $a \in A$, define a translation $\mathcal{T}_a : \mathcal{L}_{BNG} \to \mathcal{L}_{IBG}$ inductively as follows.

$$p^{\mathcal{T}_a} = p_a \qquad (\varphi \vee \psi)^{\mathcal{T}_a} = \varphi^{\mathcal{T}_a} \vee \psi^{\mathcal{T}_a} \qquad (X\varphi)^{\mathcal{T}_a} = \mathbf{X}\varphi^{\mathcal{T}_a}$$

$$(\neg\varphi)^{\mathcal{T}_a} = \neg\varphi^{\mathcal{T}_a} \qquad (\Box\varphi)^{\mathcal{T}_a} = \bigwedge_{b \in R_a} \varphi^{\mathcal{T}_b} \qquad (\varphi U \psi)^{\mathcal{T}_a} = \varphi^{\mathcal{T}_a} \mathbf{U} \psi^{\mathcal{T}_a}$$

where $p \in \mathsf{PROP}$. This translation accounts for the change to indexed propositions. Note that the replacement of $\Box\varphi$ with a conjunction indicating φ should be true at all the neighbours of a implicitly encodes R.

- Define $\mathcal{T}_g : A \to \mathcal{L}_{IBG}$ by $\mathcal{T}_g(a) = \bigwedge_{p \in g(a)} p^{\mathcal{T}_a} \wedge \bigwedge_{p \notin g(a)} \neg p^{\mathcal{T}_a}$. Recall that BNGs specify a start state (g) where IBGs do not. This function is used to ensure the start state is (weakly) met in the IBG.
- Define the iterated boolean game $\mathcal{T}(G)$ as:

$$\mathcal{T}(G) = \left(A, \Phi, \Phi_1, \dots \Phi_n, \gamma(1)^{\mathcal{T}_1} \wedge \mathcal{T}_g(1), \dots \gamma(n)^{\mathcal{T}_n} \wedge \mathcal{T}_g(n) \right)$$

Here each player has the translation of its BNG goal and also its required start state as its goal for $\mathcal{T}(G)$.

Thus we have the same agents in $\mathcal{T}(G)$ as in G. The set of propositional variables of $\mathcal{T}(G)$ is the set PROP, indexed by the agents in A. Each agent controls exactly those variables indexed by it, and so the sets Φ_a are all disjoint. Intuitively, we are using the sets Φ_a as the propositions at a's location in $\langle A, R \rangle$.

Strategy Translation. For strategy $s(a) = \langle N_{sa}, T_{sa}, I_{sa}, O_{sa} \rangle$ define

$$\mathcal{T}(s(a)) = (N_{sa}, I_{sa}, T_{sa}^{\mathcal{T}}, O_{sa}^{\mathcal{T}})$$

where $T_{sa}^{\mathcal{T}} : N_{sa} \times \mathcal{P}(\Phi) \to N_{sa}$ is defined by

$$T_{sa}^{\mathcal{T}}(v, V) = T_{sa}(v, \{\langle b, \{p \in \mathsf{PROP} \mid p^{\mathcal{T}_b} \in V\} \rangle \mid b \in R_a\})$$

(we ignore elements of V not in the neighbourhood of a and treat p_b as being p at b) and where $O_{sa}^{\mathcal{T}} : N_{sa} \to \mathcal{P}(\Phi_a)$ is defined as

$$O_{sa}^{\mathcal{T}}(v) = \{p^{\mathcal{T}_a} \mid p \in O_{sa}(v)\}$$

We define the translation of the strategy profile s as

$$s^{\mathcal{T}} = (\mathcal{T}(s(1)), \dots \mathcal{T}(s(n)))$$

So a strategy is translated by keeping the same nodes, using the same transition function (by restricting inputs to those acceptable for that function) and translating outputs.

3.2 Properties of $\mathcal{T}$

We now consider which properties are preserved under $\mathcal{T}$. We specifically consider translations of games with strategies, as this allows us to consider questions of Nash equilibria. Due to limited space, some proofs have been omitted, but we have given brief description of them where possible.

We begin by showing that $\mathcal{T}(G)$ gives the same outcomes as G. This establishes that truth of formulas is preserved by $\mathcal{T}$.

Lemma 1 (Preservation of Outcomes). *Let $G = \langle M, \gamma \rangle$ be a BNG where $M = \langle A, R, g \rangle$ and let s be a strategy profile for G. Then*

$$M, s, a, i \vDash \varphi \quad \textit{iff} \quad (\rho(s^T), i) \vDash \varphi^{T_a}$$

for every $a \in A$, formula $\varphi \in \mathcal{L}_{BNG}$ and timestep i, where the right hand side is taken with respect to $\mathcal{T}(G)$. In particular, $M, s, a \vDash \varphi$ iff $\rho(s^T) \vDash \varphi^{T_a}$.

Proof. By induction on the complexity of φ. The case $\varphi = p$ can be proved by an induction on i. The propositional and LTL cases are trivial. The case for $\Box\varphi$ remains.

$$
\begin{aligned}
M, s, a, i \vDash \Box\varphi \quad &\textit{iff} \quad M, s, b, i \vDash \varphi \text{ for all } b \in R_a \\
&\textit{iff} \quad (\rho(\mathcal{T}(s)), i) \vDash \varphi^{T_b} \text{ for all } b \in R_a \text{ by inductive hypothesis} \\
&\textit{iff} \quad (\rho(\mathcal{T}(s)), i) \vDash (\Box\varphi)^{T_a}
\end{aligned}
$$

For the particular result, recall that $M, s, a \vDash \varphi$ is defined as $M, s, a, 0 \vDash \varphi$ and $\rho(s^T) \vDash \varphi^{T_a}$ as $(\rho(s^T), 0) \vDash \varphi^{T_a}$. $\qquad\Box$

We can conclude that the truth of all $\mathcal{L}_{BNG}$ formulas is preserved under $\mathcal{T}$, where formulas are translated relative to an agent. We have successfully simulated G as an IBG, using a BNG strategy. It can be seen that the formula translation functions $\mathcal{T}_a$ have inverses. Hence we can translate back any outcomes we reach in $\mathcal{T}(G)$.

The translation of formulas has been successful, but what about our translated goals? Recall that in $\mathcal{T}(G)$, agent a's goal is $\gamma(a) \wedge \mathcal{T}_g(a)$, where $\mathcal{T}_g(a)$ is the conjunction of the propositions in a's start state. If a obtains its goal in G with s, does it in $\mathcal{T}(G)$ with $\mathcal{T}(s)$? Using Lemma 1, this reduces to asking if $\rho(s^T) \vDash \mathcal{T}_g(a)$ for every agent a. This can be shown by noting that a must use an available strategy. It follows that agents' utilities are preserved under $\mathcal{T}$. That is, a obtains its goal with s in G iff a obtains its goal with $\mathcal{T}(s)$ in $\mathcal{T}(G)$.

Let us now consider how Nash equilibria are affected by $\mathcal{T}$. If s is not a Nash equilibrium for G could $\mathcal{T}(s)$ be an Nash equilibrium for $\mathcal{T}(G)$? No. If s is not a Nash equilibrium, then some player a can do better with a different strategy $s_{a:\mathfrak{s}}$. By Lemma 1, a can do better with $\mathcal{T}(\mathfrak{s})$ in $\mathcal{T}(G)$. Hence we have Lemma 2.

Lemma 2. *Let G be a BNG. Then $\mathcal{T}(\overline{NE(G)}) \subseteq \overline{NE(\mathcal{T}(G))}$.*

But what if s *is* a Nash equilibrium in G? Will $\mathcal{T}(s)$ be a Nash equilibrium of $\mathcal{T}(G)$? In order to answer this question, we will make use of the notion of *myopic strategies*, as used in [4]. A myopic strategy is a strategy in which every node has a unique successor; for IBGs, a strategy (Q, q^0, δ, τ) is myopic iff $\delta(q, v_1) = \delta(q, v_2)$ for every node q and valuation v_1, v_2. For myopic strategies, we write the transition function as $\delta(q)$ since the valuation does not matter.

Since IBGs have finitely many players, and each strategy has finitely many nodes, there are finitely many possible configurations of players in nodes, given a strategy profile. Each configuration gives a unique subsequent configuration, so since runs are infinite they must loop. We can utilise this to build a myopic strategy which impersonates any player's strategy (assuming the other strategies are kept constant). This is summarised as Lemma 3.

Lemma 3. *Let G be an IBG and $\boldsymbol{\sigma}$ be a strategy profile for G. Then for every player a there is a myopic strategy σ'_a such that $\rho(\boldsymbol{\sigma}) = \rho(\boldsymbol{\sigma}_{-a}, \sigma'_a)$.*

We are now ready to answer our question: if s is a Nash equilibrium for G, is $\mathcal{T}(s)$ a Nash equilibrium for $\mathcal{T}(G)$?

Lemma 4. *Let G be a BNG. Then $\mathcal{T}(NE(G)) \subseteq NE(\mathcal{T}(G))$.*

Proof. By contradiction. Suppose $s \in NE(G)$ but $s^{\mathcal{T}} \notin NE(\mathcal{T}(G))$. So there is a player a and a strategy $\sigma'_a = (Q'_a, q_a^{0\,\prime}, \delta'_a, \tau'_a)$ such that $\rho(s^{\mathcal{T}}) \nvDash \gamma(a)^{T_a} \wedge T_g(a)$ and $\rho(s^{\mathcal{T}}_{-a}, \sigma'_a) \vDash \gamma(a)^{T_a} \wedge T_g(a)$. By Lemma 3 we can assume σ'_a is myopic.

Define a myopic BNG strategy $\mathfrak{s} = \langle Q'_a, T'_a, q_a^{0\,\prime}, O'_a \rangle$ for a such that $T'_a(v, g_a) = \delta'_a(v)$ and $O'_a(v) = \{p \in \mathsf{PROP} \mid p^{T_a} \in \tau'_a(v)\}$. Now $\mathcal{T}(\mathfrak{s}) = (Q'_a, q_a^{0\,\prime}, T'_a{}^{\mathcal{T}}, O'_a{}^{\mathcal{T}})$, where

$$T'_a{}^{\mathcal{T}}(v, V) = T'_a(v, \{\langle b, \{p \in \mathsf{PROP} \mid p^{T_b} \in V\}\rangle \mid b \in R_a\}) = \delta'_a(v)$$

$$O'_a{}^{\mathcal{T}}(v) = \{p^{T_a} \mid p \in O'_a(v)\} = \tau'_a(v)$$

So $\mathcal{T}(\mathfrak{s}) = \sigma'_a$. We know $\mathfrak{s}$ is available for a since $\rho(s^{\mathcal{T}}_{-a}, \sigma'_a) \vDash T_g(a)$. It follows that $\rho(s_{a:\mathfrak{s}}^{\mathcal{T}}) = \rho(s^{\mathcal{T}}_{-a}, \sigma'_a)$. Since $\rho(s^{\mathcal{T}}_{-a}, \sigma'_a) \vDash \gamma(a)^{T_a} \wedge T_g(a)$ it must be that $M, s_{a:\mathfrak{s}}, a \Vdash \gamma(a)$ by Lemma 1. Similarly, since $\rho(s^{\mathcal{T}}) \nvDash \gamma(a)^{T_a} \wedge T_g(a)$ it must be that $M, s, a \nVdash \gamma(a)$. But then $u_a(s_{a:\mathfrak{s}}) > u_a(s)$ so $s \notin NE(G)$, a contradiction.

$\square$

So Nash equilibria are preserved and non-Nash equilibria are preserved. We can summarise these results with the following theorem.

Theorem 1. *Let G be a BNG. Then $s \in NE(G)$ iff $\mathcal{T}(s) \in NE(\mathcal{T}(G))$*

Proof. Left to right is by Lemma 4. Right to left is by contrapositive, using Lemma 2.
$\square$

The reader should take care to note that Theorem 1 does *not* say $\mathcal{T}(NE(G)) = NE(\mathcal{T}(G))$. Indeed, if $s \in NE(G)$ and there are players a, b with $b \notin R_a$ then $s_a^{\mathcal{T}}$ has no transitions depending on the state of b. So we can modify $s_a^{\mathcal{T}}$ to $s_a^{\mathcal{T}'}$ by duplicating some node, and modifying the transition function so that it goes to a different duplicate depending on the state of b. We still have $s^{\mathcal{T}'} \in NE(\mathcal{T}(G))$ so in this case $\mathcal{T}(NE(G)) \subsetneq NE(\mathcal{T}(G))$.

3.3 Translation from Iterated Boolean Games to Boolean Network Games

Now we consider the opposite direction. Given an IBG G, how can we simulate it as a BNG? Again, by abuse of notation, we use T to represent any functions used in the translation.

Game Translation. Suppose $G = (A, \Phi, \Phi_1, \ldots \Phi_n, \gamma_1, \ldots \gamma_n)$ is an IBG. Set $R = \{\langle a, b \rangle \mid a \neq b\}$. Thus every player can see every other player. Use Φ for PROP. Define a translation $\mathsf{T}_a : \mathcal{L}_{IBG} \to \mathcal{L}_{BNG}$ for each player a as follows:

$$p^{\mathsf{T}_a} = \begin{cases} p & \text{if } p \in \Phi_a \\ \Diamond p & \text{if } p \notin \Phi_a \end{cases} \qquad\qquad (\neg\varphi)^{\mathsf{T}_a} = \neg\varphi^{\mathsf{T}_a}$$

$$(\mathbf{X}\varphi)^{\mathsf{T}_a} = X\varphi^{\mathsf{T}_a} \qquad\qquad (\varphi \vee \psi)^{\mathsf{T}_a} = \varphi^{\mathsf{T}_a} \vee \psi^{\mathsf{T}_a}$$

$$(\varphi \mathbf{U} \psi)^{\mathsf{T}_a} = \varphi^{\mathsf{T}_a} U \psi^{\mathsf{T}_a}$$

Define a goal profile γ as

$$\gamma(a) = \gamma_a^{\mathsf{T}_a}$$

Since BNGs require a specified initial state, and IBGs do not, we cannot define $\mathsf{T}(G)$ further until we have defined translation for strategies.

Strategy Translation. Consider the strategy $\sigma_a = (Q_a, q_a^0, \delta_a, \tau_a)$.

In the translated game, each player has control over all the variables in PROP $= \Phi$, including those they do not control in the IBG. In the translation, each player's strategy sets all the variables they "should not control" to false. That is,

$$\tau_a^{\mathsf{T}}(q_a^k) = \tau_a(q_a^k)$$

This explains our translation of formulas T_a. If player a wants p in the IBG, then they want the player controlling p to set it true. So if a controls p, then in the translation a wants p. If a does not control p, they want the player who controls p to set it true. Every player who does not control p will set it false, so a wants $\Diamond p$.

The transition function only considers the value of the variables at the players who "should be" controlling them. So we should evaluate using only values from correct players. Hence define the translation of δ_a as

$$\delta_a^{\mathsf{T}}(q_a^k, g_a) = \delta_a \left(q_a^k, \bigcup_{b \in A} g(b) \cap \Phi_b \right)$$

We can now define the translation of σ_a.

$$\mathsf{T}(\sigma_a) = \langle Q_a, \delta_a^{\mathsf{T}}, q_a^0, \tau_a^{\mathsf{T}} \rangle$$

If $\sigma = (\sigma_1, \ldots \sigma_n)$ is a strategy profile, define σ^{T} such that $\sigma^{\mathsf{T}}(a) = \mathsf{T}(\sigma_a)$.

Game Translation (continued). Take G from above and a strategy profile $\sigma = (\sigma_1, \ldots \sigma_n)$, where $\sigma_a = (Q_a, q_a^0, \delta_a, \tau_a)$ for all a. Define a global state g_σ where

$$g_\sigma(a) = \tau_a(q_a^0)$$

That is, each player's initial state is the initial state of its strategy.

Now take $\mathsf{T}(G, \sigma) = \langle \langle A, R, g_\sigma \rangle, \gamma \rangle$, where R, γ are defined as above. This gives us a BNG corresponding to both G, with start state corresponding to σ. We write $\mathsf{T}(G)$ for the set of possible translations of G, and also when it is clear which strategy is being used for translation.

3.4   Properties of T

We now consider properties of T. Our goal is to show similar properties to those we proved for $\mathcal{T}$, namely that game outcomes, player utilities and Nash equilibria are all preserved under T.

The parallel of Lemma 1 becomes trickier for T since p's translation depends on which agent we are evaluating at. To help the proof, we have the following lemma.

Lemma 5. *Given an IBG G and a strategy profile σ for G and a timestep i. For every agent a and $p \in \Phi_a$, we have $p \in \rho(\sigma)[i]$ iff $p \in g^i(a)$, where g^i is the corresponding global state in the translation.*

Proof. By induction on i.   □

We build on this result to show that truth of formulas is preserved under T.

Lemma 6 (Preservation of Outcomes). *Let $G = (A, \Phi, \Phi_1, \ldots \Phi_n, \gamma_1, \ldots \gamma_n)$ be an IBG and σ a strategy profile for G. Suppose $T(G, \sigma) = \langle M_T, \gamma_T \rangle$ Then*

$$(\rho(\sigma), i) \vDash \varphi \quad iff \quad M_T, \sigma^T, a, i \vDash \varphi^{T_a}$$

for every $a \in A$, formula $\varphi \in \mathcal{L}_{IBG}$ and timestep i.

Proof. By induction on the complexity of φ. For the case $\varphi = p$ there are two subcases. If $p \in \Phi_a$ then the case follows from Lemma 5. So suppose $p \notin \Phi_a$. There is a $b \in A$ such that $p \in \Phi_b$ since the agent-indexed sets partition Φ. By Lemma 5, $M_T, \sigma^T, b, i \vDash p$. By the structure of $T(G)$, Rab and so we have $M_T, \sigma^T, b, i \vDash \Diamond p$. So $M_T, \sigma^T, b, i \vDash p^{T_a}$. The other direction is similar. The cases when $\varphi \neq p$ are routine.   □

We have established that truth of formulas is preserved under translation. Since players' goals in $T(G)$ are simply translations of their goals in G it follows that players' utilities are preserved under translation.

Let us now consider Nash equilibria. First, if σ is not a Nash equilibrium for G, can we be sure that $T(\sigma)$ is not a Nash equilibrium for $T(G)$? Yes. As with Lemma 2, if a can do better by changing its strategy to σ'_a in G, then a can do better by changing its strategy to $T(\sigma'_a)$ in $T(G)$.

Lemma 7. *Let G be an IBG. Then $T(\overline{NE(G)}) \subseteq \overline{NE(T(G))}$.*

Non-Nash equilibria are preserved, but what about Nash equilibria? We make use of myopic strategies once again. We are still dealing with finitely many configurations over an infinite run, so we have the following lemma.

Lemma 8. *Let G be a BNG and s a strategy profile for G. Then for every player a there is a myopic strategy s for a such that $M, s_{a:s}, b, i \vDash \varphi$ iff $M, s, b, i \vDash \varphi$ for every player b and timestep i.*

In the parallel argument for $\mathcal{T}$, we next showed that $\mathcal{T}(NE(G)) \subseteq NE(\mathcal{T}(G))$ (Lemma 4). The proof hinged on our being able to find a BNG strategy $\mathfrak{s}$ to map onto the myopic σ'_a. This was a straightforward exercise. If we attempt the same for T, we reach a problem: we do not know that the myopic strategy only outputs allowable valuations. Since players in $\mathsf{T}(G)$ have control over all propositions, perhaps a's better strategy involves changing a proposition it can't change in G. In order to account for this problem, we provide the following lemma.

Lemma 9. *Let G be an IBG and σ a strategy profile for G. Then for every BNG strategy $\mathfrak{s} = \langle N, T, I, O \rangle$ for a there is an IBG strategy σ'_a for a such that*

$$M_{\mathsf{T}}, \sigma^{\mathsf{T}}_{a:\mathfrak{s}}, a, i \vDash \varphi^{\mathsf{T}_a} \quad \textit{iff} \quad M'_{\mathsf{T}}, (\sigma_{-a}, \sigma'_a)^{\mathsf{T}}, a, i \vDash \varphi^{\mathsf{T}_a}$$

for every timestep i.

Proof. By Lemma 8 we can assume $\mathfrak{s}$ is myopic. Define $\sigma'_a = (N, I, T', O')$ where $T'(q, \boldsymbol{v}) = T(q)$ for all $\boldsymbol{v}$ and $O'(q) = O(q) \cap \Phi_a$ So σ'_a is $\mathfrak{s}$ with outputs restricted to Φ_a. We claim that σ'_a satisfies the requirements of the lemma. Let the outcome behaviour of M_{T} with $\sigma^{\mathsf{T}}_{a:\mathfrak{s}}$ be $\{\langle g^i, \xi^i \rangle\}_{i \in \mathbb{N}}$ and the outcome behaviour of M'_{T} with $(\sigma_{-a}, \sigma_a)^{\mathsf{T}}$ be $\{\langle g'^i, \xi'^i \rangle\}_{i \in \mathbb{N}}$. An induction on i establishes that

$$g^i(a) \cap \Phi_a = g'^i(a) \qquad\qquad \xi^i(a) = \xi'^i(a) \tag{1}$$

$$g^i(b) = g'^i(b) \qquad\qquad \xi^i(b) = \xi'^i(b) \tag{2}$$

for every $b \in A \setminus \{a\}$ and timestep i. We now proceed by induction on the complexity of φ to show that

$$M_{\mathsf{T}}, \sigma^{\mathsf{T}}_{a:\mathfrak{s}}, a, i \vDash \varphi^{\mathsf{T}_a} \quad \textit{iff} \quad M'_{\mathsf{T}}, (\sigma_{-a}, \sigma'_a)^{\mathsf{T}}, a, i \vDash \varphi^{\mathsf{T}_a}$$

First suppose $\varphi = p$. If $p \in \Phi_a$ then $p^{\mathsf{T}_a} = p$ and $M_{\mathsf{T}}, \sigma^{\mathsf{T}}_{a:\mathfrak{s}}, a, i \vDash p$ iff $p \in g^i(a)$. By 1, this is the case iff $p \in g'^i(a)$ since $p \in \Phi_a$. But this means $M'_{\mathsf{T}}, (\sigma_{-a}, \sigma'_a)^{\mathsf{T}}, a, i \vDash p$. If $p \notin \Phi_a$ then $p^{\mathsf{T}_a} = \Diamond p$. Again we have $M_{\mathsf{T}}, \sigma^{\mathsf{T}}_{a:\mathfrak{s}}, a, i \vDash \Diamond p$ iff $p \in g^i(b)$ for some b with Rab. By 2 this means $p \in g'^i(b)$ and so $M'_{\mathsf{T}}, (\sigma_{-a}, \sigma'_a)^{\mathsf{T}}, a, i \vDash \Diamond p$.

The propositional and temporal cases follow by routine arguments. Thus σ'_a fulfils our requirements and we have our result. $\qquad\qquad\square$

With Lemma 9 proved it is now straightforward to obtain the parallel of Lemma 4.

Lemma 10. *Let G be an IBG. Then $\mathsf{T}(NE(G)) \subseteq NE(\mathsf{T}(G))$.*

Proof. Suppose $\sigma \in NE(G)$. Suppose for contradiction that $\sigma^{\mathsf{T}} \notin NE(\mathsf{T}(G))$. Then there is a player $a \in A$ and a BNG strategy $\mathfrak{s}$ for a such that $u_a(\sigma^{\mathsf{T}}_{a:\mathfrak{s}}) > u_a(\sigma^{\mathsf{T}})$.

We must have $M_{\mathsf{T}}, \sigma^{\mathsf{T}}_{a:\mathfrak{s}}, a \vDash \gamma(a)$ and $M_{\mathsf{T}}, \sigma^{\mathsf{T}}, a \nvDash \gamma(a)$ by the definition of u_a. Since $\gamma(a) = \gamma_a{}^{\mathsf{T}_a}$, by Lemma 6 (Preservation of Outcomes) we have $\rho(\sigma) \nvDash \gamma_a$. By Lemma 9 there is a strategy σ'_a for a such that

$$M_{\mathsf{T}}, \sigma^{\mathsf{T}}_{a:\mathfrak{s}}, a \vDash \varphi^{\mathsf{T}_a} \quad \textit{iff} \quad M'_{\mathsf{T}}, (\sigma_{-a}, \sigma'_a)^{\mathsf{T}}, a \vDash \varphi^{\mathsf{T}_a}$$

Since $M_{\mathsf{T}}, \boldsymbol{\sigma}_{a:s}^{\mathsf{T}}, a \vDash \gamma(a)$ *we have* $M'_{\mathsf{T}}, (\boldsymbol{\sigma}_{-a}, \sigma'_a)^{\mathsf{T}}, a \vDash \gamma(a)$. *So by Lemma 6,* $\rho(\boldsymbol{\sigma}_{-a}, \sigma'_a) \vDash \gamma_a$.

But now we have $\rho(\boldsymbol{\sigma}_{-a}, \sigma'_a) \vDash \gamma_a$ *and* $\rho(\boldsymbol{\sigma}) \nvDash \gamma_a$, *so* $\rho(\boldsymbol{\sigma}) \not\succsim_a \rho(\boldsymbol{\sigma}_{-a}, \sigma'_a)$ *and hence* $\boldsymbol{\sigma} \notin NE(G)$, *a contradiction.* □

Finally we are ready to prove the parallel of Theorem 1.

Theorem 2. *Let G be an IBG. Then $\boldsymbol{\sigma} \in NE(G)$ iff $\mathsf{T}(\boldsymbol{\sigma}) \in NE(\mathsf{T}(G))$*

Proof. Left to right is by Lemma 10. Right to left is by contrapositive, using Lemma 7.

As with Theorem 1, we note that this does not mean $\mathsf{T}(NE(G)) = NE(\mathsf{T}(G))$ and indeed basic changes to strategies can show this is the case.

It is crucial to realise that T is not an inverse of $\mathcal{T}$. Indeed, it is the case that $\mathsf{T}(\mathcal{T}(G)) \neq G$ and $\mathcal{T}(\mathsf{T}(G)) \neq G$. Since $\mathcal{T}$ indexes propositions by agent and T allows all agents control over every proposition, iterated applications of $\mathcal{T}$ and T will increase the number of propositions in the game.

3.5 Computational Complexity of Decision Problems for Boolean Network Games

The translation T allows us to make conclusions about the complexity of decision problems for Boolean network games by way of reductions to IBGs. An examination of T shows that it can be accomplished in polynomial time. Consider the following problem for testing Nash equilibria.

> BNG MEMBERSHIP
> *Given:*   BNG game G, strategy profile s.
> *Question:* Is it the case that $s \in NE(G)$?

From [4] we have that IBG MEMBERSHIP is PSPACE-complete. By Theorem 2 there is a polynomial time reduction from BNG MEMBERSHIP to IBG MEMBERSHIP, using T. Hence, we have the following proposition.

Proposition 1. BNG MEMBERSHIP *is PSPACE-hard.*

Note that since the translation $\mathcal{T}$ is exponential due to the translation of $\square$, we do not have a PSPACE-completeness result. However, if the degree of the players' goals is bounded (that is, the maximum depth of nested modalities), then we get PSPACE-completeness for each fixed bound.

4   Conclusion and Further Work

We introduced the concept of a Boolean network game and proved a two-way reduction to iterated Boolean games. This enabled us to prove theorems about determining whether a given strategy profile is a Nash equilibrium.

The problem of determining the existence of an equilibrium is also interesting. Unfortunately a reduction to IBG will not help here. Recall that while $\mathcal{T}(NE(G)) \subseteq NE(\mathcal{T}(G))$, we do not have that $\mathcal{T}(NE(G)) = NE(\mathcal{T}(G))$. $\mathcal{T}$ does not suggest an obvious method for translating strategy profiles for $\mathcal{T}(G)$ into those for G - we achieved this for individual strategies given fixed others, but not for strategy profiles. Perhaps it is the case that $NE(G) = \emptyset$ and $NE(\mathcal{T}(G)) \neq \emptyset$. Whether this is possible remains an open question, as do similar questions for $\mathcal{T}$.

A further step is to study the logic of Nash equilibria: those formulas of LTL that hold at all Nash equilibria, so as to account for social reasoning in equilibrium situations. [3] contains some initial results in this direction for IBGs.

Finally, [1] extends the Boolean game framework to epistemic games, in which the goal formula of an agent may concern the epistemic states of other agents, and the actions of players are announcements. The theory of Boolean network games promises to be a useful base from which to study epistemic Boolean games within the context of social epistemic logic [8].

References

1. Ågotnes, T., Harrenstein, P., van der Hoek, W., Wooldridge, M.: Boolean Games with Epistemic Goals. In: Grossi, D., Roy, O., Huang, H. (eds.) LORI. LNCS, vol. 8196, pp. 1–14. Springer, Heidelberg (2013), http://link.springer.com/chapter/10.1007/978-3-642-40948-6_1
2. Blackburn, P., de Rijke, M., de Venema, Y.: Modal Logic. Cambridge University Press (August 2002)
3. Gutierrez, J., Harrenstein, P., Wooldridge, M.: Iterated boolean games. In: Proceedings of the Twenty-Third International Joint Conference on Artificial Intelligence, pp. 932–938. AAAI Press (2013)
4. Gutierrez, J., Harrenstein, P., Wooldridge, M.: Iterated boolean games. Information and Computation (2015)
5. Harrenstein, P., van der Hoek, W., Meyer, J.-J., Witteveen, C.: Boolean games. In: Proceedings of the 8th Conference on Theoretical Aspects of Rationality and Knowledge, pp. 287–298. Morgan Kaufmann Publishers Inc. (2001)
6. Jackson, M.O., Zenou, Y.: Games on Networks. In: Handbook of Game Theory with Economic Applications. Handbook of Game Theory with Economic Applications, vol. 4, pp. 95–163. Elsevier (2015)
7. Kearns, M., Suri, S., Montfort, N.: An experimental study of the coloring problem on human subject networks. Science 313(5788), 824–827 (2006)
8. Seligman, J., Liu, F., Girard, P.: Logic in the community. In: Banerjee, M., Seth, A. (eds.) ICLA 2011. LNCS (LNAI), vol. 6521, pp. 178–188. Springer, Heidelberg (2011)

Symbolic Model Checking for Dynamic Epistemic Logic

Johan van Benthem[1,2], Jan van Eijck[1,3], Malvin Gattinger[1], and Kaile Su[4,5]

[1] Institute for Logic, Language & Computation (ILLC), University of Amsterdam
[2] Department of Philosophy, Stanford University
[3] Centrum Wiskunde & Informatica, Amsterdam
[4] Institute for Integrated and Intelligent Systems, Griffith University
[5] Department of Computer Science, Jinan University

Abstract. Dynamic Epistemic Logic (DEL) can model complex information scenarios in a way that appeals to logicians. However, existing DEL implementations are ad-hoc, so we do not know how the framework really performs. For this purpose, we want to hook up with the best available model-checking and SAT techniques in computational logic. We do this by first providing a bridge: a new faithful representation of DEL models as so-called knowledge structures that allow for symbolic model checking. Next, we show that we can now solve well-known benchmark problems in epistemic scenarios much faster than with existing DEL methods. Finally, we show that our method is not just a matter of implementation, but that it raises significant issues about logical representation and update.

1 Introduction

We bring together two strains in the area of epistemic model checking. On one side, there are many frameworks for symbolic model checking on interpreted systems using temporal logics [24,30]. On the other hand, there are explicit model checkers for variants of Dynamic Epistemic Logic (DEL) like DEMO [15] with inferior performance but superior usability as they allow specification in dynamic languages directly. The goal of our work is to connect the two worlds of symbolic model checking and DEL in order to gain new insights on both sides.

Existing work on model checking DEL mainly focuses on specific examples, for example the Dining Cryptographers [28], the Sum and Product riddle [26] or Russian Cards [12]. Given these specific approaches, a general approach to symbolic model checking the full DEL language is desirable. A first step is [30] which presents symbolic model checking for temporal logics of knowledge. However, it does not cover announcements or other dynamics. The framework here extends these ideas with dynamic operators and a twist on the semantics.

Our knowledge structures are similar in spirit to hypercubes from [25], but of a different type: We do not use interpreted systems and temporal relations are not part of our models. Hence also our language does not contain temporal operators but primitives for epistemic events like announcements.

W. van der Hoek et al. (Eds.): LORI 2015, LNCS 9394, pp. 366–378, 2015.
DOI: 10.1007/978-3-662-48561-3_30

Related to our work is also [13] where DEL is translated into temporal epistemic logics for which symbolic model checkers exist. However, this method has not been implemented and the complexity and performance are not known. We do not translate to a temporal logic but check DEL formulas directly.

The paper is structured as follows. In Section 2 we recall standard semantics of DEL as in [11]. We then present knowledge structures in Section 3 and discuss the famous Muddy Children example in Section 4, together with experimental results in Section 5. Section 6 is a case study of the Russian Cards problem. Our main theoretical results are in Section 7: Knowledge structures are equivalent to S5 Kripke models. Moreover, S5 action models from [1] can be described in the same way. Section 8 gives a conclusion and suggestions for further research.

All source code can be found at `https://github.com/jrclogic/SMCDEL`.

2   Dynamic Epistemic Logic on Kripke Models

Definition 1. *Fix a set of propositions V and a finite set of agents I. The DEL language $\mathcal{L}(V)$ is given by*

$$\varphi ::= p \mid \neg\varphi \mid \varphi \wedge \varphi \mid K_i\varphi \mid C_\Delta\varphi \mid [\varphi]\varphi \mid [\varphi]_\Delta\varphi$$

where $p \in V$, $i \in I$ and $\Delta \subseteq I$. We also use the abbreviations $\varphi \vee \psi := \neg(\neg\varphi \wedge \neg\psi)$ and $\varphi \to \psi := \neg(\varphi \wedge \neg\psi)$. The boolean *formulas are $\varphi ::= p \mid \neg\varphi \mid \varphi \wedge \varphi$.*

The formula $K_i\varphi$ is read as *"agent i knows φ"* while $C_\Delta\varphi$ says that φ is common knowledge among agents in Δ. The formula $[\psi]\varphi$ indicates that after a *public announcement* of ψ, φ holds. In contrast, $[\psi]_\Delta\varphi$ says that after announcing ψ to the agents in Δ, φ holds. The standard semantics for $\mathcal{L}(V)$ are given by means of Kripke models as follows.

Definition 2. *A* Kripke model *for n agents is a tuple $M = (W, \pi, \mathcal{K}_1, \cdots, \mathcal{K}_n)$, where W is a set of* worlds, *π associates with each world a truth assignment to the primitive propositions, so that $\pi(w)(p) \in \{\top, \bot\}$ for each world w and primitive proposition p, and $\mathcal{K}_1, \cdots, \mathcal{K}_n$ are binary accessibility relations on W. By convention, W^M, $\mathcal{K}_i^M$ and π^M are used to refer to the components of M. We omit the superscript M if it is clear from context. Finally, let $\mathcal{C}_\Delta^M$ be the transitive closure of $\bigcup_{i\in\Delta} \mathcal{K}_i^M$.*

A pointed Kripke model *is a pair (M, w) consisting of a Kripke model and a world $w \in W^M$. A model M is called an S5 Kripke model iff, for every i, $\mathcal{K}_i^M$ is an equivalence relation. A model M is called* finite *iff W^M is finite.*

Definition 3. *Semantics for $\mathcal{L}(V)$ on pointed Kripke models are given inductively as follows.*

1. *$(M, w) \models p$ iff $\pi^M(w)(p) = \top$.*
2. *$(M, w) \models \neg\varphi$ iff not $(M, w) \models \varphi$*
3. *$(M, w) \models \varphi \wedge \psi$ iff $(M, w) \models \varphi$ and $(M, w) \models \psi$*
4. *$(M, w) \models K_i\varphi$ iff for all $w' \in W$, if $w\mathcal{K}_i^M w'$, then $(M, w') \models \varphi$.*

5. $(M, w) \models C_\Delta \varphi$ iff for all $w' \in W$, if $w \mathcal{C}_\Delta^M w'$, then $(M, w') \models \varphi$.

6. $(M, w) \models [\psi]\varphi$ iff $(M, w) \models \psi$ implies $(M^\psi, w) \models \varphi$ where M^ψ is a new Kripke model defined by the set $W^{M^\psi} := \{w \in W^M \mid (M, w) \models \psi\}$, the relations $\mathcal{K}_i^{M^\psi} := \mathcal{K}_i^M \cap (W^{M^\psi})^2$ and the valuation $\pi^{M^\psi}(w) := \pi^M(w)$.

7. $(M, w) \models [\psi]_\Delta \varphi$ iff $(M, w) \models \psi$ implies that $(M_\psi^\Delta, (1, w)) \models \varphi$ where

   (a) $W^{M_\psi^\Delta} := \{(1, w) \mid w \in W^M \text{ and } (M, w) \models \psi\} \cup \{(0, w) \mid w \in W^M\}$

   (b) For (b, w) and (b', w') in $W^{M_\psi^\Delta}$, if $i \in \Delta$, let $(b, w)\mathcal{K}_i^{M_\psi^\Delta}(b', w')$ iff $b = b'$ and $w \mathcal{K}_i^M w'$. If $i \notin \Delta$, then let $(b, w)\mathcal{K}_i^{M_\psi^\Delta}(b', w')$ iff $w \mathcal{K}_i^M w'$.

   (c) For each $(b, w) \in W^{M_\psi^\Delta}$, $\pi^{M_\psi^\Delta}((b, w)) := \pi^M(w)$.

Note that a group announcement $[\psi]_\Delta \varphi$ is private in the sense that only the agents in Δ obtain knowledge about ψ. However, the announcement is *not secret* because the other agents still learn that the agents in Δ might have learned ψ.

3 Knowledge Structures

While the preceding semantics is standard in logic, it cannot serve directly as an input to current sophisticated model-checking techniques. For this purpose, in this section we introduce a new format, *knowledge structures*. Their main advantage is that also knowledge and results of announcements can be computed via purely boolean operations. We first recapitulate some notions and abbreviations.

Given a set of propositional variables P, we identify a *truth assignment over P* with a subset of P. We say a formula φ is a formula *over P* if each propositional variable occurring in φ is in P. For convenience, we use the logical constants $\top$ and $\bot$ which are always true and always false, respectively. We also use $\models$ to denote the usual satisfaction relation between a truth assignment and a formula.

We use substitution and quantification as follows. For any formula φ and $\psi \in \{\top, \bot\}$, and any propositional variable p, let $\varphi(\frac{p}{\psi})$ denote the result of replacing every p in φ by ψ. For any $A = \{p_1, \ldots, p_n\}$, let $\varphi(\frac{A}{\psi}) := \psi(\frac{p_1}{\psi})(\frac{p_2}{\psi})\ldots(\frac{p_n}{\psi})$, i.e. the result of substituting ψ for all elements of A. We use $\forall p \varphi$ to denote $\varphi\left(\frac{p}{\top}\right) \wedge \varphi\left(\frac{p}{\bot}\right)$. For any $A = \{p_1, \ldots, p_n\}$, let $\forall A \varphi := \forall p_1 \forall p_2 \ldots \forall p_n \varphi$.

Definition 4. *Suppose we have n agents. A* knowledge structure *is a tuple $\mathcal{F} = (V, \theta, O_1, \ldots, O_n)$ where V is a finite set of propositional variables, θ is a boolean formula over V and for each agent i, $O_i \subseteq V$.*

Set V is the vocabulary *of $\mathcal{F}$. Formula θ is the* state law *of $\mathcal{F}$. It determines the set of states of $\mathcal{F}$ and may only contain boolean operators. The variables in O_i are called agent i's* observable variables. *An assignment over V, given as the set of true propositions, that satisfies θ is called a* state *of $\mathcal{F}$. Any knowledge structure only has finitely many states. Given a state s of $\mathcal{F}$, we say that $(\mathcal{F}, s)$ is a* scene *and define the* local state *of an agent i at s as $s \cap O_i$.*

Given a knowledge structure $(V, \theta, O_1, \cdots, O_n)$ and a set $\mathcal{V}$ of subsets of V, we use $\mathcal{E}_\mathcal{V}$ to denote a relation between two assignments s, s' on V satisfying θ such that $(s, s') \in \mathcal{E}_\mathcal{V}$ iff there exists a $P \in \mathcal{V}$ with $s \cap P = s' \cap P$. We use $\mathcal{E}_\mathcal{V}^*$

to denote the transitive closure of $\mathcal{E}_\mathcal{V}$. Let $\mathcal{V}_\Delta = \{O_i \mid i \in \Delta\}$. We then have $(s, s') \in \mathcal{E}_{\mathcal{V}_\Delta}$ iff there exists an $i \in \Delta$ with $s \cap O_i = s' \cap O_i$.

We now give alternative semantics for $\mathcal{L}(V)$ on knowledge structures. Definitions 5 and 6 run in parallel, both proceeding by the structure of φ.

Definition 5. *Semantics for DEL on scenes are defined inductively as follows.*

1. $(\mathcal{F}, s) \models p$ iff $s \models p$.
2. $(\mathcal{F}, s) \models \neg\varphi$ iff not $(\mathcal{F}, s) \models \varphi$
3. $(\mathcal{F}, s) \models \varphi \wedge \psi$ iff $(\mathcal{F}, s) \models \varphi$ and $(\mathcal{F}, s) \models \psi$
4. $(\mathcal{F}, s) \models K_i\varphi$ iff for all s' of $\mathcal{F}$, if $s \cap O_i = s' \cap O_i$, then $(\mathcal{F}, s') \models \varphi$.
5. $(\mathcal{F}, s) \models C_\Delta\varphi$ iff for all s' of $\mathcal{F}$, if $(s, s') \in \mathcal{E}_{\mathcal{V}_\Delta}^*$, then $(\mathcal{F}, s') \models \varphi$.
6. $(\mathcal{F}, s) \models [\psi]\varphi$ iff $(\mathcal{F}, s) \models \psi$ implies $(\mathcal{F}^\psi, s) \models \varphi$ where $\|\psi\|_\mathcal{F}$ is given by Definition 6 and
$$\mathcal{F}^\psi := (V, \theta \wedge \|\psi\|_\mathcal{F}, O_1, \cdots, O_n)$$
7. $(\mathcal{F}, s) \models [\psi]_\Delta\varphi$ iff $(\mathcal{F}, s) \models \psi$ implies $(\mathcal{F}_\psi^\Delta, s \cup \{p_\psi\}) \models \varphi$ where p_ψ is a new propositional variable, $\|\psi\|_\mathcal{F}$ is given by Definition 6 and
$$\mathcal{F}_\psi^\Delta := (V \cup \{p_\psi\}, \theta \wedge (p_\psi \to \|\psi\|_\mathcal{F}), O_1', \cdots, O_n')$$
where $O_i' := O_i \cup \{p_\psi\}$ if $i \in \Delta$ and $O_i' := O_i$ otherwise.

Before defining the boolean equivalents of formulas, we can already explain some similarities and differences between Definitions 3 and 5. The semantics of the boolean connectives are the same. For the knowledge operators, on Kripke models we use an accessibility relation $\mathcal{K}_i$. On knowledge structures this is replaced with the condition $s \cap O_i = s' \cap O_i$, inducing an equivalence relation on the states. We can already guess that knowledge structures encode S5 Kripke models.

Definition 6. *For any knowledge structure* $\mathcal{F} = (V, \theta, O_1, \cdots, O_n)$ *and any DEL formula* φ, *we define a boolean formula* $\|\varphi\|_\mathcal{F}$.

1. *For any primitive formula, let* $\|p\|_\mathcal{F} := p$.
2. *For negation, let* $\|\neg\psi\|_\mathcal{F} := \neg\|\psi\|_\mathcal{F}$.
3. *For conjunction, let* $\|\psi_1 \wedge \psi_2\|_\mathcal{F} := \|\psi_1\|_\mathcal{F} \wedge \|\psi_2\|_\mathcal{F}$.
4. *For knowledge, let* $\|K_i\psi\|_\mathcal{F} := \forall(V \setminus O_i)(\theta \to \|\psi\|_\mathcal{F})$.
5. *For common knowledge, let* $\|C_\Delta\psi\|_\mathcal{F} := \mathbf{gfp}\Lambda$ *where* Λ *is the following operator on boolean formulas given and* $\mathbf{gfp}\Lambda$ *denotes its greatest fixed point:*
$$\Lambda(\alpha) := \|\psi\|_\mathcal{F} \wedge \bigwedge_{i \in \Delta} \forall(V \setminus O_i)(\theta \to \alpha)$$
6. *For public announcements, let* $\|[\psi]\xi\|_\mathcal{F} := \|\psi\|_\mathcal{F} \to \|\xi\|_{\mathcal{F}^\psi}$.
7. *For group announcements, let* $\|[\psi]_\Delta\xi\|_\mathcal{F} := \|\psi\|_\mathcal{F} \to (\|\xi\|_{\mathcal{F}_\psi^\Delta})(\frac{p_\psi}{\top})$.

where $\mathcal{F}^\psi$ *and* $\mathcal{F}_\Delta^\psi$ *are as given by Definition 5.*

Given these definitions, a simple induction on φ gives us the following Theorem.

Theorem 1. *Definition 6 preserves and reflects truth. That is, for any formula φ and any scene $(\mathcal{F}, s)$ we have that $(\mathcal{F}, s) \models \varphi$ iff $s \models \|\varphi\|_{\mathcal{F}}$.*

We can now explain the public and group announcements. First observe that *public* announcements only modify the state law of the knowledge structure. Moreover, the new state law is always a conjunction containing the previous one. Hence the set of states is restricted, just like public announcements on Kripke models restrict the set of possible worlds. Second, note that a group announcement adds a single observational variable and can therefore at most double the number of states, just like in the Kripke semantics in Definition 3.

4   Example 1: Muddy Children

How does our new format do in practice? For this purpose, we consider some well-known benchmarks in the epistemic agency literature. We start with how their new representations looks like. After that, we go on to actual computational experiments. The famous Muddy Children example will illustrate how announcements, both of propositional and of epistemic facts, work on knowledge structures. An early version of the puzzle are the three ladies on a train in [23]. For a standard analysis with Kripke models, see [17, p. 24-30] or [11, p. 93-96].

Let p_i stand for "child i is muddy". We consider the case of three children $I = \{1, 2, 3\}$ who are all muddy, i.e. the actual state is $\{p_1, p_2, p_3\}$. At the beginning the children do not have any information, hence the initial knowledge structure $\mathcal{F}_0$ in Figure 1 has the state law $\theta_0 = \top$. All children can observe whether the others are muddy but do not see their own face. This is represented with observational variables: Agent 1 observes p_2 and p_3, etc. Now the father says: "At least one of you is muddy." This public announcement limits the set of states by adding this statement to the state law. Note that it already is a purely boolean statement, hence the formula is added as it is, leading to $\mathcal{F}_1$.

$$\mathcal{F}_0 = \left(V = \{p_1, p_2, p_3\}, \theta_0 = \top, \begin{matrix} O_1 = \{p_2, p_3\} \\ O_2 = \{p_1, p_3\} \\ O_3 = \{p_1, p_2\} \end{matrix} \right)$$

$$\mathcal{F}_1 = \left(V = \{p_1, p_2, p_3\}, \theta_1 = (p_1 \vee p_2 \vee p_3), \begin{matrix} O_1 = \{p_2, p_3\} \\ O_2 = \{p_1, p_3\} \\ O_3 = \{p_1, p_2\} \end{matrix} \right)$$

Fig. 1. Knowledge structures before and after the first announcement.

The father now asks "Do you know if you are muddy?" but none of the children does. As it is common in the literature, we understand this as a public announcement of "Nobody knows their own state.": $\bigwedge_{i \in I} (\neg(K_i p_i \vee K_i \neg p_i))$. This is not a purely boolean formula, hence the public announcement is slightly more complicated: Using Definition 6 and Theorem 1 we find a boolean formula which on the current knowledge structure $\mathcal{F}_1$ is equivalent to the announced formula. Then this boolean equivalent is added to θ. We have

$$\|K_1 p_1\|_{\mathcal{F}_1} = \forall(V \setminus O_1)(\theta_1 \to \|p_1\|_{\mathcal{F}_1}) = \forall p_1((p_1 \vee p_2 \vee p_3) \to p_1)$$
$$= ((\top \vee p_2 \vee p_3) \to \top) \wedge ((\bot \vee p_2 \vee p_3) \to \bot) = \neg(p_2 \vee p_3)$$

$$\|K_1 \neg p_1\|_{\mathcal{F}_1} = \forall(V \setminus O_1)(\theta_1 \to \|\neg p_1\|_{\mathcal{F}_1}) = \forall p_1((p_1 \vee p_2 \vee p_3) \to \neg p_1)$$
$$= ((\top \vee p_2 \vee p_3) \to \neg\top) \wedge ((\bot \vee p_2 \vee p_3) \to \neg\bot) = \bot$$

and analogous for $K_2 p_2$, $K_2 \neg p_2$, $K_3 p_3$ and $K_3 \neg p_3$. These results make intuitive sense: In our situation where all children are muddy, a child knows it is muddy iff it sees that the other two children are clean. It can never know that it is clean itself. The announced formula becomes

$$\| \bigwedge_{i \in I} (\neg(K_i p_i \vee K_i \neg p_i)) \|_{\mathcal{F}_1} = \bigwedge_{i \in I} \|\neg(K_i p_i \vee K_i \neg p_i)\|_{\mathcal{F}_1}$$
$$= \neg(\neg(p_2 \vee p_3)) \wedge \neg(\neg(p_1 \vee p_3)) \wedge \neg(\neg(p_1 \vee p_2))$$
$$= (p_2 \vee p_3) \wedge (p_1 \vee p_3) \wedge (p_1 \vee p_2)$$

The announcement essentially says that at least two children are muddy. We get a knowledge structure $\mathcal{F}_2$ with the following more restrictive state law θ_2. Vocabulary and observational variables do not change, so we do not repeat them.

$$\theta_2 = (p_1 \vee p_2 \vee p_3) \wedge ((p_2 \vee p_3) \wedge (p_1 \vee p_3) \wedge (p_1 \vee p_2))$$

Now the same announcement ("Nobody knows their own state.") is made again. It is important that again we start with the epistemic formula $\bigwedge_{i \in I}(\neg(K_i p_i \vee K_i \neg p_i))$ and compute an equivalent formula with respect to $\mathcal{F}_2$. For reasons of space we skip tedious boolean reasoning and just note that

$$\|K_1 p_1\|_{\mathcal{F}_2} = \forall(V \setminus O_1)(\theta_2 \to \|p_1\|_{\mathcal{F}_2}) = \neg(p_3 \wedge p_2)$$

$$\|K_1 \neg p_1\|_{\mathcal{F}_2} = \forall(V \setminus O_1)(\theta_2 \to \|\neg p_1\|_{\mathcal{F}_2}) = \neg(p_2 \vee p_3)$$

which gives us $\|\neg(K_1 p_1 \vee K_1 \neg p_1)\|_{\mathcal{F}_2} = p_3 \wedge p_2$ and analogous formulas for children 2 and 3. Hence with respect to $\mathcal{F}_2$ we get the following boolean equivalent of the announcement, essentially saying that everyone is muddy.

$$\| \bigwedge_{i \in I} (\neg(K_i p_i \vee K_i \neg p_i)) \|_{\mathcal{F}_2} = (p_3 \wedge p_2) \wedge (p_3 \wedge p_1) \wedge (p_2 \wedge p_1)$$
$$= p_1 \wedge p_2 \wedge p_3$$

The resulting knowledge structure thus has the state law $\theta_3 = \theta_2 \wedge (p_1 \wedge p_2 \wedge p_3)$ which is in fact equivalent to $p_1 \wedge p_2 \wedge p_3$ and marks the end of the story: The only state left is the situation in which all three children are muddy.

5 Symbolic Model Checking: Implementation and Benchmarking

The previous section showed how epistemic operators get replaced by booleans when a new state law is computed. We could see that syntactically the state law becomes more and more complex, but semantically the same boolean function can be represented with a much shorter formula. This is where Binary Decision Diagrams (BDDs) come in extremely handy.

First presented in [5], BDDs provide an elegant data structure for boolean functions. In many cases they are less redundant and thus smaller than a corresponding truth table. Additionally, they can be manipulated efficiently: Given BDDs for φ and ψ we can compute the BDD for $\varphi \wedge \psi$, $\varphi \rightarrow \psi$ etc. Moreover, BDDs are canonical: Two formulas are equivalent iff their BDDs are identical. For an in-depth introduction, see [22, p. 202-280]. To see how BDDs can be used to describe knowledge structures, Figure 2 shows the BDDs for θ_0 to θ_3.

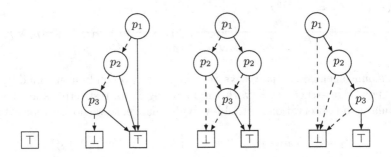

Fig. 2. Four BDDs representing the state laws θ_0 to θ_3.

Our new symbolic model checker SMCDEL works as follows: It takes two inputs, a scene $(\mathcal{F}, s)$ where the state law is given as a BDD, and a DEL formula φ. To check whether φ holds at state s we first compute the equivalent boolean formula $\|\varphi\|_{\mathcal{F}}$ according to Definition 6 and then check the boolean satisfaction $s \vDash \|\varphi\|_{\mathcal{F}}$. Alternatively, we can check whether a formula is valid on $\mathcal{F}$, i.e. true at *all* states, by checking whether $\theta \rightarrow \|\varphi\|_F$ is a tautology. The full set of states does not have to be generated and events are not executed explicitly.

We compared the performance of this method to DEMO-S5, an explicit model checker optimized for multi-agent S5 [15]. As a benchmark we used the question "For n muddy children, how many announcements of »Nobody knows their own state.« are needed until they do know their own state?". We measured how long each method takes to find and verify the correct answer, namely $n - 1$.

Figure 3 shows the results on a logarithmic scale: Explicit model checking with DEMO-S5 quickly becomes unfeasible whereas our symbolic model checker SMCDEL can deal with scenarios up to 40 agents in less than a second.

The model checker is implemented in Haskell and can be used similarly to DEMO-S5. To represent BDDs we use CacBDD [27] via the binding library

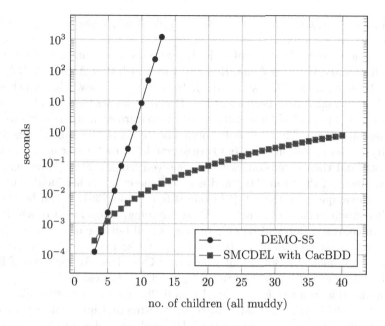

Fig. 3. Benchmark Results on a logarithmic scale.

HasCacBDD [19]. The program can also be used with CUDD [18,29] which provides very similar performance. All experiments were done using 64-bit Debian GNU/Linux 8.0 with kernel 3.16.0-4, GHC 7.8.3 and g++ 4.9 on an Intel Core i3-2120 3.30 GHz processor and 4 GB of memory.

Muddy Children has also been used to benchmark MCMAS [24] but the formula checked there concerns the correctness of behavior and not how many rounds are needed. Moreover, the interpreted system semantics of model checkers like MCMAS are very different from DEL. Still, connections between DEL and temporal logics have been studied and translations are available [3,13].

A scenario which fits nicely into both frameworks is the dining cryptographers protocol [7]. The statement "If cryptographer 1 did not pay the bill, then after the announcements are made, he knows that no cryptographers paid, or that someone paid, but in this case he does not know who did." is also checked in [24]. It can be formalized in DEL as follows where p_i says that agent i paid and ψ is the announcement: $\neg p_1 \rightarrow [\psi](K_1(\bigwedge_{i=1}^{n} \neg p_i) \vee (K_1(\bigvee_{i=2}^{n} p_i) \wedge \bigwedge_{i=2}^{n}(\neg K_1 p_i)))$. SMCDEL can check this for $n = 50$ in less than a second. Proper benchmarks and comparisons of all parameters will be done in the future.

6   Example 2: Russian Cards

As a second case study we applied our symbolic model checker to the Russian Cards Problem. One of its first logical analyses is [10] and the problem has since gained notable attention as an intuitive example of information-theoretically (in contrast to computationally) secure cryptography [9,14].

The basic version of the problem is this: Seven cards, enumerated from 0 to 6, are distributed between Anne, Bob and Crow such that Anne and Bob both receive three cards and Crow one card. It is common knowledge which cards exist and how many cards each agent has. Everyone knows their own but not the others' cards. The goal of Anne and Bob now is to learn each others cards without Crow learning them. They can only communicate via public announcements.

Many different solutions exist but here we will focus on the so-called five-hands protocols (and their extensions with six or seven hands): First Anne makes an announcement of the form "My hand is one of these: ...". If her hand is 012 she could for example take the set $\{012, 034, 056, 135, 146, 236\}$. It can be checked that this announcement does not tell Crow anything, independent of which card it has. In contrast, Bob will be able to rule out all but one of the hands in the list depending on his own hand. Hence the second and last step of the protocol is an announcement by Bob about which card Crow has. For example, if Bob's hand is 345 he would finish the protocol with "Crow has card 6.".

Verifying this protocol for the fixed deal 012|345|6 with our symbolic model checker takes less than a second. Moreover, checking multiple protocols in a row does not take much longer because the BDD package caches results. Compared to that, a DEMO implementation [12] needs 4 seconds to check one protocol.

We can not just verify but also find all 5/6/7-hands protocols, using a combination of manual reasoning and brute-force. By Proposition 32 in [10] safe announcements from Anne never contain "crossing" hands, i.e. two hands with multiple card in common. If we also assume that the hands are lexicographically ordered, this leaves us with 1290 possible lists of five, six or seven hands of three cards. Only some of them are safe announcements which can be used by Anne. We can find them by checking all the corresponding 1290 formulas. Our model checker can filter out the 102 safe announcements within 1.6 seconds, generating and verifying the same list as in [10] where it was manually generated.

7   Equivalence of S5 Kripke Models and Knowledge Structures

Having shown the computational advantage of our new knowledge models, we now look more deeply into the foundations of what we have been doing. For a start, we show that knowledge structures and standard models for DEL are equivalent from a semantic point of view. Lemma 1 gives us a canonical way to show that a knowledge structure and an S5 Kripke model satisfy the same formulas. Theorems 2 and 3 say that such equivalent models and structures can always be found. These translations are also implemented in SMCDEL.

Lemma 1. *Suppose we have a knowledge structure $\mathcal{F} = (V', \theta, O_1, \cdots, O_n)$ and a finite S5 Kripke model $M = (W, \pi, \mathcal{K}_1, \cdots, \mathcal{K}_n)$ with a set of primitive propositions $V \subseteq V'$. Furthermore, suppose we have a function $g : W \to \mathcal{P}(V')$ such that*

C1 For all $w_1, w_2 \in W$, and all i such that $1 \leq i \leq n$, we have that $g(w_1) \cap O_i = g(w_2) \cap O_i$ iff $w_1 \mathcal{K}_i w_2$.

C2 For all $w \in W$ and $v \in V$, we have that $v \in g(w)$ iff $\pi(w)(v) = \mathbf{true}$.

C3 For every $s \subseteq V'$, s is a state of $\mathcal{F}$ iff $s = g(w)$ for some $w \in W$.

Then, for every formula φ over V we have $(\mathcal{F}, g(w)) \models \varphi$ iff $(M, w) \models \varphi$.

Proof. By induction on φ: Use C2 for atomic propositions, note that the boolean semantics are the same, use C1 and C3 for the knowledge operator and show that the conditions carry over to the results of announcements.

We do not give details here because the proof does not provide any new insights: Conditions C1 to C3 describe a special case of a p-morphism between M and the Kripke model encoded by $\mathcal{F}$, see Definition 7 below. Hence their equivalence with respect to the modal language already follows from general invariance results in modal logic [4, §2.1]. The following definition and theorem show that for every knowledge structure there is an equivalent Kripke model.

Definition 7. *For any* $\mathcal{F} = (V, \theta, O_1, \cdots, O_n)$, *we define the Kripke model* $M(\mathcal{F}) := (W, \pi, \mathcal{K}_1, \cdots, \mathcal{K}_n)$ *as follows*

1. W *is the set of all states of* $\mathcal{F}$,
2. *for each* $w \in W$, *let the assignment* $\pi(w)$ *be* w *itself and*
3. *for each agent* i *and all* $w, w' \in W$, *let* $w \mathcal{K}_i w'$ *iff* $w \cap O_i = w' \cap O_i$.

Theorem 2. *For any knowledge structure* $\mathcal{F}$, *any state* s *of* $\mathcal{F}$, *and any* φ *we have* $(\mathcal{F}, s) \models \varphi$ *iff* $(M(\mathcal{F}), s) \models \varphi$.

Proof. By Lemma 1 using the identity function as g.

Vice versa, for any S5 Kripke model we can find an equivalent knowledge structure. The essential idea is to add propositions as observational variables to encode the relations of each agent. To obtain a simple knowledge structure we should add as few propositions as possible. The method below adds $\sum_{i \in I} \mathsf{ceiling}(\log_2 k_i)$ propositions where k_i is the number of $\mathcal{K}_i$-equivalence classes and $\mathsf{ceiling}(\cdot)$ denotes the smallest integer not less than the argument. This could be further improved if one were to find a general way of using the propositions already present in the Kripke model as observational variables directly.

Definition 8. *For any S5 model* $M = (W, \pi, \mathcal{K}_1, \cdots, \mathcal{K}_n)$ *we define a knowledge structure* $\mathcal{F}(M)$ *as follows. For each* i, *write* $\gamma_1, \ldots, \gamma_{k_i}$ *for the equivalence classes given by* $\mathcal{K}_i$ *and let* $l_i := \mathsf{ceiling}(\log_2 k_i)$. *Let* O_i *be a set of* l_i *many fresh propositions. This yields the sets of observational variables* $O_1, \ldots, O_n$, *all disjoint to each other. If agent* i *has a total relation, i.e. only one equivalence class, then we have* $O_i = \varnothing$. *Enumerate* k_i *many subsets of* O_i *as* $O_{\gamma_1}, \ldots, O_{\gamma_{k_i}}$ *and define the function* $g_i : W \to \mathcal{P}(O_i)$ *by* $g_i(w) := O_{\gamma(w)}$ *where* $\gamma(w)$ *is the equivalence class of* w. *Let* $V' := V \cup \bigcup_{0 < i \leq n} O_i$ *and define* $g : W \to \mathcal{P}(V')$ *by*

$$g(w) := \{v \in V \mid \pi(w)(v) = \top\} \cup \bigcup_{0 < i \leq n} g_i(w)$$

Let $\overline{V'}$ be the set of atomic propositions and their negations from V'. Finally, let $\mathcal{F}(M) := (V', \theta_M, O_1, \ldots, O_n)$ where

$$\theta_M := \bigwedge\left\{\bigvee Q \mid Q \subseteq \overline{V'} \text{ and } g(w) \models \bigvee Q \text{ for all } w \in W\right\}$$

Theorem 3. *For any finite S5 pointed Kripke model (M, w) and every formula φ, we have that $(M, w) \models \varphi$ iff $(\mathcal{F}(M), g(w)) \models \varphi$.*

Proof. By Definition 8, g_i is such that for all $w_1, w_2 \in W$, $g_i(w_1)$ and $g_i(w_2)$ are the same subset of O_i iff w_1 and w_2 are in the same equivalence class of $\mathcal{K}_i$. It is therefore easy to check the first two conditions of Lemma 1. For the "if" part of C3: If $s = g(w')$ for some $w' \in W$, then by the definition of θ_M, we have that $g(w') \models \theta_M$ and hence $g(w')$ is a state of $\mathcal{F}(M)$. For the "only if" part, assume that for every $w \in W$, $s \neq g(w)$. Then, for every $w \in W$, there is an atomic formula φ_w over V' such that $s \models \varphi_w$ but $g(w) \models \neg\varphi_w$. Therefore, $s \models \bigwedge_{w \in W} \varphi_w$. Moreover, we have for every $w' \in W$, $g(w') \models \bigvee_{w \in W} \neg\varphi_w$, and hence $\bigvee_{w \in W} \neg\varphi_w \in \Gamma_M$. Consequently, we have $s \not\models \Gamma_M$ and hence s is not a state of $\mathcal{F}(M)$. Now the theorem follows from Lemma 1.

What we have seen is how the two ways of modeling in this paper, though computationally different, are semantically equivalent. This leads us to consider how their interplay will work in more complex settings. The obvious direction to probe this is the area where DEL unleashes its full power: We now give an outlook how knowledge structures can be generalized to action models. They were first described in [1] and we do not repeat definitions here but refer to [11] for a textbook treatment. What action models are to Kripke frames, the following knowledge transformers are to knowledge structures.

Definition 9. *A* knowledge transformer *for a given vocabulary V is a tuple $\mathcal{X} = (V^+, \theta^+, O_1, \ldots, O_n)$ where V^+ is a set of atomic propositions such that $V \cap V^+ = \varnothing$, θ^+ is a possibly epistemic formula over $V \cup V^+$ and $O_i \subseteq V^+$ for all agents i. An* event *is a knowledge transformer together with a subset $x \subseteq V^+$, written as $(\mathcal{X}, x)$.*

The knowledge transformation *of a knowledge structure $\mathcal{F} = (V, \theta, O_1, \ldots, O_n)$ with a knowledge transformer $\mathcal{X} = (V^+, \theta^+, O_1^+, \ldots, O_n^+)$ for V is defined by:*

$$\mathcal{F}^{\mathcal{X}} := (V \cup V^+, \theta \wedge \|\theta^+\|_{\mathcal{F}}, O_1 \cup O_1^+, \ldots, O_n \cup O_n^+)$$

Given a scene $(\mathcal{F}, s)$ and an event $(\mathcal{X}, x)$ we define $(\mathcal{F}, s)^{(\mathcal{X}, x)} := (\mathcal{F}^{\mathcal{X}}, s \cup x)$.

The two kinds of events discussed above fit well into this general definition: The public announcement of φ is the event $((\varnothing, \varphi, \varnothing, \ldots, \varnothing), \varnothing)$ and the announcement of φ to Δ is given by $((\{p_\varphi\}, p_\varphi \to \varphi, O_1^+, \ldots, O_n^+), \{p_\varphi\})$ where $O_i^+ = \{p_\varphi\}$ if $i \in \Delta$ and $O_i^+ = \varnothing$ otherwise.

Theorem 4. *For any S5 action model there is an equivalent knowledge transformer and vice versa.*

Proof. Define translations similar to Definitions 7 and 8. Then use Lemma 1.

Finally, Definition 6 can be extended to cover event operators: Let $\|[\mathcal{X}, x]\varphi\|_{\mathcal{F}} := \|\theta_x^+\|_{\mathcal{F}} \to \|\varphi'\|_{\mathcal{F}^{\mathcal{X}}}$ where $\theta_x^+ := \theta^+\left(\frac{x}{\top}\right)\left(\frac{V^+\setminus x}{\bot}\right)$ and $\varphi' := \varphi\left(\frac{x}{\top}\right)\left(\frac{V^+\setminus x}{\bot}\right)$.

8   Conclusion and Future Work

We have achieved our goal of putting a new engine into DEL by a suitable semantic model transformation. This was shown to work well in various benchmarks, for example the Muddy Children and Russian cards. But there is obviously more to be explored now that we know this. In future work we aim to extend our theoretical framework and the implementation in different directions.

One line would be to use the same models with richer languages, and see whether the parallels that we found still persist. For example, action models with factual change [2] should also be representable as knowledge transformers. They also motivate a new notion of action equivalence which might help to solve a problem with action models where bisimulation had to be replaced with the more complicated notion of action emulation [16].

Another direction would be to extend the framework to other dynamic phenomena such as belief change or preference change which are usually non-S5. For this we can use the literature on abstraction for transition systems, starting with the seminal [8]. Moreover, BDDs have already been used to model belief change in [21]. Also abstraction ideas from the DEL literature could be implemented and their performance compared, for example the very compact modeling of Muddy Children in [20] and the mental programs from [6].

But perhaps the deepest issue that we see emerging in our approach is this. While standard logical approaches to information flow assume a sharp distinction between syntax and semantic models, our BDD-oriented approach suggests the existence of a third intermediate level of representation combining features of both that may be the right level to be at, also from a cognitive viewpoint. We leave the exploration of the latter grander program to another occasion.

Acknowledgements. This work was partially supported by NSFC grant 61472369 and carried out within the Tsinghua-UvA Joint Research Center in Logic. We thank our anonymous referees for useful comments and suggestions.

References

1. Baltag, A., Moss, L.S., Solecki, S.: The logic of public announcements, common knowledge, and private suspicions. In: Bilboa, I. (ed.) TARK 1998, pp. 43–56 (1998)
2. van Benthem, J., van Eijck, J., Kooi, B.: Logics of communication and change. Information and Computation 204(11), 1620–1662 (2006)
3. van Benthem, J., Gerbrandy, J., Hoshi, T., Pacuit, E.: Merging frameworks for interaction. Journal of Philosophical Logic 38(5), 491–526 (2009)
4. Blackburn, P., de Rijke, M., Venema, Y.: Modal Logic. In: Cambridge Tracts in Theoretical Computer Science, no. 53. CUP, Cambridge (2001)
5. Bryant, R.E.: Graph-Based Algorithms for Boolean Function Manipulation. IEEE Transaction on Computers C-35(8), 677–691 (1986)
6. Charrier, T., Schwarzentruber, F.: Arbitrary public announcement logic with mental programs. In: Proceedings of the 2015 International Conference on Autonomous Agents and Multiagent Systems, pp. 1471–1479. IFAAMAS (2015)

7. Chaum, D.: The dining cryptographers problem: Unconditional sender and recipient untraceability. Journal of Cryptology 1(1), 65–75 (1988)
8. Clarke, E.M., Grumberg, O., Long, D.E.: Model checking and abstraction. ACM Transactions on Programming Languages and Systems 16(5), 1512–1542 (1994)
9. Cordón-Franco, A., van Ditmarsch, H., Fernández-Duque, D., Soler-Toscano, F.: A geometric protocol for cryptography with cards. Designs, Codes and Cryptography 74(1), 113–125 (2015), http://dx.doi.org/10.1007/s10623-013-9855-y
10. van Ditmarsch, H.: The russian cards problem. Studia Logica 75(1), 31–62 (2003)
11. van Ditmarsch, H., van der Hoek, W., Kooi, B.: Dynamic epistemic logic, vol. 1. Springer, Heidelberg (2007)
12. van Ditmarsch, H., van der Hoek, W., van der Meyden, R., Ruan, J.: Model Checking Russian Cards. Electr. Notes Theor. Comput. Sci. 149(2), 105–123 (2006)
13. van Ditmarsch, H., van der Hoek, W., Ruan, J.: Connecting dynamic epistemic and temporal epistemic logics. Logic Journal of IGPL 21(3), 380–403 (2013)
14. Duque, D.F., Goranko, V.: Secure aggregation of distributed information. CoRR abs/1407.7582 (2014), http://arxiv.org/abs/1407.7582
15. van Eijck, J.: DEMO-S5. Tech. rep., CWI (2014)
16. van Eijck, J., Ruan, J., Sadzik, T.: Action emulation. Synthese 185(1), 131–151 (2012)
17. Fagin, R., Halpern, J.Y., Moses, Y., Vardi, M.Y.: Reasoning about knowledge, vol. 4. MIT Press, Cambridge (1995)
18. Gammie, P.: hBDD. https://github.com/peteg/hBDD (2011, updated 2014)
19. Gattinger, M.: HasCacBDD (2015), https://github.com/m4lvin/HasCacBDD
20. Gierasimczuk, N., Szymanik, J.: A note on a generalization of the Muddy Children puzzle. In: Apt, K.R. (ed.) TARK 2011, pp. 257–264. ACM (2011)
21. Gorogiannis, N., Ryan, M.D.: Implementation of Belief Change Operators Using BDDs. Studia Logica 70(1), 131–156 (2002)
22. Knuth, D.E.: The Art of Computer Programming. Combinatorial Algorithms, Part 1, vol. 4A. Addison-Wesley Professional (2011)
23. Littlewood, J.: A Mathematician's Miscellany. Methuen, London (1953)
24. Lomuscio, A., Qu, H., Raimondi, F.: MCMAS: an open-source model checker for the verification of multi-agent systems. International Journal on Software Tools for Technology Transfer, 1–22 (2015)
25. Lomuscio, A.R., van der Meyden, R., Ryan, M.: Knowledge in Multiagent Systems: Initial Configurations and Broadcast. ACM Trans. Comp. L. 1(2), 247–284 (2000)
26. Luo, X., Su, K., Sattar, A., Chen, Y.: Solving Sum and Product Riddle via BDD-Based Model Checking. In: Web Intel./IAT Workshops, pp. 630–633. IEEE (2008)
27. Lv, G., Su, K., Xu, Y.: CacBDD: A BDD Package with Dynamic Cache Management. In: Sharygina, N., Veith, H. (eds.) CAV 2013. LNCS, vol. 8044, pp. 229–234. Springer, Heidelberg (2013)
28. van der Meyden, R., Su, K.: Symbolic Model Checking the Knowledge of the Dining Cryptographers. In: CSFW, pp. 280–291. IEEE Computer Society (2004)
29. Somenzi, F.: CUDD: CU Decision Diagram Package Release 2.5.0 (2012)
30. Su, K., Sattar, A., Luo, X.: Model Checking Temporal Logics of Knowledge Via OBDDs. The Computer Journal 50(4), 403–420 (2007)

Three-Valued Plurivaluationism
of Vague Predicates

Wen-Fang Wang

Institute of Philosophy of Mind and Cognition
National Yang Ming University, Taipei, Taiwan
wfwang@ym.edu.tw

Abstract. Disagreeing with most authors on vagueness, the author proposes a solution that he calls "three-valued plurivaluationism" to the age-old sorites paradox. In essence, it is a three-valued semantics for a first-order language with identity with the additional suggestion that a vague language has more than one correct interpretation. Unlike the traditional three-valued approach to a vague language, the so-called three-valued plurivaluationism, so the author argues, can accommodate the phenomenon of higher-order vagueness. And, unlike the traditional three-valued approach to a vague language, the so-called three-valued purivaluationism, so the author argues, can also accommodate the phenomenon of penumbral connection when equipped with "suitable conditionals". The author also shows that this three-valued purivaluationism is a natural consequence of a restricted form of Tolerance Principle (T_R) and a few related ideas, and argues that (T_R) is well-motivated by considerations of how we learn, teach, and use vague predicates.

Keywords: vagueness, sorites paradox, tolerance principle, three-valued semantics, plurivationism, conditionals.

1 Vague Predicates and the Sorites Paradox

A vague predicate is a predicate that has *possible* borderline cases, i.e., possible cases such that it is semantically indeterminate whether or not the predicate applies. Examples of vague predicates abound in natural languages. Here is just a short list of examples in English: "bald", "heap", "tall", "red", "table", "a small portion of C" (where C is a class of, say, 50 students), "similar to", "identical with", and so on.

A problem about vague predicates is that they give rise to the sorites paradox. Let "a_i" be a name of someone with i hairs. Then from the apparently plausible premises "a_0 is bald" and "if a_n is bald, so is a_{n+1}, for whatever number n", one can infer the absurd conclusion that "$a_{100,000}$ is bald". Or, to take another example from [18], let b_0 be me, and suppose that there are n molecules in my body. Let $b_0, b_1, ..., b_n$ be a sequence of objects each of which is obtained from its predecessor by replacing one molecule of me with a molecule of scrambled egg, so that b_n is all scrambled egg. Let β_i be the statement that "$b_{i-1} = b_i$". Then

© Springer-Verlag Berlin Heidelberg 2015
W. van der Hoek et al. (Eds.): LORI 2015, LNCS 9394, pp. 379–391, 2015.
DOI: 10.1007/978-3-662-48561-3_31

each β_i, where $1 \leq i \leq n$, seems to be true. Yet, by n applications of the rule of transitivity of identity, we reach the absurd conclusion that I am scrambled egg. From the fact that one can "prove" almost everything s/he wants to prove by a sorties argument, of course we should conclude that some of these sorites arguments must be unsound. However, it has been proved very difficult both to pinpoint the problem(s) of these arguments and to give a plausible explanation of why we are taken in.

In the past 40 years, philosophers have witnessed a bunch of theories aiming at solving the sorites paradox[1]. A benefit/cost analysis of even a small portion of these theories will be an impossible task for a short paper like the present one. This paper suggests that we start from scratch to re-think about how we learn, teach, and use vague predicates and hopes that we will gain some insights from such an inspection.

2   Start from Scratch

Before we start, however, let me give a few preliminary comments. In the very beginning of this paper, I defined "a vague predicate" to be a predicate "having *possible* borderline cases", but why possible cases? Why not define a vague perdicate in terms of its actual borderline cases? Here is the reason. If we call a predicate "vague" only when it *actually* has some borderline cases, then some predicates that are intuitively vague will not be "vague" in the defined sense, and this seems undesirable. For example, if we define an F-snail to be a snail that walks much faster than most slow turtles, then intuitively this notion of *F-snails* is a vague one so long as notions of *much faster than, most*, and *slow* are. Yet, surely nothing in the world is an F-snail (or, if this is not the case, replace the word "turtle" by "panther"), so there is no actual borderline case for this intuitively vague predicate. As a result, the notion of *F-snails* turns out not to be "vague" in the new, defined sense, and this seems undesirable. Here is another example. If we define a baldsome male to be a male who is both very bald and very handsome, then, again, intuitively this notion of *baldsome males* is a vague one so long as both the notion of *very bald* and that of *very handsome* are. However, it may happen that there are a few men that are clearly both very bald and clearly very handsome while all others are either clearly not-very-bald (though some of them may be vaguely very handsome) or clearly not-very-handsome (though some of them may be vaguely very bald), so there will be, in this case, no actual borderline case for this intuitively vague predicate.[2] As a result, the

---

[1] To name just a few: epistemicism proposed by [3], [2], [28], and [23], gap theories proposed by [8] and [13], glut theories proposed by [18] and [10], supervaluationism proposed by [6], [14], [11], [4], [19], [1], and [12], fuzzy theories proposed by [17] and [22], plurivaluationism proposed by [27] and [16], and contextualist theories proposed by [25], [20], [5], and [21].

[2] On the other hand, it is not clear that predicates like "bald but not self-identical", "bald or self-identical", and "tall or greater-than-or-equal-to exactly four feet in height" are vague predicates, for it is impossible for these predicates to have borderline cases.

notion of *baldsome males* turns out not to be "vague" in the new, defined sense either, and this seems equally undesirable. So, if we want to characterize vague predicates as predicates having borderline cases, it seems better that we take into account all possible cases as well as all actual ones of these predicates. Or, I may even put things in this way: what I will call "the extension" of a vague predicate in this paper may be called by other philosophers its "intension", but I don't think that there will be anything important that hinges on this difference.

There are different kinds of vague predicates; especially, there are primitive ones as well as defined ones, and, among each category, there are perceptual ones as well as non-perceptual ones. So which ones will I be talking about in this paper? I intend the semantics proposed in sections 3 and 4 to be applicable to vague predicates in general, but I will restrict my discussion in this section to primitive ones to make my exposition simpler. As examples of primitive vague predicates, I take "red", "bald", "soft" (these are vague perceptual predicates), "a small portion of C" (where C is a class of 50 students), and "identity" (these are non-perceptual ones). Examples of non-primitive vague predicates include, on the other hand, "4-tall" (exactly 4 feet in height or is tall), "baldsome" (very bald but very handsome), and "F-snails" (snails that walks much faster than most slow turtles). If you think that some of these examples are wrongly classified, be my guest to adjust the classification by yourself. Again, I do not think that there will be anything important that hinges on the choice.

The notion of F-*relevant respects* of a vague predicate F will play an important role in what follows, so we'd better get a good grip of it now. Let me begin with the notion of *determination*. A set of respects (properties or relations) $R_1, ..., R_n$ determines a certain respect R (property or relation) iff, for all possible objects (or sequences of possible objects) α and β, it is necessarily the case that if α and β are exactly the same with respect to $R_1, ..., R_n$ then they are exactly the same with respect to R (and so the two sentences "Ra" and "Rb" will have the same truth value (either both are true, or both are false, or both are neither), where "a" and "b" are names of α and β (or are sequences of names of objects in α and β)). It can easily be proved that if a set S of respects determines a respect R, so does any superset S' of S, so the notion of determination is not a very useful one. For the purpose of defining "F-relevant respects" of a vague predicate F, we need a tighter notion than that of *determination*: *m-determination* (short for "minimal determination"). A set S of respects (properties or relations) $R_1, ..., R_n$ m-determines a certain respect R (property or relation) iff (a) S determines R, and (b) for any set $T = \{R'_1, ..., R'_m\}$ that also determines R, the set T "entails" the set S in the sense that, for any possible object (or sequence of possible objects) α, it is necessarily true that if α has every respect in T then it also has every respect in S. An F-relevant respect of a vague predicate F is then a respect in any set S that m-determines whether something is F. Given this definition of a relevant respect of a vague predicate, it is easy to see that even a primitive vague predicate, such as "red", may have multiple relevant respects, such as hue, value, and chroma. A semantically primitive vague predicate may therefore stand for an ontologically complex property, i.e., a property whose existence ontologically depends on the existence of several simpler properties.

With these preliminaries in mind, we are now in a position to investigate how we learn, teach, and use vague predicates. In general, I believe that the following story is a rough but faithful picture about how we learn and teach vague predicates. When learning or teaching how to use a primitive vague predicate F, we do so by means of ostension (what else can we do?), i.e., by giving or by being given examples or "paradigms", both positive and negative ones, of F. Moreover, some paradigms are introduced explicitly by pointing to them or by showing them, while others are introduced implicitly by hints or by implicatures. For examples, we may point to a few heads and call them "bald", at the same time implicitly implying or implicating that heads with fewer hairs or heads whose numbers of hairs are between those of the paradigms are also paradigms of bald heads. Another example: we may illustrate the use of "a small portion of C", where C is a class of 50 students, by saying loud that "a subset of C with 5 or less members is a small portion of C", at the same time implicitly implying or implicating that a set of C with 45 or more members are negative paradigms of the predicate. Because we all teach and learn a primitive vague predicate F in this standard ostensive way, each competent speaker of F, i.e., one who understands how to use F correctly, will have both some positive paradigms and some negative paradigms of F in his or her mind. Moreover, it seems that nothing in the process of teaching and learning F can be both a positive and a negative paradigm of F on pain of confusion.

However, in order for the teaching and learning process of a vague predicate F to be successful, the difference in F-relevant respects between any positive paradigm and any negative paradigm of F must be "salient" to the learner. Otherwise, it is hard to imagine how the learner can even re-identify a positive (or negative) paradigm of F as a positive (or negative) paradigm of F again, let alone has an idea about how to make further applications of the predicate F. We say that two paradigms of F differ saliently in F-relevant respects to a subject S in an occasion O iff the overall dissimilarity between them in F-relevant respects is easily observable for S in O or is intellectually significant for S in O. The requirement that it is easily observable for S in O is tailored especially for vague perceptual predicates, such as "red", so that, according to this requirement, the overall difference in red-relevant respects between a paradigm red patch and a paradigm not-red one must be easily observable to the learner when the predicate is learned. The requirement that it is intellectually significant for S in O, on the other hand, is tailored especially for non-perceptual vague predicates, such as "is a small portion of C", so that, according to this requirement, the overall difference in a-small-portion-of-C-relevant respects between, say, a 5-membered subset of C and a 45-membered subset of C must be intellectually significant to the learner, and presumably the intellectual significance in this case may simply consist in the fact that the difference between the ratios of the two subsets to C is close to 1 or at least much greater than a half.

So far, there is no guarantee that two competent speakers of a vague predicate F will have any common positive paradigm or any common negative paradigm in their minds, and this seems to make the publicity of a vague language problematic. Fortunately, because people have roughly, though not exactly, the same perceptual and intellectual capacities, and also because many positive and negative paradigms of a vague predicate F are implicitly introduced when teaching or learning F, all competent speakers of F ultimately share at least some common paradigms, both positive and negative ones, of F in their minds. This is not to deny that the perceptual and intellectual capacities that one has differ from person to person and from occasion to occasion. But this fact should not lead us to overlook another equally important fact that our perceptual and intellectual capacities are very similar after all. (Another important fact about our perceptual and intellectual capacities is this: we are all limited creatures; our abilities of discernment and our intellectual swiftness and astuteness are all very limited, so that, for example, no one can really discriminate a large number of "borderline shades of colors" between a positive paradigm and a negative paradigm of redness. I will not emphasize this important fact here, but I will come back to it when I consider the problem of "gradual transition" in next section.) Thus, if it makes sense at all to assign an extension F^+ and an anti-extension F^- to a vague predicate F, at least these *common* positive paradigms of F should be included in the extension F^+, and at least these *common* negative paradigms of F should be included in the anti-extension of F^-. And, from what we have said two paragraphs ago, it is also reasonable to assume that these two extensions of a vague predicate are mutually exclusive.

As I see it, the most distinguished feature of any vague predicate, in contrast with a precise predicate, is the existence of a "sorites sequence" for the predicate: for any occasion O and any two paradigms a_1 and a_n of a vague predicate F, and for any competent speaker S of F, there always is a sequence of possible cases $< a_1, ..., a_n >$ between a_1 and a_n such that any two adjacent cases in the sequence are "very similar" to S in F-relevant respects in O in the sense that the overall dissimilarity in F-relevant respects between them is not observable or is intellectually insignificant for S in O.[3] Now, a vague predicate F must allow its competent users to be able to apply and re-apply it, not only to those positive and negative paradigms that are introduced in the learning and teaching process, but also to possible cases beyond these paradigms (this is also true of most precise predicates), otherwise, it will not be a vague predicate at all but belongs to a very special kind of precise predicates. (Consider Fine's example:

---

[3] This feature does not seem to me to be owned by any precise predicate, perhaps because the F-relevant respects of a precise predicate F are just those respects specified in the definition of F, so everything falling within F differs saliently, observationally or intellectually, in F-relevant respects from everything falling out of F. (Consider the case of the precise predicate "is an even number".) As a result, even if one can fine a sequence of possible cases $< a_1, ..., a_n >$ between a positive paradigm a_1 and a negative one a_n such that any two adjacent cases in it are very similar in some respects, there still will be two adjacent cases in the sequence that differ saliently in F-relevant respects.

a number is an F if it is smaller than or equal to 13 and is not an F if it is greater than or equal to 15. Defined in this way, this predicate F will have no further possible application beyond those paradigms that are introduced in this definition, but it will not be regarded as a vague predicate by most philosophers either.) For precise predicates, the possibility of further applications is given by their definitions. But this is obviously not the case for primitive vague predicates. So, by what rule (or rules) does a competent speaker of a vague predicate F extend its use to cases other than those introduced in the learning process? To this question, I suggest[4] the following answer: every competent speaker S of a vague predicate F tacitly accepts the following "restricted tolerance principle" (T_R):

> (T_R): If it is correct for a subject S to classify x as a member of F^+ (or F^-) in an occasion O and y and x are "very similar" for S in O, then it is also correct for S to classify y as a member of F^+ (or F^-) in O, *so long as, after so classified, the difference in F-relevant respects between any member of F^+ and any member of F^- remains observationally or intellectually salient for S in O.*

In short, I believe that the following statements 1-7 jointly constitute a roughly true story about how we learn, teach and use primitive vague predicates:

1. We learn and teach how to use a vague predicate F by ostension. Some paradigms of F are introduced explicitly by pointing to them or by showing them, while others are introduced implicitly by hints or by implicatures.
2. Due to the way we learn and teach a vague predicate F, each competent speaker of F will have in mind some positive paradigms and some negative paradigms of F.
3. For any competent speaker of F, the difference in F-relevant respects between any positive paradigm and any negative paradigm of F must be either perceptually or intellectually salient.
4. We have roughly, though not exactly, the same perceptual and intellectual capacities.
5. Due to facts 1 and 4, all competent speakers of a vague predicate F share at least *some* common positive paradigms that belong to the extension F^+ of

---

[4] I do not just suggest (T_R), but also think that it is supported by at least three arguments. First, not only is (T_R) true of vague predicates, it is also true of precise predicates if we interpret "very similar" in it as "having or lacking the same defining properties". So (T_R) seems to be a principle for predicates in general. Second, (T_R) is a logically weaker principle than Wright's tolerance principle (T): If it is correct for S to classify x as a member of F^+ (or F^-) in O and y and x are "very similar" in F-relevant respects, then it is also correct to classify y as a member of F^+ (or F^-) in O. So evidences for (T) are automatically evidences for (T_R), and [29] did provide a few good evidences for (T). Finally, I believe that (T_R) can better explain, while (T) cannot, the phenomenon that is found in the "forced march sorites paradox" [9], but I will leave the justification of this explanatory power of (T_R) to another paper due to its complicated nature.

F and *some* common negative paradigms that belong to the anti-extension F⁻ of F.

6. For any occasion O and any two paradigms a_1 and a_n of a vague predicate F, and for any competent speaker S of F, there always is a sequence of possible cases $< a_1, ..., a_n >$ between a_1 and a_n such that any two adjacent cases in the sequence are "very similar" to S in F-relevant respects in O in the sense that the overall dissimilarity in F-relevant respects between them is not observable or is intellectually insignificant for S in O.

7. Every competent speaker S of a vague predicate F tacitly accepts the restricted tolerance principle (T_R): If it is correct for a subject S to classify x as a member of F^+ (or F^-) in an occasion O and y and x are "very similar" for S in O, then it is also correct for S to classify y as a member of F^+ (or F^-) in O, *so long as, after so classified, the difference in F-relevant respects between any member of F^+ and any member of F^- remains observationally or intellectually salient for S at O.*

However, if statements 1-7 are correct, then it follows that:

8. For any vague predicate F, $F^+ \cup F^-$ is not equal to the set of everything that the predicate F can meaningfully apply, for any competent speaker S of F.

9. Due to 4 and 7, the extension F^+ and the anti-extension F^- of F may be different for different competent speakers of F, though there is a common "core" for all competent speakers.

10. Although 9, so long as one's assignment of F^+ and F^- to F obeys (T_R) and some other "natural restrictions", his or her interpretation of F is correct.

11. Due to 8, a correct interpretation of a vague language L must be a three-valued interpretation and there seems to be no reason for having more than three values. Due to 10, there can be more than one correct interpretation of a vague language L.

3   Let's Get a Bit Formal

Let L be a first-order language with identity sign, vague predicates, and connectives "¬", "∧" and "∨". A model $M = <D_M, VI_M, v_M>$ for L is a triple that satisfies the following conditions:

1. D_M is a non-empty set.
2. VI_M is a subset of $D_M{}^2$, where (i) for any $< \alpha, \beta >$ that belongs to VI_M, α is not the same as β, (ii) if $< \alpha, \beta >$ belongs to VI_M, so does $< \beta, \alpha >$, and (iii) if $< \alpha_1, ..., \alpha_{i-1}, \alpha, \alpha_{i+1}, ..., \alpha_n > \in F_M{}^+$ while $< \alpha_1, ..., \alpha_{i-1}, \beta, \alpha_{i+1}, ..., \alpha_n > \in F_M{}^-$ for some n-place predicate F, $< \alpha, \beta > \notin VI_M$.
3. v_M assigns to each constant of L a member of D_M to be its value and assigns to each n-place predicate F a pair of sets $<F_M{}^+, F_M{}^->$ of n-tuples of members of D_M such that $F_M{}^+ \cap F_M{}^- = \emptyset$

Intuitively, VI_M specifies a relation of "vague identity" that is both irreflexive and symmetric on the domain D_M and never invalidates Leibiz's Law. Given a model M, we define the concept of true-in-M ($v_M(A) = 1$ in symbol), that of false-in-M ($v_M(A) = 0$ in symbol), and that of neither-true-nor-false-in-M ($v_M(A) = n$ in symbol) in the usual way:

1. $v_M(Fc_1...c_n) = 1$ if $< v_M(c_1), ..., v_M(c_n) >$ belongs to $F_M{}^+$. $v_M(Fc_1...c_n) = 0$ if $< v_M(c_1), ..., v_M(c_n) >$ belongs to $F_M{}^-$. Otherwise, $v_M(Fc_1...c_n) = n$.
2. $v_M(c_1 = c_2) = 1$ if $v_M(c_1) = v_M(c_2)$. $v_M(c_1 = c_2) = n$ if $< v_M(c_1), v_M(c_n) >$ belongs to VI_M. Otherwise, $v_M(c_1 = c_2) = 0$.
3. Truth-values of compound sentences are determined by the following strong K_3 charts:

p	$\neg p$
1	0
n	n
0	1

$\wedge$	1	n	0
1	1	n	0
n	n	n	0
0	0	0	0

$\vee$	1	n	0
1	1	1	1
n	1	n	n
0	1	n	0

4. $v_M(\forall x_i \phi) = 1$ if $v_M(\phi(c_i)) = 1$ for every constant c_i. $v_M(\forall x_i \phi) = 0$ if $v_M(\phi(c_i)) = 0$ for some constant c_i. Otherwise $v_M(\forall x_i \phi) = n$. (For simplicity, we assume that everything in the domain has a name.)

Again, the notion of validity is defined in the usual way: an argument is valid iff it preserves truth-in-M for every model M.

However, the present approach differs from most semantic theories of vagueness in that it proposes that there is more than one correct (or intended) interpretation of a vague language L, all of which differ only in how vague predicates are to be interpreted. According to what we have said in the previous section, while these intended interpretations may differ in assigning different pairs $<F_M{}^+, F_M{}^->$ to a vague predicate F, these different pairs nevertheless share a "common core". Now, let S be the set of all correct interpretations of a vague language L, it is plausible to assume (call this "Assumption (A)") that S is closed under the following relation:

> Assumption (A): Let A and B be any sentences of L. If there is a model $M \in S$ s.t. $v_M(A) = n$ and there is a model $M' \in S$ s.t. $v_{M'}(B) = n$, then there is a model $M^* \in S$ s.t. $v_{M^*}(A) = v_{M^*}(B) = n$.

In words: if it is correct to classify A as a borderline sentence and it is also correct to classify B as a borderline sentence, then it is correct both to classify A and B as borderline sentences. For the record, I also list below two more assumptions about the set S of all correct interpretations of a vague language L:

> Assumption (B): For any *atomic* sentence A, if there is a model $M \in S$ s.t. $v_M(A) \neq 1$ and there is a model $M' \in S$ s.t. $v_{M'}(A) \neq 0$, then there is a model $M^* \in S$ s.t. $v_{M^*}(A) = n$.

> Assumption (C): The cardinality of S is some finite number. (Or, at least, for any vague predicate F, there are only finite numbers of subsets of S such that the difference in the assignment to F between any two members of the same subset is perceptually or intellectually indistinguishable to everyone.)

Assumption (B) says that, in terms of the terminologies that I am about to introduce, if an *atomic* sentence is neither true *simpliciter* nor false *simpliciter*, then it is correct to interpret it as having no truth value, and I think that it is wholly justified by what we have said in the previous section. As to Assumption D, I think it can also be justified by the fact that our discriminatory and intellectual powers are very limited.

Given a vague language L and the set S of all its correct interpretations, we can define the notions of "true *simpliciter*" and "false *simpliciter*" as follows:

A sentence is true *simpliciter*, iff it is true-in-M for every M in S.

A sentence is false *simpliciter*, iff it is false-in-M for every M in S.

Borderline sentences are then sentences that are neither true *simpliciter* nor false *simpliciter*. (For a further classification of borderline sentences, see below.)

Notice that, even though the definition of the notion of truth *simpliciter* (or falsity *simpliciter*) is similar to that of the notion of supertruth (or superfalsity) of supervaluationism, these two notions differ significantly in at least two respects. First, the former does not, while the latter does, appeal to the notion of a *classical* precisification of a three-valued model for its definition. Second and more importantly, with Assumption (A) and (B), we can prove that all operators that we have met so far are truth-functional in the sense that, e.g., a disjunction is true *simpliciter* iff one of its disjunct is true *simpliciter*. (The proof is left as an exercise to readers.[5]) As a further result, the definition of truth (or falsity) *simpliciter* given here does not, while the notion of supertruth (and superfalsity) does, suffer from the problem of missing witness. For example, with Assumption (A) and (B) in hands, we can show that an existential statement is true *simpliciter* iff one of its instance is true *simpliciter* and that a conjunction is false *simpliciter* iff one of its conjunct is false *simpliciter*.

We can make a further distinction among borderline sentences if we want to. It may or may not happen that a borderline sentence is neither-true-in-M-nor-false-in-M for *every* M of S. When it happens in this way, we call such a sentence "a pure borderline sentence" and the object it mentions "a pure borderline case" of the vague predicate. We say that the kind of vagueness that these sentences and cases have is first-order. However, it may also happen that a borderline sentence is true-in-M for some but not all M of S, or false-in-M for some but not all M of S, or both. When a sentence is true-in-M for some but not all M of S, or false-in-M for some but not all M of S, or both, we call such a sentence "an impure borderline sentence" and the object it mentions "an impure borderline case". We also say that the kind of vagueness that these sentences and cases have is higher-order. Of course, we can make a further distinction among sentences of higher-order vagueness according to their fate in the set S, but there is no need to pursue this line of thought here.

---

[5] I briefly indicate how the proof should go here. With the help of Assumption (A), one can prove by induction that Assumption (B) is true not only for atomic sentences but for all sentences in general. With this general result and Assumption (A) in hands, one can then easily prove that a disjunction is true *simpliciter* iff one of its disjuncts is true *simpliciter*.

The following, then, is my formal "solution" to the sorites paradox, and I suggest the name "three-valued plurivaluationism" for it. In short, three-valued plurivaluationism asserts that a vague language L has more than one correct three-valued interpretation, and it diagnoses the fallacy of a paradoxical sorites argument as follows: in each correct interpretation M of L, there is a premise in the sorites argument that is neither-true-nor-false-in-M; so one of the premises of the sorites argument is not true *simpliciter*. The argument is still valid, as one can easily verify, but it is unsound. Why are we taken in by a paradoxical sorites argument? Traditionally, the reply to this question from a three-valued theorist is mainly this: even though one of its premises is not true *simpliciter*, none of its premises are false *simpliciter* either. Because none of the premises of a paradoxical sorites argument is false *simpliciter*, we are thereby led to think that all of them are true *simpliciter*, and this is how we are taken in. A three-valued plurivaluationist would agree with this reply, but s/he would also add to it: we are led to take the premises of a paradoxical sorites argument to be true, not only because none of them is false *simpliciter*, but also because they are often true-in-M in some, perhaps even in many or most though not in all, correct interpretation M.

4   Objections and Replies

There are two main objections to a three-valued solution to the sorites paradox. First, it may be argued that a three-valued solution overlooks what [6] called the phenomenon of "penumbral connection": logical relations exist between borderline sentences, as illustrated by the following "intuitively true sentences": "Every head is either bald or not bald", "No head is both bald and not bald", "Every head is such that if it is bald then it is bald, and if it is bald then it is either bald or shining". But this penumbral connection, says the objector, is missing in a standard three-valued semantics. Second, it may be said that a three-valued solution faces what [22] called "the jolt problem": vague predicates force a "gradual transition" from truth to falsity, but such a gradual transition cannot be accommodated in a three-valued semantics. I'll begin with the second objection first.

It is not clear to me that a three-valued semantics cannot accommodate a gradual transition from truth to falsity. After all, there are different orders, i.e., first-order and higher-order, of vagueness between truth and falsity as we have seen, so that one cannot directly jump from truth to falsity without passing by all these intermediate borderline sentences. But a three-valued plurivaluationist can actually do better than just having a few intermediaries between truth and falsity. Let S be the set of all correct interpretations of a vague language L. Let S^{A+} (S^{A-}) be the subset of S containing all and only those models such that A is true (false) in them. We define the degree of closeness to truth (or to falsity) of a sentence A $c^+(A)$ (or $c^-(A)$) simply as $|S^{A+}|/|S|(or|S^{A-}|/|S|)$. By these definitions, every sentence A will receive a pair of rational numbers $<c^+(A), c^-(A)>$ between 0 and 1 that measure its degree of closeness to truth and its degree of closeness to falsity separately. A first-order vague sentence A will then be one such that $c^+(A)=c^-(A)=0$, while a higher-order vague sentence may receive any rational number between 0 and 1 as its degree of closeness to truth (or to falsity). This already gives us both a gradual transition from truth to pure borderline cases and a gradual transition from the latter to falsity. However, if one insists that we should have a unique number for "the degree of truth" of a sentence, we may define the degree of truth* of a sentence A to be $(1+c^+(A)-c^-(A))/2$. By this final definition (or any other equally plausible definition), three-valued plurivaluationism will then allow sentences to have a gradual transition from truth* of degree 1 (positive cases) to truth* of degree 0, i.e., falisity* (negative cases). Either way, we will have an explanation of why some people, such as Smith, think that vague predicates force a gradual transition from truth to falsity. (Thank Assumption C, we don't need to worry about the troublesome possibility that $|S|$ may be an infinite cardinal number.)

Turning now to the problem of penumbral connection, the first thing to notice is that what seems to be a datum for penumbral connection to Fine may not seem so to other philosophers. As Smith points out in [21] p.86:[6]

> ...*Consider 'red'. If one indicates a point on a rainbow midway between clear red and clear orange and asks an ordinary speaker the following questions, then in my experience the responses are along the lines indicated:*
> - *"Is the point red?" Umm, well, sort of.*
> - *"Is the point orange?" Umm, well, sort of.*
> - *"But it's certainly not red and orange, right?" Well, no, it sort of is red and orange.*
> - *"OK, well it's definitely red or orange, right?" No, that's what I've been saying, it's a bit of both, the colours blend into one another.*
> *These reactions fit with the recursive assignments of truth values, not the supervaluationist assignments.*

The right thing to conclude from these remarks, I think, is that some of the claimed data for penumbral connection, especially those involving truth-functional connectives, are not genuine data at all. But this is not to deny that some data *are* still genuine, especially those involving conditionals, such as "Every head is such that if it is bald then it is bald" and "Every head is such that if it is bald then it is either bald or shining". However, the fact that these conditionals are indeed true shows only that the connective "if ... then ..." should, as many philosophers think it should, be construed as a non-truth-functional connective for a theorist who prefers a three-valued treatment of a vague language.

[24] and [15] have proposed a very popular way of treating the connective "if ... then ..." as a non-truth-functional connective. According to this line of treatment, a conditional "if A then B" asserts that, to simplify a bit, every closest A-world is also a B-worlds. Following this line of thought, we can define a model for a vague language L to be a 5-tuple $<W_M, D_M, f_M, VI_M, v_M>$, where W_M

---

[6] I also found such a reaction in [26].

is a non-empty set of possible worlds and f_M is a selection function from a world and a sentence (or a proposition) to a set of worlds satisfying a few conditions. What kind of logic we will have for conditionals will then depend on the formal properties we put on the selection function f_M. In most semantic systems that have been proposed, sentences of the forms "If A then A" and "If A then A or B" are valid, as desired. However, here is another simpler suggestion: we may take a conditional to be a claim not in the object language but in the meta-language, so leave the object language L intact. According to this suggestion, a meta-claim "if A then B" ("A $\rightarrow$ B" in symbol) is true *simpliciter* (or false *simpliciter*) iff S^{A+} is a subset of S^{B+} (or a subset of S^{B-}). Otherwise, "A $\rightarrow$ B" is a borderline sentence. Either way, we will have the desired penumbral connection.

References

1. Bennett, B.: Modal Semantics for Knowledge Bases Dealing with Vague Concepts. In: Cohn, A.G., Schubert, L., Shapiro, S. (eds.) Principles of Knowledge Representation and Reasoning: Proceedings of the 6th International Conference, pp. 234–244. Morgan Kaufmann, San Mateo (1998)
2. Campbell, R.: The Sorites Paradox. Philosophical Studies 26, 175–191 (1974)
3. Cargile, J.: The Sorites Paradox. In: Keefe, R., Smith, N.J.J. (eds.) Vagueness: A Reader. MIT Press, Cambridge (1997)
4. Dummett, M.A.E.: Wang's Paradox. Synthese 30, 301–324 (1975)
5. Fara, D.G.: Shifting Sands: An Interest-Relative Theory of Vagueness. Philosophical Topics 28, 45–81 (2000)
6. Fine, K.: Vagueness, Truth and Logic. Synthese 30, 265–300 (1975)
7. Goguen, J.A.: The Logic of Inexact Concepts. Synthese 19, 325–373 (1969)
8. Halldén, S.: The Logic of Nonsense. Uppsala, Uppsala Universitets Arsskrift (1949)
9. Horgan, T.: Robust Vagueness and the Forced-March Sorites Paradox. Philosophical Perspective 8, 159–188 (1994)
10. Hyde, D.: From Heaps and Gaps to Heaps of Gluts. Mind 108, 641–660 (1997)
11. Kamp, H.: Two Theories about Adjectives. In: Keenan, E.L. (ed.) Formal Semantics of Natural Language, pp. 123–155. Cambridge University Press, Cambridge (2000)
12. Keefe, R.: Theories of Vagueness. Cambridge University Press, Cambridge (2000)
13. Körner, S.: Conceptual Thinking. Cambridge University Press, Cambridge (1955)
14. Lewis, D.K.: General semantics. Synthese 22, 18–67 (1983), Reprinted in his Philosophical Papers, vol. 1. Oxford University Press, Oxford
15. Lewis, D.K.: Counterfactuals. Basil Blackwell Ltd., OxOxford (1973)
16. McGee, V.: Kilimanjaro. Canadian Journal of Philosophy (sup. 23) 141–63 (1997)
17. Priest, G.: Fuzzy Identity and Local Validity. The Monist 81(2), 331–342 (1998)
18. Priest, G.: A Site for Sorites. In: Beall, J.C. (ed.) Liars and Heaps: New Essays on Paradox. Oxford University Press, Oxford (2003)
19. Przelecki, M.: Fuzziness as Multiplicity. Erkenntnis 10, 371–380 (1976)
20. Raffman, D.: Vagueness without Paradox. Philosophical Review 103, 41–74 (1994)
21. Shapiro, S.: Vagueness in Context. Clarendon Press, Oxford (2006)
22. Smith, N.J.J.: Vagueness and Degrees of Truth. Oxford University Press, Oxford (2008)
23. Sorensen, R.: Vagueness and Contradiction. Clarendon Press, Oxford (2001)

24. Stalnaker, R.: A Theory of Conditionals. In: Rescher, N. (ed.) Studies in Logical Theory, American Philosophical Quarterly, Monograph, vol. 2, pp. 98–112. Blackwell, Oxford (1968)
25. Tappenden, J.: The Liar and Sorites Paradoxes: Toward a Unified Treatment. J. of Philosophy. 90, 551–577 (1993)
26. van Inwagen, P.: How to Reason about Vague Objects. Philosophical Topics 16, 255–284 (1988)
27. Varzi, A.C.: Supervaluationism and Its Logics. Mind 112, 295–299 (2001)
28. Williamson, T.: Vagueness. Routledge, London (1994)
29. Wright, C.: On the Coherence of Vague Predicates. Synthese 30, 325–365 (1973)
30. Wright, C.: Further Reflections on the Sorites Paradox. Philosophical Topics 15, 227–290 (1987)

A Logic of Knowing How

Yanjing Wang*

Department of Philosophy, Peking University, Beijing, China
y.wang@pku.edu.cn

Abstract. In this paper, we propose a single-agent modal logic framework for reasoning about goal-direct "knowing how" based on ideas from linguistics, philosophy, modal logic and automated planning. We first define a modal language to express "I know how to guarantee φ given ψ" with a semantics not based on standard epistemic models but labelled transition systems that represent the agent's knowledge of his own abilities. A sound and complete proof system is given to capture the valid reasoning patterns about "knowing how" where the most important axiom suggests its compositional nature.

1 Introduction[1]

1.1 Background: Beyond "Knowing That"

Von Wright and Hinttika laid out the syntactic and semantic foundations of epistemic logic respectively in their seminal works [1] and [2]. The standard picture of epistemic logic usually consists of: a modal language which can express "an agent knows that φ"; a Kripke semantics incarnates the slogan "knowledge (information) as elimination of uncertainty"; a proof system syntactically characterizes a normal modal logic somewhere between S4 and S5 subjective to different opinions about the so-called introspection axioms. Despite the suspicions from philosophers in its early days, the past half-century has witnessed the blossom of this logical investigation of propositional knowledge with applications in epistemology, theoretical computer science, artificial intelligence, economics, and many other disciplines besides its birth place of modal logic.[2]

However, the large body of research on epistemic logic mainly focuses on propositional knowledge expressed by "knowing that φ", despite the fact that in everyday life knowledge is expressed by also "knowing how", "knowing why", "knowing what", "knowing whether", and so on (knowing?X below for brevity).

---

* The author thanks Frank Veltman for his insightful comments on an earlier version of this paper. The author is also gratful to the support from NSSF key projects 12&ZD119 and 15AZX020.

[1] To impatient technical readers: this rather philosophical introduction will help you to know *how* the formalism works in the later sections. A bit of philosophy can lead us further.

[2] For an excellent survey of the early history of epistemic logic, see [3, Chapter 2]. For a contemporary comprehensive introduction to its various topics, see [4].

© Springer-Verlag Berlin Heidelberg 2015
W. van der Hoek et al. (Eds.): LORI 2015, LNCS 9394, pp. 392–405, 2015.
DOI: 10.1007/978-3-662-48561-3_32

Linguistically, these expressions of knowledge share the common form consisting of the verb "know" followed by some embedded questions.[3] It is natural to assign a high-level uniform truth condition for these knowledge expressions in terms of knowing an answer of the corresponding question [6]. In fact, in the early days of epistemic logic, Hinttika has elaborate discussions on knowing?X and its relation with questions in terms of first-order modal logic [2], which also shapes his later work on *Socratic Epistemology* [7]. For example, "knowing who Frank is" is rendered as $\exists x \mathcal{K}(Frank = x)$ in [2]. However, partly because of the then-infamous philosophical and technical issues regarding the foundation of first-order modal logic (largely due to Quine), the development of epistemic logics beyond "knowing that" was hindered.[4] In the seminal work [10], the first-order epistemic logic is just briefly touched without specific discussion of those expressions using different embedded questions. A promising recent approach is based on *inquisitive semantics* where propositions may have both informative content and inquisitive content (cf. e.g.,[11]). An inquisitive epistemic logic which can handle "knowing that" and "knowing whether" is proposed in [12].

Departing from the linguistically motivated compositional analysis on knowing?X, some researchers took a knowing?X construction as a whole, and introduce a new modality instead of breaking it down by allowing quantifiers, equalities and other logical constants to occur freely in the language [13,14,15]. For example, "knowing what a password is" is rendered by "*Kv password*" in [13] instead of $\exists x \mathcal{K} \; password = x$, where Kv is the new modality. This move seems promising since by restricting the language we may avoid some philosophical issues of first-order modal logic, retain the decidability of the logic, and focus on special logical properties of each particular knowing-?X construction at a high abstraction level. A recent line of work results from this idea [16,17,18,19,20]. Besides the evident non-nomality of the resulting logics,[5] a 'signature' technical difficulty in such an approach is the apparent mismatch of syntax and semantics: the modal language is relatively weak compared to the models which contain enough information to facilitate a reasonable semantics of knowing?X, and this requires new techniques.

---

[3] There is a cross-lingual fact: such knowing?X sentences become meaningless if the verb "know" is replaced by "believe", e.g., I believe how to swim. This may shed some shadow on philosophers' usual conception of knowledge in terms of strengthened belief. Linguistically, this phenomenon occurs to many other verbs which can be roughly categorized using factivity, cf., e.g, [5].

[4] Nevertheless Hintikka addressed some of those issues about first-order modal logic insightfully in the context of epistemic logic, see, e.g., a wonderful survey paper [8]. Many of those issues are also elegantly addressed in intensional first-order modal logic cf. e.g., [9].

[5] For example, *knowing whether* $p \to q$ and knowing whether p together does not entail knowing whether q. Likewise, *knowing how* to p and knowing how to q does not entail knowing how to $p \land q$. Moreover, you may not *know why* a tautology is a tautology which contradicts necessitation.

1.2 Knowing How

Among all the knowing?X expressions, the most discussed one in philosophy and AI is "knowing how". Indeed, it sounds the most distant from propositional knowledge (knowledge-that): knowing how to swim seems distinctly different from knowing that it is raining outside. One question that keeps philosophers busy is whether knowledge-how (the knowledge expressed by "knowing how") is reducible to knowledge-that. Here philosophers split into two groups: the intellectualists who think knowledge-how is a subspecies of knowledge-that (e.g., [21]), and the anti-intellectuallists who do not think so (e.g., [22]). The anti-intellectualism may win your heart at the first glance by equating knowledge-how to certain ability, but the linguistically and logically well-versed intellectualists may have their comebacks at times (think about the previously mentioned interpretation of knowing?X as knowing an answer).[6] In AI, starting from the early days [24,25,26], people have been studying about representation and reasoning of *procedural knowledge* which is often treated as synonym for knowledge-how in AI, in particular about knowledge-how based on specifiable procedures such as coming out of a maze or winning a game. However, there is no common consensus on how to capture the logic of "knowing how" formally (cf. the excellent surveys [27,28]). In this paper we presents an attempt to formalize an important kind of "knowing how" and lay out its logic foundation, inspired by the aforementioned perspectives of linguistics, philosophy, and AI.

Some clarifications have to be made before mentioning our ideas and their sources:

- We will focus on the logic of *goal-direct* "knowing how" as Gochet puts it [27], such as knowing how to prove a theorem, how to open the door, how to bake a cake, and how to cure the disease, i.e., linguistically, mainly about knowing how followed by a *achievement verb* or an *accomplishment verb* according to the classification of Vendler [29].[7] On the other hand, we will not talk about the following "knowing how": I know how the computer works (explanation), I know how happy she is (degree of emotion), I know how to speak English (rule-direct) and so on.
- The goal of this paper is *not* to address the philosophical debate between intellectualism and anti-intellectualism which we did discuss in [30,31]. However, to some extent, we are inspired by the ideas from both stands,

---

[6] See [23] for a survey of the debate. A comprehensive collection of the related papers (200+) can be found at http://philpapers.org/browse/knowledge-how, edited by John Bengson.

[7] Here knowing how to maintain something or to do an activity (like swimming) are *not* typical examples for our treatment, although we hope our formalism captures some common features shared also by them. As discussed in [27], "knowing how" plus activities, though more philosophically interesting, is less demanding in logic rendering than others.

and combine them in the formal work which may in turn shed new light on this philosophical issue. [8]

— We focus on the single-agent case without probability, as the first step.

1.3 Basic Ideas Behind the Syntax and Semantics

Different from the cases on "knowing whether" and "knowing what", there is nothing close to a consensus on what would be the syntax and semantics of the logic of "knowing how". Various attempts were made using Situation Calculus, ATL, or STIT logic to express different versions of "knowing how", cf. e.g., [26,32,33,34,35,27]. However, as we mentioned before, we do not favour a compositional analysis using powerful logical languages. Instead, we would like to take the "knowing how" construction as a single (and only) modality in our language. It seems natural to introduce a modality $\mathcal{K}h\varphi$ to express the goaldirect "knowing how to achieve the goal φ". It sounds similar to "having the ability to achieve the goal φ", as many anti-intellectualists would agree. It seems harmless to go one step further as in the AI literature to interpret this type of "knowing how" as that the agent *can* achieve φ. However, it is crucial to note the following problems of such an anti-intelectualistic ability account:

1. Knowing how to achieve a goal may not entail that you *can* realize the goal now. For example, as intellectualists would remark, a broken-arm pianist may still know how to play piano even if he cannot play right now, and a chef may still know how to make cakes even when the sugar is run out (cf. e.g., [21]).

2. Even when you have the ability to win a lottery by luckily buying the right ticket (and indeed win it in the end), it does not mean you know how to win the lottery, since you cannot *guarantee* the result (cf. e.g., [36]).

To reconcile our intuition about the ability involved in "knowing how" and the first problem above, it is observed in [30] that "knowing how" expressions in context often come with implicit preconditions.[9] For example, when you claim that you know how to go to the city center of Taipei from the airport, you are talking about what you can do under some implicit preconditions: e.g., the public transportation is still running or there is no strike of the taxi drivers. Likewise, it sounds all right to say that you know how to bake a cake even when you do not have all the ingredients right now: you can do it *given* you have all the ingredients. In our logical language, we make such context-dependent preconditions explicit by introducing the modality $\mathcal{K}h(\psi, \varphi)$ expressing that the agent knows how to achieve φ given the precondition ψ.[10] Actually, we used a similar

---

[8] Our logic is more about knowing how than knowledge-how though they are clearly related and often discussed interchangeably in the philosophy literature. The full nature of knowledge-how may not be revealed by the inference structure of the linguistic construction of knowing how.

[9] Such conditions are rarely discussed in the philosophical literature of "knowing how" with few exceptions such as [37].

[10] By using the condition, one can be said to know better how to swim than another if he can do it in a more hostile environment (thus weakening the condition) [30].

conditional knowing what operator in [17] to capture the conditional knowledge such as "I would know what my password for this website is, given it is 4-digit" (since I only have one 4-digit password ever).[11] In [17], this conditionaliztion is proved to be also useful to encode the potential dynamics of knowledge. We will come back to this at the end of the paper.

Now, to reconcile the intuition of ability with the second problem above, we need to interpret the ability more precisely to exclude the lucky draws. Our main idea comes from *conformant planning* in AI which is exactly about *how* to achieve a goal by a linear plan which can never fail given some initial uncertainty (cf. e.g., [39]). For example (taken from [40,41]), consider the following map of a floor, and suppose that you know you are at a place marked by p but do not know exactly where you are. Do you know how to reach a safe place (marked by q)? Note that the marks are only on the map.

Example 1.

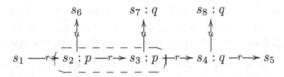

It is not hard to see that there exists a plan to *guarantee* your safety from any place marked by p, which is to move r first then move u. On the contrary, the plan rr and the plan u may fail sometimes depending on where you are actually. The locations in the map can be viewed as states of affairs and the labelled directional edges between the states can encode your own "knowledge map" of the available actions and their effects.[12] Intuitively, to know how to achieve φ requires that you can guarantee φ. Consider the following examples which represent the agent's knowledge about his own abilities.

Example 2.

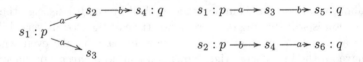

The graph on the left denotes that you know you can do a at the p-state s_1 but you are not sure what the consequence is: it may lead to either s_2 or s_3, and the exact outcome is out of your control. Therefore, this action is *non-deterministic* to you. In this case, ab is not a good plan since it may fail to be executable. Thus it sounds unreasonable to claim that you know how to reach q given p.

---

[11] Such conditionals are clearly not simple (material) implications and they are closely related to conditional probability and conditional belief (cf. e.g., [38]).

[12] The agent may have more abilities *de facto* than what he may realize. It is important to make sure the agent can *knowingly* guarantee the goal in terms of the ability he is aware of, cf. [24,34,28].

Now consider the graph on the right. Let ab and ba be two treatments for the same symptom p depending on the exact cause (s_1 or s_2). As a doctor, it is indeed true that you can cure the patient (to achieve q) if you are told the exact cause. However, responsible as you are, can you say you know how to cure the patient given only the symptom p? A wrong treatment may kill the patient. These planning examples suggest the following truth condition for the modal formula $\mathcal{K}h(\psi, \varphi)$ w.r.t. a graph-like model representing the agent's knowledge about his or her abilities (available actions and their possibly non-deterministic effects):

> There *exists* a sequence σ of actions such that from *all* the ψ-states in the graph, σ will *always* succeed in reaching φ-states.

Note that the nesting structure of quantifiers in the above truth condition is $\exists\forall\forall$.[13] The first $\exists$ fixes a unique sequence, the first $\forall$ checks all the possible states satisfying the condition ψ, and the second $\forall$ make sure the goal is guaranteed.

There are several points to be highlighted: 1. $\exists$ cannot be swapped with the first $\forall$: see the discussion about the second graph in Example 2, which amounts to the distinction between *de re* and *de dicto* in the setting of "knowing how" (cf. also [26,35,33,28] and uniform strategies in imperfect information games); 2. There is no explicit "knowing that" in the above truth condition, which differs from the truth conditions advocated by intellectualism [43] and the linguistically motivated $\exists x \mathcal{K}\varphi(x)$ rendering.[14] On the other hand, the graph model represents the agent's knowledge of his actions and their effects (cf. [44]). 3. The truth condition is based on a Kripke-like model without epistemic relations as in the treatment of (imperfect) procedure information in [44]. As it will become more clear later on, it is not necessary to go for neighbourhood or topological models to accommodate non-normal modal logics if the truth condition of the modality is non-standard (cf. also [45,19,16]); 4. Finally, our interpretation of "knowing how" does not fit the standard scheme "knowledge as elimination of uncertainty", and it is not about possible worlds indistinguishable from the "real world". The truth of $\mathcal{K}h(\psi, \varphi)$ does not depend on the actual world: it is "global" in nature.

In the next section, we will flesh out the above ideas in precise definitions and proofs: first a simple formal language, then the semantics based on the idea of planning, and finally a sound and complete proof system. We hope our formal theory can clarify the above informal ideas further. In the last section, we summarize our novel ideas beyond the standard schema of epistemic logic, and point out many future directions.

Note: Due to the lack of space, we omit the proofs.[15]

---

[13] In [42], the author introduced a modality for *can* φ with the following $\exists\forall$ schema over neighbourhood models: *there is* a relevant cluster of possible worlds (as the outcomes of an action) where φ is true in *all* of them.

[14] This also distinguishes this work from our earlier philosophical discussion [30] where intellectualism was defended by giving an $\exists x \mathcal{K}\varphi(x)$-like truth condition informally.

[15] For detailed proofs, see http://arxiv.org/abs/1505.06651.

2   The Logic

Definition 1. *Given a countable non-empty set of proposition letters* **P**, *the language* **L$_{\mathrm{Kh}}$** *is defined as follows:*

$$\varphi \quad ::= \quad \top \mid p \mid \neg\varphi \mid (\varphi \wedge \varphi) \mid \mathcal{K}h(\varphi, \varphi)$$

where $p \in \mathbf{P}$. *As discussed in the previous section,* $\mathcal{K}h(\psi, \varphi)$ *expresses that the agent knows how to achieve* φ *given* ψ. *We use the standard abbreviations* $\bot, \varphi\vee\psi$ *and* $\varphi \to \psi$, *and define* $\mathcal{U}\varphi$ *as* $\mathcal{K}h(\neg\varphi, \bot)$. *The meaning of* $\mathcal{U}$ *will become more clear after the semantics is defined.*

Definition 2. *Given the set of proposition letters* **P** *and a countable non-empty set of action symbols* $\boldsymbol{\Sigma}$. *An* ability map *is essentially a labelled transition system* $(\mathcal{S}, \mathcal{R}, \mathcal{V})$ *where:*

- $\mathcal{S}$ *is a non-empty set of states;*
- $\mathcal{R} : \boldsymbol{\Sigma} \to 2^{\mathcal{S}\times\mathcal{S}}$ *is a collection of transitions labelled by actions in* $\boldsymbol{\Sigma}$;
- $\mathcal{V} : \mathcal{S} \to 2^{\mathbf{P}}$ *is a valuation function.*

We write $s \xrightarrow{a} t$ *if* $(s,t) \in \mathcal{R}(a)$. *For a sequence* $\sigma = a_1 \dots a_n \in \boldsymbol{\Sigma}^*$, *we write* $s \xrightarrow{\sigma} t$ *if there exist* $s_2 \dots s_n$ *such that* $s \xrightarrow{a_1} s_2 \xrightarrow{a_2} \cdots \xrightarrow{a_{n-1}} s_n \xrightarrow{a_n} t$. *Note that* σ *can be the empty sequence* ϵ *(when* $n = 0$*), and we set* $s \xrightarrow{\epsilon} s$ *for any* s. *Let* σ_k *be the initial segment of* σ *up to* a_k *for* $k \leq |\sigma|$. *In particular let* $\sigma_0 = \epsilon$. *We say that* $\sigma = a_1 \dots a_n$ *is* strongly executable *at* s *if: for any* $0 \leq k < n$ *and any* t, $s \xrightarrow{\sigma_k} t$ *implies that* t *has at least one* a_{k+1}*-successor. It is not hard to see that if* σ *is strongly executable at* s *then it is executable at* s, *i.e.,* $s \xrightarrow{\sigma} t$ *for some* t.

Note that, according to our above definition, ab is not strongly executable from s_1 in the left-hand-side model of Example 2, since s_3 has no b-successor but it can be reached from s_1 by $a = (ab)_1$.

Definition 3 (Semantics of L$_{\mathrm{Kh}}$)

$\mathcal{M}, s \vDash \top$	*always*
$\mathcal{M}, s \vDash p$	$\Leftrightarrow p \in V(s)$
$\mathcal{M}, s \vDash \neg\varphi$	$\Leftrightarrow \mathcal{M}, s \nvDash \varphi$
$\mathcal{M}, s \vDash \varphi \wedge \psi$	$\Leftrightarrow \mathcal{M}, s \vDash \varphi$ *and* $\mathcal{M}, s \vDash \psi$
$\mathcal{M}, s \vDash \mathcal{K}h(\psi, \varphi)$	$\Leftrightarrow$ *there exists a* $\sigma \in \boldsymbol{\Sigma}^*$ *such that for all* s' *such that* $\mathcal{M}, s' \vDash \psi$: σ *is strongly executable at* s' *and for all* t *such that* $s' \xrightarrow{\sigma} t, \mathcal{M}, t \vDash \varphi$

Note that the modality $\mathcal{K}h$ is *not local* in the sense that its truth does not depend on the designated state where it is evaluated. Thus it either holds on all the states or none of them. It is not hard to see that the schema of $\exists\forall\forall$ appears in the truth condition for $\mathcal{K}h$ where the last $\forall$ actually consists of two parts: the strong executability (there is a $\forall$ in its definition) and the guarantee of the goal. These two together make sure the plan will never fail to achieve φ. It is a simple exercise to see that $\mathcal{K}h(p, q)$ holds in the model of Example 1, but not in the models of Example 2. Moreover, the operator $\mathcal{U}$ defined by $\mathcal{K}h$ is actually a *universal modality*:[16]

$$\mathcal{M}, s \vDash \mathcal{U}\varphi \Leftrightarrow \text{ for all } t \in \mathcal{S}, \mathcal{M}, t \vDash \varphi$$

To see this, check the following:

$$
\begin{aligned}
\mathcal{M}, s \vDash \mathcal{K}h(\neg\psi, \bot) \Leftrightarrow\ & \text{there exists a } \sigma \in \Sigma^* \text{ such that for every } \mathcal{M}, s' \vDash \neg\psi : \\
& \sigma \text{ is strongly executable at } s' \text{ and if } s' \xrightarrow{\sigma} t \text{ then } \mathcal{M}, t \vDash \bot \\
\Leftrightarrow\ & \text{there exists a } \sigma \in \Sigma^* \text{ such that for every } \mathcal{M}, s' \vDash \neg\psi : \\
& \sigma \text{ is strongly executable at } s' \text{ and there is no } t \text{ such that } s' \xrightarrow{\sigma} t \\
\Leftrightarrow\ & \text{there exists a } \sigma \in \Sigma^* \text{ such that for every } \mathcal{M}, s' \vDash \neg\psi : \bot \text{ holds} \\
\Leftrightarrow\ & \text{there exists a } \sigma \in \Sigma^* \text{ such that there is no } s' \text{ such that } \mathcal{M}, s' \nvDash \psi \\
\Leftrightarrow\ & \text{for all } t \in \mathcal{S}, \mathcal{M}, t \vDash \psi
\end{aligned}
$$

Proposition 1. *The following are valid:*

1 $\ \mathcal{U}p \wedge \mathcal{U}(p \to q) \to \mathcal{U}q$ 2 $\ \mathcal{K}h(p, r) \wedge \mathcal{K}h(r, q) \to \mathcal{K}h(p, q)$

3 $\ \mathcal{U}(p \to q) \to \mathcal{K}h(p, q)$ 4 $\ \mathcal{U}p \to p$

5 $\ \mathcal{K}h(p, q) \to \mathcal{U}\mathcal{K}h(p, q)$ 6 $\ \neg\mathcal{K}h(p, q) \to \mathcal{U}\neg\mathcal{K}h(p, q)$

The validity of (2) above actually captures the intuitive compositionality of "knowing how", as desired. Note that $\mathcal{K}h(p, q) \wedge \mathcal{K}h(p, r) \to \mathcal{K}h(p, q \wedge r)$ is not valid, as desired.

Based on the above axioms, we propose the following proof system $\mathbb{SKH}$ for $\mathbf{L_{Kh}}$ (where $\varphi[\psi/p]$ is obtained by uniformly substituting p in φ by ψ):

System $\mathbb{SKH}$

Axioms		**Rules**
TAUT	all axioms of propositional logic	MP $\quad \dfrac{\varphi, \varphi \to \psi}{\psi}$
DISTU	$\mathcal{U}p \wedge \mathcal{U}(p \to q) \to \mathcal{U}q$	NECU $\quad \dfrac{\varphi}{\mathcal{U}\varphi}$
COMPKh	$\mathcal{K}h(p, r) \wedge \mathcal{K}h(r, q) \to \mathcal{K}h(p, q)$	SUB $\quad \dfrac{\varphi(p)}{\varphi[\psi/p]}$
EMP	$\mathcal{U}(p \to q) \to \mathcal{K}h(p, q)$	
TU	$\mathcal{U}p \to p$	
4KU	$\mathcal{K}h(p, q) \to \mathcal{U}\mathcal{K}h(p, q)$	
5KU	$\neg\mathcal{K}h(p, q) \to \mathcal{U}\neg\mathcal{K}h(p, q)$	

Proposition 1 plus some reflection on the usual inference rules should establish the soundness of $\mathbb{SKH}$. For completeness, we first get a taste of the deductive power of $\mathbb{SKH}$ by proving the following formulas which play important roles in the later completeness proof. In the rest of the paper we use $\vdash$ to denote $\vdash_{\mathbb{SKH}}$.

---

[16] Note that $\mathcal{U}$ is a very powerful modality in its expressiveness when combined with the standard $\square$ modality, cf. [46].

Proposition 2. *We can derive the following in* $\mathbb{SKH}$ *(names are given to be used later in the proofs):*

TRI	$\mathcal{K}h(p,p)$
WSKh	$\mathcal{U}(p \to r) \wedge \mathcal{U}(o \to q) \wedge \mathcal{K}h(r,o) \to \mathcal{K}h(p,q)$
4U	$\mathcal{U}p \to \mathcal{U}\mathcal{U}p$
5U	$\neg\mathcal{U}p \to \mathcal{U}\neg\mathcal{U}p$
COND	$\mathcal{K}h(\bot,p)$
UCONJ	$\mathcal{U}(\varphi \wedge \psi) \leftrightarrow (\mathcal{U}\varphi \wedge \mathcal{U}\psi).$
PREKh	$\mathcal{K}h(\mathcal{K}h(p,q) \wedge p, q).$
POSTKh	$\mathcal{K}h(r, \mathcal{K}h(p,q) \wedge p) \to \mathcal{K}h(r,q)$

Moreover, the following rule NECKh *is admissible:* $\vdash \varphi \implies \vdash \mathcal{K}h(\psi,\varphi).$

Interestingly, PREKh says that you know how to guarantee q given both p and the fact that you know how to guarantee q given p. POSTKh says that you know how to achieve q given r if you know how to achieve a state where you know how to continue to achieve q.[17]

Remark 1. From the above proposition and the system $\mathbb{SKH}$, we see that $\mathcal{U}$ is indeed an S5 modality which can be considered as a version of "knowing that": you know that φ iff it holds on all the relevant possible states under the current restriction of attention (not just the epistemic alternatives to the actual one). The difference is that here the knowledge-that expressed by $\mathcal{U}\varphi$ refers to the "background facts" that you take for granted for now, rather than contingent but epistemically true facts in the standard epistemic logic. Another interesting thing to notice is that WSKh actually captures an important connection between "knowing that" and "knowing how", e.g., you know how to cure a disease if you know that it is of a certain type and you know that you know how to cure this type of the disease in general. We will come back to the relation between "knowing how" and "knowing that" at the end of the paper.

It is crucial to establish the following replacement rule to ease the later proofs.

Proposition 3. *The replacement of equivalents* $(\vdash \varphi \leftrightarrow \psi \implies \vdash \chi \leftrightarrow \chi[\psi/\varphi])$[18] *is an admissible rule in* $\mathbb{SKH}$.

In the rest of the paper we often use the above rule of replacement implicitly.

Here are some notions before we prove the completeness. Given a set of $\mathbf{L_{Kh}}$ formulas Δ, let $\Delta|_{\mathcal{K}h}$ be the collection of its $\mathcal{K}h$ formulas:

$$\Delta|_{\mathcal{K}h} = \{\chi \mid \chi = \mathcal{K}h(\psi,\varphi) \in \Delta\}.$$

Now for each maximal consistent set of $\mathbf{L_{Kh}}$ formulas we build a canonical model.

---

[17] This is an analog of a requirement of the modality Can in [26].

[18] Here the substitution can apply to some (not necessarily all) of the occurrences.

Definition 4. *Given a maximal consistent set Γ w.r.t. $\mathbb{SKH}$, let $\boldsymbol{\Sigma}_\Gamma = \{\langle\psi,\varphi\rangle \mid \mathcal{K}h(\psi,\varphi) \in \Gamma\}$, the canonical model for Γ is $\mathcal{M}_\Gamma^c = \langle \mathcal{S}_\Gamma^c, \mathcal{R}^c, \mathcal{V}^c\rangle$ where:*

- *$\mathcal{S}_\Gamma^c = \{\Delta \mid \Delta$ is a maximal consistent set w.r.t. $\mathbb{SKH}$ and $\Gamma|_{\mathcal{K}h} = \Delta|_{\mathcal{K}h}\}$;*
- *$\Delta \xrightarrow{\langle\psi,\varphi\rangle}_c \Theta$ iff $\mathcal{K}h(\psi,\varphi) \in \Gamma, \psi \in \Delta$, and $\varphi \in \Theta$;*
- *$p \in \mathcal{V}^c(\Delta)$ iff $p \in \Delta$.*

Clearly Γ is a state in $\mathcal{M}_\Gamma^c$. We say that $\Delta \in \mathcal{S}_\Gamma^c$ is a φ-state if $\varphi \in \Delta$.

The following two propositions are immediate:

Proposition 4. *For any Δ, Δ' in $\mathcal{S}_\Gamma^c$, any $\mathcal{K}h(\psi,\varphi) \in \mathbf{L_{Kh}}$, $\mathcal{K}h(\psi,\varphi) \in \Delta$ iff $\mathcal{K}h(\psi,\varphi) \in \Delta'$ iff $\mathcal{K}h(\psi,\varphi) \in \Gamma$.*

Proposition 5. *If $\Delta \xrightarrow{\langle\psi,\varphi\rangle} \Theta$ for some $\Delta, \Theta \in \mathcal{S}_\Gamma^c$ then $\Delta \xrightarrow{\langle\psi,\varphi\rangle} \Theta'$ for any Θ' such that $\varphi \in \Theta'$.*

Based on Proposition 4 and $\mathbb{S}5$ axioms for $\mathcal{U}$, we can prove a crucial proposition to be used later.

Proposition 6. *If $\varphi \in \Delta$ for all $\Delta \in \mathcal{S}_\Gamma^c$ then $\mathcal{U}\varphi \in \Delta$ for all $\Delta \in \mathcal{S}_\Gamma^c$.*

Now we are ready to establish another key proposition for the truth lemma.

Proposition 7. *Suppose that there are $\psi', \varphi' \in \mathbf{L_{Kh}}$ such that for each ψ-state $\Delta \in \mathcal{S}_\Gamma^c$ we have $\Delta \xrightarrow{\langle\psi',\varphi'\rangle} \Theta$ for some $\Theta \in \mathcal{S}_\Gamma^c$, then $\mathcal{U}(\psi \to \psi') \in \Delta$ for all $\Delta \in \mathcal{S}_\Gamma^c$.*

Now we are ready to prove the truth lemma based on the above two propositions.[19]

Lemma 1 (Truth lemma). *For any $\varphi \in \Gamma : \mathcal{M}_\Gamma^c, \Delta \vDash \varphi \iff \varphi \in \Delta$*

Now due to a standard Lindenbaum-like argument, each $\mathbb{SKH}$-consistent set of formulas can be extended to a maximal consistent set Γ. Due to the truth lemma, $\mathcal{M}_\Gamma^c, \Gamma \vDash \Gamma$. The completeness of $\mathbb{SKH}$ follows immediately.

Theorem 1. *$\mathbb{SKH}$ is sound and strongly complete w.r.t. the class of all models.*

3   Conclusions and Future Work

In this paper, we propose and study a modal logic of goal-direct "knowing how". The highlights of our framework are summarized below with connections to our earlier ideas on non-standard epistemic logics:

---

[19] The proof is quite non-trivial. Please refer to the online version:
http://arxiv.org/abs/1505.06651.

- The "knowing how" construction is treated as a whole similar to our works on "knowing whether" and "knowing what" [16,19]. We would like to keep the language neat.
- Semantically, "knowing how" is treated as a special conditional: *being able to guarantee a goal given a precondition*, partly inspired by the conditionalization in [17].
- The *ability* involved is further interpreted as having a plan that never fails to achieve the goal under the precondition, inspired by the work on conformant planning [41] where we used the epistemic PDL language to encode the planning problem.
- The semantics is based on labelled transition systems representing the agent's knowledge of his own abilities, inspired by the framework experimented in [44].
- Compared to the standard semantic schema of knowledge-that: true in *all* indistinguishable alternatives, our work has a more existential flavour: knowing how as having at least *one* good plan. Our modal operator is not local to the indistinguishable alternatives but it is about all the possible states even when they are distinguishable from the current world. Thus a cook can still be said to know how to cook a certain dish even if he knows that the ingredients are not available right now.

There are a lot more to explore. We conjecture the logic is decidable and leave the model-theoretical issues to the full version of this paper. Moreover, it is a natural extension to introduce the standard knowing-that operator $\mathcal{K}$ into the language and correspondingly add a set $\mathcal{E} \subseteq \mathcal{S}$ in the model to capture the agent's *local* epistemic alternatives. Then we can define the local version of "knowing how" $\mathcal{K}h\varphi$ as $\mathcal{K}\psi \wedge \mathcal{K}h(\psi, \varphi)$ for some ψ. Other obvious next steps include probabilistic and multi-agent versions of $\mathcal{K}h$. It also makes good sense to consider group notions of "knowing how" which may bring it closer to the framework of ATEL where a group of agents may achieve a lot more together (cf. [28]). More generally, we may consider program-based "knowing how" where conditional plans and iterated plans are allowed, which can be used to *maintain* a goal. It is also interesting to add the dynamic operators to the picture, i.e., the public announcements $[\varphi]$. In particular, it is interesting to see how new knowledge-how is obtained by learning new knowledge-that e.g., $\mathcal{K}h(p, q) \rightarrow [p](\mathcal{U}p \wedge \mathcal{K}h(\top, q))$ may be a desired valid formula.[20]

There are also interesting philosophical questions related to our formal theory. For example, a new kind of logical omniscience may occur: if there is indeed a good plan to achieve φ according to the agent's abilities then he knows how to achieve φ. To the taste of philosophers, maybe an empty plan is not acceptable to witness knowledge-how, e.g., people would not say I know how to digest (by doing nothing). We can define a stronger modality $\mathcal{K}h^+(\psi, \varphi)$ as $\mathcal{K}h(\psi, \varphi) \wedge$

---

[20] Note that $\mathbf{L_{Kh}}$ may not have the enough pre-encoding power for announcements in itself, similar to the case of PALC discussed in [47]. In particular, $[\chi]\mathcal{K}h(\psi, \varphi) \leftrightarrow \mathcal{K}h([\chi]\psi, [\chi]\varphi)$ may not be valid due to the lack of control in the syntax for the intermediate stages of the execution path of a plan.

$\neg\mathcal{U}(\psi \to \varphi)$ to rule out such cases.[21] Note that although $\mathcal{U}$ is definable by $\mathcal{K}h$ in our setting, it does not have the philosophical implication that knowledge-that is actually a subspecies of knowledge-how, as strong anti-intellectulism would argue. Nevertheless, our axioms do tell us something about the interactions between "knowing how" and "knowing that", e.g., WSKh says some background knowledge may let us know better how to reach our goal.

References

1. Von Wright, G.H.: An Essay in Modal Logic. North-Holland, Amsterdam (1951)
2. Hintikka, J.: Knowledge and Belief: An Introduction to the Logic of the Two Notions. Cornell University Press, Ithaca (1962)
3. Wang, R.J.: Timed Modal Epistemic Logic. PhD thesis, City University of New York (2011)
4. van Ditmarsch, H., Halpern, J.Y., van der Hoek, W., Kooi, B. (eds.): Handbook of Epistemic Logic. College Publications (2015)
5. Egré, P.: Question-embedding and factivity. Grazer Philosophische Studien 77(1), 85–125 (2008)
6. Harrah, D.: The logic of questions. In: Gabbay, D. (ed.) Handbook of Philosophical Logic, vol. 8, Springer (2002)
7. Hintikka, J.: Socratic Epistemology: Explorations of Knowledge-Seeking by Questioning. Cambridge University Press (2007)
8. Hintikka, J., Hintikka, M.: Reasoning about knowledge in philosophy: The paradigm of epistemic logic. In: The Logic of Epistemology and the Epistemology of Logic. Synthese Library, vol. 200, pp. 17–35. Springer, Netherlands (1989)
9. Fitting, M., Mendelsohn, R.L.: First-order modal logic. Synthese Library. Springer (1998)
10. Fagin, R., Halpern, J., Moses, Y., Vardi, M.: Reasoning about knowledge. MIT Press (1995)
11. Ciardelli, I., Groenendijk, J., Roelofsen, F.: Inquisitive semantics: A new notion of meaning. Language and Linguistics Compass 7(9), 459–476 (2013)
12. Ciardelli, I., Roelofsen, F.: Inquisitive dynamic epistemic logic. Synthese 192(6), 1643–1687 (2015)
13. Plaza, J.A.: Logics of public communications. In: Emrich, M.L., Pfeifer, M.S., Hadzikadic, M., Ras, Z.W. (eds.) Proceedings of the 4th International Symposium on Methodologies for Intelligent Systems, pp. 201–216 (1989)
14. Hart, S., Heifetz, A., Samet, D.: Knowing whether, knowing that, and the cardinality of state spaces. Journal of Economic Theory 70(1), 249–256 (1996)
15. van der Hoek, W., Lomuscio, A.: Ignore at your peril - towards a logic for ignorance. In: Proceedings of AAMAS 2003, pp. 1148–1149. ACM (2003)
16. Wang, Y., Fan, J.: Knowing that, knowing what, and public communication: Public announcement logic with kv operators. In: Proceedings of IJCAI, 2013 (2013)
17. Wang, Y., Fan, J.: Conditionally knowing what. In: Advances in Modal Logic, vol. 10, pp. 569–587 (2014)
18. Fan, J., Wang, Y., van Ditmarsch, H.: Almost necessary. In: Advances in Modal Logic, vol. 10, pp. 178–196 (2014)

---

[21] The distinction between $\mathcal{K}h$ and $\mathcal{K}h^+$ is similar to the distinction between STIT and deliberative STIT.

19. Fan, J., Wang, Y., van Ditmarsch, H.: Contingency and knowing whether. The Review of Symbolic Logic 8, 75–107 (2015)
20. Xiong, S.: Decidability of **ELKv^r**. Bachelaor thesis (2014)
21. Stanley, J., Williamson, T.: Knowing how. The Journal of Philosophy, 411–444 (2001)
22. Ryle, G.: Knowing how and knowing that: The presidential address. In: Proceedings of the Aristotelian Society, vol. 46, pp. 1–16 (1946)
23. Fantl, J.: Knowing-how and knowing-that. Philosophy Compass 3(3), 451–470 (2008)
24. McCarthy, J., Hayes, P.J.: Some philosophical problems from the standpoint of artificial intelligence. In: Machine Intelligence, pp. 463–502. Edinburgh University Press (1969)
25. McCarthy, J.: First-Order theories of individual concepts and propositions. Machine Intelligence 9, 129–147 (1979)
26. Moore, R.C.: A formal theory of knowledge and action. In: Hobbs, J.R., Moore, R.C. (eds.) Formal Theories of the Commonsense World. Ablex Publishing Corporation (1985)
27. Gochet, P.: An open problem in the logic of knowing how. In: Hintikka, J. (ed.) Open Problems in Epistemology. The Philosophical Society of Finland (2013)
28. Ågotnes, T., Goranko, V., Jamroga, W., Wooldridge, M.: Knowledge and ability. In: van Ditmarsch, H., Halpern, J., van der Hoek, W., Kooi, B. (eds.) Handbook of Epistemic Logic, pp. 543–589. College Publications (2015)
29. Vendler, Z.: Linguistics in Philosophy. Cornell University Press (1967)
30. Lau, T., Wang, Y.: Formalizing "knowing how". (unpublished manuscript) (2015)
31. Lau, T.: Formalizing "knowing how". Master's thesis, Peking University (2015) (in Chinese)
32. Morgenstern, L.: A first order theory of planning, knowledge, and action. In: Proceedings of the 1986 Conference on Theoretical Aspects of Reasoning About Knowledge, pp. 99–114. Morgan Kaufmann Publishers Inc., San Francisco (1986)
33. Herzig, A., Troquard, N.: Knowing how to play: uniform choices in logics of agency. In: Proceedings of AAMAS 2006, pp. 209–216 (2006)
34. Broersen, J.: A logical analysis of the interaction between 'obligation-to-do' and 'knowingly doing'. In: van der Meyden, R., van der Torre, L. (eds.) DEON 2008. LNCS (LNAI), vol. 5076, pp. 140–154. Springer, Heidelberg (2008)
35. Jamroga, W., van der Hoek, W.: Agents that know how to play. Fundamenta Informaticae 63(2-3), 185–219 (2004)
36. Carr, D.: The logic of knowing how and ability. Mind 88(351), 394–409 (1979)
37. Noë, A.: Against intellectualism. Analysis, 278–290 (2005)
38. Tillio, A.D., Halpern, J.Y., Samet, D.: Conditional belief types. Games and Economic Behavior 87, 253–268 (2014)
39. Smith, D.E., Weld, D.S.: Conformant graphplan. In: AAAI 1998, pp. 889–896 (1998)
40. Wang, Y., Li, Y.: Not all those who wander are lost: Dynamic epistemic reasoning in navigation. In: Advances in Modal Logic, vol. 9, pp. 559–580 (2012)
41. Yu, Q., Li, Y., Wang, Y.: A dynamic epistemic framework for conformant planning. In: Proceedings of TARK 2015, pp. 249–259 (2015)
42. Brown, M.A.: On the logic of ability. Journal of Philosophical Logic 17(1), 1–26 (1988)
43. Stanley, J.: Know how. Oxford University Press (2011)

44. Wang, Y.: Representing imperfect information of procedures with hyper models. In: Banerjee, M., Krishna, S.N. (eds.) ICLA. LNCS, vol. 8923, pp. 218–231. Springer, Heidelberg (2015)
45. Kracht, M., Wolter, F.: Simulation and transfer results in modal logic - A survey. Studia Logica 59(1), 149–177 (1997)
46. Goranko, V., Passy, S.: Using the universal modality: Gains and questions. Journal of Logic and Computation 2(1), 5–30 (1992)
47. van Benthem, J., van Eijck, J., Kooi, B.: Logics of communication and change. Information and Computation 204(11), 1620–1662 (2006)

A Dynamic Epistemic Logic with a Knowability Principle

Michael Cohen

Munich Center for Mathematical Philosophy, LMU
michaelco13@gmail.com

Abstract. A dynamic epistemic logic is presented in which the single agent can reason about his knowledge stages before and after announcements. The logic is generated by reinterpreting multi agent private announcements in a single agent environment. It is shown that a knowability principle is valid for such logic: any initially true φ can be known after a certain number of announcements.

In recent years a novel explication of *knowability* has been studied in the framework of dynamic epistemic logic (DEL) [4], in which 'knowable' is read as 'can be known after an announcement' [1], [5]. Under this explication it has been shown that the *knowability principle* (KP), the principle according to which all truths are knowable, does not hold in public announcement logic (PAL) [1] nor in its extension arbitrary public announcement logic (APAL) [5]. Instead of using public announcements, this contribution focuses on the investigation of knowability using the logical structure of private announcements [4]. For this purpose, private announcements are reinterpreted in a single agent environment.

1 Agents in Private Announcements as Stages of Knowledge

One can reinterpret the logical structure of private announcements to a set of epistemic agents as an update of a single agent who can reason about past and present stages of knowledge.

For instance, consider the situation in which two agents, 0 and 1, don't know p, and the private announcement of p which is given to agent 1. Let $[\mathcal{E}, e]$ denote this action of private announcement. Then $\mathcal{M}, w \models [\mathcal{E}, e]K_1(p \wedge \neg K_0 p)$ is the case, which says that after the private announcement, agent 1 knows p and she also knows that agent 0 does not know p. This situation can be reinterpreted as one containing a single agent with two sequential stages of knowledge, stage 0 as the initial stage, and stage 1 as the stage after the first announcement. Then K_0 and K_1 represent the knowledge stages before and after the first announcement, respectively. Under this interpretation, $[\mathcal{E}, e]K_1(p \wedge \neg K_0 p)$ is read 'after the first update, the agent knows (K_1) that p and that before the update she didn't know $(\neg K_0)$ p'. The above example is depicted in Fig. 1. For a proper exposition of private announcements in DEL, see ([4], p. 173).

© Springer-Verlag Berlin Heidelberg 2015
W. van der Hoek et al. (Eds.): LORI 2015, LNCS 9394, pp. 406–410, 2015.
DOI: 10.1007/978-3-662-48561-3_33

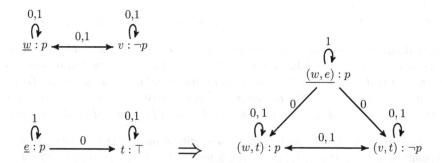

Fig. 1. An example of a private announcement to agent 1 in action model logic (AML). The upper left epistemic model $\mathcal{M}$ describes the initial ignorance of agents 0 and 1; the lower left model $\mathcal{E}$ describes the private announcement of p to 1; the right model $\mathcal{M} \times \mathcal{E}$ is the epistemic model after the announcement. 0 and 1 can be reinterpreted as stages of knowledge of a single agent.

The aim below is to present such a dynamic epistemic logic of stages of knowledge, as a modification of PAL.

2   Stages of Knowledge Logic SKL

The alphabet of stages of knowledge logic (SKL) is the same as of PAL. For the inductive definition of the language of SKL $\mathcal{L}_{SKL}$, we have the following restricted form of a PAL language

$$\varphi := \ | \ p \ | \ \neg\varphi \ | \ \varphi \wedge \psi \ | \ K_i\varphi \ | \ [\alpha]\psi \ \text{ s.t.}$$

$$\alpha := \ | \ p \ | \ \neg\alpha \ | \ \alpha_1 \wedge \alpha_2 \ | \ K_i\alpha \ | \ \alpha_1 \wedge [\alpha_1]\alpha_2$$

Let $\mathcal{L}_{EL}$ and $\mathcal{L}_{PL}$ denote the languages of epistemic logic without announcements and of propositional logic, respectively. In SKL, the K_i operator represents the single agent's knowledge after the i-th announcement. For that we define a degree function d that assigns a natural number to each occurrence of $[\varphi]\psi$ in an SKL formula, s.t. $d([\varphi]\psi) = i$ is read as 'φ is the ith announcement'. When clear by context, instead of writing $d([\varphi]\psi)$, we write $d(\varphi)$.

Definition 1. *Let $d : \mathcal{L}_{SKL} \to \mathbb{N}$, be a degree function, assigning every formula and sub-formula in $\mathcal{L}_{SKL}$ a natural number. Let $d_{max}(\alpha)$ assign for each α of the language the highest d of its sub-formulae. To determine the d of $\alpha \in \mathcal{L}_{SKL}$ and all of its sub-formulae, one applies the following tree rules, where at each node d is applied according to the below specification. For the root of the tree let $n = 0$.*

$$\varphi \wedge \psi \ n$$
$$\varphi \ n' = n + m(\varphi \wedge \psi) \qquad \psi \ n' = n + m(\varphi \wedge \psi)$$

$$\star\varphi \ n$$
$$|$$
$$\varphi \ n' = n + m(\star\varphi)$$

$$[\psi]\varphi \ n$$
$$\psi \ n' = n \qquad \varphi \ n' = n + m([\psi]\varphi)$$

(Where $\star = \neg, K_i$)
$d(\alpha) = n + m(\alpha)$. $m(\alpha) = 0$ if $\alpha \neq [\varphi]\psi$. Otherwise, $m([\varphi]\psi) = 1 + d_{max}(\varphi)$

Definition 2. *Given an announcement* $[\varphi]\psi$ *s.t.* $d(\varphi) = n$ *define its* φ*-list to be a list* $< \alpha_1'...\varphi >$ *s.t. the i-th member* $(1 \leq i \leq n)$ *of the list is the announcement of degree i that has, or is within an announcement that has,* ψ *in its scope. Given a* φ*-list, a* φ*-sequence is obtained by replacing any member* α_i *in the* φ*-list of the form* $\beta \wedge [\beta]\gamma$ *to* γ *s.t. the* φ *sequence contains only formulae of epistemic logic.*

The semantics of SKL can be seen as a modification of PAL semantics (for an exposition of the latter, see [4]). Given a model $\mathcal{M}$, after the announcement of φ, instead of moving to the model $\mathcal{M}_{|\varphi}$ in which all the φ states are eliminated as we do in PAL, in SKL we move to a certain *union* of the existing model $\mathcal{M}$ and its PAL update $\mathcal{M}_{|\varphi}$. This requires defining an initial SKL model, its PAL update model, and the union of the two.

Definition 3. *Given a single agent S5 epistemic model* $\mathcal{M} = (W, R, V)$ *its SKL initial model is a structure* $\mathcal{M}_0 = \langle W^0, \{R_i^0 | i \leq n\}, V^0 \rangle$, *where:*
$W^0 = \{(w, 0) | w \in W\}$.
$R_i^0 = R$ *for any* $0 \leq i \leq n$.
$(w, 0) \in V^0(p)$ *iff* $w \in V(p)$.

Definition 4. *Given a model* $\mathcal{M}_i$ *and an announcement* φ *s.t.* $d(\varphi) = i + 1$, *the PAL' model of* $\mathcal{M}_i$, *written* $\mathcal{M}_{i|\varphi}$ *is:*
$W_{|\varphi}^i = \{(w, i+1) : w \in W$ *and* $\mathcal{M}_i, (w, i) \models \varphi\}$
$(w, i+1)R_{j|\varphi}^i(v, i+1)$ *iff* $j \geq d(\varphi)$ *and* $(w, i)R_j^i(v, i)$
$(w, i+1) \in V_{|\varphi}^i(p)$ *iff* $w \in V(p)$.
We abbreviate (w, i) *as* w_i.

Definition 5. *Given an initial* $\mathcal{M}_0$ *model and an announcement* φ *s.t.* $d(\varphi) = n$ *and* $< \alpha_1...\alpha_n >$ *is the sequence of* φ, *define the model* $\mathcal{M}_n$ *to be:*
$W^n = W^{n-1} \cup W_{|\alpha_n}^{n-1}$
$R_j^n = R_j^{n-1} \cup R_{j|\alpha_n}^{n-1}$ *and if* $j < n$, *then* $w_n R_j^n v_j$ *iff* $w_j R_j^j v_j$
$V^n = V^{n-1} \cup V_{|\alpha_n}^{n-1}$

We read $w_x R_i^j u_y$ as 'in the model $\mathcal{M}_j$, w_x is related to u_y with relation i'. In the R clause of the definition we specify that a state w_n can 'look down' at a state u_j $(j < n)$ only with the relation j: $w_n R_j^n u_j$.

Definition 6. *Given an SKL model* $\mathcal{M}_i$ *we define satisfaction as usual with the following change*
$\mathcal{M}_i, w_j \models [\varphi]\psi$ *iff, if* $\mathcal{M}_i, w_j \models \varphi$, *then* $\mathcal{M}_{d(\varphi)}, w_{d(\varphi)} \models \psi$
$\mathcal{M}_i, w_j \models \langle\varphi\rangle\psi$ *iff* $\mathcal{M}_i, w_j \models \varphi$, *and* $\mathcal{M}_{d(\varphi)}, w_{d(\varphi)} \models \psi$

For a simple example of the SKL semantics, consider an agent who initially does not know p. After announcing p the agent knows p and that before the announcement he didn't know p: $\mathcal{M}_0, w_0 \models [p]K_1(p \wedge \neg K_0 p)$. This update is depicted in Fig. 2. Note that $d(p) = 1$. As was implied earlier, SKL can be reinterpreted as a multi agent logic with n agents s.t. the first announcement is given to agents 1 to n and excludes agent 0, the second is given to agents 2 to n and excludes agents 0 and 1, and so on. Similarly to PAL, SKL contains reduction axioms for announcements which allow the translation of each SKL formula to an epistemic logic formula, and which make the system complete [3].

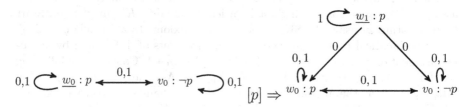

Fig. 2. The single agent who initially does not know p learns p. The left model is $\mathcal{M}_0$, the initial model. The right model is $\mathcal{M}_1$, the model after the first announcement. Note that in $\mathcal{M}_1$ relation R_1 is reflexive. In general, R_i is always reflexive in $\mathcal{M}_i$ [3].

3   SKL and the Knowability Principle

In the philosophical literature the KP is regularly formulated as $\varphi \rightarrow \Diamond K\varphi$. The *knowability paradox* connected to this formulation is the modal derivation showing that if the KP holds, then all truths are actually known: $\varphi \rightarrow K\varphi$ [2]. The KP in PAL is formulated as $\varphi \rightarrow \langle \psi \rangle K\varphi$ for some ψ, and it is invalid in the latter logic and in its relevant extensions [1], [5]. In this section, we show that for any φ, the formula $\varphi \rightarrow \langle \psi \rangle K_n \varphi$ is true for a 'high enough' n. To prove so, we first define the notion of bisimilarity (relevant to SKL).

Definition 7. *A non-empty relation $Z \subseteq W^n \times W^m$ is called a bisimulation from 0 to k between $\mathcal{M}_n$ and $\mathcal{M}_m$ iff the following conditions are satisfied:*
Atoms: If $w_i Z w_j$ then w_i and v_j satisfy the same propositional letters.
Forth: If $w_i Z w_j$ and $w_i R_l^n v_h$ (s.t. $l \leq k$), then there is a $v_x \in W^m$ s.t. $v_h Z v_x$ and $w_j R_l^m v_x$.
Back: If $w_i Z w_j$ and $w_j R_l^m v_h$ (s.t. $l \leq k$), then there is a $v_x \in W^n$ s.t. $v_x Z v_h$ and $w_i R_l^n v_x$. We write $(\mathcal{M}_n, w_i) \leftrightarrow (\mathcal{M}_m, w_j)$ for bisimilar states. If $(\mathcal{M}_n, w_i) \leftrightarrow (\mathcal{M}_m, w_j)$ from 0 to k, then $\mathcal{M}_n, w_i \models \varphi$ iff $\mathcal{M}_m, w_j \models \varphi$ for any $\varphi \in \mathcal{L}_{EL}$ with epistemic modalities $K_0...K_k$.

Lemma 1. $(\mathcal{M}_n, w_n) \leftrightarrow (\mathcal{M}_{n+1}, w_{n+1})$ *from 0 to n.*

proof sketch: Define a relation $Z \subseteq W^n \times W^{n+1}$ s.t. for any w_i in W^n and W^{n+1} ($i \leq n$) respectively, $w_i Z w_i$, and for $w_{n+1} \in W^{n+1}$, let $w_n Z w_{n+1}$. Then Z is bisimulation for the modalities $K_0...K_n$. For a full proof, see the extended version of this contribution ([3], p. 61).

Given the bisimilarity result one can show that a KP is valid in SKL.

Theorem 1. *For an arbitrary initial SKL model* $(\mathcal{M}_0, w_0)$,

$$\mathcal{M}_0, w_0 \models \varphi \to \langle\psi\rangle K_n \varphi$$

for some ψ *and a* K_n *s.t.* $n \geq 1$.

Proof. The proof goes by the construction of a formula ψ and stage n for each given φ of the language of SKL. If $\varphi \in \mathcal{L}_{PL}$ then let $\varphi = \psi$ and $n = 1$. Then for such φ, $\mathcal{M}_0, w_0 \models \varphi \to \langle\varphi\rangle K_1 \varphi$ can be easily checked to be true. Otherwise, assume $\varphi' \in \mathcal{L}_{SKL}$ s.t. the highest knowledge modality K_m in φ' is m. Start by translating φ' using the SKL announcement axioms to a formula $\varphi \in \mathcal{L}_{EL}$. Let ψ be identical to a sequence of m announcements of $\top$ followed by an announcement of φ, i.e. $\psi = \langle\top\rangle...\langle\top\rangle\langle\varphi\rangle$ s.t. $d(\varphi) = m+1$. Consider the following KP for φ: $\mathcal{M}_0, w_0 \models \varphi \to \langle\top\rangle...\langle\top\rangle\langle\varphi\rangle K_{m+1}\varphi$. Assume $\mathcal{M}_0, w_0 \models \varphi$. Then $\mathcal{M}_m, w_m \models \varphi$ as the announcements of $\top$ do not change the truth value of φ [3]. Hence, in order to show that $\mathcal{M}_m, w_m \models \langle\varphi\rangle K_{m+1}\varphi$ it remains to show that $\mathcal{M}_{m+1}, w_{m+1} \models K_{m+1}\varphi$. Pick an arbitrary u_x s.t. $w_{m+1} R_{m+1}^{m+1} u_x$. By Definition 5 it must be that $u_x = u_{m+1}$ and by the assumption of the existence of u_{m+1} it follows that $\mathcal{M}_m, u_m \models \varphi$. By Lemma 1 $(\mathcal{M}_m, u_m) \leftrightarrow (\mathcal{M}_{m+1}, u_{m+1})$ from 0 to m, and since by assumption φ contains modalities up to m, it follows that $\mathcal{M}_{m+1}, u_{m+1} \models \varphi$. Therefore, $\mathcal{M}_{m+1}, w_{m+1} \models K_{m+1}\varphi$ and so $\mathcal{M}_m, w_m \models \langle\varphi\rangle K_{m+1}\varphi$. Note that sequence of announcements $\langle\top\rangle...\langle\top\rangle\langle\varphi\rangle$ is equivalent to one nested announcement $\langle\psi'\rangle$ in SKL as in PAL.

Unlike other epistemic logics that can express a KP, SKL avoids the knowability paradox: all truths are knowable but not all truths are known. I note that while the standard exposition of the paradox assumes the necessitation rule for the possibility operator $\Diamond$ to derive the paradox [2], the necessitation rule for announcements is unsound in SKL [3].

References

1. van Benthem, J.: What one come to know. Analysis 64(282), 95–105 (2004)
2. Brogaard, B., Salerno, J.: Fitch's Paradox of Knowability. In: Zalta, E.N. (ed.) The Stanford Encyclopedia of Philosophy (2013), http://plato.stanford.edu/archives/win2013/entries/fitch-paradox
3. Cohen, M.: Dynamic Knowability: The Knowability Paradox in Dynamic Epistemic logic. MA Thesis, LMU Munich (2015)
4. van Ditmarsch, H.P., van der Hoek, W., Kooi, B.: Dynamic Epistemic Logic. Springer, Heidelberg (2007)
5. van Ditmarsch, H.P., van der Hoek, W., Iliev, P.: Everything is knowable – How to Get to Know Whether a Proposition Is True. Theoria 78(2), 93–114 (2012)

Reflective Oracles: A Foundation for Game Theory in Artificial Intelligence*

Benja Fallenstein[1], Jessica Taylor[1], and Paul F. Christiano[2]

[1] Machine Intelligence Research Institute, Berkeley, USA
{benja,jessica}@intelligence.org
[2] UC Berkeley, Berkeley, USA
paulfchristiano@eecs.berkeley.edu

Abstract. Game theory treats players as special: A description of a game contains a full, explicit enumeration of all players. This isn't a realistic assumption for autonomous intelligent agents. In this paper, we propose a framework in which agents and their environments are both modelled as probablistic oracle machines with access to a "reflective" oracle, which is able to answer questions about the outputs of other machines with access to the same oracle. These oracles avoid diagonalization problems by answering some queries randomly. Agents make decisions by asking the oracle questions about their environment, which they model as an arbitrary oracle machines. Since agents are themselves oracle machines, the environment can contain other agents as non-distinguished subprocesses, removing the special treatment of players in the classical theory. We show that agents interacting in this way play Nash equilibria.

1 Introduction

Classical game theory treats players as special: A description of a game contains a full, explicit enumeration of all players [5]. This isn't a realistic assumption for autonomous intelligent agents. Ideally, such agents would be based on a decision-theoretic foundation for game theory in which their coplayers are a non-distinguished part of the agent's environment, but it is non-trivial to find such a foundation. Attempts to model both players and the environment as Turing machines, for example, fail for standard diagonalization reasons [1].

In this paper, we consider oracle machines with access to a "reflective" oracle, which is able to answer questions about the outputs of other machines with access to the same oracle. These oracles avoid diagonalization by answering some queries randomly. We show that machines with access to a reflective oracle can be used to define rational agents using causal decision theory [8]. These agents model their environment as a probabilistic oracle machine, which may contain other agents as a non-distinguished part.

We show that if such agents interact, they will play a Nash equilibrium, with the randomization in mixed strategies coming from the randomization in the

---

* An extended version of this paper is available as a technical report [3].

© Springer-Verlag Berlin Heidelberg 2015
W. van der Hoek et al. (Eds.): LORI 2015, LNCS 9394, pp. 411–415, 2015.
DOI: 10.1007/978-3-662-48561-3_34

oracle's answers. This can be seen as providing a foundation for classical game theory in which players aren't special.

This method can be applied to *Solomonoff induction* [7], a method for sequence prediction that can learn sequences generated by arbitrary computer programs, and Marcus Hutter's AIXI [4], a "universally intelligent" agent that can learn to interact with arbitrary computable environments. Using reflective oracles, it is possible to define variants of Solomonoff induction and AIXI whose hypothesis space contains environments inhabited by other predictors or agents of the same type [2]. This is not true of the original versions, since they are uncomputable but consider only computable hypotheses.

Let $\mathcal{M}$ be the set of *probabilistic oracle machines,* defined here as Turing machines which can execute special instructions to (i) flip a coin that has an arbitrary rational probability of coming up heads, and to (ii) call an oracle O, whose behavior might itself be probabilistic.

Roughly speaking, the oracle answers questions of the form: "Is the probability that machine M returns 1 greater than p?" Thus, O takes two inputs, a machine $M \in \mathcal{M}$ and a rational probability $p \in [0, 1] \cap \mathbb{Q}$, and returns either 0 or 1. If M is guaranteed to halt and to output either 0 or 1 itself, we want $O(M, p) = 1$ to mean that the probability that M returns 1 (when run with O) is at least p, and $O(M, p) = 0$ to mean that it is at most p; if it is equal to p, both conditions are true, and the oracle may answer randomly. In summary,

$$
\begin{aligned}
\mathbb{P}(M^O() = 1) > p &\implies \mathbb{P}(O(M, p) = 1) = 1 \\
\mathbb{P}(M^O() = 1) < p &\implies \mathbb{P}(O(M, p) = 0) = 1
\end{aligned}
\tag{1}
$$

where we write $\mathbb{P}(M^O() = 1)$ for the probability that M returns 1 when run with oracle O, and $\mathbb{P}(O(M, p) = 1)$ for the probability that the oracle returns 1 on input (M, p). We assume that different calls to the oracle are stochastically independent events (even if they are about the same pair (M, p)); hence, the behavior of an oracle O is fully specified by the probabilities $\mathbb{P}(O(M, p) = 1)$.

Definition 1. *A query (with respect to a particular oracle O) is a pair (M, p), where $p \in [0, 1] \cap \mathbb{Q}$ and $M^O()$ is a probabilistic oracle machine which almost surely halts and returns an element of $\{0, 1\}$. An oracle is called* reflective on R, *where R is a set of queries, if it satisfies the two conditions displayed above for every $(M, p) \in R$. It is called* reflective *if it is reflective on the set of all queries.*

Theorem 1. *(i) There is a reflective oracle. (ii) For any oracle O and every set of queries R, there is an oracle O' which is reflective on R and satisfies $\mathbb{P}(O'(M, p) = 1) = \mathbb{P}(O(M, p) = 1)$ for all $(M, p) \notin R$.*

Proof. For the proof of (ii), see Appendix B of the extended version of this paper [3]; see also Theorem 5.1 in [3], which gives a more elementary proof of a special case. Part (i) follows from part (ii) by choosing R to be the set of all queries and letting O be arbitrary.

As an example, consider the machine given by $M^O() = 1 - O(M, 0.5)$, which implements a version of the liar paradox by asking the oracle what it will return and then returning the opposite. By the existence theorem, there is an oracle which is reflective on $R = \{(M, 0.5)\}$. This is no contradiction: We can set $\mathbb{P}(O(M, 0.5) = 1) = \mathbb{P}(O(M, 0.5) = 0) = 0.5$, leading the program to output 1 half the time and 0 the other half of the time.

2   From Reflective Oracles to Causal Decision Theory

We now show how reflective oracles can be used to implement a perfect Bayesian reasoner. We assume that each possible environment that this agent might find itself in can likewise be modeled as an oracle machine; that is, we assume that the laws of physics are computable by a probabilistic Turing machine with access to the same reflective oracle as the agent. For example, we might imagine our agent as being embedded in a Turing-complete probabilistic cellular automaton, whose laws are specified in terms of the oracle.

Here, we consider agents implementing causal decision theory (CDT) [8], which evaluates actions according to the consequences they cause: For example, if the agent is a robot embedded in a cellular automaton, it might evaluate the expected utility of taking action 0 or 1 by simulating what would happen in the environment if the output signal of its decision-making component were replaced by either 0 or 1. We will assume that the agent's model of the counterfactual consequences of taking different actions a is described by a machine $W_A^O(a)$; e.g.,

$$W_A^O(a) = \begin{cases} \$20 & \text{if } a = 0 \\ \$15 & \text{otherwise} \end{cases} \tag{2}$$

We assume that the agent has a utility function over outcomes, $u(\cdot)$, implemented as a lookup table, which takes rational values in $[0, 1]$.[1] Furthermore, we assume that both $W_A^O(0)$ and $W_A^O(1)$ halt almost surely and return a value in the domain of $u(\cdot)$. Causal decision theory then prescribes choosing the action that maximizes expected utility; in other words, we want to the agent to be a machine $A^O()$, returning the agent's action, such that

$$A^O() = \operatorname*{argmax}_a \mathbb{E}\left[u\left(W_A^O(a)\right)\right] \tag{3}$$

In the case of ties, any action maximizing utility is allowed, and it is acceptable for $A^O()$ to randomize.

We cannot compute this expectation by simply running $u(W_A^O(a))$ many times to obtain samples, since the environment might contain other agents of the same type, potentially leading to infinite loops. However, we can find an optimal action by making use of a reflective oracle. This is easiest when the agent has only two

---

[1] Since the meaning of utility functions is invariant under affine transformations, the choice of the particular interval $[0, 1]$ is no restriction.

actions (0 and 1), but similar analysis extends to any number of actions. Define a machine

$$E^O() := \mathrm{flip}\big(\big(u(W_A^O(1)) - u(W_A^O(0)) + 1 \big)/2 \big) \tag{4}$$

where $\mathrm{flip}(p)$ is a probabilistic function that returns 1 with probability p and 0 with probability $1 - p$.

Theorem 2. *O is reflective on $\{(E, 1/2)\}$ if and only if $A^O() := O(E, 1/2)$ returns a utility-maximizing action.*

Proof. The demand that $A^O()$ return a utility-maxmizing action is equivalent to

$$\begin{aligned} \mathbb{E}[u(W_A^O(1))] > \mathbb{E}[u(W_A^O(0))] &\implies A^O() = 1 \\ \mathbb{E}[u(W_A^O(1))] < \mathbb{E}[u(W_A^O(0))] &\implies A^O() = 0 \end{aligned} \tag{5}$$

We have

$$\mathbb{P}(E^O() = 1) = \mathbb{E}\left[\big(u(W_A^O(1)) - u(W_A^O(0)) + 1 \big)/2 \right] \tag{6}$$

It is not difficult to check that $\mathbb{E}[u(W_A^O(1))] \gtrless \mathbb{E}[u(W_a^O(0))]$ iff $\mathbb{P}(E^O() = 1) \gtrless 1/2$. Together with the definition of $A^O()$, we can use this to rewrite the above conditions as $\mathbb{P}(E^O() = 1) > 1/2 \implies O(E, 1/2) = 1$ and similarly, $\mathbb{P}(E^O() = 1) < 1/2 \implies O(E, 1/2) = 0$. But this is precisely the definition of "O is reflective on $\{(E, 1/2)\}$".

In order to handle agents which can choose between more than two actions, we can compare action 0 to action 1, then compare action 2 to the best of actions 0 and 1, then compare action 3 to the best of the first three actions, and so on. Adding more actions in this fashion does not substantially change the analysis.

3   From Causal Decision Theory to Nash Equilibria

Since we have taken care to define our agents' world models $W_A^O(a)$ in such a way that they can embed other agents,[2] we need not do anything special to pass from single-agent to multi-agent settings. We represent the environment as a machine $F^O(a_1, \ldots, a_n)$ which takes each agent's action and returns an outcome. Then, the true distribution of outcomes is given by $F^O(A_1^O(), \ldots, A_n^O())$, and the causal counterfactuals of agent i are given by $W_i^O(a_i) := F^O(a_i, A_{-i}^O()) := F(A_1^O(), \ldots, A_{i-1}^O(), a_i, A_{i+1}^O(), \ldots, A_n^O())$.

We assume that each agent has a utility function $u_i(\cdot)$ of the same type as in the previous subsection. Hence, we can define the agent programs $A_i^O()$ just as before: We set $A_i^O() = O(E_i, 1/2)$, where

$$E_i^O() = \mathrm{flip}\big(\big(u_i(W_i^O(1)) - u_i(W_i^O(0)) + 1 \big)/2 \big) \tag{7}$$

---

[2] More precisely, we have only required that $W_A^O(a)$ always halt and produce a value in the domain of the utility function $u(\cdot)$. Since all our agents do is to perform a single oracle call, they always halt, making them safe to call from $W_A^O(a)$.

Here, each $E_i^O()$ calls $W_i^O()$, which calls $A_j^O()$ for each $j \neq i$, which refers to the source code of $E_j^O()$, but Kleene's second recursion theorem shows that this kind of self-reference poses no theoretical problem [6].

This setup very much resembles the setting of normal-form games. In fact:

Theorem 3. *Given an oracle O, consider the n-player normal-form game in which the payoff of player i, given the pure strategy profile $(a_1, \ldots, a_n)$, is $\mathbb{E}[u_i(F^O(a_1, \ldots, a_n))]$. The mixed strategy profile given by $s_i := \mathbb{P}(A_i^O() = 1)$ is a Nash equilibrium of this game if and only if O is reflective on $\{(E_1, 1/2), \ldots, (E_n, 1/2)\}$.*

Proof. $(s_1, \ldots, s_n)$ is a Nash equilibrium iff a pure strategy a_i is only assigned positive probability if it maximizes $\mathbb{E}[u_i(F^O(a_i, A_{-i}^O()))] = \mathbb{E}[u_i(W_i^O(a_i))]$. By Theorem 2, this is equivalent to O being reflective on $\{(E_i, 1/2)\}$.

Note that, in particular, any normal-form game with rational-valued payoffs can be represented in this way. The theorem shows that then, every reflective oracle (which exists by Theorem 1) gives rise to a Nash equilibrium. Theorem 3 together with Theorem 1(ii) show that for any Nash equilibrium $(s_1, \ldots, s_n)$ of the normal-form game, there is a reflective oracle such that $\mathbb{P}(A_i^O() = 1) = s_i$.

4 Conclusions

In this paper, we have introduced *reflective oracles*, which are able to answer questions about the behavior of oracle machines with access to the same oracle. We've shown that such oracle machines can implement a version of causal decision theory. We have focused on answering queries about oracle machines that halt with probability 1, but the reflection principle presented in Section 1 can be modified to apply to machines that do not necessarily halt [3]. This can be used to define reflective variants of Solomonoff induction and AIXI [2].

References

1. Binmore, K.: Modeling rational players: Part I. Economics and Philosophy 3(02), 179–214 (1987)
2. Fallenstein, B., Soares, N., Taylor, J.: Reflective variants of Solomonoff induction and AIXI. In: Bieger, J., Goertzel, B., Potapov, A. (eds.) AGI 2015. LNCS, vol. 9205, pp. 60–69. Springer, Heidelberg (2015)
3. Fallenstein, B., Taylor, J., Christiano, P.: Reflective oracles: A foundation for classical game theory. Tech. rep., Machine Intelligence Research Institute (2015), http://intelligence.org/files/ReflectiveOracles.pdf
4. Hutter, M.: Universal Artificial Intelligence. Texts in Theoretical Computer Science. Springer (2005)
5. Kreps, D.M.: A Course in Microeconomic Theory. Princeton University Press (1990)
6. Rogers, H.: Theory of Recursive Functions and Effective Computability. McGraw-Hill (1967)
7. Solomonoff, R.J.: A formal theory of inductive inference. Part I 7(1), 1–22 (1964)
8. Weirich, P.: Causal decision theory. In: Zalta, E.N. (ed.) The Stanford Encyclopedia of Philosophy (2012)

Infinite Ordinals and Finite Improvement

Aaron Hunter

British Columbia Institute of Technology
aaron_hunter@bcit.ca

Abstract. Ordinal Conditional Functions assign every state an ordinal value representing its plausibility. In most existing work, plausibility values are restricted to the set of natural numbers; however, the fully general theory allows infinite values as well. In this paper, we explore simple arithmetical approaches to belief revision with ordinal conditional functions that might take infinite plausibility values. We suggest that infinite values need not be seen as a mathematical artifact of the theory; they provide a natural tool for a generalized form of conditional reasoning.

1 Introduction

The theory of belief change is concerned with the way agents incorporate new information. Many formal models of belief change require an agent to have some form of *ordering* or *ranking* that gives the relative plausibility of possible states. One well-known tool for representing plausibility is an Ordinal Conditional Function (OCF), which is just a function from states to ordinals [Spo88, Wil94]. While the original definition allows the range of an OCF to be any ordinal, in existing work it is common to restrict the range to the natural numbers, possibly with an additional symbol ∞ representing impossibility. In this paper, we demonstrate the utility of a wider range of ordinal plausibility values.

2 Preliminaries

Belief revision is the form of belief change that occurs when new information is presented to an agent with some a priori beliefs. We assume an underlying propositional signature $\mathbf{P}$. An interpretation over $\mathbf{P}$ is called a *state*, while a logically closed set of formulas over $\mathbf{P}$ is called a *belief set*. A belief revision operator is a function that combines the initial belief set and a formula to produce a new belief set. While OCFs use a quantitiative ranking function to represent plausibility, other formal approaches often rely on an underlying ordering over states [AGM85, KM92, DP97].

We refer the reader to [Dev93] for an excellent introduction to infinite ordinals and ordinal arithmetic. For our purposes, it is sufficient to note that ordinals are sets defined by an "order type." The finite ordinals are the natural numbers, and the first infinite ordinal is the set ω of all natural numbers. It is easy to construct a countably infinite set that is not order isomorphic to ω: just add

W. van der Hoek et al. (Eds.): LORI 2015, LNCS 9394, pp. 416–420, 2015.
DOI: 10.1007/978-3-662-48561-3_35

another symbol ∞ at the end that is larger than every natural number. The ordinal that defines the order type of this set is written $\omega + 1$. Similarly, there exists a distinct ordinal $\omega + n$ for any natural number n. If we add a complete copy of the natural numbers, then we have the ordinal $\omega + \omega$ which is normally written as $\omega \cdot 2$. By taking powers, we can get even more order types; we will not delve further into this topic.

There has been some recent work on the use of infinite valued ordinals in OCFs [Kon09]. Our approach is different in that we explicitly use the ordering on limit ordinals to represent infinite leaps in plausibility.

3   Algebra on Ordinal Conditional Functions

The following definition allows us to define conditional functions over sets of ordinals.

Definition 1. *Let Γ be a collection of ordinals. A Γ-CF (Γ conditional function) over a set S is a function $r : S \to \Gamma$ such that $r(s) = 0$ for some state s.*

In this paper, we are primarily interested in the class of ω^2-CFs. Every element of ω^2 can be written as $\omega \cdot k + c$ for some k and c.

Definition 2. *Let r be an ω^2 ranking with $\min(f) = \omega \cdot k + c$. Then k is the degree of f and c is the finite shift, written $deg(f)$ and $fin(f)$ respectively.*

We think of these conditional functions as having countably many infinite levels of implausibility. If r is a Γ-CF, we define $Bel(r) = \{x \mid r(x) = 0\}$.

We use the term Γ ranking to refer to an arbitrary function from S to Γ. For Γ rankings r_1 and r_2, we write $r_1 \sim r_2$ just in case $r_1(s) < r_1(t) \iff r_2(s) < r_2(t)$ holds for all s, t. Clearly $\sim$ is an equivalence relation.

Definition 3. *Let r be an ω^2 ranking with $deg(r) = k$ and $fin(r) = c$. Define $\bar{r}$ as follows. Let s be a state with $r(s) = \omega \cdot m + p$.*

1. If $m > k$, then $\bar{r}(s) = \omega \cdot (m - k) + c$.
2. If $m = k$, then $\bar{r}(s) = (p - c)$.

We call $\bar{r}$ the *finite zeroing* of r. Intuitively, elements at the "lowest level" are normalized to zero and elements at higher levels are shifted down by the degree of r. The following result is easy to prove.

Proposition 1. *If r is an ω^2 ranking, then $\bar{r}$ is a ω^2-CF and $r \sim \bar{r}$.*

Hence, the finite zeroing of any ranking is an equivalent ω^2-CF.

Definition 4. *Let r_1, r_2 be ω^2-CFs. Then $r_1 * r_2 = \overline{r_1 + r_2}$.*

We think of $*$ as a revision operator. Let r_1 be a ω-CF representing the initial beliefs of an agent. Let ϕ be a formula, let d be a positive integer, and let r_2 be the ranking function defined as follows:

$$r_2(s) = \begin{cases} 0 \text{ if } s \models \phi \\ d \text{ otherwise} \end{cases}$$

Then $r_1 * r_2$ is equivalent to Spohn's conditionalization of r_1 by ϕ with strength d. Similarly, if r_2 takes only two values and the degree of strength of r_2 is strictly larger than the degree of strength of r_1, then $r_1 \mp r_2$ is AGM revision. Morever, in the ω-CF case, $*$ is equivalent to addition followed by minimization; hence the algebra defined by $*$ for finite valued functions is an abelian group. In the ω^2 case, the algebra is different.

Proposition 2. *The class of ω^2-CFs is a non-abelian group under $*$. (i.e. it is closed, associative, and every element has an inverse, but it is not commutative).*

The fact that $*$ is not commutative has interesting consequences, notably the fact that infinite jumps in plausibility outweigh concerns of primacy and recency.

We consider the relationship with the *improvement operators* of [KP08], which are belief change operators that satisfy a set of postulates that includes the following:

(I1) *There exists $n \in \mathbf{N}$ such that $B(\Psi \circ^n \phi) \vdash \phi$.*

Here Ψ is an epistemic state, and $B(\cdot)$ maps an epistemic state to the minimal elements of the underlying ordering. An analogous statement can be formulated in our context.

We say that an OCF r is a ϕ-strengthening iff $Bel(r) = \{s \mid s \models \phi\}$. If r_ϕ denotes a ϕ-strengthening, we can express the content of **(I1)** as follows.

(I*) *There exists $n \in \mathbf{N}$ such that $Bel(r *^n r_\phi) \models \phi$.*

It turns out that this property holds for ω-CFs.

Proposition 3. *If r is an ω-CF and r_ϕ is a ϕ-strengthening with finite strength, then $\mathbf{I^*}$ holds.*

In fact, it turns out that finite strengthenings define so-called *weak improvement* operators in this context. However, if we move to ω^2-CFs, this is no longer the case.

Proposition 4. *If r is an ω^2-CF and r_ϕ is a ϕ-strengthening with finite strength, then $\mathbf{I^*}$ need not hold.*

This result essentially states that no finite sequence of improvements at level d will ever impact the actual beliefs at lower levels.

4   Nearly Counterfactual Reasoning

Suppose that we initially believe that it is impossible for dogs to fly. Now suppose we are told that flying things have hollow bones. This does not give us any new information about dogs. Suppose we subsequently become convinced of the existence of flying dogs. Ideally, we should incorporate the fact about hollow bones: we should believe that flying dogs, unlike regular dogs, have hollow bones.

We refer to the reasoning in the preceding example as *nearly-counterfactual* reasoning. This kind of reasoning falls somewhere between conditional reasoning and counterfactual reasoning. Kern-Isberner has previously addressed conditional reasoning with (finite-valued) OCFs [Ker99]. In this section, we discuss the way that infinite values can help when reasoning about highly unlikely states. In our example, we would like to keep the information about flying dogs (that they have hollow bones), for the unlikely case where they happen to exist. Hence, we suggest that our beliefs should not be unchanged; they should be changed in sort of an infinitesimally small way. While our beliefs about the actual world do not change, our beliefs about some (nearly) impossible worlds do, in fact, change.

In order to capture nearly counterfactual reasoning, We follow the basic intuition of Lewis: the truth of a counterfactual sentence is determined by its truth in alternative worlds [Lew73]. We can represent this idea with ω^2-CFs. At each limit ordinal $\omega \cdot k$, we essentially have an entirely new plausibility ordering. As k increases, each such ordering represents an increasingly implausible world. However, a sufficiently strong observation can force our beliefs to jump to any of these unlikely worlds. As such, these are not counterfactual worlds, because we admit the possibility that they may eventually be believed possible.

The important property that we can capture with ω^2-CFs is the following: there are some formulas that may be true, yet we can not be convinced to believe them based on any finite number of pieces of "weak evidence."

Definition 5. *If r is an ω^2-CF, a formula ϕ is nearly counterfactual with respect to r just in case there is no ω-CF r' such that $Bel(r * r') \models \phi$.*

The following is an immediate consequence of this definition.

Proposition 5. *If ϕ is nearly counterfactual with respect to r, then there is no finite sequence $r_1, \ldots, r_n$ of ω-CFs such that $Bel(r * r_1 * \cdots * r_n) \models \phi$.*

We are interested in modelling revision by nearly counterfactual conditionals. Let r be an ω^2-CF and let B, A be formulas. Let $deg(A)$ be the least natural number k such that there is some state s such that $s \models A$ and $r(s) = \omega \cdot k + c$. For any natural number n let $r(n, B|A)$ be the function defined as follows:

$$r(n, B|A)(s) = \begin{cases} r(s), \text{ if } deg(s) \neq k \\ r_k * r(B, n) \text{ otherwise} \end{cases}$$

We call this function the n-stengthening of B conditioned on A. This function finds the least level of r containing an A-state, and then essentially revises that level by B. In the flying-dog example, this approach allows us to conclude that hollow bones are more possible in cases where dogs can fly. This may be important for counterfactual reasoning, and it may be important if we actually find a flying dog. Roughly speaking, after revising by $(hollow|fly)$, we now believe that hollow bones are more plausible in all states where a dog can fly. However, our beliefs about the actual world have not changed.

5   Conclusion

We have explored the use of infinite ordinals for reasoning about belief change, focusing on the case where plausibility values are drawn from ω^2. Here we have multiple infinite levels of implausibility that can represent situations where stubbornly held beliefs are resistant to evidence. This allows us to capture a notion of belief improvement, in which no finite number of improvements will actually lead to a change in the belief state. We also discussed the suitability of this model for "nearly counterfactual" revision, where we incorporate information that is conditional on a highly unlikely statement. In future work, we intend to move beyond ω^2-CFs, to completely characterize the relationship with improvement operators, and to consider practical applications of nearly counterfactual reasoning.

References

[AGM85]   Alchourrón, C.E., Gärdenfors, P., Makinson, D.: On the logic of theory change: Partial meet functions for contraction and revision. Journal of Symbolic Logic 50(2), 510–530 (1985)

[Dev93]   Devlin, K.: The Joy of Sets. Springer (1993)

[DP97]   Darwiche, A., Pearl, J.: On the logic of iterated belief revision. Artificial Intelligence 89(1-2), 1–29 (1997)

[Ker99]   Kern-Isberner, G.: Postulates for conditional belief revision. In: Proceedings of the Sixteenth International Joint Conference on Artificial Intelligence (IJCAI), pp. 186–191 (1999)

[KM92]   Katsuno, H., Mendelzon, A.O.: Propositional knowledge base revision and minimal change. Artificial Intelligence 52(2), 263–294 (1992)

[Kon09]   Konieczny, S.: Using transfinite ordinal conditional functions. In: Sossai, C., Chemello, G. (eds.) ECSQARU 2009. LNCS, vol. 5590, pp. 396–407. Springer, Heidelberg (2009)

[KP08]   Konieczny, S., Péréz, R.P.: Improvement operators. In: Eleventh International Conference on Principles of Knowledge Representation and Reasoning (KR 2008), pp. 177–186 (2008)

[Lew73]   Lewis, D.: Counterfactuals. Harvard University Press (1973)

[Spo88]   Spohn, W.: Ordinal conditional functions. A dynamic theory of epistemic states. In: Harper, W.L., Skyrms, B. (eds.) Causation in Decision, Belief Change, and Statistics, vol. II, pp. 105–134. Kluwer Academic Publishers (1988)

[Wil94]   Williams, M.A.: Transmutations of knowledge systems. In: Proceedings of the Fourth International Conference on the Principles of Knowledge Representation and Reasoning (KR 1994), pp. 619–629 (1994)

Solving the HI-LO Puzzle

Brian Kim

Oklahoma State University

Abstract. I defend classical decision theory from the challenge posed
to it by the HI-LO game.

The HI-LO game poses a problem for classical decision theory and game theory.
In the two player game, both players have the option of selecting either HI or LO.
If both select HI, they both receive a higher payoff than if they both select LO. If
the two players select different options, they receive nothing. For the discussion
that follows, let's assume a version of the game with a 10:1 ratio between payoffs.

When faced with this game, human players will almost always select HI.[1] And
it is also clear that if playing this game, one ought to select HI. Unfortunately,
traditional decision theory and game theory has no obvious explanation of why
HI is the uniquely rational choice. And some have gone so far as to argue that
with its focus on individual rationality and reasoning, the traditional views are
simply unable to produce the right result.[2]. In defense of the traditional account,
I will propose one explanation of why HI is the uniquely rational choice from
a traditional point of view. And since HI appears to not only be the uniquely
rational choice but also a dominant choice, my aim will be to provide this type
of solution.

The challenge posed by the HI-LO game is to provide some explanation of
how players, who only possess a common knowledge of rationality and whose
only aim is to maximize their own payoffs, can reason to the conclusion that
they both ought to choose HI. The task is made all the more difficult because
when restricted to a common knowledge of rationality, each player is initially
completely ignorant of what the other player is going to do. For assigning any
determinate probability to the other player's choice is tantamount to deciding
what one ought to do.[3] And so one cannot assume such a probability assignment
without already possessing an argument for what one ought to do. So whatever
the argument for choosing HI is, it must begin with an assumption of ignorance.
But how can we come to the judgment that some action is best when we are
operating under such massive uncertainty? To solve this problem, I will appeal
to the principle of strong dominance. After all, this principle is one of the least
controversial principles governing decisions under uncertainty.

To capture how Player 1 could reason, we begin at the initial state where he
is ignorant of how Player 2 will choose. And while Player 1 lacks any evidence

---

[1] [3], fn. 3
[2] [1]
[3] See [2] for further discussion.

© Springer-Verlag Berlin Heidelberg 2015
W. van der Hoek et al. (Eds.): LORI 2015, LNCS 9394, pp. 421–425, 2015.
DOI: 10.1007/978-3-662-48561-3_36

about how and what Player 2 will choose, he does know that a choice must made. This is, after all, a forced choice scenario. So if Player 2 must make a choice, then there must be some way that she chooses.[4]

I believe the key to solving the puzzle is by considering how one might reason about what to do given that the other player adopts some way or other of choosing. So Player 1 can start by imagining a variety of choice procedures that Player 2 uses. A choice procedure is any procedure that someone can use to choose between options. For the sake of simplicity, I will only consider choice procedures for picking between HI and LO. For example, Player 2 might pick the option that is mentioned first. Alternatively, she can ask her aunt, pick the option that is aesthetically pleasing, or draw straws.

We can abstract away from the particular details of choice procedures since the only feature of a choice procedure that matters for determining what one ought to do is the likelihood that such a procedure will result in the choice of HI or LO. Thus, for any choice procedure c, let us define $P_c(\text{HI})$ and $P_c(\text{LO})$ as the probability that a player's use of c will result in respectively selecting HI or LO. Therefore, choice procedures can be understood as flips of coins whose bias determines P_c. Of course, I am not proposing that we imagine Player 2 actually uses a coin. Rather, I am proposing to model all possible choice procedures that Player 2 could use as if they were flips of a biased coin.

Suppose that Player 1 begins by imagining that Player 2 chooses as if she is flipping some particular coin. On this supposition, he can judge the likelihood of HI and LO by considering the bias of the coin and then calculate the expected payoff of his own choices. Of course, while Player 1 can engage in such suppositional reasoning, he would not be justified in assuming that Player 2 has selected one or another choice procedure. After all, Player 1 does not yet possess any information about how Player 2 will choose. Furthermore, in our modeling, it is clear that Player 1's ignorance should extend to how the coin is used as a choice procedure. Suppose that C is a choice procedure that uses this particular coin. In order to use a coin flip to make a choice, the two sides of the coin must be associated with the two choices. So let us suppose that C is the choice procedure where HEADS is associated with HI and TAILS is associated with LO. Of course, Player 1 has no reason to think that the coin would be used in one way rather than another. After all, Player 2 could have associated HEADS with LO and TAILS with HI. Thus, given Player 1's ignorance and uncertainty when considering the possibility that Player 2 uses some arbitrary choice procedure c, he also ought to consider the possibility that Player 2 uses the mirror image of c – the procedure in which the two sides of the coin are associated with different options. Let us call this c* – the choice procedure whose probabilities are given by P_{c*}, where $P_{c*}(\text{HI})=P_c(\text{LO})$. Call c and c* a *matching pair* of choice procedures.

---

[4] I am not assuming that there is a reason Player 2 chooses one option rather than the other. Player 2 may simply pick. Paradigmatic cases of picking are those where one picks between indistinguishable options (c.f. [4]).

It may at first seem as though the appeal to matching pairs makes sense only if we are talking about coin flips. However, it's easy to see how the notion of matching pairs applies to any type of choice procedure. For example, suppose I decide to ask my aunt whether to choose HI or LO. I could choose HI if she asserts HI but I could also choose LO if she asserts HI. To construct a matching pair, one first identifies a way of differentiating the two choices - in terms of flips of a coin, assertions of an aunt, lexical ordering, etc. And then one can identify two ways of using this differentiation to make one's choice. Choosing the HI side of the coin if it lands face up or face down. Choosing what one's aunt says or the opposite. Choosing what alphabetically comes first or second. Thus, the notion of matching pairs applies to all choice procedures.

In order to capture Player 1's initial ignorance of Player 2's choice, I propose that Player 1 must always consider matching pairs of choice procedures together. Now how might Player 1 reason given this way of considering his ignorance of Player 2's choice? As I noted above, the least controversial principles that are relevant in cases of uncertainty are dominance principles. For our purposes, we will only need a strong dominance principle, and for the sake of simplicity, I will only discuss a principle that applies in the HI-LO game. The following, of course, can easily be generalized.

Strong Dominance*: HI≻LO if (1) for all matching pairs of choice procedures, $EU_c(\text{HI}) > EU_c(\text{LO})$ or $EU_{c^*}(\text{HI}) > EU_{c^*}(\text{LO})$ and (2) for some matching pair of choice procedures, $EU_c(\text{HI}) > EU_c(\text{LO})$ and $EU_{c^*}(\text{HI}) > EU_{c^*}(\text{LO})$

Strong dominance states that HI is strictly better than LO if for every possible matching pair of choice procedures that one can consider, HI expects a better outcome than LO according to at least one of the pair and there is at least one matching pair that one can consider where HI expects a greater payoff then LO according to both choice procedures in a matching pair.

If this is a valid principle for decision making under uncertainty, then we may conclude that HI is strictly preferable to LO. After all, for all coins with a bias of less than $\frac{10}{11}$, both ways of labeling the coin will result in the expected payoff of HI being greater than that of LO. If the coin's bias is greater than or equal to $\frac{10}{11}$, then for one of the matching pair of choice procedures, the expected payoff of HI is greater. Therefore, HI strongly dominates* LO.

In order to defend the proposed solution, I must defend Strong Dominance*. To do so, I will show that given two assumptions, Strong Dominance* is just a special case of the standard strong dominance principle in decision theory. Along the way, I will defend the two assumptions.

The first assumption is that Player 1's decision problem may be framed such that the set of states are represented by an exhaustive and exclusive set of matching pairs of choice procedures. Let us use $\{MP_1, MP_2, \ldots, MP_n\}$ to represent this set. In addition, $\{O_1, O_2, \ldots, O_n\}$ and $\{O'_1, O'_2, \ldots, O'_n\}$ respectively represent the outcomes of choosing HI or LO. As I argued above, given Player 1's ignorance, he cannot differentiate matching pairs of choice procedures. That is, for any way one can distinguish the two choices, Player 1 has no reason to think that Player 2 uses one or another way of using this distinction to make a choice. Thus, matching pairs should always be considered together.

The second assumption makes explicit how we are determining our preference over outcomes in the special cases where states are individuated by matching pairs of choice procedures. We can state the assumption with the following two principles (where c and c* are the relevant matching pair of choice procedures in any given state).[5]

Weak Preference: $O_i \succcurlyeq O_i'$ just in case $EU_c(\text{HI}) > EU_c(\text{LO})$ or $EU_{c*}(\text{HI}) > EU_{c*}(\text{LO})$

Strong Preference: $O_i \succ O_i'$ just in case $EU_c(\text{HI}) > EU_c(\text{LO})$ and $EU_{c*}(\text{HI}) > EU_{c*}(\text{LO})$

Given the two assumptions that we have made, it can easily be shown that our principle of strong dominance can be derived from the standard principle of strong dominance. The standard principle of strong dominance entails that HI≻LO just in case for every i , $O_i \succcurlyeq O_i'$ and for some i, $O_i \succ O_i'$. So if each state of the world is a matching pair of choice procedures and Player 2's preferences over the possible outcomes is defined in terms of the expected utilities relative to these matching pairs, then strong dominance* is simply a special case of standard strong dominance.

Why should we accept weak and strong preference? I propose that we can derive these from standard expected utility theory given a certain judgment of symmetry. Let me explain.

I mentioned above that Player 1 has no reason to differentiate Player 2s labeling of the coin in one way or the other. And for that reason, Player 1 cannot differentiate choice procedures within a matching pair. This means, in part, that one does not have any evidence that a player would use one or another of a matching pair. As we have seen, such ignorance does not provide any useful information. However, there is an intuitive symmetry judgment that one may have about matching pairs of choice procedures. When we consider matching pairs, we not only think that we don't know how a player picks between one or the other of a pair, but we also think that the question of determining whether to use on or another of a matching pair of choice procedures is perfectly symmetrical. Or to put the point more concretely, we think that insofar as a player is just trying to pick a way of choosing, the question of determining how to label the two sides of a coin is perfectly symmetrical.

This symmetry judgment is akin to the symmetry judgments that underlie many of our probability judgments.[6] For example, when it comes to the flip of certain coins, it is not sufficient to simply state that we do not know whether the coin will land heads or tails. After all, we often think that the question of how likely it is that the coin lands heads is in every respect the same as the question of how likely it is that the coin lands tails. And if we think that these

---

[5] Since the players' aims are solely to maximize their own payoffs, I will assume that their utilities are identical to their payoffs.

[6] For a nice discussion of probability judgments and symmetry judgments, see [5].

questions are alike in every respect, then these questions must have the same answer. Thus, we judge that it is equally likely that the coin lands heads or tails.

Similarly, if we judge that the question of how likely it is that HEADS is associated with HI is just like the question of how likely it is that HEADS is associated with LO, then we must think that these two question have the same answer. And if they have the same answer, then the likelihood of using one or another of a matching pair of choice procedures is identical.[7]

If the use of c or c* is equiprobable for every matching pair, then it is clear to see why Weak and Strong Preference are valid principles. If the use of c and c* is equiprobable and according to one HI is strictly better and according to the other LO is strictly better, then there is an equal chance at doing better with HI or LO. Thus, HI is at least as good as LO, and LO is at least as good as HI. If according to both, HI is strictly better, then HI ought to be strictly preferred.

In providing an explanation of why HI is the uniquely rational choice, I have been motivated by the intuition that HI dominates LO even for players who are motivated solely to maximize their own payoffs. On my view, what an adequate solution to this game requires is an explanation of how players might frame their decision problems in a way that, despite their uncertainty, HI is nevertheless the dominant choice. I hope to have offered one such framing and line of reasoning.

References

1. Bacharach, M.: Beyond individual choice: teams and frames in game theory. Princeton University Press (2006)
2. Colman, A.M.: Cooperation, psychological game theory, and limitations of rationality in social interaction. Behavioral and Brain Sciences 26, 139–153 (2003)
3. Gold, N., Sugden, R.: Theories of team agency. In: Peter, F., Schmid, H.B. (eds.) Rationality and Commitment, pp. 280–312. Oxford University Press (2008)
4. Ullmann-Margalit, E., Morgenbesser, S.: Picking and choosing. Social Research, 757–785 (1977)
5. Vasudevan, A.: On the a priori and a posteriori assessment of probabilities. Journal of Applied Logic 11(4), 440–451 (2013)

---

[7] The reader may object that while I have rejected any determinate probability judgment for Player 2's choice of HI or LO, I am now arguing that the use of one or another choice procedure in a matching pair is equiprobable. However, only the latter has a reasonable justification. In this case, it is justified on the basis of a symmetry judgment.

Epistemic Updates on Bilattices

Zeinab Bakhtiarinoodeh[1],* and Umberto Rivieccio[2],**

[1] LORIA, CNRS-Université de Lorraine, France
[2] Delft University of Technology, Delft, The Netherlands

The Logic of Epistemic Actions and Knowledge (EAK) has been introduced by Baltag, Moss and Solecki [1] as a framework for reasoning about knowledge in a dynamic setting. It is thus a language expansion of (classical) modal logic having, besides the usual modal operators that represent knowledge and beliefs of agents, dynamic operators used to represent the epistemic change that can be brought about by epistemic actions such as, e.g., announcements.

Formally, epistemic changes are modeled via the so-called *product update* construction on the Kripke-style models that constitute the relational semantics of EAK. Through the product update, a Kripke model encoding the current epistemic setup of a group of agents is replaced by an updated model, which encodes the setup of the agents after an epistemic action has taken place.

In [3,4] product updates are dually characterized as a construction (called *epistemic update*) that transforms the complex algebra associated with a given Kripke model into the complex algebra associated with the model updated by means of an action structure; in this way EAK is endowed with an algebraic semantics that is dual to the relational one via a Jónsson-Tarski-type duality. Moreover, the methods of [3,4] can be used to define a logic of Epistemic Actions and Knowledge on a propositional basis that is weaker than classical logic. This provides us with a more flexible logical formalism, which can be applied to a variety of contexts where classical reasoning is not suitable. This line of research has been further pursued in [5,6], which extends the mechanism of updates to the bilattice modal logic of [2], obtaining a *bilattice public announcement logic*.

In the present contribution we report on ongoing research that aims at further extending the methods of [5,6] to introduce a suitable notion of product update on relational and algebraic models of bilattice modal logic, thus providing a semantics and a complete axiomatization for a bilattice-based Logic of Epistemic Action and Knowledge (BEAK).

Bilattice modal logic is a logic defined by Kripke models $\langle W, R, v \rangle$ in which both valuations and the accessibility relation $R: W \times W \to$ FOUR take values into the four-element Belnap bilattice FOUR. The language of bilattice modal logic $\langle \wedge, \vee, \to, \neg, \Diamond, \mathsf{t}, \top, \mathsf{f}, \bot \rangle$ is essentially the same as that of classical modal logic (augmented with constants representing elements of FOUR), but the propositional connectives as well as the modal operator $\Diamond$ are interpreted using the

---

* Support by European Research Council grant EPS 313360 is gratefully acknowledged.

** The research of the second author has been supported by the Netherlands Organization for Scientific Research (NWO) Vidi grant 016.138.314.

© Springer-Verlag Berlin Heidelberg 2015
W. van der Hoek et al. (Eds.): LORI 2015, LNCS 9394, pp. 426–428, 2015.
DOI: 10.1007/978-3-662-48561-3_37

algebraic operations of FOUR. This logic can be extended to define bilattice-based epistemic logics, for example a four-valued analogue of modal logic S5 (we refer to [2] for further details and motivation on bilattice modal logic).

We obtain the language of (single-agent)[1] BEAK by expanding that of bilattice modal logic with a dynamic modal operator $\langle \alpha \rangle$, where α is an action structure defined as below. Thus, for every formula $\varphi \in Fm$, we have that $\langle \alpha \rangle \varphi$ is also a formula. In our four-valued setting, an epistemic action is a structure $\alpha = (K, k, R_\alpha, \mathsf{Pre}_\alpha)$ where K is a finite non-empty set, $k \in K$, $R_\alpha : K \times K \to$ FOUR and $\mathsf{Pre}_\alpha : K \to Fm$ is a map taking each point in K to a formula of BEAK (the precondition of the action).

Drawing inspiration from [3,4], we introduce an algebraic semantics for BEAK via *intermediate structures*. For every modal bilattice **B** (modal bilattices are the algebraic semantics of the bilattice modal logic introduced in [2]) and every action structure $\alpha = (K, k, R_\alpha, \mathsf{Pre}_\alpha)$, the intermediate structure $\prod_\alpha \mathbf{B}$ is given by the direct power $\mathbf{B}^K$, which is obviously an algebra in the same variety. A special quotient of $\prod_\alpha \mathbf{B}$ is then taken, as an instance of the general construction introduced in [5,6] to account for public announcements in a bilattice setting. This is called *pseudo-quotient*, because it is obtained by means of a relation that is compatible with all the bilattice connectives except for the $\Diamond$ operator. We note that the pseudo-quotient definition from [3] does not work in the bilattice setting, for produces a relation that is already not compatible with one non-modal connective (the bilattice negation), and has therefore to be adapted as indicated in [5,6].

The above product and pseudo-quotient constructions allow us to define a suitable notion of algebraic models of BEAK. We then use the duality developed in [2] to obtain a relational semantics for the logic. Given a four-valued Kripke model $M = (W, R, V)$ and an action structure α, the intermediate structure $M \times \alpha$ is given by the coproduct $\coprod_\alpha M := (\coprod_K W, R \times R_\alpha, \coprod_K V)$, where $\coprod_K W$ is the $|K|$-fold coproduct of W (which is set-isomorphic to $W \times K$), $R \times R_\alpha$ is a four-valued relation on $\coprod_K W$ and $(\coprod_K V)(p) := \coprod_K V(p)$ for every atomic formula p. Finally, the update of M with the action structure α is the submodel $M^\alpha := (W^\alpha, R_\alpha, V^\alpha)$ of $\coprod_\alpha M$ the domain of which is the subset

$$W^\alpha := \{(w, j) \in \coprod_K W : M, w \models \mathsf{Pre}_\alpha(j)\}.$$

The constructions sketched above allow us to devise suitable interaction axioms between the dynamic modality and the other connectives of bilattice modal logic, which give us a Hilbert-style axiomatization of BEAK. Completeness with respect to algebraic models is obtained, as in [3,4], via reduction to the static fragment of the logic; completeness with respect to the relational models then follows by duality.

---

[1] The multi-agent version of BEAK results from indexing modal operators with agents and interpreting relations (both on models and on action structures) over a set of agents.

References

1. Baltag, A., Moss, L., Solecki, A.: The logic of public announcements, common knowledge, and private suspicions. CWI technical report SEN-R9922 (1999)
2. Jung, A., Rivieccio, U.: Kripke semantics for modal bilattice logic. In: Proceedings of the 28th Annual ACM/IEEE Symposium on Logic in Computer Science, pp. 438–447. IEEE Computer Society Press (2013)
3. Kurz, A., Palmigiano, A.: Epistemic Updates on Algebras. Logical Methods in Computer Science 9(4:17), 1–28 (2013)
4. Ma, M., Palmigiano, A., Sadrzadeh, M.: Algebraic semantics and model completeness for Intuitionistic Public Announcement Logic. Annals of Pure and Applied Logic 165, 963–995 (2014)
5. Rivieccio, U.: Algebraic semantics for bilattice public announcement logic. In: Indrzejczak, A., Kaczmarek, J., Zawidzki, M. (eds.) Proceedings of Trends in Logic XIII, Lodz, Poland, July 2-5, pp. 199–215. Lodz University Press (2014)
6. Rivieccio, U.: Bilattice public announcement logic. In: Goré, R., Kooi, B., Kurucz, A. (eds.) Advances in Modal Logic, vol. 10, pp. 459–477. College Publications (2014)

On the Complexity of Input/Output Logic

Xin Sun[1] and Diego Agustín Ambrossio[1,2]

[1] Faculty of Science, Technology and Communication, University of Luxembourg,
Luxembourg
xin.sun@uni.lu
[2] Interdisciplinary Centre for Security, Reliability and Trust, University of
Luxembourg, Luxembourg
diego.ambrossio@uni.lu

Abstract. The complexity of input/output logic has been sparsely developed. In this paper we study the complexity of four existing input/output logics. We show that the lower bound of the complexity of the fulfillment problem of these input/output logics is coNP, while the upper bound is either coNP, or P^{NP}.

1 Introduction

In the first volume of the handbook of deontic logic and normative systems [4], input/output logic [6,7,8,9] appears as one of the new achievements in deontic logic in recent years. Input/output logic takes its origin in the study of conditional norms. Unlike the modal logic framework, which usually uses possible world semantics, input/output logic adopts mainly operational semantics: a normative system is conceived in input/output logic as a deductive machine, like a black box which produces normative statements as output, when we feed it descriptive statements as input. For a comprehensive introduction to input/output logic, see Parent and van der Torre [9]. A technical toolbox to build input/output logic can be found in Sun [12].

While the semantics and application of input/output logic has been well developed in recent years, the complexity of input/output logic has not been studied yet. In this paper we fill this gap. We show that the lower bound of the complexity for the fulfillment problem of four input/output logics is coNP, while the upper bound is either coNP or P^{NP}.

The structure of this paper is as follows: we present a summary of basic concepts and results in input/output logic and some notes in complexity theory, in Section 2. In Section 3 we study the complexity of input/output logic. We point out some directions for future work and conclude this paper in Section 4.

2 Background

2.1 Input/Output Logic

Makinson and van der Torre introduce input/output logic as a general framework for reasoning about the detachment of obligations, permissions and institutional

© Springer-Verlag Berlin Heidelberg 2015
W. van der Hoek et al. (Eds.): LORI 2015, LNCS 9394, pp. 429–434, 2015.
DOI: 10.1007/978-3-662-48561-3_38

facts from conditional norms. Strictly speaking input/output logic is not a single logic but a family of logics, just like modal logic is a family of logics containing systems K, KD, S4, S5, ... We refer to the family as the input/output framework. The proposed framework has been applied to domains other than normative reasoning, for example causal reasoning, argumentation, logic programming and non-monotonic logic, see Bochman [2].

Let $\mathbb{P} = \{p_0, p_1, \ldots\}$ be a countable set of propositional letters and PL be the propositional language built upon $\mathbb{P}$. Let $N \subseteq PL \times PL$ be a set of ordered pairs of formulas of PL. We call N a normative system. A pair $(a, x) \in N$, call it a *norm*, is read as "given a, it ought to be x". N can be viewed as a function from 2^{PL} to 2^{PL} such that for a set A of formulas, $N(A) = \{x \in PL : (a, x) \in N$ for some $a \in A\}$. Intuitively, N can be interpreted as a *normative code* composed of conditional norms and the set A serves as explicit input representing factual statements.

Makison and van der Torre [6] define the semantics of input/output logics from O_1 to O_4 as follows:

- $O_1(N, A) = Cn(N(Cn(A)))$.
- $O_2(N, A) = \bigcap\{Cn(N(V)) : A \subseteq V, V \text{ is complete}\}$.
- $O_3(N, A) = \bigcap\{Cn(N(B)) : A \subseteq B = Cn(B) \supseteq N(B)\}$.
- $O_4(N, A) = \bigcap\{Cn(N(V) : A \subseteq V \supseteq N(V)), V \text{ is complete}\}$.

Here Cn is the classical consequence operator of propositional logic, and a set of formulas is *complete* if it is either *maximal consistent* or equal to PL. These four operators are called *simple-minded output, basic output, simple-minded reusable output* and *basic reusable output* respectively. For each of these four operators, a *throughput* version that allows inputs to reappear as outputs, defined as $O_i^+(N, A) = O_i(N_{id}, A)$, where $N_{id} = N \cup \{(a, a) \mid a \in PL\}$. When A is a singleton, we write $O_i(N, a)$ for $O_i(N, \{a\})$.

Input/output logics are given a proof theoretic characterization. We say that an ordered pair of formulas is derivable from a set N iff (a, x) is in the least set that extends $N \cup \{(\top, \top)\}$ and is closed under a number of derivation rules. The following are the rules we need to define O_1 to O_4^+:

- SI (strengthening the input): from (a, x) to (b, x) whenever $b \vdash a$. Here $\vdash$ is the classical entailment relation of propositional logic.
- OR (disjunction of input): from (a, x) and (b, x) to $(a \vee b, x)$.
- WO (weakening the output): from (a, x) to (a, y) whenever $x \vdash y$.
- AND (conjunction of output): from (a, x) and (a, y) to $(a, x \wedge y)$.
- CT (cumulative transitivity): from (a, x) and $(a \wedge x, y)$ to (a, y).
- ID (identity): from nothing to (a, a).

The derivation system based on the rules SI, WO and AND is called D_1. Adding OR to D_1 gives D_2. Adding CT to D_1 gives D_3. The five rules together give D_4. Adding ID to D_i gives D_i^+ for $i \in \{1, 2, 3, 4\}$. $(a, x) \in D_i(N)$ is used to denote the norms (a, x) derivable from N using rules of derivation system D_i. In Makinson and van der Torre [6], the following soundness and completeness theorems are given:

Theorem 1 ([6]). *Given an arbitrary normative system N and formula a,*

- $x \in O_i(N, a)$ *iff* $(a, x) \in D_i(N)$, *for* $i \in \{1, 2, 3, 4\}$.
- $x \in O_i^+(N, a)$ *iff* $(a, x) \in D_i^+(N)$, *for* $i \in \{1, 2, 3, 4\}$.

2.2 Complexity Theory

Complexity theory is the theory to investigate the time, memory, or other resources required for solving computational problems. In this subsection we briefly review those concepts and results from complexity theory which will be used in this paper. More comprehensive introduction of complexity theory can be found in [11,1]

We assume the readers are familiar with notions like Turing machine and the complexity class P, NP and coNP. Oracle Turing machine and one complexity class related to oracle Turing machine will be used in this paper.

Definition 1 (oracle Turing machine). *An* oracle *for a language L is device that is capable of reporting whether any string w is a member of L. An (resp. non-deterministic) oracle Truing machine M^L is a modified (resp. non-deterministic) Turing machine that has the additional capability of querying an oracle. Whenever M^L writes a string on a special oracle tape it is informed whether that string is a member of L, in a single computation step.*

Definition 2 ($\mathbf{P}^{NP}$). *P^{NP} is the class of languages decidable with a polynomial time oracle Truing machine that uses oracle $L \in NP$.*

3 Complexity of Input/Output Logic

The complexity of input/output logic has been sparsely studied in the past. Although the reversibility of derivations rules as a proof re-writing mechanism has been studied for input/output logic framework [6], the length or complexity of such proofs have not been developed. We approach the complexity of input/output logic from a semantic point of view.

We now start to study the complexity of the following input/output logics: $O_1, O_1^+, O_3,$ and O_3^+. We focus on three different problems:
Given a finite set of norms N, a finite set of formulas A and a formula x:

> (1) *Fulfillment problem:*   is $x \in O(N, A)$?
> (2) *Violation problem:*   is $\neg x \in O(N, A)$?
> (3) *Compatibility problem:* is $\neg x \notin O(N, A)$?

The aim of the fulfillment problem is to check whether the formula x appears among the obligations detached from the normative system N and facts A. The intuitive reading of the violation problem is: if the obligation to fulfill $\neg x$ exists, then x is a violation. Finally, the compatibility problem says if $\neg x$ is not obligatory, then x is compatible with the normative system N, given facts A.

The compatibility problem is often referred as a *negative permission* [8,3], and corresponds to what is called weak permission.[1] It can be proven that the other two problems can be reduced to the comliance problem. Therefore we focus on the compliance problem.

3.1 Simple-Minded O_1

Theorem 2. *The fulfillment problem of simple-minded input/output logic is coNP-complete.*

Corollary 1. *The violation problem of simple-minded input/output logic is coNP-complete. The compatibility problem of simple-minded input/output logic is NP-complete.*

3.2 Simple-Minded Throughput O_1^+

Theorem 3. *The fulfillment problem of simple-minded throughput input/output logic is coNP-complete.*

Corollary 2. *The violation problem of simple-minded throughput input/output logic is coNP-complete. The compatibility problem of simple-minded throughput input/output logic is NP-complete.*

3.3 Simple-Minded Reusable O_3

Theorem 4. *The fulfillment problem of simple-minded reusable input/output logic is between coNP and P^{NP}.*

Corollary 3. *The violation problem of simple-minded reusable input/output logic is between coNP and P^{NP}. The compatibility problem of simple-minded reusable input/output logic is between NP and P^{NP}.*

3.4 Simple-Minded Reusable Throughput O_3^+

Theorem 5. *The fulfillment problem of simple-minded reusable throughput input/output logic is between coNP and P^{NP}.*

Corollary 4. *The violation problem of simple-minded reusable throughput input/output logic is between coNP and P^{NP}. The compatibility problem of simple-minded reusable throughput input/output logic is between NP and P^{NP}.*

---

[1] "An act will be said to be permitted in the weak sense if it is not forbidden ..." [13].

4   Conclusion and Future Work

In this paper we develop complexity results of input/output logic. We show that four input/output logics (O_1, O_1^+, O_3, O_3^+) have lower bound coNP and upper bound either coNP or P^{NP}. There are several natural directions for future work:

1. What is the complexity of other input/output logic?
2. What is the complexity of constraint input/output logic? Constraint input/output logic [7] is developed to deal with the inconsistency of output. The semantics of constraint input/output logic is more complex than those input/output loic discussed in this paper. This might increase the complexity of the compliance problem. Constraint input/output logic based on O_3^+ has close relation with Reiter's default logic [10]. Gottlob [5] presents some complexity results of Reiter's default logic, which will give us insights on the complexity of constraint input/output logic.
3. What is the complexity of different types of permission? three different of permissions are introduced in Makinson and van der Torre [8]. In this paper we study the complexity of only one of them (namely, negative permissions) as the compatibility problem. The semantics of these three logics are different, which suggests different complexity for the new problems related to permissions.

References

1. Arora, S., Barak, B.: Computational Complexity: A Modern Approach. Cambridge University Press, New York (2009)
2. Bochman, A.: A causal approach to nonmonotonic reasoning. Artificial intelligence 160(1-2), 105–143 (2004)
3. Boella, G., van der Torre, L.W.N.: Permissions and obligations in hierarchical normative systems. In: ICAIL, pp. 109–118 (2003), http://dl.acm.org/authorize?859450
4. Gabbay, D., Horty, J., Parent, X., van der Meyden, R., van der Torre, L.: Handbook of Deontic Logic and Normative Systems. College Publications, London (2013)
5. Gottlob, G.: Complexity results for nonmonotonic logics. J. Log. Comput. 2(3), 397–425 (1992), http://dx.doi.org/10.1093/logcom/2.3.397
6. Makinson, D., van der Torre, L.: Input-output logics. Journal of Philosophical Logic 29, 383–408 (2000)
7. Makinson, D., van der Torre, L.: Constraints for input/output logics. Journal of Philosophical Logic 30(2), 155–185 (2001)
8. Makinson, D., van der Torre, L.: Permission from an input/output perspective. Journal of Philosophical Logic 32, 391–416 (2003)
9. Parent, X., van der Torre, L.: I/O logic. In: Horty, J., Gabbay, D., Parent, X., van der Meyden, R., van der Torre, L. (eds.) Handbook of Deontic Logic and Normative Systems. College Publications (2013)
10. Reiter, R.: A logic for default reasoning. Artif. Intell. 13(1-2), 81–132 (1980), http://dx.doi.org/10.1016/0004-37028090014-4
11. Sipser, M.: Introduction to the theory of computation, 3rd edn. Cengage Learning, Boston (2012)

12. Sun, X.: How to build input/output logic. In: Bulling, N., van der Torre, L., Villata, S., Jamroga, W., Vasconcelos, W. (eds.) CLIMA 2014. LNCS, vol. 8624, pp. 123–137. Springer, Heidelberg (2014),
http://dx.doi.org/10.1007/978-3-319-09764-0_8,
doi:10.1007/978-3-319-09764-0_8
13. von Wright, G.: Norm and Action. Routledge and Kegan (1963)

Translating a Counterpart Theory into a Quantified Modal Language with Descriptors

Chi-Her Yang

National Taiwan university, Taipei, Taiwan
d99124008@ntu.edu.tw

Abstract. Following Fitting's method, a translation of a Lewis-style counterpart theory in the language L(NI) into the language L(NId) is provided.

Keywords: singular term, quantified modal logic, rigidity, counterpart theory.

We start from L(NI)[1]. L(NI) is a quantified modal logic language, which has its variables, constants, quantifiers, connectives, modal operators, and formation rules in a standard way.

Definition 1. *A counterpart frame or a cF is $\langle W, D, R, C \rangle$ in which W is a non-empty set of worlds, D is a non-empty domain, and R is a binary accessibility relation on W. C is a function mapping each member of $W \times W$ to a counterpart relation on D.*

David Lewis in [1] accepts possible worlds but he argues against transworld identity for the reason that objects cannot have being or be identifiable across possible worlds, just as an object cannot exist in different places at the same time. He develops the counterpart theory: an object in a possible world has counterparts in other worlds rather than existing by itself in them. In order to capture his idea, we have C in the counterpart frame. For example, if $\langle x, y \rangle \in C(w, w')$, then y in w' is a counterpart of x in w.[2]

Definition 2. *A counterpart model on a counterpart frame or a cF-model is $\langle W, D, R, C, v \rangle$. v is a function such that: if c is a constant, $v(c) \in D$; if P_w is an n-place predicate for a world w, $v_w(P) \subseteq D^n$.*

As to the semantics, truth values are assigned to all closed formulas. We consider non-modal cases first. For atomic sentence, $v_w(P_{a_1 \ldots a_n}) = 1$ iff

---

[1] I use this notation because this language works well with the notion of rigid designators to validate necessary identity. Although necessary identity does not hold with respect to the semantics for the counterpart theory I introduced later.

[2] I also suggest that we make R follow C in the sense that if there is a counterpart relation between objects of two worlds, then there is a accessibility relation between these two worlds.

© Springer-Verlag Berlin Heidelberg 2015
W. van der Hoek et al. (Eds.): LORI 2015, LNCS 9394, pp. 435–438, 2015.
DOI: 10.1007/978-3-662-48561-3_39

$\langle v(a_1)...v(a_n)\rangle \in v_w(P)$. We expand the language to ensure that every member of D has a name; for all $d \in D$, we add a constant k_d to the language. For the quantifiers, $v_w(\forall x \Psi(x)) = 1$ iff for all $d \in D, v_w(\Psi(k_d) = 1)$; $v_w(\exists x \Psi(x)) = 1$ iff for some $d \in D, v_w(\Psi(k_d) = 1$. The truth conditions for the connectives are in a standard way. For modal cases, we define counterpart-sentences first.

Definition 3. *For each $d \in D$, we add a constant k_d as usual. We pick up all n-tuples of k_d such that $\{v(a_i), v(k_{d_i})\} \in C(w, w')$ for every k_{d_i} in $\Phi_{w'}$ and its corresponding a_i in Φ_w, and for all $w' \in W$ such that wRw'. For each of these n-tuples, $\Phi_{w'}(k_{d_1}...k_{d_n})$ is a counterpart-sentence of $\Phi_w(a_1...a_n)$.*

Hence, $v_w(\Box\Phi) = 1$ iff every counterpart-sentence of Φ is true; $v_w(\Diamond\Phi) = 1$ iff some counterpart-sentence of Φ is true. Notice that we might have different objects in the domain as candidates for k_{d_i} in different n-tuples, because an object might have two counterparts in one possible world.

Now let us turn to the language L(NId) which is based on L(NI) and extended by adding descriptors. We need more definitions before we go further.

Definition 4. $\langle W, D, R\rangle$ *is a L(NId) frame or dF; $\langle W, D, R, v\rangle$ is a dF-model.*

Descriptors are non-rigid. We assign each descriptor a denotation $v_w(\alpha)$ at each world. Hence, the truth conditions of closed atomic sentences would be $v_w(P_{t_1...t_n}) = 1$ iff $\langle v_w(t_1)...v_w(t_n)\rangle \in v_w(P)$, and a term t is either a rigid constant or a descriptor. For modal formulae, $v_w(\Box\Phi(\alpha)) = 1$ iff for any world w' such that wRw', the denotation of α in w' satisfies Φ. And similarly for $\Diamond$.

In order to do the translation, we need to collect some special functions. Here we consider a loop-free case in which our collecting would not lead to many-valued functions.

Definition 5. *A frame is simply connected if there is only one path, if any, between any two worlds. A path in the frame is a sequence of worlds (with no repetition) such that each of them is related to the next. w and w' is related if wRw' or $w'Rw$.*

Definition 6. *Let f be a function defined on a frame, mapping worlds to members of D. f is counterpart frame compatible or cF-compatible if for any w and w' such that wRw', $f(w')$ is a counterpart of $f(w)$.*

If a cF is simply connected and we have a cF-model for the cF, then we have a corresponding dF-model which has the same W, D, R with the cF-model, and has v expanded by adding valuations of descriptors such that each descriptor corresponds to each cF-compatible function. I explain this by an example. (Cf. Fig. 1.) Suppose there is a cF:

$W : \{w_0, w_1, w_2\}$,
$D : \{O_{01}, O_{02}, O_{11}, O_{12}\, O_{21}, O_{22}\}$,
$R : \{\langle w_0, w_1\rangle\, \langle w_0, w_2\rangle, \langle w_0, w_0\rangle, \langle w_1, w_1\rangle, \langle w_2, w_2\rangle\}$,
$C(w_0, w_1) : \{\langle O_{01}, O_{11}\rangle, \langle O_{02}, O_{12}\rangle\}$,
$C(w_0, w_2) : \{\langle O_{01}, O_{21}\rangle, \langle O_{01}, O_{22}\rangle, \langle O_{02}, O_{22}\rangle\}$,
Also for any w, $C(w, w)$ is an identity function $\{\langle O, O\rangle\}$ for all objects. [3]

---

[3] I assume that an object is the counterpart of itself in its world.

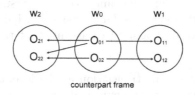

counterpart frame

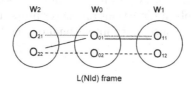

L(NId) frame

Fig. 1.

Then we have a corresponding dF-model:
W, D, R is the same, and we add the following valuations into v. (Since v in cF-model is only about rigid constants and predicates, this adding does not cause contradiction.)
$v(\alpha_1)$ is a function: $\{\langle w_0, O_{01}\rangle, \langle w_1, O_{11}\rangle, \langle w_2, O_{21}\rangle\}$
$v(\alpha_2)$ is a function: $\{\langle w_0, O_{01}\rangle, \langle w_1, O_{11}\rangle, \langle w_2, O_{22}\rangle\}$
$v(\alpha_3)$ is a function: $\{\langle w_0, O_{02}\rangle, \langle w_1, O_{12}\rangle, \langle w_2, O_{22}\rangle\}$

I suggest that for any Φ in a counterpart theory in the language L(NI), there is a corresponding Φ^* in the language L(NId) such that Φ is true in a cF-model iff Φ^* is true in a corresponding dF-model. The translation from Φ to Φ^* is defined in this way,

1. If A is atomic, $A^* = A$.
2. $(A \wedge B)^* = (A^* \wedge B^*)$, and similarly for other connectives.
3. $(\forall x A)^* = \forall x A^*$, and similarly for existential quantifiers.
4. Suppose that constants of A are $a_1...a_n$. For each cF-compatible function we add a descriptor into the language, so we got a set of these descriptors Δ and the Cartesian product Δ^n. We collect every n-tuple $\{\alpha_{i1}...\alpha_{in}\} \subseteq \Delta^n$ such that $a_1 = \alpha_{i1}$ is true, $a_2 = \alpha_{i2}$ is true, and etc. The number of this kind of n-tuple is i, $(\Box A(a_1...a_n))^* = \bigwedge_{1 \leq j \leq i} \Box A^*(\alpha_{j1}...\alpha_{jn})$. And similarly for $\Diamond$.

For example, using the previous cF-frame, suppose Φ is $\Box Pa$ which is true in w_0 in a cF-model when $v(a) = O_{01}$, then we have $\Box P\alpha_1 \wedge \Box P\alpha_2$ which is true in w_0 in a corresponding dF-model.

Remark 1. For the translation, a crucial point is to make the non-rigid descriptors in a dF-model play the role of C in a cF model.

Remark 2. Modal operators for the counterpart theory still works locally. If we just want to do the translation for modal formulae which do not contain iterated modal operators, then we do not need to assume that the cF frame is simple connected.

Remark 3. If we want to do the translation at the level of cF frames and dF frames without assuming a simple connected cF, we need to avoid assigning many-valued functions to descriptors. Fitting suggests that we can 'duplicate' possible worlds. For modal realists[4], it might be more reasonable, even not easier in technical details, to duplicate functions and descriptors (unless you also accept the existence of individual concepts or intensional objects).

References

1. Lewis, D.: Counterpart theory and quantified modal logic. Journal of Philosophy 65, 113–126 (1968)
2. Fitting, M., Mendelsohn, R.L.: First Order Modal Logic. Kluwer, Dordrecht (1999)
3. Fitting, M.: First-order intensional logic. Annals of Pure and Applied Logic 127, 191–193 (2004)
4. Fitting, M.: Intensional Logic. The Stanford Encyclopedia of Philosophy (Spring 2014 Edition) (2011)
5. Kripke, S.: Semantical Considerations on Modal Logic. Acta Philosophica Fennica 16, 83–94 (1963)
6. Kripke, S.: Naming and Necessity. Harvard University Press, Cambridge (1980)
7. Priest, G.: An Introduction to Non-Classical Logic. Cambridge University Press, Cambridge (2008)

---

[4] I suppose that they prefer a simpler reality, but I might be wrong.

Author Index

Printed in the United States
By Bookmasters